Practical Quantitative Vibrational and Electronic Spectroscopy

Practical Quantitative Vibrational and Electronic Spectroscopy

A Guide for Developing, Optimizing, and Validating Procedures

Edited by

Deborah A. Peru
DP Spectroscopy and Training, LLC
Lebanon, New Jersey
United States

WILEY

Published by John Wiley & Sons, Inc., Hoboken, New Jersey.

Library of Congress Cataloging-in-Publication Data Applied for:

Hardback ISBN: 9781394227228

Cover Design: Wiley
Cover Image: Courtesy of Deborah A. Peru and John Wasylyk

Set in 9.5/12.5pt STIXTwoText by Straive, Chennai, India

To my husband, Jim, and children, Cora and Nicholas. You have provided me with your patience and companionship as I navigated through my career, and now, during the preparation of this book. Each of you has inspired me in so many creative and unique ways. Your love is the foundation that gives meaning to my research and teaching career.

Contents

List of Contributors

Albert Arias
Virginia Commonwealth University School of Medicine
Richmond, VA
USA

Pierre Beckers
ViBra-Sante Hub
Laboratory of Pharmaceutical Analytical Chemistry
Department of Pharmacy
University of Liege (ULiege)
Liege
Belgium

Julie Y. Chen
HORIBA Instruments Inc.
Irvine, CA
USA

Emil W. Ciurczak
Doramaxx Consulting LLC
Goldens Bridge, NY
USA

Charlotte De Bleye
ViBra-Sante Hub
Laboratory of Pharmaceutical Analytical Chemistry
Department of Pharmacy
University of Liege (ULiege)
Liege
Belgium

Stuart Farquharson
Real-Time Analyzers, Inc.
Middletown, CT
USA

Adam Gilmore
Horiba Instruments Inc.
Piscataway, NJ
USA

Philippe Hubert
ViBra-Sante Hub
Laboratory of Pharmaceutical Analytical Chemistry
Department of Pharmacy
University of Liege (ULiege)
Liege
Belgium

Benoît Igne
Manufacturing Science and Technology
Vertex Pharmaceutical Inc
Boston, MA
USA

Sanga Kim
HORIBA Korea Ltd.
Manan-Gu, Anyang-Si
Gyeonggi-Do
Republic of Korea

Maritza Reyna Liriano
University of Puerto Rico at Mayagüez
Mayagüez
Puerto Rico

Eunah Lee
HORIBA Reno Technology Center
Reno, NV
USA

Ryan Lerud
Process Insights Inc.
Sacramento, CA
USA

Charles E. Miller
ChEM Analytics LLC
Lansdale, PA
USA

Deborah A. Peru
DP Spectroscopy and Training LLC
Lebanon, NJ
USA

Brian G. Rohrback
Infometrix, Inc.
Bothell, WA
USA

Rodolfo J. Romañach
University of Puerto Rico at Mayagüez
Mayagüez
Puerto Rico

Pierre-Yves Sacré
Research Support Unit in Chemometrics
Department of Pharmacy
University of Liege (ULiege)
Liege
Belgium

Doreen Schober
Concha y Toro
Pencahue
Maule
Chile

and

Organic Agricultural Sciences, Organic Food Quality
University of Kassel
Witzenhausen, Hessen
Germany

Chetan Shende
Real-Time Analyzers, Inc.
Middletown, CT
USA

Zhenqi Shi
Synthetic Molecule Pharmaceutical Science
Genentech Research and Early Development
Genentech
A Member of Roche Company
South San Francisco, CA
USA

Wayne Smith
WSmithConsulting, Inc.
South Glastonbury, CT
USA

Luis Torres
Quality Control Analyst
Avara Pharmaceuticals
Arecibo
Puerto Rico

John Wasylyk
Bristol Myers Squibb
Yardville, NJ
USA

Alexis R. Weber
PerkinElmer
Albany, NY
USA

Eric M. Ziemons
ViBra-Sante Hub
Laboratory of Pharmaceutical Analytical Chemistry
Department of Pharmacy
University of Liege (ULiege)
Liege
Belgium

Preface

The field of spectroscopy continues to advance in both hardware and software, and how it is applied in biological, pharmaceutical, medical, electronic, and environmental industries, among others. Much of the previous literature published in the 1980s and 1990s pertains to using quantitative spectroscopy in the agricultural and chemical industries. With the rapid improvement in computer speed, process analytical technology (PAT) applications rapidly became a "hot topic" at conferences. PAT continues to be an important area of focus for spectroscopists and engineers in the 21st century for improving product quality and manufacturing cost savings. Chemometrics (i.e. principal component analysis) and machine learning for extracting information from large quantities of redundant and correlated data is inarguably one of the most widely used and published topics in spectroscopy today.

The approach taken in this book is to offer a broad look at how and why spectroscopy is currently being used across multiple industries. The intent is to provide an instructional guide for developing methods that are not theoretical but practical to implement and fit for purpose. In many industrial sectors, quantitative spectroscopy procedures are used throughout the product's life cycle, from the determination of contaminants in raw materials to problem-solving when product performance is compromised. Validated methods that adhere to FDA and ICH guidelines may need to be developed and implemented quickly to support rapidly changing business priorities. The traditional approach of creating a "training set" consisting of matrix-matched calibration standards may not always be needed or plausible. A good example is when light-based devices are used in longitudinal studies to measure the color change from a baseline. This book addresses real-world topics like short development time, training set design, wavelength selection strategies, and best practices for the lifecycle management of procedures, equipping you with the necessary tools to succeed in your projects.

Part I captures years of knowledge gained by the authors when developing, validating, and implementing a quantitative procedure. It covers several quantitative approaches used in industry and academia to extract quantitative information from spectral data. The chapters are written from the perspective that many creative and useful spectroscopy procedures can be developed and used for many different purposes. This includes patent applications, clinical studies, and environmental testing both on Earth and in space. Many laboratories using spectroscopy instrumentation are implementing procedures for routine quality testing of samples using CFR 21 Part 11 compliant software as part of the initiative to improve laboratory efficiency.

Part II of this book includes commercial ultraviolet, visible, near-infrared, mid-infrared, Raman, and SERS instrumentation applications. The applications were chosen to highlight important aspects of method development presented in Part I. The authors discuss how specific challenges were overcome to enhance method performance in various relevant applications. These include the measurement of microplastics, petrochemicals, wine, pharmaceuticals, forensics, and illicit drugs. These practical examples are included to reassure you of your understanding and boost your confidence in applying the knowledge in your work.

Part III of this book contains several "Vignettes" describing examples of problems solved while developing and implementing spectroscopy instrumentation for quantitative analysis. This valuable information is being shared, so developers can learn from these experienced authors and apply it in future applications.

We aim to provide the reader with an understanding of managing quantitative spectroscopy projects. Part I of the book provides a mix of theory and experience-based knowledge on the various stages of building quantitative procedures. Part II and Part III are intended to broaden the readers' understanding of how quantitative methods are applied and help to spark creativity for developing new quantitative spectroscopy procedures.

While preparing these chapters, I gathered feedback from various colleagues. I am grateful to the following individuals for their insightful reviews of different portions of the book: Emil W. Ciurczak, Pierre-Yves Sacré, Brian Rohrback, and Ryan Lerud. I also thank my creative colleague, Nicholas Minadeo, for enhancing the quality and resolution of many figures and images in this book. Finally, I am grateful for the fruitful discussions with Gary Ritchie, who identified two outstanding authors for this book.

Deborah A. Peru
Lebanon, NJ, USA
February 14, 2025

About the Companion Website

This book is accompanied by a companion website:

www.wiley.com/go/peruelectronicspectroscopy

Scan this QR code to visit the companion website

This website include:

- Appendix
- Tables
- Guidelines for UV-VIS-EEM Measurements
- Standard Operating Procedure Manual
- SERS Spectra from All patients in the study

Part I

Fundamentals of Quantitative Spectroscopy

1

An Introduction to Quantitative Spectroscopy

Deborah A. Peru

DP Spectroscopy and Training LLC, Lebanon, NJ, USA

1.1 Introduction

The fundamental purpose of quantitative spectroscopy is to determine the concentration of an analyte, molecule, ingredient, or property of interest in an unknown sample using the spectral signals obtained when samples interact with light. A method can be developed using the entire spectrum, one or more specific regions, or with one or several wavelengths or frequencies. Many spectroscopic methods that determine concentration utilize light energy to measure absorbance, reflectance, excitation, or emission of specific molecules in all forms of matter (solid, liquid, and gas). In this book, we will limit our discussions to liquid and solid forms of matter, although the principles and techniques apply to the measurement of gas-phase molecules.

Applications using mid-infrared and other spectroscopy techniques have seen a rapid increase since the computer revolution began in the early 1980s. With the advent of more affordable computers and improved computational speeds, real-time processing of the data collected became a reality [1]. This led to applying chemometric algorithms to extract more useful information from the spectral fingerprint. The potential for using these techniques is no longer limited to the chemical, petrochemical, agricultural, and pharmaceutical industries [2], but found everywhere. In the last decade, quantitative applications have been applied more broadly in many diverse areas, including nutraceuticals, biomedical, environmental, forensics, biologicals, and clinical for testing and quality analysis.

As depicted in Figure 1.1, applications of spectroscopy are being applied in all stages of the manufacturing supply chain from early research through quality control testing and for problem solving. During the early stages of development and innovation, quantitative procedures can provide insight to enhance the chemical understanding and quality of a product and process for improved performance. For example, when raw materials arrive at a manufacturing plant, quantitative analysis provides a safeguard against contaminants entering the plant and for checking purity levels.

During research and development and scale-up, methods can be applied to optimize new products or processes in the laboratory and pilot plant. In manufacturing, quantitative methods provide information for making critical business decisions, quality assurance, and quality control. Spectroscopy is routinely used in forensic investigations and to solve performance problems even after the products and materials reach the consumer. Often, these problems are solved using a variety of instrumental analysis techniques in combination with spectroscopy, which typically begins the investigation because of the ability to measure samples "as is."

Practical Quantitative Vibrational and Electronic Spectroscopy: A Guide for Developing, Optimizing, and Validating Procedures, First Edition. Edited by Deborah A. Peru.

Companion website: www.wiley.com/go/peruelectronicspectroscopy

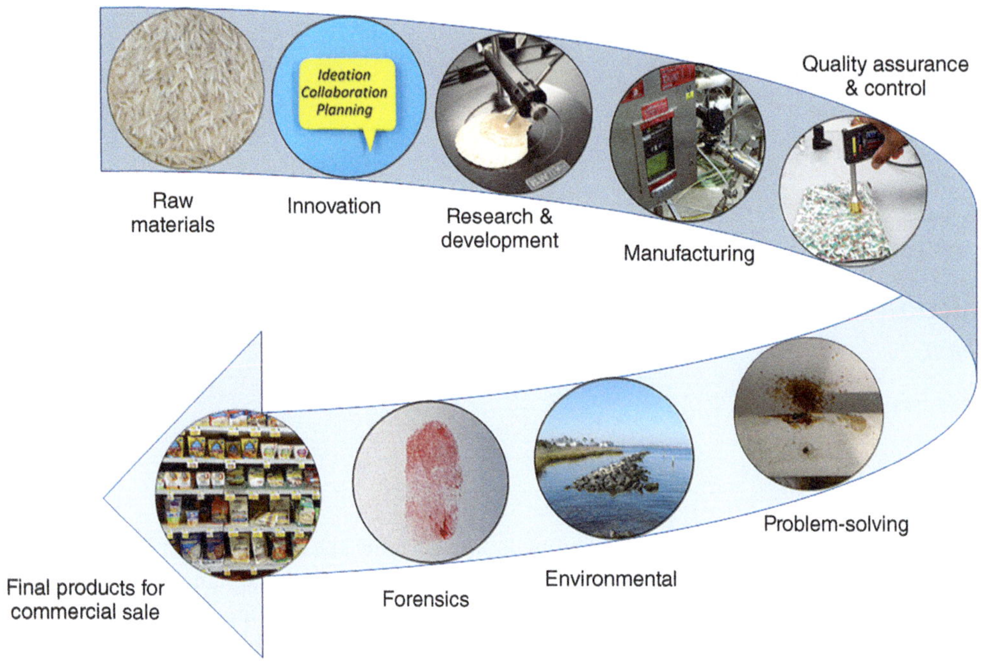

Figure 1.1 Stages in product lifecycle when spectroscopy procedures can be used. *Source*: Deborah A. Peru.

What is driving the use of spectroscopy in industry? Companies invest time and resources in developing spectroscopic-based applications because this approach often requires little or no sample preparation. Analysis time is quick, and results are available within minutes. With proper knowledge, experience, and training in method development, quantitative methods can be designed to be as reliable and sometimes more reliable than compendial methods because errors introduced from sample preparation, solvent interactions, and analyst errors are often reduced.

In addition, spectroscopy eliminates the analyte recovery problem associated with extraction methods. These methods can be classified as green technology [3]. It reduces the need for harsh solvents and thus is safer for operators. It also reduces disposal costs, solvent waste, and contamination of the environment [4]. With all these benefits, why are not all analytical procedures developed using spectroscopy techniques? Well, the short answer is the fact that many molecules absorb or emit light at similar frequencies, and thus, developing methods that are highly specific for the molecular or ingredient of interest requires a good understanding of the mathematical tools available to extract that specific information desired, particularly when measuring complex samples and formulations containing many components. Additionally, compendial and primary methods are essential in developing spectroscopy methods. The concentrations assigned to calibration standards are often obtained using primary methods such as high-performance liquid chromatography (HPLC), gas chromatography, Karl Fischer moisture determination, and various wet chemical methods. These methods must be accurate, reliable, and validated before spectroscopy methods can be implemented with full confidence that the information is reliable. Yet, as you will see in this book, there are other ways of using spectral information for quantitative analysis that may not always involve using a primary method or developing a calibration regression model.

The first principle in quantitative spectroscopy began in 1729 when Pierre Bouguer found that the intensity transmitted by light through a solution diminishes logarithmically as the depth of the solution or pathlength increased [5]. Unfortunately, this work was not discovered until much later in history, and his name is often not mentioned. In 1760, Johann Heinrich Lambert created a mathematical expression Eq. (1.1) related to this finding, known as the Lambert law [6].

$$I = I_o e^{-\varepsilon l} \tag{1.1}$$

where

ε – intrinsic ability of the solution (or molecule) to absorb light at a particular frequency or wavelength
l – pathlength of the light through the solution
I – intensity after passing through a finite thickness of matter
I_o – intensity of incident radiation.

In 1852, the German scientist August Beer conducted an experiment to determine the absorption of red light in colored liquids. Based on the measurement of transmitted light, he proposed a law relating transmittance to concentration [7]. Most quantitative spectroscopy techniques are based on these two laws. Molecules absorb energy from a light source at specific wavelengths. The transmission of light through a liquid exponentially declines as the sample pathlength thickens (at constant concentration) and as the concentration increases at constant pathlength (Figure 1.2).

Absorbance is related to the logarithmic ratio of the transmittance through a material and is linearly proportional to concentration at constant pathlength as shown in Figure 1.3.

The relationship in Eq. (1.2) is known as the *Beer–Lambert law*. The absorptivity constant (ε) is an intrinsic property of each molecule and defines the amount of light energy absorbed at each wavelength. Other names are used for ε, such as "molar absorption coefficient," "absorbancy index," and "attenuation coefficient" [8]. Strongly absorbing molecules will have a high absorptivity, and weak absorbers will have a much lower absorptivity constant.

$$\text{Abs} = \log\left(\frac{I_o}{I}\right) = \varepsilon lc \tag{1.2}$$

If the instrument measures transmittance as a percentage, conversion to absorbance (Abs) is 2-log % T. Most instrumental software automatically converts the spectral data from T or % T into absorbance and

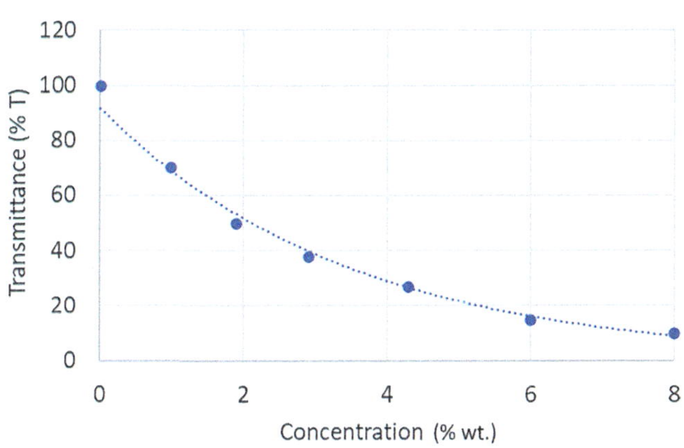

Figure 1.2 Relationship between transmittance of light through matter as concentration increases at a constant pathlength. *Source:* Deborah A. Peru.

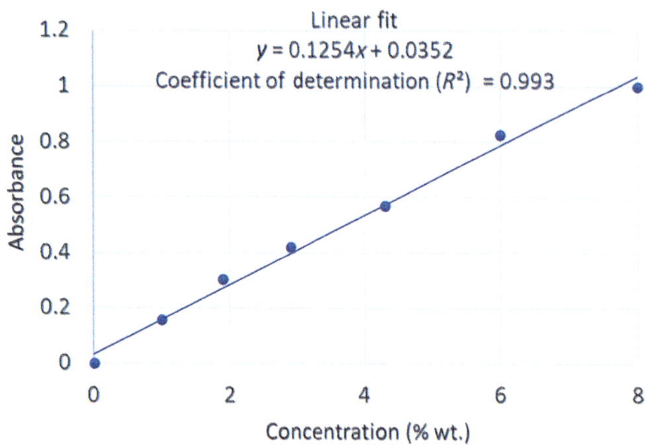

Figure 1.3 Generalized relationship between absorbance of light and concentration. *Source:* Deborah A. Peru.

vice versa. The only unit that changes is related to the Y axis of the spectrum. The wavelength or wavenumber positions remain the same. For quantitative applications, it is often more convenient to convert the spectral data into absorbance units because the relationship between absorbance and concentration is linear, whereas the relationship between transmittance and concentration is logarithmic and requires complex mathematical solutions to be developed. The Beer–Lambert law of absorption states that the amount of light absorbed by the molecule is directly proportional to concentration. This is fundamental for building a linear calibration model to predict the concentration of unknown samples.

The principle that electromagnetic energy is absorbed by molecules at specific frequencies forms the basis for ultraviolet, visible, near-infrared, and mid-infrared spectroscopy. The absorbed energy from the light source can cause the molecule to undergo an electronic, vibrational, rotational, and/or translational transition. When measured with appropriate instrumentation, these molecular transitions yield a unique spectral fingerprint providing molecular structure and concentration information.

The ultraviolet and visible frequencies are located on the blue end of the electromagnetic spectrum as shown in Figure 1.4. Light absorption in this region generally excites electronic transitions within the molecule. Light radiation in this region is characterized as having a higher frequency, higher energy, and shorter wavelengths. The mid-infrared and Raman vibrational frequencies are located on the red end of the electromagnetic spectrum. This region has lower frequency, lower energy, and longer wavelengths. Absorption and emission of light in this region predominantly cause vibration changes within the molecule.

Depending on the molecules involved, the energy of the light source, the state of matter, and the chemical and physical condition of the sample, the spectral fingerprint may have features caused by one or more types of molecular transition [9]. For example, liquids are capable of vibrational motion, but at the

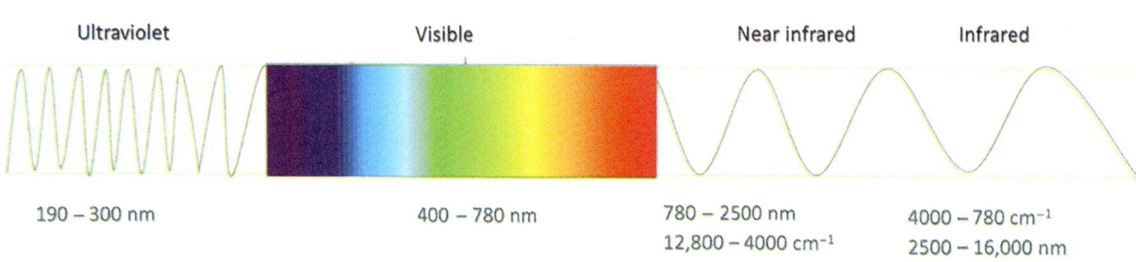

Figure 1.4 Regions of the electromagnetic spectrum covered in this book. *Source:* Deborah A. Peru.

Table 1.1 Types of molecular transition(s) found in regions of the electromagnetic spectrum.

Spectroscopy technique	Primary type of molecular transition	Wavelength, nm (wavenumber, cm^{-1})	
Ultraviolet	Electronic (valence shell electrons)	190–400 nm 52,632–25,000 cm^{-1}	
Visible/fluorescence	Electronic (valence shell electrons)	400–780 nm 25,000–12,820 cm^{-1}	
		Visible wavelength and color observed	
		410 nm	Violet
		470 nm	Blue
		510 nm	Cyan
		540 nm	Green
		580 nm	Yellow
		610 nm	Orange
		685 nm	Red
Near-infrared	Vibration Rotation	780–2500 nm 12,820–4000 cm^{-1}	
Mid-infrared/Raman	Vibration	2500–16,000 nm 4000–625 cm^{-1}	

same time, they can also show rotational and translational motions due to weak intermolecular forces between molecules. Rotational transitions occur mostly between rotational levels of the same vibrational state. For light molecules, many examples exist where combinations of vibrational and rotational transitions can be measured [9]. Table 1.1 summarizes the type of molecular transition within the molecule and the wavelength associated with each region.

1.2 Basic Principles of Ultraviolet and Visible Spectroscopy

The UV–vis region of the electromagnetic spectrum is found between 190 and 780 nm. When the energy of the incident light source matches a specific quantized energy level, an electronic transition will occur, promoting one or more of the lower-energy electrons to a higher-energy state. For each electronic state of the molecule, there may also be several vibration states and rotational states [10]. The absorption of light follows "selection rules" that depend on the molecule's symmetry, types of atoms, and bonding involved [11]. Because of these constraints, an electronic transition does not occur for many substances. However, it is possible in unsaturated and conjugated molecules (chromophores), dienes, trienes, and molecules containing π-bonded atoms (e.g., aromatics) and atoms with non-bonding orbitals. For example, molecules containing nitrogen (electron acceptor), and oxygen (electron donor), hydrated cations, and halogen.

Electronic transitions occur when the energy of the incident light matches or is greater than the quantum energy separating the ground state and an excited state. The transition between the $\sigma \rightarrow \sigma^*$ and $n \rightarrow \sigma^*$ molecular orbitals is very energetic and occurs only in the vacuum ultraviolet and far-ultraviolet region. Whereas the transitions between $\pi \rightarrow \pi^*$ and $n \rightarrow \pi^*$ molecular levels are the least energetic and occur with the absorption of light in the ultraviolet and visible regions [11]. The "n" electrons are lone-pair electrons that are non-bonded, sigma (σ) electrons are involved in the formation of signal bonds, and pi (π) electrons form double bonds. When light passes through a compound, some of the energy is used to promote an electron from one of the bonding or non-bonding orbitals into one of the anti-bonding orbitals. The energy gaps between these levels determine the frequency (or wavelength) of the light absorbed, and these gaps are often distinct in different compounds.

The area of the molecule where the electronic transitions happen is called a chromophore [12]. When the molecule is transparent in this region, derivatization and complexation reactions can significantly enhance absorption [10, 13]. When derivatization is used, the sensitivity and sometimes the selectivity of the UV–vis methodology increase [14].

For UV–vis measurements of liquids, the light passes through a sample. The determination begins with analyzing a reference sample, often called the "blank sample." A cuvette is filled with a similar solvent used to prepare the sample. For example, when an aqueous buffered solution containing the sample is used for measurement, the buffered solution without the analyte of interest is used as the reference. The instrument uses the reference spectrum to remove the contributions from the solvent to obtain the true absorbance values of the analytes.

The UV–vis method can be developed to determine the optical properties of solids by measuring in reflectance mode and converting to absorbance or Kubelka–Munk units that are more correlated to concentration [15]. The optical and electronic properties of solids, films, and powders can be characterized in the UV–vis region. Measurements are frequently used for quantitative applications in many industries, including pharmaceutical, food and beverage, and chemical [16]. In some applications, auxochromes may intensify the color produced by molecules or cause a red or blue shift in the absorption band [17]. These molecules are attached to a chromophore and can modify the ability of the chromophore to absorb light.

Molecules that absorb ultraviolet and visible light generally have very large absorptivity constants, and thus, sensitivity using this spectroscopy is very high. Methods can be developed to measure molecules at the ppm concentration level. The main disadvantage is that only certain molecules will absorb light in this region, and specificity can be poor in some cases [16]. Molecules can have very different molecular structures, and yet their absorption bands are found in similar locations, making interpretation challenging. However, once the samples are well characterized, very efficient quantitative applications can be developed and used for routine testing of aromatics and other absorbers.

Like other spectroscopies, UV–vis spectroscopy procedures can be developed for non-destructive determination, allowing the sample to be reused for further processing or analyses by other techniques. Measurements can be made quickly, allowing easy integration for manufacturing or rapid determinations when information is needed quickly. Instruments are inexpensive and easy to operate requiring minimal user training prior to use.

1.3 Theory of Colorimetry (Visible Spectroscopy)

When we pass light through a prism, white light is separated into the colors of the rainbow. The spectrum is arranged in the order of indigo, blue, green, yellow, orange, and red, from shortest to longest wavelengths. In 1666, Isaac Newton, a 23-year-old Cambridge graduate, performed experiments with sunlight (white light) using a simple prism to separate the light into individual wavelengths of color to describe what the prism was doing to light [18].

The light reflected from solids and transmitted through liquids becomes the color of the object that is perceived. The specific wavelengths of light radiated stimulate the eye's retina, which in turn transmits this information to the brain [19]. A spectrophotometer can also measure the same colors observed by the human eye using color-matching functions $\bar{x}(\lambda)$, $\bar{y}(\lambda)$, and $\bar{z}(\lambda)$ applied to the transmittance or reflectance spectrum measured from 380 to 780 nm [19–23]. The appearance of color is different under different lighting

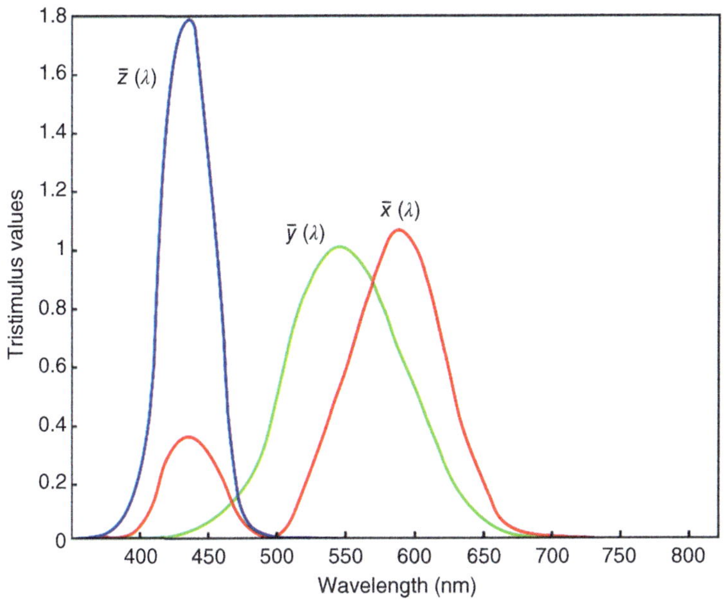

Figure 1.5 CIE 1931 Guild and Wright color-matching functions for the *X* (red line), *Y* (green line), and *Z* (blue line) from visible spectra. *Source*: Recreated by Deborah A. Peru.

conditions, so the color temperature of the illuminant must be specified along with the viewing angle when describing color from spectra.

The Commission Internationale de l'éclairage (CIE) developed the color-matching equations in 1931 using the specific regions of the visible spectrum as shown in Figure 1.5.

At each wavelength, the standard observer functions $\bar{x}(\lambda)$, $\bar{y}(\lambda)$, $\bar{z}(\lambda)$ are multiplied by the CIE illuminant and the reflectance or transmittance value of the sample at each wavelength [20, 23]. The values for all the wavelengths are then summed and integrated to determine the CIE X, Y, Z color values as depicted in Eqs. (1.3)–(1.5):

$$X = k \int S(\lambda)\bar{x}(\lambda)R\lambda \tag{1.3}$$

$$Y = k \int S(\lambda)\bar{y}(\lambda)R\lambda \tag{1.4}$$

$$Z = k \int S(\lambda)\bar{z}(\lambda)R\lambda \tag{1.5}$$

S is the relative spectral distribution power of the illuminant, $\bar{x}(\lambda)$, $\bar{y}(\lambda)$, and $\bar{z}(\lambda)$ are the color-matching functions for the standard observers. $R(\lambda)$ is the sample's spectral reflectance or transmittance value [20, 23]. *k* is a normalizing factor determined with a white reflectance standard to obtain a *Y* value of 100. ASTM E 308 describes these procedures and contains tables of the CIE Standard Observer functions for 2° established in 1931 and 10° established in 1964 [21].

Many applications require converting the visible spectrum into parameters that describe how humans perceive color. Color is a useful parameter to measure in many quality control applications. Simple transforms can convert the CIE XYZ values to other common color spaces, such as RGB, CIE L*C*h, and CIE L*a*b* color space models [20]. In the latter, L* represents lightness; 0 is perfect black; and 100 is perfect white. Positive a* represents redness, and negative a* represents greenness. Positive b* is yellowness, and negative b* represents blueness. Many near-infrared and ultraviolet instruments extend into the visible region, and thus, it may be useful to use the spectrum collected to track changes in color and chemical changes in your samples.

Quantitative applications can be built to track a specific change in color or hue between two measurement points. This type of measurement may be needed during stability studies, clinical studies, or to measure product performance and quality compared to a gold standard. The total color difference is determined as the distance between two measurements using all the color parameters [19, 23]. The delta E (ΔE) parameter defines a change between two measurements of color using the CIE L*a*b* parameters:

$$\Delta E = \sqrt{\Delta L^2 + \Delta a^2 + \Delta b^2} \tag{1.6}$$

In some industries, a weighting factor can also be applied to the components of ΔE to create industry-specific indexes, such as the whitening index or yellowness index used in the dental and textile industries, respectively.

1.4 Basic Principles of Fluorescence Spectroscopy

Two other spectroscopies in the visible region are used for qualitative and quantitative assessments of molecules. Fluorescence and phosphorescence measure the emission or release of light energy following an absorption event. Aleksander Jabłoński, a Polish physicist regarded as the father of fluorescence spectroscopy, devoted his life to the study of molecular absorbance and light emission. He is famously known for developing the Jablonski diagram, often cited and used to explain the kinetics of fluorescence, delayed fluorescence, and phosphorescence [24].

Fluorescence occurs in conjugated molecules that often contain a ring structure [25]. When molecules interact with light energy in the visible region, electrons in the lowest-energy level (S_0 ground state) become excited and transition to an excited energy level (S_1). Within the excited state, electrons may undergo any number of processes that can deactivate the excited state (S_1 energy level), resulting in loss of excess energy and return to the ground state (S_0). These processes include fluorescence emission ($S_1 \rightarrow S_0 + h\nu'$), internal conversion ($S_1 \rightarrow S_0 + \text{heat}$), vibrational relaxation, intersystem crossing ($S_1 \rightarrow T_1$), and collisional quenching ($S_1 + Q \rightarrow S_0 + \text{heat}$) [11]. This emission of light photons occurs at much longer wavelengths and with lower energy than the incident wavelength used for excitation because energy is transferred or lost as heat within the molecule.

Fluorescence spectrometers are often designed to measure the excitation and emission spectrum of samples, providing valuable insights into fluorophore behavior. Excitation spectra measurements reveal the optimal excitation wavelength for a particular fluorophore to produce the greatest fluorescence response, whereas the emission spectrum shows the relative intensity of fluorescence as a function of wavelength.

Simple and advanced applications have been developed to determine the concentration of the analytes and other unique properties of materials. The fluorescence intensity is directly proportional to the concentration of the fluorescent molecule. Thus, a calibration curve can be created to determine the concentration of the same molecule in unknown samples.

Quenching experiments can also be performed to track the decrease in intensity of the fluorophore signal as additives are added systematically. Quantitative methods can track structural changes in proteins, such as folding/unfolding and association/dissociation, often used in structure-based medicine development studies [26]. Time-resolved fluorescence measures fluorophore response as a function of time after excitation by a pulse of light [26]. Quantitative methods combining the emission and absorbance spectra using chemometric multiblock algorithms have been used in food, cannabis, and other industries [27]. For clinical studies, fluorescence spectroscopy has the advantages of being sensitive and selective, provided a fluorophore is present. Many applications exist, including detecting microbial biofilms in clinical applications [28]. Additional resources on fluorescence theory and applications are available elsewhere [24, 28–30].

1.5 Basic Principles of Near-Infrared Spectroscopy

The near-infrared region is between the visible and infrared regions and was largely ignored up until the mid-1960s because it was considered to lack relevant chemical information. The bands consist of overtone absorption bands appearing at frequency positions equal to two, three, and four times the fundamental band frequency and combination bands, which are the sum of two or more fundamental vibrational frequencies within the molecule involving the same functional group with the same symmetry [31]. Near-infrared vibrational bands are often broad and severely overlapped, making interpretation more difficult. The spectral intensity is much weaker than the sharp, intense, fundamental bands in the mid-infrared region. However, there are several unique spectral features measured in the near-infrared region that occur due to a property called anharmonicity. Anharmonicity is a deviation from the classical mechanic harmonic oscillation of molecular bonds, causing overtones and combination bands to occur at inexplicit wavelength positions. The non-idealities that appear have made near-infrared a powerful technique, particularly for measurements in aqueous environments. Owing to anharmonicity, the effect of hydrogen bonding, a predominant feature in aqueous samples, is much different in the near-infrared region than in the mid-infrared spectrum. For example, when hydrogen bonding occurs, an electron-donor molecule often exhibits a decrease in intensity in its respective overtone and combination bands. In contrast, overtones and combination bands from non-hydrogen-bonded molecules become very intense [31].

Near-infrared has become a very powerful method in many industries because of the unique fingerprint caused by small changes in the chemical and physical environment of the sample. Similar to mid-infrared spectroscopy, the factors that affect band intensity and band position include whether there is a dipole moment created, the symmetry of the molecule, atomic mass, bond strength, and the type of vibration mode allowed. The most intense vibrations in the near-infrared region are due to molecular bonds involving C—H, N—H, and O—H stretching and bending vibrations along with C—O, C—N, and N—O stretching owing to a larger dipole moment.

Because the spectral fingerprint is rich with hydrogen bonding information, many useful applications can be developed using this technique to exploit subtle changes in the fingerprint resulting from interactions between polar molecules and bound and unbound forms of water [32, 33]. Among the early adopters for *in vivo* measurements was the application of measuring moisture content in skin [34, 35]. Owing to the excellent signal-to-noise of most near-infrared instruments and robust hardware design, extensive use of near-infrared for routine quantitative determinations exists in most industries. Process analytical technology (PAT) applications have also dominated the use of near-infrared spectroscopy in manufacturing environments using fiber-optic probes for monitoring during production, including biopharmaceutical processes [36].

1.6 Basic Principles of Mid-Infrared Spectroscopy

When light from a mid-infrared source, such as a tungsten or incandescent lamp, is absorbed by molecules, the bonds absorb energy, which causes them to vibrate with frequencies equal to the energy absorbed. The frequencies or positions of the vibrational bands are used for molecular identification, whereas the intensity is used to determine concentration. The bonds between atoms can undergo stretching vibrations, both symmetric and asymmetric, and any number of bending vibrations (i.e., scissoring, wagging, twisting, and rocking). This yields a unique vibrational fingerprint for each molecule in a sample.

Hooke's law describes the relationship between the frequency of oscillation and bond strength [37]. The stronger the bond force constant, the higher the vibration frequency. For example, O—H, N—H, and C—H have strong bond force constants and thus appear at higher frequencies than the weaker C—C or C—O bond vibrations. Chemical environment and physical interactions can also influence the vibrational frequency and intensity.

Molecule symmetry will also influence the number of active vibrational modes. In theory, linear molecules have 3N-5 vibrational modes, and non-linear molecules have 3N-6 vibrational modes or vibrational degrees of freedom, where N is the number of atoms in the molecule. For example, H_2O, a non-linear molecule, has three vibrational modes that appear in the mid-infrared spectrum. A broad band appears near 3500 cm^{-1} caused by two unresolved bands near 3404 cm^{-1} assigned to the symmetric stretching (ν_1) near 3280 cm^{-1} and the asymmetric stretching (ν_3) near 3490 cm^{-1}, and a third band is observed at 1643 cm^{-1} due to the ν_2 bending vibration [38]. The water molecule does not have rocking, wagging, and twisting modes of vibration because these types of vibrations would involve the whole molecule rather than just the atomic bonds within it.

Mid-infrared instrumentation is widely used for both qualitative and quantitative applications. Over the past 60 years, mid-infrared spectroscopy has matured from a slow and time-consuming laboratory technique to a rapid go-to technique for most chemical investigations. Many highly sensitive and robust quantitative applications have been developed and used in the medical/clinical, industrial/manufacturing, and environmental industries, among others [1]. With the rapid improvement and growing popularity of miniaturized devices, future trends are field deployable and *in situ* instrumentation for diagnostics and point-of-care applications.

1.7 Basic Principles of Raman Spectroscopy

Raman spectroscopy is a technique used to analyze the vibrational and rotational modes of molecules. It is based on the inelastic scattering of light, known as Raman scattering or the Raman effect [39]. As illustrated in Figure 1.6, when a molecule is irradiated with a monochromatic laser, most of the scattered light undergoes elastic scattering, known as Rayleigh scattering, which does not change the energy of the scattered

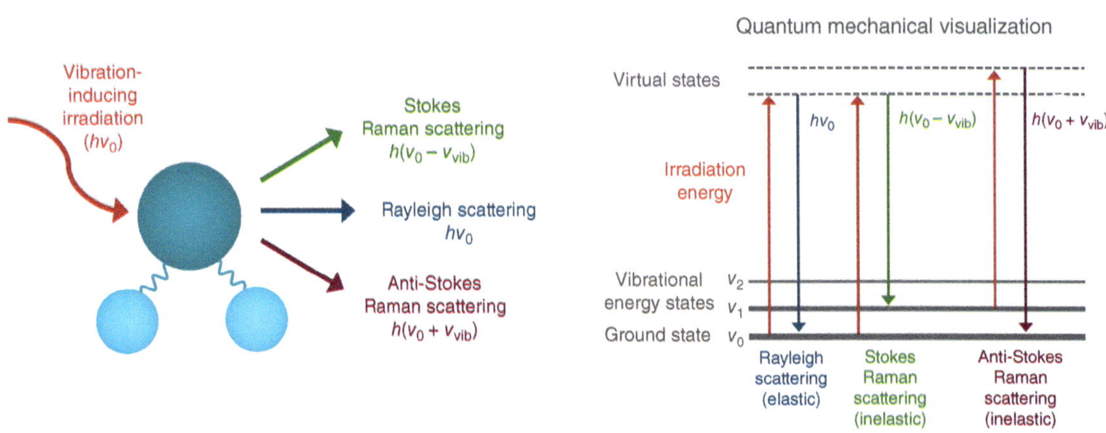

Figure 1.6 Illustration of Raman scattering and its electronic diagram. *Source*: With permission from Metrohm.

photon. However, a small fraction of the scattered light undergoes inelastic scattering, leading to a shift in energy. This shift can be either lower in energy (Stokes scattering) or higher in energy (anti-Stokes scattering) than the incident light source. The energy difference corresponds to the energy associated with the vibrational or rotational modes of the molecule.

In Stokes scattering, the incident photons lose energy and are scattered with longer wavelengths, while the energy lost is transferred to the molecule, resulting in molecular vibration. Anti-Stokes scattering occurs when the molecules are already in an excited state. In this case, the scattered light photons gain energy, resulting in shorter wavelengths and higher energy than incident light. By detecting the vibrational frequency and intensity of the scattered light, the Raman spectroscopy technique can provide information about a sample's molecular composition, structure, and concentration.

Raman spectroscopy has a wide range of applications, including quantitative analysis and material characterization [39, 40]. It is often used in biomedical, pharmaceutical, and clinical research. It has widely been used in biopharmaceutical manufacturing in all stages of production, from the earliest stages of the manufacturing processes (e.g., raw materials) to online and offline monitoring of bioprocesses and final product characterization [40]. Among its strengths is the ability to develop procedures for direct analysis through plastic, glass, and turbid media, and with fiber-optic probes, it can be used *in situ* and process environments. Its biggest challenge is the overall lack of sensitivity because of the weakness of the Raman scattering process. However, specific applications involving the use of gold or silver nanoparticles in close proximity to the analyte of interest result in a localized plasmon-mediated amplification of an electric field resulting in an enhancement in Raman scattering [41]. This technique, surface-enhanced Raman spectroscopy (SERS), has significantly improved the sensitivity for low-level detection [41]. As you will learn throughout this book, with a good experimental design in hand, robust analytical methods can be developed and implemented that often overcome many limitations and challenges.

1.8 Analysis of Liquids

Liquids are the easiest type of sample to measure. Samples can simply be poured into a plastic or quartz cuvette and placed into a spectrophotometer for an absorption measurement. In transmission mode, light radiation passes through a sample, and photons that are not absorbed by the sample are measured with a detector. The pathlength is fixed based on the measurement cuvette or vial size, as shown in Figure 1.7. For ultraviolet, visible, fluorescence, and near-infrared spectroscopy the analyst often can measure samples "as is." During method development, the analyst will need to prepare samples using a variety of cells to identify the optimal pathlength and cuvette material to yield a spectrum with the highest sensitivity and no detector saturation. Equally important is the ease of sample preparation and cleanup procedures.

A powerful advantage of transmission sampling techniques is that the beam passes through the entire sample. Thus, bulk chemistry is predominantly measured compared with other sampling accessories that only provide chemical information on the surface of the sample.

Figure 1.7 Transmission measurement of liquid. *Source*: Deborah A. Peru.

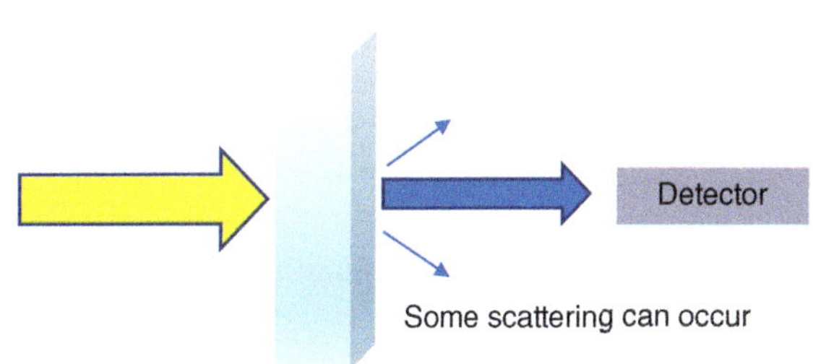

The two disadvantages of transmission sampling are the possibility of detector saturation when absorbancy is higher than 1.5 units, and when dissolved particles scatter light. Both can lead to non-linearity in the relationship between absorbance and concentration. When the sample is opaque, the transmission of light may be difficult, unreproducible, or impossible. In ultraviolet spectroscopy, dilution with a non-interfering solvent may be required when the absorptivity or concentration of the molecule is high. However, dissolution of solids in certain solvents may result in interactions, such as hydrogen bonding, which can cause band distortion or shifting, particularly at lower concentrations [42]. Notwithstanding, Dang et al. developed an ultraviolet procedure for the determination of ibuprofen, paracetamol, and caffeine during dissolution testing of commercial tablets and capsules in buffered solutions [43]. This procedure provided comparable precision and accuracy to HPLC and was more eco-friendly, cost effective, and offered considerable time savings.

Some vendors may offer extended-range detectors that are linear above two absorbance units, which can help extend linearity in some applications. However, the linear relationship between absorbance and concentration can change at very high and low levels, requiring multiple calibration curves for the most accurate determination over a wide concentration range.

The transflection (a.k.a interactance immersion) probe is a fiber-optic-based sampling accessory used to measure liquid samples remotely. The liquid moves into the cavity between two measurement plates, carefully adjusted for the proper path length. Light is sent to the sample through a fiber bundle (one or more fibers). A mirror or highly polished stainless-steel surface reflects the remaining light after sample absorption back through another set of collection fibers to the detector. In a transflection measurement, the light travels twice through the sample. Once going toward the mirror and again going back to the detector. Thus, the pathlength is twice the distance between the plates. This sample accessory is useful for measurements of emulsions, gels, and turbid liquids such as milk or inline measurement of fermentation slurries in a production environment [36].

In mid-infrared spectroscopy, most materials are highly absorbing. Thus, transmission often requires diluting the sample with a low-absorbing solvent or measuring a very thin layer of sample, as small as 1–20 μm. In the modern laboratory, attenuated total reflectance (ATR) is often the most convenient sample accessory for analyzing liquid samples. This type of sampling cell eliminates the problem of detector saturation because the accessory severely reduces light throughput [1]. A small sample (several drops) is placed on the ATR sensor's surface. The light interacts with the sample *via* an evanescent electromagnetic wave generated at the interface between the sensor's surface and the adjacent sample [1, 44]. The crystal's refractive index must be significantly greater than the sample's, or internal reflectance will not occur [44]. Several crystal materials are used in ATR accessories. The most frequently used crystal for a traditional ATR sensor is diamond because it is the most durable and chemical resistant. Other common crystal materials used are zinc selenide (ZnSe) and germanium (Ge).

Many vendors offer advice and a variety of liquid sample accessories to help the analyst optimize the proper liquid accessory for analysis. Table 1.2 provides a summary of different types of liquid sampling accessories.

1.9 Analysis of Solids

The interaction of light with solids, powders, grains, pellets, thin films, etc., is more complex and may require more careful attention to details to create robust methods. Light interacting with a solid can be reflected, refracted internally, or scattered in all directions. The incident light may be absorbed by particles within the sample or emitted by the combined process of reflection, refraction, and diffraction [45].

Table 1.2 Common types of liquid sampling cells and accessories used for quantitative applications.

Sampling cell type	Technique	Advantage (A) Disadvantage (D)
Cuvette/microcuvette/ sipper cell/flow cell	Ultraviolet Visible Fluorescence Near-infrared Raman	(A) Microfluidic measurements for real-time analysis (D) UV-grade quartz cells is expensive
Glass dram vials and scintillation vials Plastic cuvettes	Visible Near-infrared Raman	(A) Low-cost cells (A) Disposable eliminates clean up
Sight glass	Near-infrared Visible	(A) Variety of windows available (quartz, low OH glass, or plastic) for process monitoring
Interactance immersion probe	Near-infrared Visible	(A) Real-time analysis of liquids (A) Variable pathlength (D) Prone to air bubbles on window
Transflection (diffuser used for reflectance analysis of liquids)	Near-infrared	(A) Ideal for thick samples that are difficult to pour into cuvettes (A) Variable pathlength diffusers (D) Clean up required
Sealed window cells	Mid-infrared	(A) Variety of cell materials (D) Difficult to clean
Salt window plates (NaCl or KBr)	Mid-infrared	(A) Easy to use and clean (D) Organic materials only (not for aqueous samples)
Attenuated total reflectance (ATR)	Mid-infrared	(A) Chemical inert for most samples (A) Easy to clean after use

The emitted light can then be absorbed or emitted by other particles in the sample. The measurement's accuracy often depends on the instrument design, particularly the detector collection efficiency. The sample accessories are designed for measuring different types of reflectance properties (specular, internal, and diffuse). In the last decade, transmission through solid materials using a special high-energy light source has become routinely used for measuring pharmaceutical tablets in the near-infrared region [46, 47].

Quantitative procedures for analyzing solid samples is difficult, but possible using an ATR accessory for mid-infrared measurements. ATR uses the principle of internal reflectance where light is introduced into a prism at an angle exceeding the critical angle for internal reflection. To ensure good contact with the diamond surface of an ATR, a calibrated pressure arm is placed on top of the sample to deliver a constant amount of force [44]. For obvious reasons, ATR is typically not used online or inline for solids determination. However, with care, quantitative offline applications can be developed following the Lambert–Beer relationship.

A homogeneous sample is required for analysis because of the small surface area measured. Because measurement depth is only one to a few microns into the material, consideration must be taken when interpreting the results since surface chemistry will predominate the spectral fingerprint. Surface effects such as oxidation or other chemical changes can also cause misinterpretation of the spectrum, a common occurrence in plastics [48]. Germanium has the highest refractive index, meaning that the effective penetration depth is approximately 1 μm compared with several microns for diamond [44]. For most solid samples, this will result in a very weak spectrum. However, this is generally an advantage when analyzing highly absorbing materials such as black-carbon-filled rubber [44].

Mid-infrared and near-infrared instruments can measure solid samples in diffuse reflectance mode. Diffuse reflectance measures reflected light from solid matter at many angles. The amount of light absorbed by a material at each wavelength is directly proportional to the log(10) 1/R, where R is the intensity ratio of reflected light (I) to incident light (I_0).

$$\text{Absorbance} = \log(10)\frac{1}{R} = -\log R; \quad \text{where } R = \left(\frac{I}{I_0}\right) \tag{1.7}$$

It is important to properly characterize and control packing density and particle size during method development in diffuse reflectance mode since this can affect the scattering properties and reproducibility in quantitative measurements for all spectroscopy techniques.

Mid-infrared diffuse reflectance, sometimes referred to as DRIFTS (diffuse reflectance infrared Fourier transform spectroscopy), is commonly used to analyze both organic and inorganic samples. Mid-infrared radiation interacts with the particles and then reflects off their surfaces, causing the light to diffuse, or scatter, as it moves throughout the sample. Some samples can be measured "as is," but most are too absorbing and must be ground into a fine powder (less than 10 μm) and mixed with an inert diluent such as potassium bromide [49]. The samples are loosely and evenly packed in the sample cup to increase the penetration depth contribution of the diffusely scattered components while minimizing the contribution of the specular reflection components [50]. To properly correlate the peak intensities with concentration, a Kubelka–Munk correction is often required.

An integrating sphere accessory can be used with near-infrared, visible, and ultraviolet instruments. The accessory directs a broad beam of parallel light onto a sample positioned in the sample viewing port. The reflected light is well distributed within the sphere, resulting in multiple interactions with the sample [51]. *In vivo* collection of reflectance spectra of human skin spectra collected in the UV-to-near-infrared region (250–2500 nm) has been published by NIST using an integrated sphere [52].

In the visible region, the integrating sphere measures both diffuse reflection (color) and specular reflection (glossiness or shine) components by opening or closing the specular port [51]. Total reflectance is the sum of the specular reflectance and diffuse reflectance components. Because the perceived color of an object may change depending on the surface shine, the measurement is often made with the specular port closed, a.k.a. specular-included (SCI) mode. In specular-excluded mode (SCE), the specular reflectance is excluded from the measurement, and only the diffuse reflectance is measured. When light is specularly reflected from the surface, it appears at the same but opposite angle as the incident light and no absorbance occurs by the sample.

Some visible instruments are specifically designed to measure the % specularity of a sample quantitatively. Two measurements are made to measure specular reflection: in SCI mode and in SCE mode [23]. The % specularity is determined using the following equation:

$$\%\ \text{Specularity} = \frac{\text{SCI–SCE}}{\text{SCI}} \times 100 \tag{1.8}$$

1.10 Analysis of Viscous Materials and Slurries

Many of the same solid sampling accessories discussed can also be used to measure pastes, slurries, emulsions, and viscous polymers. A diamond ATR accessory is ideal for measuring viscous materials by mid-infrared spectroscopy. Care must be taken, however, when selecting the type of ATR crystal. For example, zinc selenide (ZnSe) is a lower-cost ATR material than diamond, but it scratches easily and degrades at certain pH levels.

For Raman determination, samples can be prepared in glass dram vials or scintillation vials for direct analysis through the glass wall with proper adjustment of the measurement focal point. This approach simplifies sample preparation and clean-up time.

An integrating sphere is a good option for slurries, viscous samples, or scattering samples using a sealed sample cup or quartz cell containing the sample mounted within the sample port [53]. An integrating sphere has been used to demonstrate the optical properties of dense algal and turbid milk samples where multi-scattering processes occur [53, 54]. The integrating sphere also has the advantage of measuring a larger sample size (700 mm^2) compared with a fiber-optic diffuse reflectance probe (9–10 mm^2) [55].

A diffuse reflectance accessory can be used in transflection mode with near-infrared instrumentation to analyze viscous liquids. Samples are poured into a quartz beaker and placed directly on the measurement window (typically sapphire) for analysis. A gold or polished stainless-steel diffuser reflects light that is not absorbed back to the detectors. The diffuser is configured with the appropriate spacing to achieve the best pathlength to avoid detector saturation. Alternatively, samples can be measured through high-quality flat-bottom glass vials with a diffuser.

An established procedure in the visible region is the determination of total suspended solids (TSS) content in liquid slurries [56]. The presence of dissolved particles blocks the transmission of light to the detector caused by light-scattering properties of the particulates. At low levels, the turbidity of samples is linearly related to particle concentration. Turbidity can be indirectly measured using a visible spectrophotometer for material characterization in research and development projects and quality assurance testing in some industries. However, variances in particle size and matrix effects can disrupt this relationship [57].

1.11 Instrument Calibration and More

Most types of quantitative applications require the development of a calibration curve to predict analyte concentration in samples. This is referred to as "instrument calibration." In spectroscopy, we typically refer to the determination of concentration as a prediction, particularly when calibration standard properties are assigned using a primary method. When developed correctly, quantitative procedures using spectroscopy techniques are often accurate and, in some instances, more reliable than other instrumental or wet chemical quantitative procedures.

A calibration algorithm is developed using one or more peak heights (measured in absorbance or intensity units), a peak area, or a spectral region or a combination of spectral regions of the spectrum. This is known as the independent or [X] block variable set. The concentration or property of interest is the dependent or [Y] block variable set. An example of a calibration curve developed using partial least squares (PLS) regression is shown in Figure 1.8. The replicates for each sample are included in the calibration to demonstrate precision in the method throughout the calibration range of concentrations.

The samples used for developing a regression equation are often referred to as the "training set" because we essentially use these standards, with known properties, to teach the spectrophotometer, using math, to accurately predict the property of interest in the unknown sample. The calibration standards can be prepared

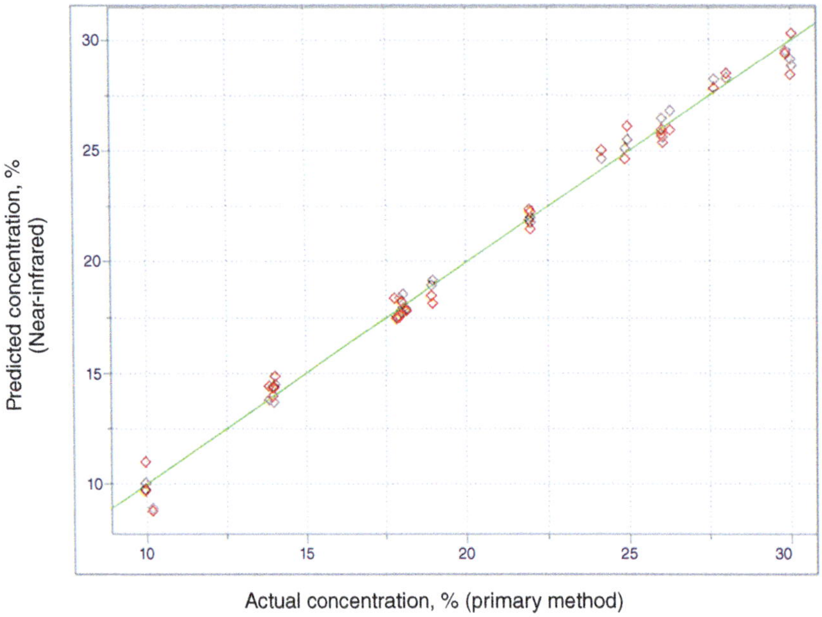

Figure 1.8 Example near-infrared calibration curve developed using partial least squares (PLS) regression. *Source*: Deborah A. Peru.

in the laboratory or pilot plant. Concentrations can be assigned based on the analytical mass or volume [58], provided no material losses occur. However, these values are generally determined by a primary method and assigned to the corresponding spectrum(s).

In Chapter 4, the author provides an in-depth look at how to extract quantitative information from spectral data using a variety of direct and inverse modeling methods, including classical least squares (CLS), multiple linear regression (MLR), and PLS regression, among others. One thing you will find is that most quantitative applications involve the use of more than one wavelength. This is almost always true in near-infrared, ultraviolet, and, frequently, mid-infrared spectroscopy since one response variable (wavelength) in the spectrum generally does not have sufficient selectivity to monitor a specific property without experiencing interferences from other components or variables. Using more of the [X] variables provides a signal average benefit and provides sufficient wavelengths to selectively remove matrix variances unrelated to the component of interest using preprocessing techniques. Once an instrument is calibrated and validated, it can predict the concentration or property of interest in unknown samples. However, the calibration curve must capture the variances that influence the relationship between the spectral response and concentration.

As you will learn in this book, quantitative applications can be developed using other approaches that do not involve developing a calibration curve beforehand. Figure 1.9 describes the various mathematical techniques used for converting spectral data into quantitative information.

In addition to developing a regression equation for predicting a property of interest, a "change" in spectral response, such as in a peak area from a baseline value, to track a statistical change in the property of interest can be used (i.e., parametric modeling). This is likely the approach used in clinical trials involving spectroscopy.

Additionally, a quantitative method can be developed simply using the relative peak area or relative peak height. The ratio of the band height, in absorbance or intensity units, or peak area, between the unknown sample and a reference material multiplied by the concentration of the reference material is often used to determine the relative concentration of the unknown sample. This approach can be used to monitor

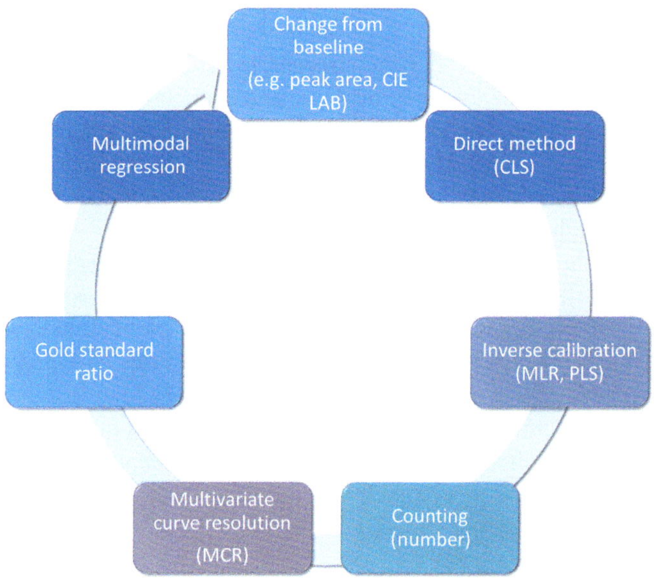

Figure 1.9 Mathematical techniques used to extract quantitative information from spectral data. *Source*: Deborah A. Peru.

the progression of yield during chemical synthesis, consumption of reagents, endpoint detection, and to determine purity. The approach is often used in early-stage research to optimize reaction conditions, provided a pure material or gold standard exists. The same approach can be used in root cause analysis investigations when spectroscopy is used to solve product performance problems. There are examples found in the biotechnology arena when a calibration curve is not necessary, such as optical density measurements for bacterial culturing [16].

Multivariate curve resolution (MCR) is often used in the laboratory, pilot plant, and sometimes during manufacturing when spectral data is collected over a time domain. The method is designed to extract the two underlying response profiles, i.e., the pure-component spectral fingerprint and their respective concentration versus time profile. MCR determines the composition of a complex mixture without *a priori* knowledge of the number of components, pure-component fingerprint, or reaction order [59].

Lastly, these same spectroscopy techniques can be used for spectroscopic imaging of samples. One application discussed in this book involves using Raman for microplastic analysis. The purpose of including this application is to introduce the reader to other avenues for using light-based instruments for quantitative applications that go beyond developing a regression curve.

1.12 Where Quantitative Methods are Used

Spectroscopists continue to find new ways to improve this technology and new ways to apply procedures both on Earth and in space. Spectroscopy techniques will undoubtedly play a significant role in addressing our ever-increasing global challenges [60]. The following sections review recent trends in using spectroscopy for the product development lifecycle and beyond in clinical, environmental, and forensic arenas. Appendix I, found in the *Wiley Companion Website*, provides additional references for quantitative applications by technique.

1.12.1 Early Research

Spectroscopy is ubiquitously being used in all industries for a variety of purposes. In the early stages of the product development cycle, spectroscopy techniques can be used to provide an understanding of the interactions between ingredients. This is part of the FDA's quality-by-design principles [61]. *In situ* monitoring at the beginning of the product development lifecycle can provide a better understanding of product chemistry and helps to identify critical processing points. The learnings are used to create methods for monitoring and controlling critical points to maintain quality throughout production.

Methods can be developed for performance testing of new prototypes, such as dissolution testing of pharmaceutical drugs or efficacy testing of new products that deliver a unique benefit [62]. Spectroscopy is useful for describing how a new product works in patent applications. One example provided is the use of ultraviolet and near-infrared spectroscopy for measuring the deposition of an active ingredient in dentifrice on hydroxyapatite disks, a model substrate typically used in developing new dentifrice formulations [63]. Many patents are granted that pertain to new products or process innovation, but few are commercialized or put into use yet demonstrate a company's strong innovative pipeline. Additionally, quantitative information using a spectroscopy procedure may provide the necessary evidentiary proof of how product works to support new product claims and in legal challenges.

1.12.2 Research and Development (R&D)

In research and development (R&D), quantitative methods can be developed to characterize raw materials for purity level or to rapidly measure the presence of contaminants such as water [64]. Water can quickly degrade food products and pharmaceuticals or create an environment suitable for rapid microbial growth [65]. Analysis of water content in materials for optimization of a lyophilization process and for quality control is critical for product safety and quality [66]. Control of moisture content in pharmaceutical solids, both raw materials and solid dosage forms, is a challenge to pharmaceutical development and manufacturing. Measurement of pharmaceutical solids often requires lengthy sample preparation methods for moisture determination. A non-invasive spectroscopy approach for moisture determination provides a more rapid and efficient approach for assessing moisture variability and for continual process qualification [67].

During accelerated aging (stability) studies, spectroscopy procedures can measure and track the concentration of key ingredient(s) in new products, when both temperature and humidity are at extreme levels. Spectroscopy is a faster approach for monitoring variation over time, requires fewer resources, and at a lower cost than chromatography methods [68]. The non-destructive nature of spectroscopy is a unique feature and reason for its use in stability studies since the inter-sample variability is eliminated, and each sample can be its own control. Additionally, the instrumental determination of product color attributes during stability studies is more precise and often can quantify changes sooner than visual examination [69].

In the pharmaceutical industry, the purity and dose levels in drugs must be controlled throughout the product lifecycle [70]. The additional regulatory demand for "evidence of control" emphasizes risk-based specifications, testing composition, and stability [70]. Spectroscopy plays an important role in assisting compliance with the generation of stability data and the evaluation of risks associated with a manufacturing process.

In consumer products, near-infrared coupled with chemometric data analysis methods have demonstrated the ability to predict the shelf-life of body lotion. The same approach has potential application in other industries, including cosmetics, pharmaceuticals, and foods [71].

1.12.3 Scale-Up of New Products and Processes

Often the scale-up of reactions from laboratory to pilot plant or full-size manufacturing plant can present many challenges. During the initial stages of scale-up, *in situ* spectroscopic monitoring is invaluable when labile intermediates are generated, equilibriums are perturbed by sampling, fast reactions obliterate the use of offline testing, dealing with expensive reagents and limited volumes, and when analytes are sensitive to atmospheric oxygen and moisture [72]. Samples can be analyzed "as is" without adulteration of the native chemistry. This provides a more accurate and objective determination of the chemical changes occurring within the batch or making tank [73]. During the technology transfer stage from laboratory scale to commercial scale, the formulation developed at the laboratory scale is generally fixed; however, the process parameters are changed. For example, as the size of the container used in the granulation process increases, the powder's volume or weight increases, and the impeller size and range of operating speeds can change [73]. These changes may lead to a change in the overall product quality. Successful scale-up is achieved when the process parameters are appropriately adjusted to achieve the same quality observed in the laboratory scale [73]. For chemical synthesis and polymerization processes, heat removal, degree of homogeneity, pressure, temperature, and reaction time should be nearly the same after scale-up [73, 74].

Concentration profiling can help optimize process parameters such as mixing speed, dwell time, order of addition, and drying procedures for best product performance and quality attributes. For example, Maltese et al. established a method for analyzing phenol-to-insulin ratio content, a critical quality attribute (CQA), by mid-infrared and near-infrared spectroscopy. The study focused on determining the chemical differences when the product was exposed to different dehydration (freeze drying versus spray drying) procedures [75].

1.12.4 Manufacturing

In manufacturing, spectroscopy can be used for in-process testing to verify the process is running within the design specification range. Process applications have been used for decades in pharmaceutical unit operations to ensure stages such as blending, tableting, or coating operations remain in control [76]. Spectral data and offline moisture content are often measured using spectroscopy procedures during batch operation of fluid-bed dryers in pharmaceuticals for endpoint determination and evaluation of mass transfer performance [77].

Hybrid operations utilizing batch and continuous manufacturing operations may strategically isolate smaller premix tanks for critical ingredient blends to reduce costs when an out-of-specification test result is found. Premix tanks are useful in consumer product making to isolate specific ingredients to ensure they are fully dissolved and to detect signs of degradation when held for long periods, often at elevated temperatures. In these processes, spectroscopy procedures are useful to ensure the premix is in specification before it is released to avoid contamination of the upstream process.

In continuous operations, a quantitative method in place during manufacturing provides an electronic "eyeball" for monitoring and controlling product chemistry as it is being made. The use of Raman spectroscopy in continuous manufacturing environments has been demonstrated since the early 1990s. Applications of Raman for bioprocess monitoring and control of ethanol, glucose, water, glycerol, and lactic acid in a laboratory scale through a production scale have been widely demonstrated [78].

Kim et al. demonstrated how PAT, including near-infrared and Raman spectroscopy, can improve the quality of drug products [79]. With continuous monitoring of unit operations, the framework for establishing a control strategy for real-time release testing was proposed [78].

1.12.5 Quality Assurance and Control

Spectroscopy methods have become the workhorse in many industries. They are replacing many compendial and primary methods owing to their speed of analysis, ability to test more samples quickly, and faster access to information needed for making critical decisions. Transmission Raman spectroscopy is a cost-effective, secondary method to HPLC, for quantifying active pharmaceutical ingredients in coated tablets [80]. The quality control procedure for the determination of nitrate (NO_3) in water and wastewater is based on the ultraviolet measurement of water at 220 nm, and a second reading at 275 nm corrects for interferences due to dissolved organic matter [81]. Mid-infrared spectroscopy is widely used for the detection of adulteration in foods, spices, and oils and for the determination of quality and safety [82]. And, finally the use of ATR mid-infrared instrumentation for the rapid determination of drug content in tablet dosage forms to eliminate substandard and falsified medicines [83].

1.12.6 Clinical Studies, Forensics, Environmental

Quantitative spectroscopy has made its way into the clinical testing world. Procedures are being applied for testing products and monitoring human and animal tissue samples in the dental, medical, and cosmetic fields. For example, tissue evaluation by fluorescence for tracking cancerous tumor growth by monitoring the ratio in the tryptophan and collagen bands [84]. A handheld filter-based fluorimeter has been used in clinical imaging of dental plaque on the teeth of subjects [85]. In this application, a blue light (avg. 404 nm) is used to excite fluorophores in the bacteria found in plaque. The detection is made with an imaging camera to quantitatively measure the release of red-color light (avg. 640 nm) in the visible region [85]. Demonstrating agreement between the clinical evaluation by dentists and a fluorescence imaging camera establishes the potential for monitoring oral hygiene with at-home devices using optical devices.

Researchers have shown both near-infrared and mid-infrared can be used to measure moisture on human skin [35, 86, 87]. Both the methods are non-invasive and can be applied in clinical assessment of new product performance. Using peak area extraction of several marker bands for moisture and proteins, researchers have demonstrated use of spectroscopy methods for the analysis of moisture in a wide variety of skin regions and with various skin conditions. Applications for using mid-infrared instrumentation with an ATR accessory have provided quantitative evidence of change in the stratum corneum of human skin as moisturizing treatments are applied [87]. Martin uses near-infrared reflectance measurements of skin to identify the presence of at least four types of water molecules: water associated with the lipid bilayers (@1875 nm), primary and secondary water molecules connected to protein groups (@1909 and 1923 nm), and a fourth type of water molecule @1890 nm assigned to bulk water content below the stratum corneum [35].

1.13 Defining the Problem

An important part of method development is to properly define the problem. Figure 1.10 outlines the preliminary questions to ask when discussing new opportunities for developing spectroscopic procedures with your customers, management team, and colleagues.

1.13.1 Background and Purpose

Gain as much background information as possible before beginning your journey in developing a quantitative method. Establish the criticality of your spectroscopy project and how it aligns with the overall business priorities. Most spectroscopists and engineers manage multiple projects simultaneously, so it is helpful to prioritize your projects and time appropriately.

Figure 1.10 Stages in asking questions to define the parameters of the application. *Source*: Deborah A. Peru.

1.13.2 Interfacing and Sampling Requirements

How will the samples be introduced to the spectrophotometer? At-line and offline analyses refer to the analysis of samples taken from a process. The instrument is located either near the location where the sample was taken or in a remote laboratory. If the sample is being measured through a product container, the variability in the container material will need to be assessed.

Inline refers to the analysis of samples in real time often with a fiber-optic probe inserted directly into the process. Online refers to the analysis of materials using a slipstream that diverts a portion of the material through a sensor or probe. Inline and online applications are typically used when the information from the spectrophotometer is used to take immediate action such as to adjust the process.

Identify the best approach for interfacing the sample with the instrument. If a new type of sample accessory or holder is required, it will need to be well characterized before use. In a cGMP environment, reusable equipment in direct contact with the product may be subject to cleaning and sanitization procedures and often requires the use of 316 L stainless-steel materials.

1.13.3 Chemical and Physical Properties

It is important to have a good understanding of the chemical composition of the unknown samples. When practicable, obtain samples of each ingredient for fingerprint analysis. Store the spectral fingerprints in a separate folder from the calibration "training set." The spectral fingerprint of the individual components provides information for:

- Interpretation of mixed standards and unknown samples
- Identification of important wavelengths and ranges for modeling
- Characterization of the specificity of the method.

The physical properties of the sample should be known when selecting the hardware and ancillary accessories used in a new procedure and can assist in sample preparation. Ask whether the sample is homogeneous or heterogeneous in nature. You may need to introduce a sample preparation method to improve the reproducibility of the measurements, such as filtering small particles in liquids, or perhaps vortexing to improve mixing of heterogeneous liquids. If the sample is corrosive, it may be safer to analyze the sample remotely using a fiber-optic probe with a chemically resistant housing (i.e., tantalum or Hastelloy). Answers will help create better procedures for sampling and "training set" development.

1.13.4 Concentration

The concentration range for unknown samples, calibration standards, and the validation set of standards should be identified. The target and range are important parameters for defining what type of spectroscopy technique(s) can be used. The choice of detector may also require knowledge of the concentration range and whether the sample needs to be diluted to avoid detector saturation or concentrated to enhance sensitivity.

1.13.5 Quantity and Frequency of Testing

The quantity and frequency of testing influence the type of interfacing required. Large numbers of samples to be measured daily might be a good candidate for implementing an autosampler or perhaps a process instrument, particularly if the determinations are used to control the process.

1.14 Plan for Development

Once you and your partners have defined the problem, the next step is to plan for development. Table 1.3 describes some of the early steps involved with planning and developing a new procedure.

Define the modeling components of the procedure. Does the procedure require only quantitative determination or both a qualitative and quantitative determination? In some applications, a qualitative component is often incorporated before a quantitative method to verify the spectrum is not an outlier or belonging to a different class of materials. This is often the case in the pharmaceutical industry when testing solid dosage forms [83].

Statistical requirements should be defined early on in development. Business-critical applications should be validated in accordance with industry standards. It is important that you identify the accuracy and precision required for your industry, risk tolerance level, and how and where in the process the information gathered will be used. Accuracy is defined as the closeness of agreement between the determined value and truth. Precision is defined as the degree of scatter between a series of measurements.

Knowing the history of the samples is important to avoid losing valuable time because of separation or chemical change. How the samples were obtained can dramatically affect the outcome of the analysis and the interpretation made. Proper sample handling procedures will ensure a representative well-preserved portion of sample is used for instrumental analysis providing true information.

Table 1.3 Steps to plan the development of a new procedure.

Define modeling components of procedure

- Only quantitative model required. Or, both a qualitative model, combined with a quantitative method
- Qualitative models identify, qualify, or classify spectrum based on a comparison with a library or reject a spectral fingerprint based on statistical parameters
- Quantitative procedures (predict concentration of ingredient or property) or apply statistical methods.

Statistical requirements

- Business critical procedures must be fully validated according to industry guidelines
- Procedures used for information gathering, and having little compliance risk, may be validated in accordance with company guidelines to ensure results are reliable
- Analytical method validation is a "best practice" (refer to Chapter 7).

History of samples

- Know history of samples to ensure composition remains intact during development
- Identify proper storage conditions
- Develop a proper sample handling procedure defined as the process of pulling a representative sample for analysis while preserving the chemistry.

Fit-for-purpose

- Select best technology (or combination of technologies) and accessories for project
- Modify prior methods to reduce development time
- Testing multiple spectroscopy techniques maybe required to fully understand the benefits and limitations of each approach.

Create a team and timeline

- Some projects require a team approach (e.g., PAT)
- Establish buy-in with partners early on
- Utilize project management tools to identify roles and responsibilities, key milestones, and timeline.

A fit-for-purpose framework for quality and reliability is an important aspect of ICH Q2 (R2) and the companion guideline ICH Q14. Identification of the appropriate combination of instrumentation to use for quality risk management, appropriate preventive maintenance, and sampling procedures are all part of development, as described in ICH Q9. Consult with your instrument vendor and colleagues and use external and internal literature to help advance method development stages. Pharmaceutical companies must establish, submit, and maintain evidence that a new analytical testing procedure is suitable.

Create project tasks, milestones, and a timeline. PAT projects that use instrumentation inline or online will unquestionably require a team approach with members from different disciplines such as engineering, R&D, analytical, quality, manufacturing, and compliance. Meet often and regularly to ensure members have a good understanding of the technology, the goal for using the procedure, and its vulnerabilities, and metrics for success. Many process applications require extra programming to use the information provided for process adjustments, and thus programmable logic control (PLC) engineers are essential.

1.15 Conclusion

This introductory chapter reviewed the basic theory of molecular spectroscopy techniques and how they can be used for the quantitative determination of key attributes in materials and products. Quantitative spectroscopy provides information for making critical decisions that can support business goals. Procedures can be used throughout the entire supply chain but may require creative thinking to deliver methods quickly. Defining the problem requires knowing what questions to ask your customers and colleagues. Developing robust procedures during their lifecycle of use requires a good understanding of how to build procedures that minimize the pernicious effects caused by measurement, sampling, and process variances. This book is intended to provide guidance on these subjects and provide the reader with the depth of knowledge to develop and implement quantitative procedures that remain successful.

References

1 Haas, J. and Mizaikoff, B. (2016). Advances in mid-infrared spectroscopy for chemical analysis. *Annual Review of Analytical Chemistry* 9 (1): 45–68.

2 Santos, M.C., Nascimento, P., Guedes, W.N. et al. (2019). Chemometrics in analytical chemistry – an overview of applications from 2014 to 2018. *Eclética Química Journal* 44 (2): 11–25.

3 He, Y., Tang, L., Xi, W. et al. (2007). Spectroscopy: the best way toward green analytical chemistry? *Applied Spectroscopy Reviews* 42 (2): 119–138.

4 Migaszewski, Z. (2013). The 12 principles of green analytical chemistry and the SIGNIFICANCE mnemonic of green analytical practices. *Trends in Analytical Chemistry* 50: 78–84.

5 Bouguer, P. (1729). *Essai d'optique sur la gradation de la lumière*. Notre-Dame: Claude Jombert.

6 Lambert, J.H. (1760). Photometria sive de mensura et gradibus luminis, colorum et umbrae. Book from the collections of Harvard University published 1892. https://archive.org/details/lambertsphotome00lambgoog.

7 Beer, A. (1852). Bestimmung der Absorption des rothen Lichts in farbigen Flüssigkeiten. *Annalen der Physik* 162: 78–88.

8 Oshina, I. and Spigulis, J. (2021). Beer–Lambert law for optical tissue diagnostics: current state of the art and the main limitations. *Journal of Biomedical Optics* 26 (10): 100901.

9 Nave, R. (2001). *Properties of Molecules*. Department of Physics and Astronomy, Georgia State University, Atlanta. http://hyperphysics.phy-astr.gsu.edu/hbase/molecule/molec.html (accessed 6 March 2024).

10 Sommer, L. (1989). *Analytical Absorption Spectrophotometry in the Visible and Ultraviolet: The Principles*, vol. 8, 2–311. New York: Elsevier Science Publishing Company, Inc.

11 Bauer, H., Christian, G., and O'Reilly, J. (1979). *Instrumental Analysis*, 3e. Boston, MA: Allyn and Bacon Chemistry Series.

12 Shukla, R., Dubey, A., Pandey, V. et al. (2017). Chromophore—an utility in UV spectrophotometer. *Inventi Rapid: Pharm Analysis & Quality Assurance* 2012 (3): 334.

13 Adegoke, O. (2012). Chemical derivatization methodologies for UV-visible spectrophotometric determination of pharmaceuticals. *International Journal of Pharmaceutical Sciences Review and Research* 14: 6–24.

14 Passos, M. and Saraiva, M.L. (2018). Detection in UV-visible spectrophotometry: detectors, detection systems, and detection strategies. *Measurement* 135: 896–904.

15 Mark, H. and Workman, J. (2003). *Statistics in Spectroscopy*, 2e. San Diego, CA: Academic Press.

16 Tom, J. (2021). UV-vis spectroscopy: principle, strengths and limitations and applications. https://www.technologynetworks.com/analysis/articles/uv-vis-spectroscopy-principle-strengths-and-limitations-and-applications-349865 (accessed 2 March 2024).

17 Noori, Z., Moreira, I.d.P.R., Bofill, J.M., and Poater, J. (2024). Adjusting UV-vis spectrum of alizarin by insertion of auxochromes. *ChemistryOpen* 13 (8): e202400030.

18 Ball, P. (2023). Refraction is then all there is to it: how Isaac Newton's experiments revealed the mystery of light. LiveScience. https://www.livescience.com/physics-mathematics/refraction-is-then-all-there-is-to-it-how-isaac-newtons-experiments-revealed-the-mystery-of-light#:~:text=The%20colors%20themselves%20are%20already%20in%20the%20white,%27tis%20generally%20believed%29%2C%20but%20Original%20and%20connate%20properties.%22 (accessed 9 March 2024).

19 Gilchrist, A. and Nobbs, J. (2000). *Colorimetry, Theory. Encyclopedia of Spectroscopy and Spectrometry*, 328–333. Cambridge, MA: Academic Press.

20 Hunter Lab AN 1002.00 (2022). CIE standard observers and calculation of CIE X, Y, Z color values. https://support.hunterlab.com/hc/en-us/articles/203420099-CIE-Standard-Observers-and-calculation-of-CIE-X-Y-Z-color-values-AN-1002b (accessed 22 March 2024.

21 ASTM E308-18 (2001). *Standard Practice for Computing the Colors of Objects by Using the CIE System*. West Conshohocken, PA: ASTM International.

22 Westland, S. (2016). The CIE system. In: *Handbook of Visual Display Technology* (ed. J. Chen, W. Cranton, and M. Fihn), 161–169. Cham: Springer.

23 Minolta (1994). Precise color communication. Osaka, Japan. https://segc.konicaminolta.com/instruments/registration_index/.

24 Lakowicz, J. (2006). *Principles of Fluorescence Spectroscopy*. New York, NY: Springer.

25 Zwinkels, J.C., DeRose, P.C., Leland J.E., (2014) Chapter 7 – Spectral fluorescence measurements. In *Experimental Methods in the Physical Sciences* Germer, T. A., Zwinkels, J. C., and Tsai, B.K. Academic Press, Cambridge, MA. 46: 221–290

26 Brader, M.L. (2020). Chapter 5 – UV-absorbance, fluorescence and FT-IR spectroscopy in biopharmaceutical development. In: *Biophysical Characterization of Proteins in Developing Biopharmaceuticals*, 2e. Elsevier, MA.

27 Hernández-Sánchez, N., Lleó, L., Diezma, B. et al. (2021). Multiblock analysis applied to fluorescence and absorbance spectra to estimate total polyphenol content in extra virgin olive oil. *Food* 10 (11): 2556.

28 Shakibaie, F., Lamard, L., Rubinsztein-Dunlop, H., and Walsh, L.J. (2018). Application of fluorescence spectroscopy for microbial detection to enhance clinical investigations. https://www.intechopen.com/chapters/59541 (accessed 24 March 2024).

29 Thermo Fisher Scientific (2011). Fluorescence fundamentals. https://www.thermofisher.com/us/en/home/references/molecular-probes-the-handbook/introduction-to-fluorescence-techniques.html (accessed 10 February 2024).

30 Horiba Scientific. Fluorescence applications. https://www.horiba.com/int/scientific/products/fluorescence-spectrometers/fluorescence-applications/ (accessed 27 March 2024).

31 Miller, C.E., (2001) Chemical Principles of Near-Infrared Technology. *Near-Infrared Technology in the Agricultural and Food Industry*. Williams, P., and Norris, K. H. American Association of Cereal Chemists, St. Paul.

32 Malegori, C., Muncan, J., Mustorgi, E. et al. (2022). Analyzing the water spectral pattern by near-infrared spectroscopy and chemometrics as a dynamic multidimensional biomarker in preservation: rice germ storage monitoring. *Spectrochimica Acta Part A: Molecular and Biomolecular Spectroscopy* 265: 120396.

33 Zhou, G.X., Ge, Z., Dorwart, J. et al. (2003). Determination and differentiation of surface and bound water in drug substances by near infrared spectroscopy. *Journal of Pharmaceutical Sciences* 92 (5): 1058–1065.

34 Walling, P. and Dabney, J.M. (1989). Moisture in skin by near-infrared reflectance spectroscopy. *Journal of the Society of Cosmetic Chemists* 40: 151–171.

35 Martin, K. (1998). in vivo measurements of water in skin by near-infrared reflectance. *Applied Spectroscopy* 52 (7): 1001–1007.

36 Sales, K.C., Rosa, F., Sampaio, P.N. et al. (2015). In situ near-infrared (NIR) versus high-throughput mid-infrared (MIR) spectroscopy to monitor biopharmaceutical production. *Applied Spectroscopy* 69 (6): 760–772.

37 Hollas, M.J. (2004). *Modern Spectroscopy*, 4e. England: Wiley.

38 Freda, M., Piluso, A., Santucci, A., and Sassi, P. (2005). Transmittance Fourier transform infrared spectra of liquid water in the whole mid-infrared region: temperature dependence and structural analysis. *Applied Spectroscopy* 59 (9): 1155–1159.

39 Carron, K. and Besli, M.M. (2015). *Monograph: Introduction to Raman Spectroscopy*. Metrohm.

40 Buckley, K. and Ryder, A.G. (2017). Applications of Raman spectroscopy in biopharmaceutical manufacturing: a short review. *Applied Spectroscopy* 71 (6): 1085–1116.

41 Stiles, P., Dieringer, J., Shah, N., and Van Duyne, R. (2008). Surface-enhanced Raman spectroscopy. *Annual Review of Analytical Chemistry* 1: 601–626.

42 Nyquist, R.A. (1990). Infrared study of maleic anhydride in solvent systems. *Applied Spectroscopy* 44 (3): 438–442.

43 Dang, H.V., Truong Thi Thu, H., Dong Thi Ha, L., and Nguyen, M.H. (2020). RP-HPLC and UV spectrophotometric analysis of paracetamol, ibuprofen, and caffeine in solid pharmaceutical dosage forms by derivative, Fourier, and wavelet transforms: a comparison study. *Journal of Analytical Methods in Chemistry* 2020: 810757.

44 Perkin Elmer Technical Note (2005). FT-IR spectroscopy attenuated Total reflectance (ATR). https://diverdi.colostate.edu/C431/experiments/infrared%20spectroscopy%20-%20transmission%20and%20attenuated%20total%20reflectance/references/ATR%20-%20Perkin%20Elmer.pdf.

45 Dahm, D.J. and Dahm, K.D. (1999). Representative layer theory for diffuse reflectance. *Applied Spectroscopy* 53 (6): 647–654.

46 Sánchez-Paternina, A., Román-Ospino, A.D., Martínez, M. et al. (2016). Near-infrared spectroscopic transmittance measurements for pharmaceutical powder mixtures. *Journal of Pharmaceutical and Biomedical Analysis* 123: 120–127.

47 Sparén, A., Hartman, M., Fransson, M. et al. (2015). Matrix effects in quantitative assessment of pharmaceutical tablets using transmission Raman and near-infrared (NIR) spectroscopy. *Applied Spectroscopy* 69 (5): 580–589.

48 Miseo, E. and Briggs, J. (2021). Common problems with FT-IR instruments and how to avoid them. Spectroscopy supplements. *The SciX 2021 Preview and Companion Guide* 36 (S9A): 15–16.

49 Bak, J. and Kindl, B. (1997). Quantitative analysis of mineral powders by DRIFTS: determination of $SrCO_3$ in superconductor precursor powders. *Applied Spectroscopy* 51 (11): 1730–1735.

50 Fahelelbom, K.M.S., Saleh, A., Al-Tabakha, M., and Ashames, A. (2022). Recent applications of quantitative analytical FTIR spectroscopy in pharmaceutical, biomedical, and clinical fields: a brief review. *Reviews in Analytical Chemistry* 41: 21–33.

51 Yeo, R. (2022). Theory & applications a white paper by Robert Yeo, pro-lite technology. https://www.electrooptics.com/sites/default/files/content/white-paper/pdfs/White%20paper%20-%20Integrating%20Spheres%20-%20Pro-Lite%20-%20August%202022.pdf.

52 Cooksey, C.C., Allen, D.W., and Tsai, B.K. (2017). Reference data set of human skin. *Journal of Research of the National Institute of Standards and Technology* 122: 26.

53 Bellini, S., Bendoula, R., Le Floc'h, E. et al. (2016). Simulation method linking dense microalgal culture spectral properties in the 400–750 nm range to the physiology of the cells. *Applied Spectroscopy* 70 (6): 1018–1033.

54 Stocker, S., Foschum, F., Krauter, P. et al. (2017). Broadband optical properties of Milk. *Applied Spectroscopy* 71 (5): 951–962.

55 Meissl, K., Smidt, E., Schwanninger, M., and Tintner, J. (2008). Determination of humic acids content in composts by means of near- and mid-infrared spectroscopy and partial least squares regression models. *Applied Spectroscopy* 62 (8): 873–880.

56 Lourenço, N.D., Paixão, F., Pinheiro, H.M., and Sousa, A. (2010). Use of spectra in the visible and near-mid-ultraviolet range with principal component analysis and partial least squares processing for monitoring of suspended solids in municipal wastewater treatment plants. *Applied Spectroscopy* 64 (9): 1061–1067.

57 Huber, E. and Frost, M. (1998). Light scattering by small particles. *Journal of Water Supply: Research and Technology-Aqua* 47 (2): 87–94.

58 Tchounga, C.A.W., Marini, R.D., Nga, E.N. et al. (2023). In-field implementation of near-infrared quantitative methods for analysis of medicines in tropical environments. *Applied Spectroscopy* 77 (11): 1264–1279.

59 Garrido, M., Rius, F.X., and Larrechi, M.S. (2008). Multivariate curve resolution–alternating least squares (MCR-ALS) applied to spectroscopic data from monitoring chemical reactions processes. *Analytical and Bioanalytical Chemistry* 390: 2059–2066.

60 MacRae, M. (2015). Trends in spectroscopy: a big-picture look at recent innovations and future directions. *Spectroscopy* 30 (3): 50–57.

61 Panzitta, M., Calamassi, N., Sabatini, C. et al. (2021). Spectrophotometry and pharmaceutical PAT/RTRT: practical challenges and regulatory landscape from development to product lifecycle. *International Journal of Pharmaceutics* 601: 120551.

62 Rignall, A. (2018). Chapter 1: ICHQ1A(R2) stability testing of new drug substance and product and ICHQ1C stability testing of new dosage forms. In: *ICH Quality Guidelines: An Implementation Guide*, 1e (ed. A. Teasdale, D. Elder, and R.W. Nims). John Wiley & Sons, Inc.

63 Gronlund, J. Xu, G., Peru, D.A., and Subramanyam, R. Analysis of substrates having agents deposited thereon. US Patent 8,803,095, filed 6 April 2009 and issued 12 August 2014.

64 Patel, A., Jin, C., Handzo, B., and Kalyanaraman, R. (2023). Measurement of moisture content in pharmaceutical tablets by handheld near-infrared spectrometer: adopting quality by design approach to analytical method lifecycle management. *Journal of Pharmaceutical and Biomedical Analysis* 229: 115381.

65 Nielsen, S.S. (2010). Determination of moisture content. In: *Food Analysis Laboratory Manual*, Food Science Texts Series (ed. S.S. Nielsen). Boston, MA: Springer. https://doi.org/10.1007/978-1-4419-1463-7_3.

66 Simtra BioPharm Solutions (2017). *Process Development for Lyophilized Products*, White Papere. Deerfield, IL: Baxter Healthcare Corp. https://biopharmasolutions.baxter.com/sites/g/files/ebysai3806/files/2021-07/Baxter_ProcessDevelopmentWhitepaper_Sacha.pdf.

67 Jameel, F., Alexeenko, A., Bhambhani, A. et al. (2021). Recommended best practices for Lyophilization validation part II: process qualification and continued process verification. *AAPS PharmSciTech* 22 (8): 266.

68 Boltia, S.A., Abdelkawy, M., Mohammed, T.A., and Mostafa, N.N. (2018). Validated stability-indicating spectrophotometric methods for the determination of silodosin in the presence of its degradation products. *Spectrochimica Acta Part A: Molecular and Biomolecular Spectroscopy* 202: 131–145.

69 Sakiroff, L.M., Chennell, P., Yessaad, M. et al. (2022, 2022). Evaluation of color changes during stability studies using spectrophotometric chromaticity measurements versus visual examination. *Scientific Reports* 12: 8959. https://doi.org/10.1038/s41598-022-13025-3.

70 Hammond, J.P. (2014). Chapter 11 – The use of spectrophotometry in the pharmaceutical industry. In: *Experimental Methods in the Physical Sciences*, vol. 46 (ed. T.A. Germer, J.C. Zwinkels, and B.K. Tsai), 409–456. Oxford, UK: Academic Press.

71 Pedro, A.M.K. and Ferreira, M.M.C. (2009). The use of near-infrared spectroscopy and chemometrics for determining the shelf-life of products. *Applied Spectroscopy* 63 (11): 1308–1314.

72 Chen, X. (2019). Seven essential steps for in situ reaction monitoring. *Spectroscopy Supplements Special Issue* 34 (8): 8–10. https://www.spectroscopyonline.com/view/seven-essential-steps-situ-reaction-monitoring (accessed 10 February 2024).

73 Jang, E.H., Park, Y.S., Kim, M.S., and Choi, D.H. (2020). Model-based scale-up methodologies for pharmaceutical granulation. *Pharmaceutics* 12 (5): 453.

74 Meyer, T. (2003). Scale-up of polymerization process: a practical example. *Organic Process Research and Development* 7 (3): 297–302.

75 Maltesen, M.J., Bjerregaard, S., Hovgaard, L. et al. (2011). Multivariate analysis of phenol in freeze-dried and spray-dried insulin formulations by NIR and FTIR. *AAPS PharmSciTech* 12 (2): 627–636.

76 Panzitta, M., Calamassi, N., Sabatini, C. et al. (2021). Spectrophotometry and pharmaceutical PAT/RTRT: practical challenges and regulatory landscape from development to product lifecycle. *International Journal of Pharmaceutics* 601: 120551.

77 Avila, C.R., Ferré, J., de Oliveira, R.R. et al. (2020). Process monitoring of moisture content and mass transfer rate in a fluidised bed with a low-cost inline MEMS NIR sensor. *Pharmaceutical Research* 37 (5): 84.

78 Esmonde-White, K.A., Cuellar, M., Uerpmann, C. et al. (2017). Raman spectroscopy as a process analytical technology for pharmaceutical manufacturing and bioprocessing. *Analytical and Bioanalytical Chemistry* 409: 637–649. https://doi.org/10.1007/s00216-016-9824-1.

79 Kim, E.J., Kim, J.H., Kim, M.S. et al. (2021). Process analytical technology tools for monitoring pharmaceutical unit operations: a control strategy for continuous process verification. *Pharmaceutics* 13 (6): 919.

80 Liu, Y., Doddi, J., Zheng, Y. et al. (2020). Transmission Raman spectroscopic quantification of active pharmaceutical ingredient in coated tablets of hot-melt extruded amorphous solid dispersion. *Applied Spectroscopy* 74 (1): 108–115.

81 Standard Methods For the Examination of Water and Wastewater (2018). 4500-NO_3 *– Nitrogen (Nitrate)*, 23e. American Water Works Association Public Health Association. https://doi.org/10.2105/SMWW.2882.089.

82 Rathnakumar, K., Pandiselvam, R., Mohammed, M. et al. (2020). Application of infrared spectroscopy techniques for the assessment of quality and safety in spices: a review. *Applied Spectroscopy Reviews* 55: 1520–1569.

83 Lawson, G., Ogwu, J., and Tanna, S. (2018). Quantitative screening of the pharmaceutical ingredient for the rapid identification of substandard and falsified medicines using reflectance infrared spectroscopy. *PLoS One* 13 (8): e0202059.

84 Pu, Y., Wang, W., Yang, Y., and Alfano, R.R. (2012). Stokes shift spectroscopy highlights differences of cancerous and normal human tissues. *Optics Letters* 37: 3360–3362.

85 Liu, Z., Gomez, J., Khan, S. et al. (2017). Red fluorescence imaging for dental plaque detection and quantification: pilot study. *Journal of Biomedical Optics* 22 (9): 1–10. https://doi.org/10.1117/1.JBO.22.9.096008.

86 Shen, X., Lan, S., Zhao, Y. et al. (2023). Characterization of skin moisture and evaluation of cosmetic moisturizing properties using miniature near-infrared spectrometer. *Infrared Physics & Technology* 132: 104759.

87 Takada, S., Naito, S., Sonoda, J., and Miyauchi, Y. (2012). Noninvasive in vivo measurement of natural moisturizing factor content in stratum Corneum of human skin by attenuated Total reflection infrared spectroscopy. *Applied Spectroscopy* 66 (1): 26–32.

2

Types of Sampling and Measurement Errors

Deborah A. Peru[1] *and Emil W. Ciurczak*[2]

[1] *DP Spectroscopy and Training, LLC, Lebanon, NJ, USA*
[2] *Doramaxx Consulting LLC, Goldens Bridge, NY, USA*

2.1 Introduction

This chapter is dedicated to unraveling the different types of sampling and measurement errors that can significantly impact the performance of quantitative determinations in spectroscopy. Understanding these errors is crucial as they can profoundly affect the accuracy and reliability of our results. One of the primary objectives of procedure development is to identify conditions that cause bias, distort the spectra, and introduce unwanted noise and interferences in the determination. Sampling errors, for instance, can compromise the information gained from the determination when the materials analyzed are not representative or have chemically changed during or prior to analysis. To develop a robust and reproducible measurement, it is essential to have a comprehensive understanding of the errors that can affect the sample, instrument, and measurement conditions at the time of measurement.

Let's start with some basic concepts in error analysis. Random errors are sometimes called unsystematic errors, system noise, or random variation. Random errors are typically caused by factors that change by chance and display no apparent pattern. This causes the results to fluctuate arbitrarily from the "true" value. One way to visualize random errors is to plot the concentration residuals. These variances are calculated as the difference between the known (actual) minus the predicted concentration using the quantitative method. Figure 2.1 graphs the concentration residuals (on the y-axis) vs. the actual concentration (on the x-axis). When random errors exist, the residuals will be evenly distributed above and below zero (0), indicating little bias as illustrated in the figure.

Truly "random" errors are expected to have a normal distribution from the "true" value with roughly 68% of the results one standard deviation from the mean difference, 95% of the samples two standard deviations from the mean, and 99.7% of the samples lie within 3 standard deviations of the mean as shown in Figure 2.2. For measurements containing only random sources of error, the sum of errors is near zero, and the results have no significant bias over the range of concentrations measured.

The sources of random errors are sometimes difficult to isolate and predict because they can fluctuate constantly without a repeating pattern. The classic example is the normal variation in temperature and humidity of the environment and sample. In some cases, a random error could occur when an intermittent or sudden deviation occurs from normal conditions, such as a sudden vibration of the instrument, an unusual deviation in the measurement container, or a misplacement of the sample in the sampling chamber. Other plausible reasons include an intermittent change in particle size, shape, packing density, or any

Practical Quantitative Vibrational and Electronic Spectroscopy: A Guide for Developing, Optimizing, and Validating Procedures, First Edition. Edited by Deborah A. Peru.

Companion website: www.wiley.com/go/peruelectronicspectroscopy

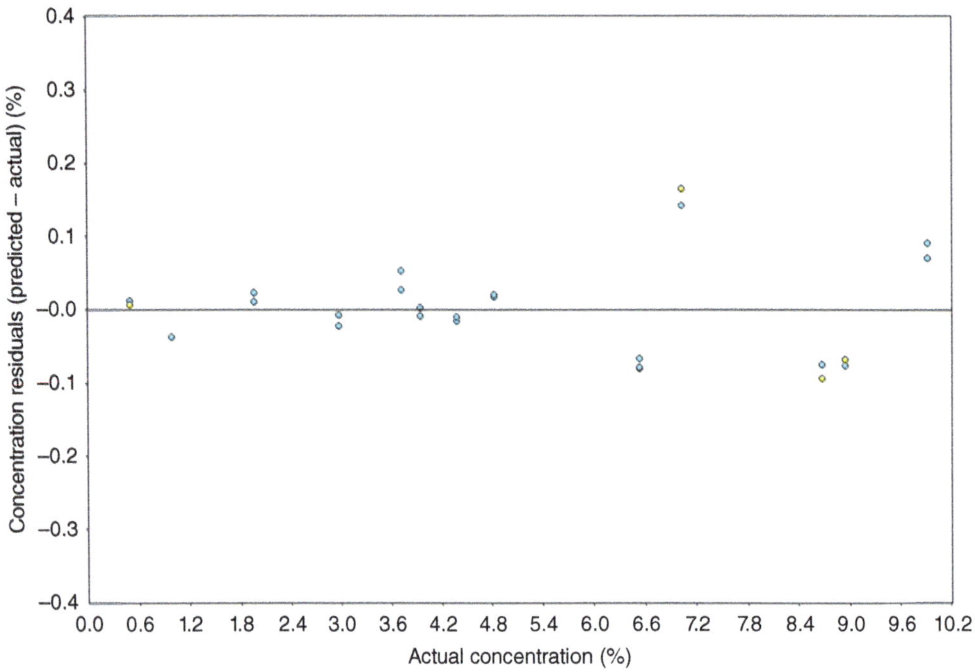

Figure 2.1 Residual plot demonstrates random errors in residuals. *Source*: Deborah A. Peru.

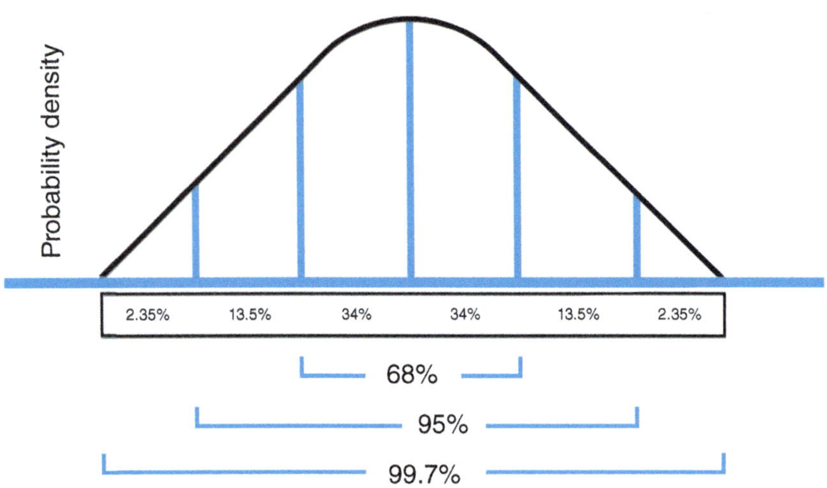

Figure 2.2 Normal distribution in determinations containing random sources of error. *Source*: Deborah A. Peru.

other unforeseen temporary change in the measurement condition. Most of these aberrations will cause the spectral fingerprint to deviate and can be detected using statistical metrics such as Hotellings T^2 and Q diagnostics (refer to Chapters 3 and 7).

Developing a robust calibration curve should, to some degree, include the expected variations caused by random sources. Collecting standard spectra over a period of time (typically weeks or months) compared to within a single day will ensure sources of random error are captured within the "training set" of spectral data. Certain mathematical approaches used to develop the regression model, e.g. partial least squares (PLS) regression, can, to some degree, identify and account for random sources of variance within the loadings, and these are typically removed from the final model.

Systematic errors, also known as determinate errors, are not caused by chance but have a defined value and an assignable cause. These errors are consistently the same for every determination made in the same manner. The effect on the accuracy of the determination can be small or large depending on the type of variance. This error could be as simple as not tarring the vessel or weighing paper when weighing a sample.

Once a systematic error is introduced into the process or measurement condition, it will ultimately make its way to the final determination. A fixed systematic error introduces an error of the same magnitude and direction, and this causes a bias in the results [1]. If the bias is negative (−), it causes results to be low; if it is positive (+), the result will be high. In Figure 2.3, a determination is made that shows a consistent bias of +0.85% magnitude in the predicted concentration (y axis) vs. the true concentration (x axis).

The source may not be obvious initially but can often be identified once a root cause analysis is completed. Most systematic errors are man-made and, when identified, can be corrected because they are reproducible. Sometimes, a bias in one direction may be due to a combination of errors.

A potential source of systematic error that often occurs in manufacturing is the seasonal change in raw materials and suppliers. Many companies have preferred vendors they use for raw materials, but sources can change seasonally due to weather-related changes in farming conditions, shipping costs, supply chain problems, and price. A change in a raw material supplier can have a subtle and, in some instances, large effect on the spectral fingerprint, particularly in the case of near-infrared spectroscopy, where small chemical and physical changes in ingredients can change the shape and the spectral fingerprint [2].

Depending on the application, it may be beneficial to include these types of variances in the calibration "training set" to avoid a future measurement bias. However, adding variables to a calibration model has the potential for degrading the accuracy of the determination. But, when proper calibration math is applied, it is

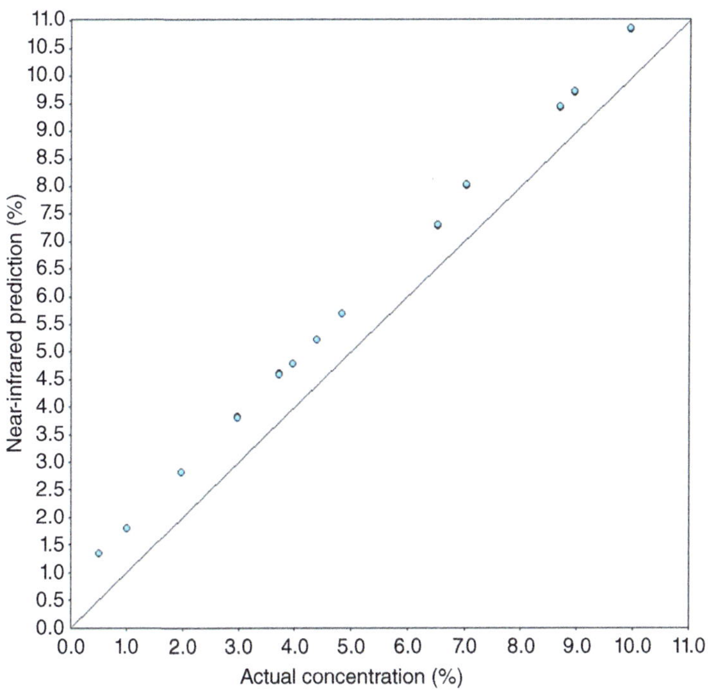

Figure 2.3 Example of systematic (determinant) error bias in quantitative prediction of analyte concentration. *Source*: Deborah A. Peru.

possible to account for these variances in the final regression equation and minimize their pernicious effect. If the variable cannot be properly modeled to yield acceptable performance, another approach would be to have separate calibration curves for seasonal changes and possibly incorporate a qualitative model to qualify the "unknown" spectral fingerprint as belonging to a particular class of standards before a quantitative model is applied. Some software packages have "indicator variables" that allow you to tag sets of data; e.g. "summer" and "winter". The software ignores interferences and only looks for common bands [3].

Figure 2.4 describes some common sources of random and systematic errors that can affect the degree of accuracy and precision in quantitative procedures. Random errors can sometimes become systematic when they don't normalize to zero over time or when proper hardware or environment maintenance procedures are not followed.

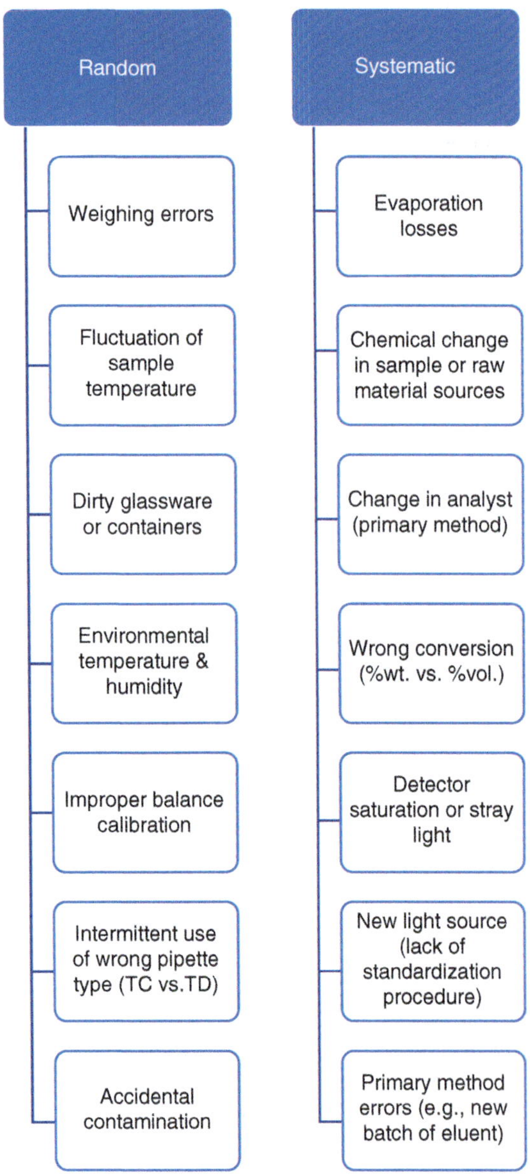

Figure 2.4 Common sources of random and systematic errors in quantitative analysis. *Source*: Deborah A. Peru.

2.2 Basic Concepts in Sampling

Sampling is the selection of a subset from within a statistical population to estimate or infer the quality characteristics of the whole population. Bulk sampling refers to a large amount of material taken at various locations within a lot. A lot is the total amount of substance you want to measure using the analytical testing procedures. For analytical testing, a small amount of material "sample" is taken from the bulk sample, which can be further partitioned into aliquots before analysis. Even the most carefully developed spectroscopy procedure can yield inaccurate results if poor sampling occurs. A sampling plan should align and support the goals of the analysis. For example, if the procedure is intended to measure surface moisture, the sampling procedure must involve the analysis of a freshly prepared surface rather than a surface exposed to the atmosphere for several hours.

Developers should be able to understand and apply basic concepts in statistical sampling. Proper sampling procedures are needed when creating a representative set of samples for instrumentation calibration and validation, selecting samples for product or process optimization, process validation, process monitoring and control, and for nonconformity and quality assurance testing of commercial products.

Implementing a sampling plan usually involves three steps: physically removing the sample from its target population, preserving it, and preparing it for analysis. Poor sample handling at any stage can lead to anomalousness and possibly bias in the results. Regardless of the cause, the information obtained from the determination may be misleading and lead to actions that have little benefit or cause detriment to the process or decision-making. Some of the more common problems in sampling are described below:

- Mass loss (evaporation or electrostatic charge on one or more components)
- Heat generation during grinding can lead to moisture loss or gain
- Changes in polymorphism structure
- Packing errors during sample repacking, such as a change in surface, squeezing out liquids, leaving spaces or gaps, and stratification
- Poorly mixed or improperly mixed samples. Some powders can undergo segregation or deblending if too much mixing is applied
- Sample size that is too small or not taken in the right location may not fully capture the average composition.

Analysis of a heterogeneous mixture in which the composition is not uniform throughout will require extra development time and experimentation along with statistical diagnostics to identify the proper sampling protocols (i.e. number of samples, location, and pretreatment) to get a representative result. To obtain a more reproducible measurement signal, the developer may need to incorporate a pretreatment step by grinding the sample before analysis [4]. Vortexing liquids and some types of solids or suspensions before analysis may help to ensure representative sampling. Some vendors offer a rotating sampling accessory to facilitate the collection of a more representative sample during the measurement. This approach works well with larger sample sizes and when spectra are acquired with longer acquisition times and co-added scan procedures. In all cases, the goals are to increase the likelihood that the spectral measurement provides a good representation of the sample population and that the information obtained from the determination is truthful.

In the field of microplastics, for example, industry leaders are working toward the adoption of standardized and globally harmonized sampling procedures to enhance the reliability and reproducibility of the data acquired to permit better comparison across studies and facilitate meta-analyses [5].

Work with the entire team to review the chemical and physical characteristics of the samples and sampling requirements to meet the objectives. Then, devise a plan for sampling, including location, weight, volume, how often sampling should occur, and the optimal storage conditions.

2.2.1 Bulk Sampling

In a process or industrial setting, a quantitative spectroscopy method may be implemented for a number of reasons, including detection of raw material contaminants at receiving, release testing of premix batches or unit operation, continuous operation monitoring and control, effluent monitoring, and the list goes on. Gaining a representative analytical sample may be challenging. A stepwise process that may involve stratified sampling to yield a representative sample of the whole can be used. In stratified sampling, the target population is divided into strata and then random samples from each stratum are analyzed [6]. The data are averaged and pooled to give an overall mean for the target population.

The number of samples and location may depend on factors such as the type of sample (solid, liquid, pellets, slurry, grains, etc.), the volume of sample that can be analyzed by spectroscopy and primary method, and degree of uniformity (homogeneous vs. heterogeneous). A heterogeneous sample would require more sampling at representative locations to achieve a representative result. A granular chemical mixture or powder can be taken with a sample thief in a symmetric pattern at various depths and at specified time intervals. Thief containers may also be used.

Grab sampling of discrete units can be taken at specified timeline intervals to capture variability within a process. Compounding is the process of combining one or more grab samples, mixing them together, and possibly grinding before analysis. Aliquots are taken from these combined sample composites for both spectroscopy and laboratory analysis. Much has been documented in the food and agricultural industry based on evidence-based testing by early adopters of near-infrared spectroscopy testing procedures [7].

The number of samples is often specified in a compendial method or may be designed for a certain degree of precision. When the samples are collected, they must not sit long before analysis to minimize compositional changes. Ideally, the spectroscopic analysis is completed first on the same portion sent to the laboratory for compendial method testing. If samples are stored for a period of time, changes may occur due to exposure to atmospheric moisture or light, oxidation, settling, chemical change, and possible reaction with a containment vessel or enclosure.

There are many excellent references on the subject of sampling. Esbensen and Romanach review the theory and practice of representative sampling to guide developers in obtaining relevant, meaningful, and reliable analytical results for various materials [8].

2.2.2 Sampling for Regulatory Purposes

Sampling for nonconformities in commercial products and environmental contamination is set by product type and the regulatory agency that oversees the product and industry. For example, the U.S. Food and Drug Administration (FDA) has established appropriate written testing standards, specifications, and sampling plans for components, in-process materials, containers, and finished products. The Current Good Manufacturing Practice (cGMP) guidance documents contain comprehensive information on when and how many samples should be tested during process validation, stability testing, and release testing. The Code of Federal Regulations (CFR) Title 21 requirements ensure products are safe for use and have the proper ingredients and concentrations [9]. The EPA SW-846 compendium comprises sampling protocols and analytical test methods, including colorimetry and infrared procedures for analyzing hazardous waste and other matrices [10].

The International Organization for Standardization (ISO) is an international organization responsible for creating standards for various industries that import and export products commercially. The standard ISO 2859 describes sampling procedures for quality inspections of materials by attributes [11]. The inspection process for attributes uses statistical sampling plans to determine whether the batch has too many nonconforming products.

2.2.3 Sampling for Laboratory vs. Process Measurements

The significant differences between laboratory testing and process analysis are the speed at which samples move past the sensors and the testing frequency. In a process, every measurement is considered a different time point. Multiple time points may be averaged in some applications, and the average determination is used for making process decisions. For process optimization, the spectral data collected at each time point is used to obtain information for process improvement and to control critical quality attributes. This is part of the 2004 Process Analytical Technology (PAT) quality-by-design initiative by the FDA for innovative pharmaceutical development [12].

Inline sampling probes are inserted directly into the process. In contrast, online probes measure products using a slipstream to divert a small amount of material from the process through the measurement sensor. Both inline and online spectroscopy probes can be used in the pilot plant to provide chemical information to help scale up new processes. The frequency of collecting spectra is application-specific and often depends on mixing conditions and reaction rates in the case of chemical synthesis and whether intermediate detection is required. In batch processing, in situ monitoring with spectroscopy instrumentation located adjacent to the batch can be used, or grab samples can be taken for rapid analysis with instruments located atline. It must be stated that freshly pressed tablets are warmer and denser than tablets allowed to "age" for several hours or more. Therefore, calibration models developed on "older" tablets may need a bias adjustment when used for the determination of process samples in real time.

In continuous manufacturing, process spectroscopy instrumentation is used to monitor and control processes. The location of where the probe is mounted in the process and where representative sampling can be obtained for corresponding offline testing are critical for understanding process chemistry and validating feedback control systems. The residence time distribution is an important concept and pertains to the amount of time before process adjustments are reflected in the spectrum [13]. Proper mixing during scale-up will require testing and evaluation to identify the best measurement locations for monitoring. Mixing during transportation and using baffles will facilitate blending but may lengthen the necessary time for chemical reactions and interactions between molecules to complete compared with aggressive mixing. The relationship between chemical changes and where process monitoring occurs is important for successful implementation.

During the early stages of method development, samples can be taken from the process at specific time points after spectroscopy measurement and sent to the laboratory for offline analytical testing. The spectral data is assigned a concentration or physical parameter for building quantitative methods that can be used for future real-time analysis. The benefit of sampling from a process is that it will contain variables that can affect the spectral fingerprint that may not otherwise be captured in the laboratory or pilot plant. A disadvantage of only sampling from a process is that the variability may be too limited for generating a regression β coefficient that explicitly describes the property of interest.

2.2.4 Chemical Changes in Sample

The two common reasons sample chemistry may change are improper storage conditions and time spent waiting for analysis. Exposure to atmospheric conditions (oxygen, carbon dioxide, temperature, and pressure) can cause unwanted chemical reactions, surface oxidation, moisture loss, settling, curing, and volatile loss.

If mixing or grinding the sample is done to ensure homogeneity, these approaches should be balanced to prevent volatile losses or chemical changes within the sample. Once the bulk sample is ready for analytical sampling, this step should be done quickly to avoid further chemical changes. Samples measured by the spectroscopic technique can be weighed and quenched in solvent for primary method determination to ensure the highest degree of agreement between the methods.

An important reality of sampling is that chemical changes can alter the outcome of the determination by both the primary and the spectroscopy methods. During the development phase, take extra time to optimize the amount of sample (volume or weight) and preserve the chemical integrity of the samples measured by both techniques. This is often easier for spectroscopy because the measurements are often made "as is," but it may require careful consideration to avoid chemical changes before primary method determinations.

Sampling inconsistency is a common problem experienced when developing and validating applications involving process dryers. During the time it takes to collect a specific number of co-added scans, several pounds of material can move past a measurement window, which is then signal averaged to obtain a final spectrum. The speed and amount of product moving past a fiber-optic probe during a process measurement can be significantly greater than the amount of sample used in a single offline Karl Fisher determination. The discrepancy in the sample size measured by the primary method can degrade agreement between the methods due to the chemical differences in what is measured. Therefore, consider making a procedural change in the primary method, such as taking a larger sample size and introducing a serial dilution step, especially during the calibration and validation stages. This will ensure the sample measured is more representative and chemically comparable to the sample measured with the spectroscopy procedure. This approach can improve agreement between the methods and create a much more accurate and robust spectroscopy calibration curve.

A fluid-bed dryer unit is often equipped with a fluid-bed "spoon" probe inserted directly into the dryer for sampling before a near-infrared measurement. After collection, an air purge exiting through the ports in the probe tip clears the "spoon" for a new sample [14]. Other applications may involve a diffuse reflectance probe for solid sampling or use of an interactance immersion probe for liquid sampling (see Chapter 3 for more details). Swift and rapid sampling near the location where these probes are installed is important for good agreement between methods.

For some applications, aseptic sampling may be needed. This is the process of pulling samples for analysis, with procedures in place to prevent contamination of the batch or remaining product.

Once a sample is taken, steps should be taken to ensure chemical stability or to re-establish original concentration levels if degradation does occur [15]. Suspensions and samples involving mixtures are prone to changes in opacity or transparency caused by precipitation or separation, settling of solids in a suspension, emulsion inversion, refractive index, or density changes with temperature or pressure change once a sample is pulled from a process. For example, two-dimensional correlation spectroscopy applied to the visible and near-infrared spectra collected on chicken meat during cold storage found chemical changes occurring due to the interactions between water molecules and meat components. Proteolysis, denaturation, and tenderization during storage and prolonged aging were identified. The asynchronous component also found discoloration of the meat occurring before all other chemically identifiable changes [16].

2.2.5 Sampling for Method Development

Three basic types of samples are used to develop the "training set" during instrument calibration.

1) Synthetic standards made in the laboratory contain the analyte of interest over a wide concentration range and include minor variances in matrix ingredients. This approach is used to develop procedures for ultraviolet, visible, and fluorescence spectrophotometers. In vibrational spectroscopy, process nuances and physical variances can cause spectral shape changes, which can cause bias or degrade performance in real-world applications. Thus, use of only laboratory-prepared samples may not fully capture the range of spectral behavior.
2) Pilot-plant-scale samples that vary the ingredient of interest and other ingredients within the process design space are often used to create standards for calibrating vibrational spectroscopy instrumentation.

When applicable, using the Design of Experiments (DOE) to incorporate chemical and process variables is ideal because it establishes the spectral response of the extreme levels while minimizing collinear relationships. This is the optimum approach for maximizing the relationship between spectral response and the parameter of interest using a minimal number of calibration samples.

3) Grab samples taken directly from the process can further supplement the training set with a real-process fingerprint. While this approach captures the truest relationship between process variability and spectral response, it often does not extend over a wide enough concentration range for calibration and validation purposes. This can lead to very lengthy calibration times (often years) or calibration models that only perform over a small range and are very difficult to validate due to the narrow range in concentration.

For pharmaceutical tablets, one approach is to grind tablets thoroughly and weigh two aliquots of the powder mix, each roughly equivalent to a single dosage form. Powder rifflers and splitters can disperse powdered material into several statistically equal aliquots for further testing. Weighed samples can be measured "as is" for solid analysis or dissolved in a solvent for liquid analyzers. The spectrophotometer can analyze the ground and dispersed samples, and then the samples can be sent for laboratory testing during the calibration and validation stages. Alternatively, if the spectroscopy method is being developed for whole tablet analysis, the method may require analysis of several tablets before grinding and mixing the tablets for offline primary method analysis. It is recommended that the spectral data for each tablet be included separately in the calibration curve using the same averaged laboratory value [17].

When assigning a concentration or parameter value to a spectrum, the best approach is to ensure the laboratory and spectroscopy techniques measure the same representative sample. Ideally, the sample should be taken and prepared for "wet chemical" analysis after completing the spectroscopy analysis. Timeliness matters to avoid chemical changes in the sample. We often forget that proper sampling affects both the primary method and the spectroscopy techniques. Optimization on both sides of the equation -the input block [X] and the output response block [Y] - is required during development to create calibration models that meet performance expectations. Proper sampling procedures may require days or weeks of experimentation to identify the right conditions for representative sampling without altering the chemical composition.

In some applications, calibration standards can be prepared gravimetrically or volumetrically. In these instances, the spectroscopy method may be considered a primary method, but an assumption is made that there are no material losses. When preparing organic liquid standards, the volume of material can be measured with a pipette or "to deliver" burette, and the corresponding weight can be recorded with an analytical balance. Using the density of the organic ingredients measured under the same conditions can validate the weight fraction added and determine the true volume fraction of each ingredient in a mixture [18].

A good example of the use of designed standards for the quantitative determination of polymorph form in pharmaceutical tablets by Raman spectroscopy is described by Feng et al. [19]. For this application, the calibration standards were prepared using a three-factor central composite design (CCD). The three factors were polymorph form, mannitol/microcrystalline cellulose ratio, and compaction force. The materials were mixed using a blender and then subdivided. Tableting was done at three levels of compaction force, and the final tablets were used to develop the calibration curve [19].

2.3 Instrument Qualification

Maintaining the integrity of the spectral data collected is crucial to ensure the calibration models remain accurate and precise throughout use. Data integrity refers to the assurance that the spectral information remains accurate and consistent. An underlying assumption is made when quantitative methods are developed that the wavelengths or frequency positions and photometric response remain stable and constant during the entire life cycle of the method [8].

Over the years, instrument vendors have included diagnostic tests that help developers and users of spectroscopic instrumentation monitor instrument performance parameters before use. These include wavelength accuracy and repeatability, photometric linearity, photometric noise, intensity or absorbance accuracy, and repeatability [8]. Some instrument software allows the user to make fine adjustments to the wavelength position and photometric scale intensity using well-characterized materials called Standard Reference Materials (SRMs). Characterization of the instrument is important before the instrument is used for collecting spectra on samples, particularly when decisions are being made based on the results. Instrument performance is one of several factors directly affecting the accuracy and repeatability of quantitative measurements. Hence, it is important that instrument performance be regularly monitored and stored so that documentary evidence can be provided.

The USP instrument-specific General Chapters describe technique-specific tests that should be successfully performed before instrumentation is commissioned for use. Remember that these guidelines are intended to meet cGMP requirements for pharmaceutical and food and drug testing of finished products. Most vendors have a full validation package that can be implemented when new instrumentation is installed. On-going performance qualification (PQ) testing procedures often include a combination of instrument performance testing diagnostics and application-specific testing protocols to demonstrate fitness to use the spectroscopy procedure for quantitative determination over the operational range (see Chapter 7 on validation for more details).

2.4 Types of Measurement Errors

Once the instrument is stable and the corresponding qualification documentation is completed and approved, it can be used to generate data. At this point, measurement errors can still occur due to several obstacles that result in random or systematic errors. Most of these errors are caused by the operator or instrument when measurement conditions are not optimized or haphazard changes occur during the measurement. Table 2.1 shows some of the typical problems that can occur.

Check for excessive or unusual baseline noise in the spectra or an unusual artifact, such as the intermittent appearance of sharp bands. A sudden baseline shift during a liquid measurement often indicates the presence of air bubbles. When measuring liquids or slurries in a process, air bubbles caused by vigorous mixing often become trapped on the surface of the optical sensor or measurement window. Attenuated total reflectance (ATR) sensors used in mid-infrared spectroscopy only penetrate a few microns into the sample. The presence of air bubbles will severely reduce the measurement signal and increase noise. In process applications, when using an ATR probe, it is beneficial to move the probe away from the mixing vortex and to angle the probe so that when bubbles collect, they can escape quickly.

When using an interactance immersion probe for inline measurement of liquid samples and slurries, bubbles may collect within the space (gap) between the parallel plates. This can reduce the signal to noise and affect the spectral baseline (first-order effect) and effective pathlength (second-order or multiplicative scatter effect) [7].

Figures 2.5 and 2.6 show the effect of entrained air in liquid soap measured by near-infrared and Raman instruments. Due to process foam formation, the S/N is severely reduced in both instrumental measurements when air is present in the sample. The Raman scattering at most wavenumbers is not even detectable when measuring a sample containing foam as shown by the flat response in the green spectrum.

In diffuse reflection analysis of solids by near-IR, the spectral fingerprint is dependent on several factors including particle size, shape and uniformity, sample chemistry, and depth of penetration of incident

Table 2.1 Common instrument problems that reduce the performance of the quantitative procedure.

Typical instrument problems	Description of symptoms
Temperature control (instrument/laboratory)	Unusual fluctuation in the laboratory or environment or overheating of instruments due to faulting cooling
General carelessness	Poorly trained or new analysts, lack of instrument understanding
Excessive noise in the spectrum	Stray light, vibration, missing sample
Relative humidity (instrument/laboratory)	Change in humidity or lack of instrument purge or desiccant
Poorly collected reference spectrum, dirty ATR sensor	Spectrum does not represent the sample, negative bands, excessive noise
Instrument–instrument differences	Calibration does not take into account systematic instrument differences, and calibration transfer algorithms are not used or are insufficient to correct for differences
Sample presentation (or transport) apparatus	The accessory has changed position, misaligned, or misused, wrong speed.
Missing sample or partially covered sensor	Fast-moving production flow or a small amount of material, crumbling problems, etc.
Misaligned interferometer	A fringe pattern in the spectrum (sinusoidal waves)
Measurement not representative of the bulk sample	Remove the surface layer or change the depth of penetration or wavelengths that measure deeper into the sample.
Sample temperature	Sample pulled from process measured before stabilized. Sample temperature changes sitting inside the instrument
Broken or bent fiber-optics	Reduces light collection, decreases signal
Poor reference collection	Changes sample spectrum and affects accuracy
Dirty sample cell/optical window fouling	Distorted spectrum, weak spectrum, reduction in light transmission to the detector
Air bubbles	Reduces efficiency of light collected, raises baseline, reduces precision and accuracy
Sample door left open	Straylight effects decrease linear range, precision, and accuracy
Poor sample contact	A common issue with ATR sampling of solids and films is improperly using a back-pressure gauge.
Fluorescence in Raman background	Reduced S/N in Raman measurement

Source: Deborah A. Peru.

light [20]. Smaller particles don't allow light to penetrate as deeply as the larger particles, and internal scattering is subsequently reduced. The baseline offset increases from shorter to higher wavelength as does the slope. These nuances are clearly illustrated in the diffuse reflectance measurement of ground and whole rice grains shown in Figure 2.7. Notice that the baseline shift is more prominent in the whole rice spectrum (red trace) than in the ground rice spectrum (blue trace). On the other hand, the higher variability between replicates of ground rice is due to the lack of consistency in particle size and shape after grinding.

When using ATR sensors in mid-infrared spectroscopy, it is a common practice to check the cleanliness of the sensor before a reference scan is taken. Most spectroscopists call this a 100% line or "clean scan" performed after cleaning and before the next background spectrum is taken. If cleaning is not done correctly,

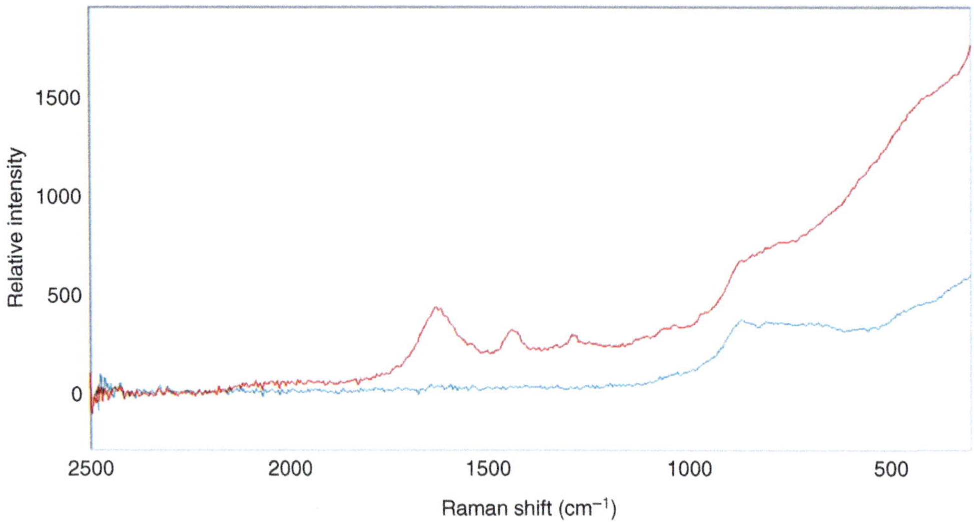

Figure 2.5 Raman spectra of liquid soap with entrained air (green) and without entrained air (red trace) due to foam formation. Source Deborah A. Peru.

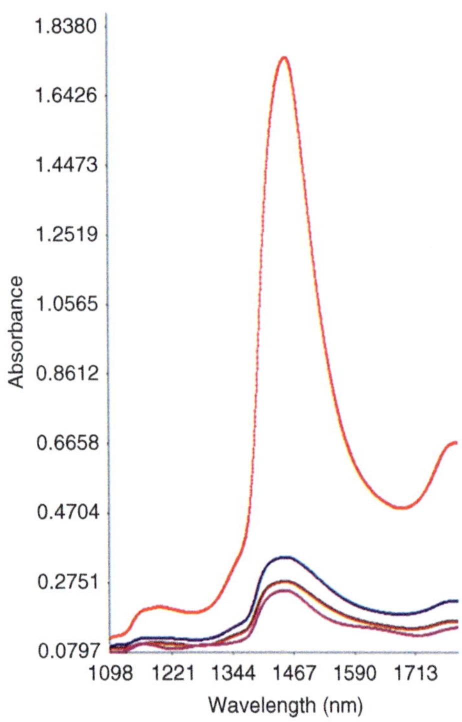

Figure 2.6 Near-infrared spectra of liquid soap sample with entrained air causing foam formation (blue, magenta, and brown replicates) and without air (red spectrum). *Source*: Deborah A. Peru.

the next measured reference spectrum will become contaminated with residue, affecting the chemical interpretation of future spectra. When a new reference scan is taken, the contamination is effectively subtracted from the measured sample as shown in Figure 2.8a. This has the effect of reducing absorbance bands and introducing negative features where bands don't exist. Figure 2.8b shows the fingerprint of a clean background and a dirty background. While the difference in intensity of the backgrounds is minor,

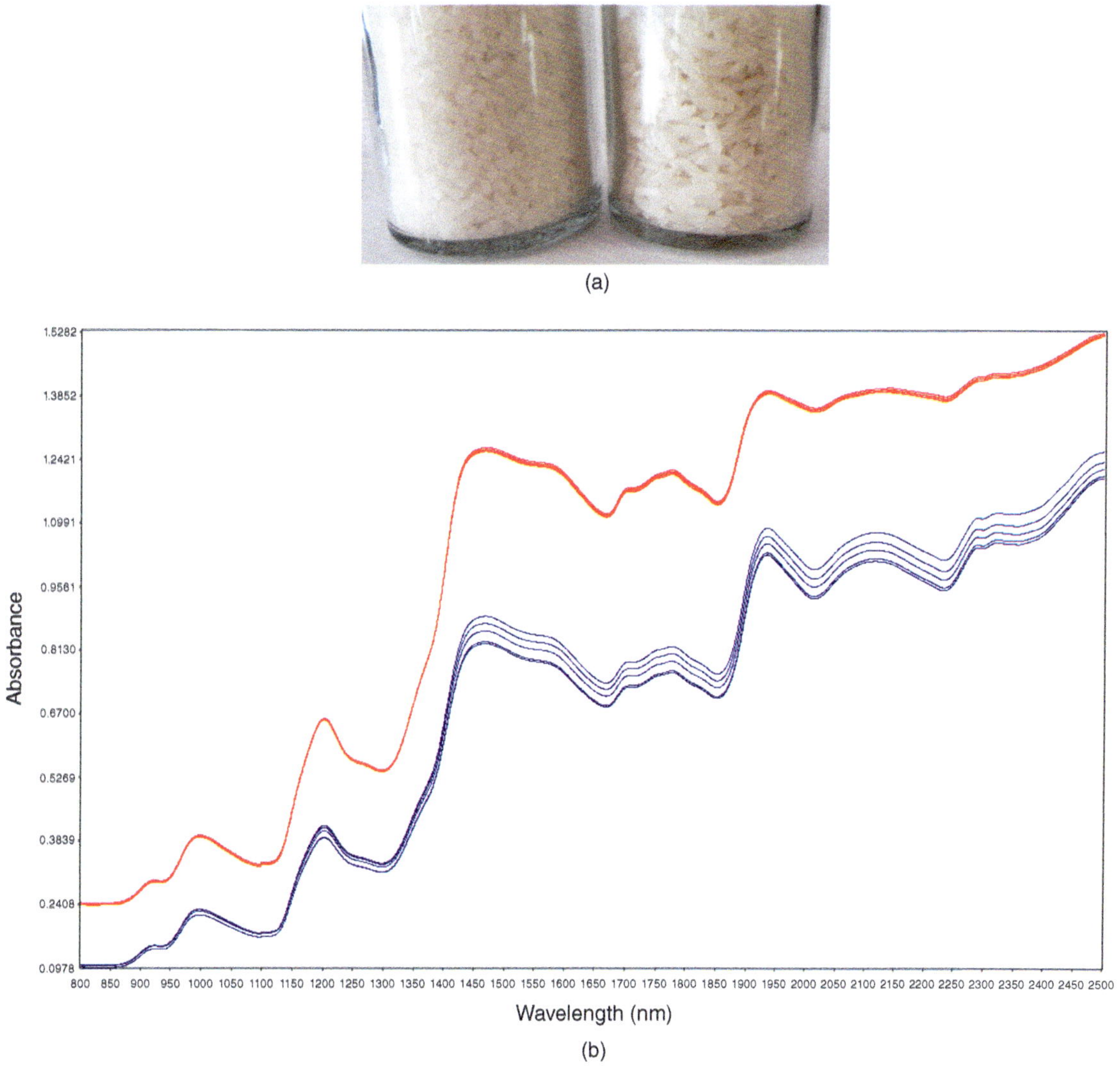

Figure 2.7 (a) Photos of ground rice (left) and whole rice (right) show packing density differences. *Source*: Deborah A. Peru. (b) Diffuse reflectance NIR spectrum of ground rice (Blue) and whole rice (Red). *Source*: Deborah A. Peru.

the difference spectrum (blue trace) clearly shows the difference in the CH functional group region of the spectrum 3000–2800 cm^{-1}, accounting for the negative bands observed in the sample spectrum.

Another important consideration during procedure development is the laser wavelength used for the excitation of Raman. The selection is most often based on maximizing the Raman signal strength while minimizing the photoluminescence that obscures the Raman bands [21]. Photoluminescence from either the sample or the substrate produces detector noise that can be as large as the Raman signal, thus obscuring the measured component [21]. Selecting an excitation source that generates little or no photoluminescence, thereby reducing the associated detector noise, will always produce superior results. The typical excitation sources are the visible range (532, 638, and 785 nm) or near-infrared range (1064 nm).

The amount of Raman signal generated from a given sample is often limited by physical characteristics such as particle size, density, scattering coefficient (elastic and Raman), and the absorption coefficient [22].

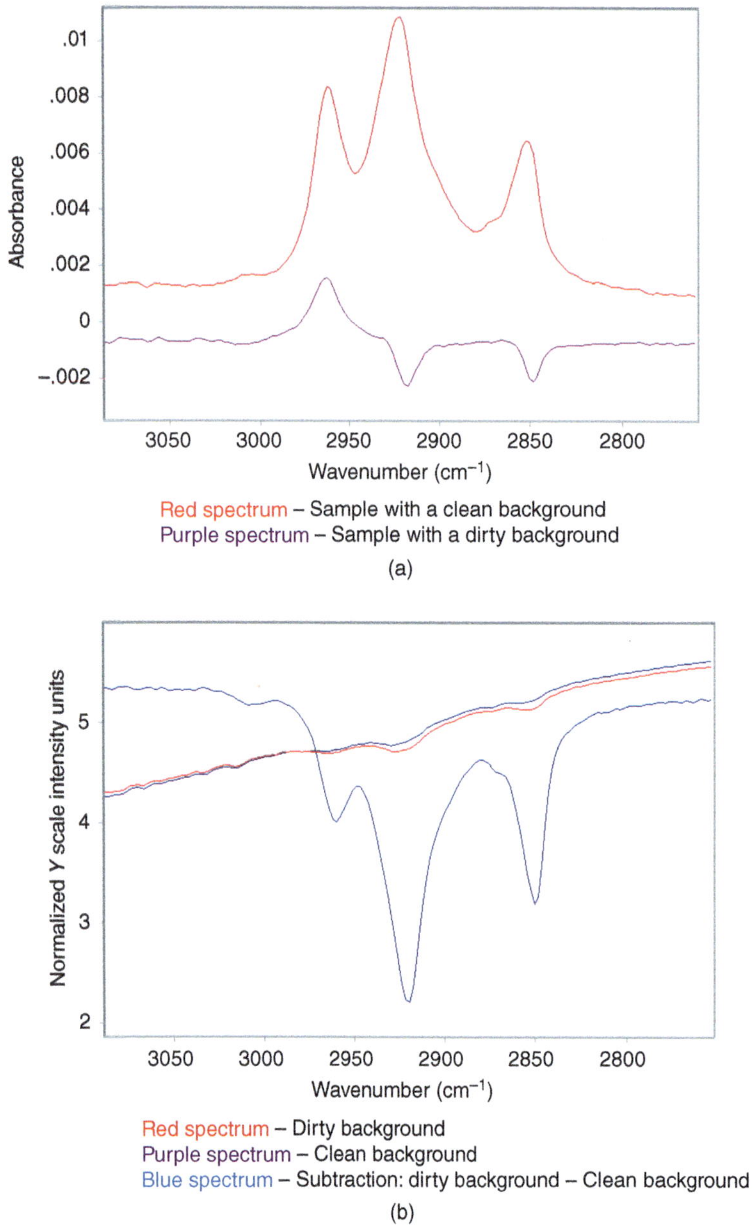

Figure 2.8 (a) Comparison of sample spectrum with clean ATR sensor and dirty ATR sensor. (b) Comparison of background spectrum of a properly clean ATR sensor and a dirty ATR sensor and the difference spectrum between them. *Source*: Spectra Courtesy of Dr. Ellen Miseo, Instrumentation and Applications Consultant in Spectroscopy.

2.5 Inconsistencies in Data Collection Measurement Parameters

Data collection inconsistencies can occur, especially when collected over long periods. Check the data collection parameters to ensure the same instrument settings are used for all the spectra collected. This is particularly important for users in large facilities where multiple scientists may use the same instrumentation. The spectral collection parameters should be optimized to achieve the best-quality spectra consistent with the measurement goals and performance specifications. This includes the laser power, integration times, number of coadded scans, spot size, resolution, spectral range, intensity correction, and use of reference standards. Reference standardization refers to using a NIST-traceable CRM (Certified Reference Material) to fine-tune the absorbance scale (y-axis) after installing a new light source. This is typically used in near-infrared spectroscopy. Similar quantitative correction factors can be applied in other spectroscopies to reduce fluctuations in the y-axis (intensity/absorbance scale), particularly when applications have a long lifetime.

When data is being collected and combined from multiple locations, it may be useful to verify the accessory used. For example, combining mid-infrared spectral data collected with an ATR accessory and a transmission accessory can often degrade performance owing to the differences in depth of penetration and differences in spectral band shape with each accessory. The penetration depth for an ATR crystal depends on the wavelength, refractive index of the ATR crystal, and on the sample itself, as well as the angle of the entering light beam [23]. The distortion of water bands using ATR accessories compared to a transmission spectrum is reportedly caused by the chemical changes of water at the interface between the optical element and the bulk solution [24].

Intermittent problems can occur over time for a variety of reasons. Optical window fouling, or residual particles can cover a sensor window and reduce the quality of the spectra. Sampling accessories can be modified to enhance spectral quality. For example, a polymer coating can prevent fouling of the optical sensor used for the analysis of blood, polluted water samples, or slime [25].

Excessive noise in a spectrum and spectral distortion can occur for multiple reasons. A missing sample can happen in a fast-moving measurement system; the sample can become misaligned during measurement, and spectral distortion can occur due to vibration. Lastly, sinusoidal features in the spectrum called fringes can occur due to multiple internal reflections that should be corrected before analysis.

Raman signal intensities are highly dependent on the hardware used. Signal strength is affected by the laser (power and band), collection efficiency, sampling pathlength, cross-section, and concentration. Since the intensity response is so unique to each spectrometer set-up, it is very important that a correction be applied to account for these intensity variations between instruments [26].

The FDA requires data collection software for pharmaceutical applications and clinical studies to comply with CFR Part 11 requirements. Many instrument vendors do provide an option to collect and analyze spectral data with 21 CFR Part 11 features enabled during standard operating procedures. This software provides authentication of the spectral data collected by tagging with date and time stamps, multi-level user permissions, audit trails, and e-signatures. CFR Part 11 compliance is further discussed in Chapter 7 on Validation.

When developing methods, check the literature and discuss your application with the vendor to see if any new advancements or advice are available that may enhance the quality of the spectral data collected. If you spot-check the spectra early in development, you may be able to stop the propagation of errors and save time-modeling-irrelevant data.

2.6 Uncontrolled Measurement Conditions

Other sources of error can occur due to the spectroscopy measurement itself. Uncontrolled experimental conditions often cause these errors. Conditions that may unexpectedly change between sample measurements include stray light, poor contact between a sample and the measurement window, particle size and shape, and variances in low-cost disposable sample containers.

When commercial instrumentation is used, errors due to stray light sources are generally caused by procedural problems resulting from a lack of training or knowledge about spectral data collection.

For example, when the instrument lid is not fully closed. Stray light reduces the linear range and reduces signal to noise by interfering with the incident light and collection efficiency. Stray light may become an issue when application-specific hand-held devices are mass-produced for clinical studies or field use.

Poor sample contact can occur when sample sizes are small or in process environments when the material only partially covers a sensor. This type of error can be controlled by improving the training provided to operators or changing how a probe is inserted into the process. Likewise, deviations in particle size can be handled either by (1) tightening specifications for the procedure, (2) introducing a sample pretreatment step to unify the particle size, or (3) building robustness to particle size variances by introducing particle size as a parameter that is varied within the "training set" design.

Applications that measure the product directly through finished product packaging may be affected if the properties of the packaging change over time, such as in its composition, thickness, or optical properties. Deviations in the measurement container thickness, chemical composition, or surface can cause a change in the spectral fingerprint. This is a particular problem in diffuse reflectance near-infrared spectroscopy when the point of measurement is through the bottom of the container. While dram vials are often used in Raman measurements, the concave shape of certain types of these containers (see Table 2.2c) and manufacturer-specific engravings (Table 2.2d) at the bottom can reduce light throughput or distort the spectrum when measurements are conducted by diffuse reflectance NIR. This type of error may be controlled by ensuring packaging is within a narrow window of tolerance or using spectroscopic-grade glass and ensuring the spectral difference between containers is smaller than the change in the property of interest. In addition, the calibration standards can be measured with several lots of disposable containers to improve robustness to vial variances. This works well if calibration modeling is done with chemometric algorithms, such as PLS regression since the effects of these variances can be modeled in the principal component loadings.

Poor sample contact can lead to distortion of the spectrum, a raised baseline, and a severely weakened measurement. Figure 2.9 shows the analysis of Olay soap bar chips when the sensor is only partially covered with the product. This type of problem might occur in soap dryer applications using a diffuse reflectance probe if the sensor is not installed correctly, not in direct contact with the product, or when crumbling occurs at low moisture. Using a near-infrared designed to extend into the visible region, a sharp offset at 1100 nm may occur, which is the point when the software stitches the spectra acquired using two detectors, i.e. the visible region measured with a silicon detector and the near-infrared region measured with an indium gallium arsenide (InGaAs) detector. Fortunately, the interface became an excellent diagnostic to catch poor contact. The problem is often solved with a better interfacing design.

Table 2.2 General comparison of glass vials and quality of bottoms for spectroscopy applications.

(a) Glass dram vial (4 mm)	**(b) Glass scintillation vial**	**(c) Glass dram vial (3 mm) for sidewall Raman measurements**	**(d) Glass container with embossed bottom**
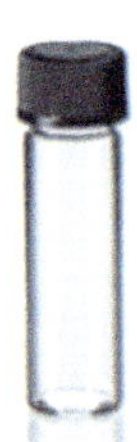			

Source: Deborah A. Peru.

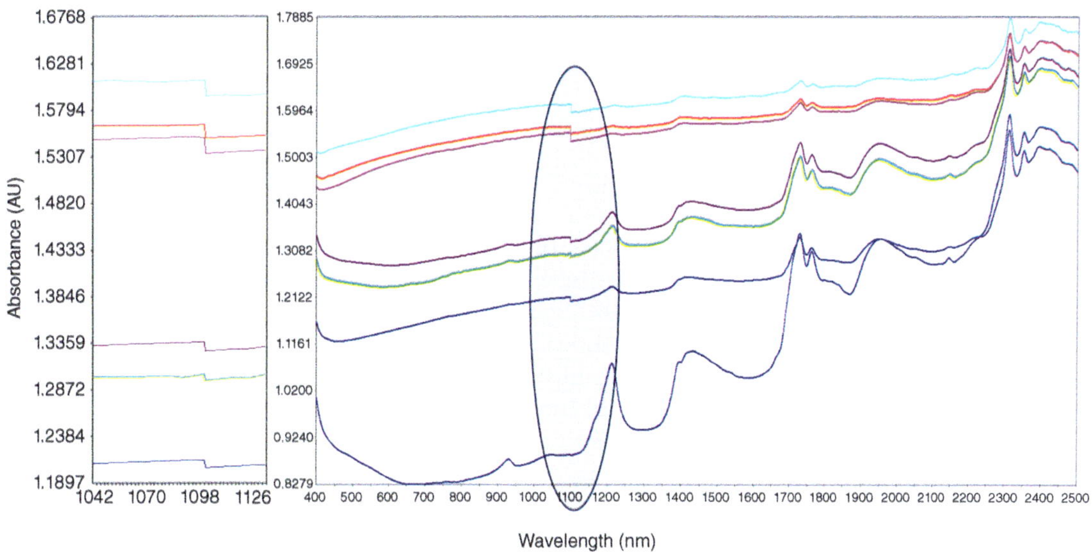

Figure 2.9 Analysis of bar soap in poor contact with the measurement sensor. The left inset shows a sharp offset at 1100 nm, the point where spectra from two detectors used in a dispersive NIR instrument are stitched together for display. *Source*: Deborah A. Peru.

In reaction monitoring the potential for carbon dioxide formation (CO_2) can cause bubbles to occur and foam formation during vigorous mixing. Carbon dioxide can also cause a change solution pH which will also affect spectral band shape and response. Water vapor within the instrument can often be observed in the mid-infrared and near-infrared spectrum. However, it can be corrected using a desiccant, air purge, or an air vapor suppression software feature.

2.7 Examine the Spectral Data

Take the necessary time during method development to overlay and visualize the data before model building begins. This is an important step that many spectroscopists forget to do. If the data set is large, data reduction methods such as principal component analysis (PCA) might be required to identify potential outliers that require further inspection. This is discussed in more detail in Chapters 3 and 5. When visualizing your data in an overlay or using PCA scores, try to develop an understanding of where and why variances are occurring and how they can have an effect on the spectrum.

Changes in the light source power or measurement container thickness can increase baseline noise. When fluorescence occurs in Raman measurements, the quality and information obtained from the spectrum can be greatly reduced [21]. A lower-energy visible laser (785 nm) or a near-infrared laser line (1064 nm) can be used to minimize or eliminate molecular fluorescence. Another approach is to use an ultraviolet laser line at 248 nm. In this instance, fluorescence emission can still occur but is widely separated from the Raman signal [27].

In Raman imaging, the charge-coupled device (CCD) detector is vulnerable to random spikes caused by atmospheric cosmic ray bands. These random cosmic spikes are superimposed on chemically meaningful spectra [28]. Due to their high intensity and potential influence on variance, they should be removed before post-analysis of the data. Many vendors include some form of removal or zap function in their software.

In fluorescence, the signal can become quenched by self-absorption or matrix poisoning. In many applications involving high concentrations of solvents or particular ingredients, absorbance or intensities can cause detector saturation depending on the type of detector used and may require incorporating a sample preparation procedure to avoid saturation and improve the linear relationship between the response and concentration.

2.8 Concepts in Propagation of Error

An important topic in statistics that directly influences the performance of a spectroscopy procedure is called "propagation of error." Measurement error is always propagated whenever a quantity is derived from two or more measurements, each containing its own uncertainty [1]. In quantitative spectroscopy, the final error in a determination is a culmination of errors occurring at multiple stages. The most prominent are (1) sampling, (2) the reference (primary) method for assigning quantitative values, and (3) the spectroscopy measurement (instrumentation and sample presentation). During method development, it is important to review each variance source and optimize the procedure to ensure the highest degree of accuracy and precision. Most importantly, the errors that introduce the largest degree of error in the final measurement should be minimized [1].

How do these sources of error affect our final determination? It depends on whether these errors are systematic or random and on the type of math operations used during the postprocessing of the spectral data. For the simplest inverse method using multiple linear regression (MLR) with one independent variable, these three sources of error will have a direct effect on the standard error (SE) determined for both the slope (β_1) and the intercept (β_0). The SE equation for the slope is shown below. The residual error term ($yi - \hat{y}i$) is defined as the predicted minus the actual value for all samples in the calibration (training) set. To determine the significance (t), divide the slope by the SE. The t value is compared with a t-distribution table at the 95% confidence interval (CI) for the appropriate degrees of freedom (DOF). The calculated t needs to be greater than the value found in the two-tail t-distribution table. Alternatively, the corresponding P-value can be determined with a t score to P-value calculator. A significant P-value is less than (<0.05) @ 95% CI.

$$\text{SE}\ (\beta_1) = \text{sqrt}\ [\Sigma(yi - \hat{y}i)^2/(n-2)]/[\Sigma(xi - x)^2] \tag{2.1}$$

When MLR modeling is extended to include multiple independent variables $x_1 + x_2 + x_3$ etc., the corresponding slopes (β_1, β_2, β_3, etc.) are determined. The formula for the SE for each regression coefficient is determined using the variance–covariance matrix of the betas which is solved using linear algebra equations where s^2 is the mean square error (MSE). Significance testing (as discussed above) is determined for each slope (i.e. wavelength) and used to determine the variables important for inclusion in the final regression equation.

$$\text{VCOV}(\beta) = s^2(X'X)^{-1} \tag{2.2}$$

For even more complex regression models involving chemometrics (i.e. PLS or PCR), the errors propagated errors in the final determination can be observed when reviewing diagnostic parameters involving the sum of squares error (SSE):

$$\text{SSE} = \Sigma(y_i - y_{\text{pred}})^2 \tag{2.3}$$

When the SSE is close to zero, we say the model has a very good fit, and the coefficient of determination, R^2, is close to 1. As SSE increases, R^2 declines below 1. The standard error of calibration (SEC) incorporates all the sources from the instrumental measurement, primary method, and sampling. It is calculated using the SSE for all training samples, using the final model, and corresponding degrees of freedom ($n - 1$) for one analyte determination.

$$\text{SEC} = \sqrt{\Sigma(y_{\text{pred.}} - y_{\text{act.}})^2/(n-1)} \tag{2.4}$$

The same type of equation can also be used to estimate the SE of prediction during cross-validation SEPV. The SEPV is estimated during development using the "leave-one-out" cross-validation procedure discussed in more detail in Chapter 5.

At the beginning of calibration development, it is recommended to review all sources of error for each step involved and try to improve precision in each analysis phase. It is especially important to focus on reducing the source of error that introduces the largest error in the determination.

2.9 The Signal-to-Noise Ratio (SNR)

Noise is a collective term used to describe unwanted sources of measurement signal that do not pertain to the material's physical or chemical properties. Some noise sources introduce systematic error, such as dark current, and can be corrected. Other noise sources introduce random error that can be averaged out over several acquisitions. The relevant types of noise in spectroscopy include dark current, shot noise, readout noise, stray light, and electronic noise. Factors such as optical and electronic components contribute to the systematic noise component and can cause a bias when calibration models are transferred from one instrument to another [8]. This is why some vendors offer calibration transfer algorithms to help transfer equations from one instrument to another.

The signal-to-noise ratio (SNR) is a metric used to measure the sensitivity of an instrument and is often used to compare instrumentation and optimize measurement conditions. It is typically measured by determining the ratio between the intensity of an on-peak measurement signal and the intensity at the noise level. The procedure involves measuring an SRM or stable material for a specific number of scans. The numerator of the equation is determined as the average peak height for a particular band in the sample. The noise is determined by calculating the average baseline noise over a range of wavelengths/wavenumbers, or the variation of the on-peak signal is determined when measuring a blank sample.

The baseline variance is sensitive to instrumental noise, including electrical noise (both dark current and fixed noise), and optical sources, such as stray light. The choice of detector, intensity of the light source, and efficient design of the light collection optics are all factors that can affect the signal. A strong light source can increase the on-peak signal but may also increase baseline noise. Instruments designed with highly sensitive detectors, such as the mercury cadmium telluride (MCT), will improve sensitivity and SNR. Meanwhile, deuterated triglycine sulfate (DTGS) detectors operate at room temperature and may require longer scanning times to achieve a good SNR for quantitative purposes. Other sources of noise that can appear in the spectrum may be caused by temperature fluctuations, vibration in the environment or tabletop where the instrument is located, and possibly shifting of the sampling accessory during normal use.

Cut-off filters can remove unwanted stray light in handheld devices. In gas-phase applications, specialty windows pass IR light into and out of the sampling cell while maintaining a leak-tight seal between the gaseous samples and the electronics compartments [29]. Some vendors include software correction algorithms to remove stray light, noise, CO_2, or moisture from the measurement, all of which will improve the measurement's SNR.

Often, the most influential factor for improving SNR is increasing the number of scans collected (co-added) and averaged in a single measurement. The SNR improves proportionately with the square root of the number of co-added scans owing to the reduction in random error sources by averaging. Figure 2.10 shows the relationship between the signal at 1439 cm^{-1} to the standard deviation in baseline noise from 1900 to 1600 cm^{-1} for Olay bar soap measured by Raman spectroscopy using 1064 nm laser. The measurement band at 1439 cm^{-1} is due to the CH_3, and CH_2 vibration of the alkanes in fatty acids.

Table 2.3 summarizes the empirical SNR for these measurements using the peak height @ 1439 cm^{-1} to the average baseline noise in the region from 1900 to 1600 cm^{-1}. The linear relationship between SNR and

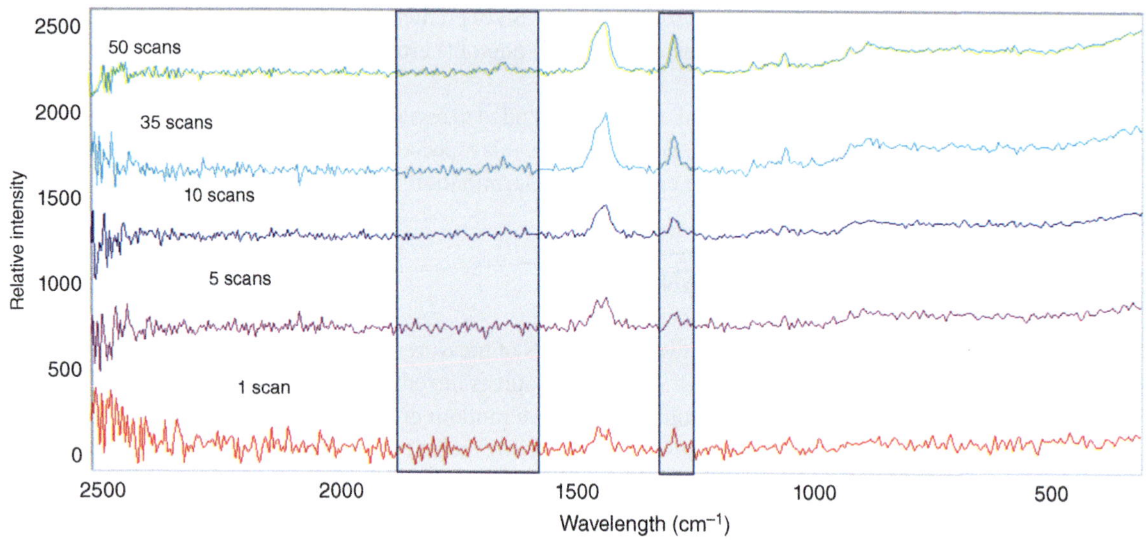

Figure 2.10 Improvement in SNR with increasing scan number for Raman measurements of Olay bar soap. *Source*: Deborah A. Peru.

Table 2.3 Determination of signal to noise with increasing number of scans.

Parameter	1 scan	5 scans	10 scans	25 scans	50 scans
Peak height @ 1439 cm^{-1}	98.61	120.41	123.42	129.06	117.61
Noise (1900–1600 cm^{-1}), std. (σ)	30.36	14.05	10.01	7.33	6.47
SNR empirical	3.25	8.57	12.33	17.62	18.19
$\sqrt{\#scan}$	1	2.24	3.16	5	7.07

Source: Deborah A. Peru.

the square root of the number of scans is ~0.94 (Pearson correlation coefficient). The measured SNR will vary between Raman instruments due to differences in laser sources and hardware configuration, but this relationship will remain the same.

Increasing the number of co-added scans has a practical advantage. A higher S/N will translate into lower detection limits; thus, lower concentrations can be measured. While increasing the number of co-added scans will lengthen measurement time, it can improve the S/N and reduce the detection limit. In some FTIR applications, collection parameters may necessitate 256, 512, or even 1024 co-added scans to detect concentrations less than 500 ppm Measurement time will increase proportionately from one minute to almost eight minutes per scan.

2.10 Conclusion

Prior to modeling the data, the three main sources of error affecting the outcome of quantitative methods are related to poor sampling, uncontrolled measurement conditions, and reference method errors. Random error is unavoidable. The best defense is to capture calibration standards over time to ensure random fluctuations are captured so that signal averaging and modeling are optimized to minimize their effect.

Systematic errors can be identified with the proper investigation. Work with your team, ask the right questions to discover the cause, and make corrections accordingly. In some instances, the range of chemical and physical (e.g. temperature, density, or particle size) variances expected in unknown samples should be incorporated within the training set to enhance the long-term robustness of the regression models developed.

A signal-to-noise measurement evaluates an instrument's intrinsic sensitivity. It can be used as a gauge to help optimize procedures during development, compare instrument hardware performance, and monitor instrument performance and long-term stability.

References

1 Caulcutt, R. and Boddy, R. (1983). *Statistics for Analytical Chemists*. London; New York: Chapman and Hall.

2 Wang, T., Ibrahim, A., Potts, A.R., and Hoag, S.W. (2015). Application of multivariate strategies to the classification of pharmaceutical excipient manufacturers based on near-infrared (NIR) spectra. *Applied Spectroscopy* 69 (11): 1257–1270.

3 Sarathjith, M.C., Das, B.S., Wani, S.P., and Sahrawat, K.L. (2016). Variable indicators for optimum wavelength selection in diffuse reflectance spectroscopy of soils. *Geoderma* 267: 1–9.

4 Gao, Y., Cui, L., Lei, B. et al. (2014). Estimating soil organic carbon content with visible–near-infrared (Vis-NIR) spectroscopy. *Applied Spectroscopy* 68 (7): 712–722.

5 Brander, S.M., Renick, V.C., Foley, M.M. et al. (2020). Sampling and quality assurance and quality control: a guide for scientists investigating the occurrence of microplastics across matrices. *Applied Spectroscopy* 74 (9): 1099–1125.

6 Gholizadeh, A., Borůvka, L., Saberioon, M., and Vašát, R. (2013). Visible, near-infrared, and mid-infrared spectroscopy applications for soil assessment with emphasis on soil organic matter content and quality: state-of-the-art and key issues. *Applied Spectroscopy* 67 (12): 1349–1362.

7 Williams, P. and Norris, K. (ed.) (1987). *Near-infrared Technology in the Agricultural and Food Industries*. American Association of Cereal Chemists.

8 Ciurczak, E.W., Igne, B., Workman, J. Jr., and Burns, D.A. (2021). *Handbook of Near-Infrared Analysis*, 4e. CRC Press.

9 FDA (2023). Current good manufacturing practice (CGMP) regulations. Maryland: US Food and Drug Administration. https://www.fda.gov/drugs/pharmaceutical-quality-resources/current-good-manufacturing-practice-cgmp-regulations (accessed 31 July 2023).

10 EPA SW-846. (1986). Hazardous waste test methods. Washington, D.C: United States Environmental Protection Agency. https://www.epa.gov/hw-sw846/sw-846-compendium (accessed 30 July 2023).

11 ISO 28590 (2017). Introduction to the ISO 2859 series of standards for sampling for inspection by attributes. International Organization for Standardization. https://www.iso.org/standard/64622.html (accessed 31 July 2023).

12 FDA (2004). A framework for innovative pharmaceutical development, manufacturing, and quality assurance guidance for industry October 2004. Maryland: US Food and Drug Administration. https://www.fda.gov/regulatory-information/search-fda-guidance-documents/pat-framework-innovative-pharmaceutical-development-manufacturing-and-quality-assurance (accessed 31 July 2023).

13 De Los Santos, P., Hofer, J., Jayjock, E. et al. (2019). Sampling considerations in continuous manufacturing. https://ispe.org/pharmaceutical-engineering/may-june-2019/sampling-considerations-continuous-manufacturing (accessed 31 July 2023).

14 Metrohm USA Application Note AN-PAN-1050. (2023). Inline moisture analysis in fluid bed dryers by near-infrared spectroscopy. https://www.metrohm.com/en_us/applications/application-notes/prozess-applikationen-anpan/an-pan-1050.html (accessed 11 July 2023).

15 Voraberger, H., Ribitsch, V., Janotta, M., and Mizaikoff, B. (2003). Application of mid-infrared spectroscopy: measuring hydrogen peroxide concentrations in bleaching baths. *Applied Spectroscopy* 57 (5): 574–579.

16 Liu, Y. and Chen, Y.-R. (2000). Two-dimensional correlation spectroscopy study of visible and near-infrared spectral variations of chicken meats in cold storage. *Applied Spectroscopy* 54 (10): 1458–1470.

17 Kenyon, A. S., Kirchhofer, R.D., and Layloff, T.P. (1992). The 1992 analytical methods for pharmaceutical analysis. Missouri: US Food and Drug Administration, Div. of Drug Analysis. http://www.layloff.net/fda/1992%20STL-FDA%20Drug%20Analyst%20Training%20Manual.pdf.

18 García-González, D.L., Sedman, J., and van de Voort, F.R. (2013). Principles, performance, and applications of spectral reconstitution (SR) in quantitative analysis of oils by Fourier transform infrared spectroscopy (FT-IR). *Applied Spectroscopy* 67 (4): 448–456.

19 Feng, H., Bondi, R.W., Anderson, C.A. et al. (2017). Investigation of the sensitivity of transmission Raman spectroscopy for polymorph detection in pharmaceutical tablets. *Applied Spectroscopy* 71 (8): 1856–1867.

20 Padalkar, M.V. and Pleshko, N. (2015). Wavelength-dependent penetration depth of near-infrared radiation into cartilage. *The Analyst* 140 (7): 2093–2100.

21 Tuschel, D. (2016). Selecting an excitation wavelength for Raman spectroscopy. *Spectroscopy* 31 (3): 14–23.

22 Barrera, B.A. and Sommer, A.J. (1998). Sampling considerations for FT-Raman spectroscopy of polymers and solutions. *Applied Spectroscopy* 52 (11): 1483–1487.

23 Bruker Optics (2013). Attenuated total reflection (ATR) mode – Advantages for FT-IR spectroscopy. https://www.azom.com/article.aspx?ArticleID=5958 (accessed 24 July 2023).

24 Hancer, M., Sperline, R.P., and Miller, J.D. (2000). Anomalous dispersion effects in the IR-ATR spectroscopy of water. *Applied Spectroscopy* 54 (1): 138–143.

25 Tenjimbayashi, M., Komatsu, H., Akamatsu, M. et al. (2016). Determination of blood potassium using a fouling-resistant PVDF–HFP-based optode. *RSC Advances* 6: 14261–14265.

26 BWTek Metrohm USA Technical Note 410000055-A (2021). The importance of relative intensity correction of Raman data. https://www.metrohm.com/en_us/applications/bw-tek-applikationen/410000055-a.html (accessed 2 January 2024).

27 Asher, S.A., Calum, H.M., and Zhenhuan, C. (1997). UV lasers revolutionize Raman spectroscopy. *Laser Focus World* 33 (7): 99–109.

28 Zhang, L. and Henson, M.J. (2007). A practical algorithm to remove cosmic spikes in Raman imaging data for pharmaceutical applications. *Applied Spectroscopy* 61 (9): 1015–1020.

29 Institute of Clean Air Companies, Inc. (2013). Guide to FTIR technology for compliance testing, performance specifications, and continuous emissions monitoring of target gases. https://montrose-env.com/wp-content/uploads/2017/09/Guide-to-FTIR-Technology-for-Compliance-Testing-Performance-Specification-Continuous-Emissions-Monitoring-CEM-of-Target-Gases.pdf (accessed 25 Aug 2023).

3

Procedure Design and Data Analytics

Deborah A. Peru

DP Spectroscopy and Training LLC, Lebanon, USA

3.1 Introduction

Data analytical tools are essential in the early stages of method development to better understand the information in the spectral data collected for instrument calibration and how to optimize the design of a quantitative procedure. Designing a procedure that is "fit for purpose" is paramount. Fit for purpose means the analytical procedure used to measure a key attribute has the desired specificity, accuracy, and precision over the specified range [1]. Optimizing and selecting the best instrument and measurement conditions for the analysis objectives is important to achieve this goal. In spectroscopy, how the spectrum is collected, i.e. sampling accessories, measurement parameters, sample presentation, and choice of detector, can have a large effect on the quality of the spectrum, depth of penetration, signal strength, and measurement area. This chapter includes a closer look at some important characteristics and differences among sampling approaches and provisions for use.

Methods for evaluating and characterizing the dependent variables, or [Y] block of properties, including range, variable distribution, and collinearity among variables, are discussed along with the fundamentals of design of experiments (DoE). Moreover, this chapter will describe how statistics and chemometrics work together to characterize the spectral data set, known as the independent variables or [X] block. Principal component analysis (PCA) is commonly performed to provide a better understanding of the chemical differences within the spectra. PCA reduces a large set of spectra into smaller subsets that can be plotted and visualized in many ways, providing valuable insight during the early stages of development.

3.2 Elements of Procedure Design

The United States Pharmacopoeia (USP) General Chapter <1220> describes the critical role that knowledge management has in analytical procedure development and lifecycle management [2, 3]. Knowledge management starts with defining the problem. Establish an analytical target profile (ATP) and conduct a risk assessment to understand better the factors contributing to the variability in the attribute of interest [4]. The ATP includes a description of the intended purpose, product attributes, quantity and frequency of testing, and relevant performance criteria [3, 4]. An evaluation of the expected variability based on knowledge of the manufacturing process and regulatory requirements for the analyte to be determined must be known. Additional information should be gathered regarding the factors that affect spectral variability, such

Practical Quantitative Vibrational and Electronic Spectroscopy: A Guide for Developing, Optimizing, and Validating Procedures, First Edition. Edited by Deborah A. Peru.

Companion website: www.wiley.com/go/peruelectronicspectroscopy

as measurement conditions, where the instrument is located relative to the sample, and product attributes, including the chemical and physical properties. ICH Q14 applies to analytical testing procedures used for the release and stability testing of commercial drug substances and products and lays the foundation for real-time testing [3, 5].

Once the problem (or process) is well understood, the developer can begin the strategic design of an analytical procedure that sufficiently satisfies the objectives and level of understanding to meet the analytical goals. Whether the goal is to create a new procedure or refine an existing method, three necessary design steps are to (1) optimize the measurement conditions, (2) use data analytics, and (3) implement analytical controls. Understanding the interrelationship between data analytics and procedure design is a fundamental aspect of quantitative spectroscopy. These processes often occur simultaneously as the effects of procedure design continue to be evaluated during development when the method is used for experimental predictions, and results are used to further improve procedure attributes.

3.3 Measurement Conditions

3.3.1 Maintaining Data Integrity and Robustness

Experimentation is required to identify the right instrumental technique, accessories, and measurement conditions to optimize performance. This stage may involve testing and comparing results to identify the best conditions for measurement. How the samples are prepared for analysis will depend on the sample type (solid, liquid, slurry, mixture, paste, etc.) and the sensitivity or response of the instrumentation being used. Proper care should be taken to fully understand any effects of sample compaction, particle-size variation on the spectral response, and sample presentation to the spectrometer must be reliable.

During early development, experiments are conducted to optimize measurement conditions such as the integration time and number of co-added to maximize the signal-to-noise ratio (SNR). When evaluating the performance of different types of accessories and measurement conditions, the SNR and standard deviation for n number of replicate scans are useful diagnostic metrics to demonstrate improvement in reproducibility (see Chapter 2, Figure 2.11). Fourier transform instruments need to be optimized for resolution and integration time. Barra et al. showed an improvement in the coefficient of determination (R^2) and a reduction in root mean square error during cross-validation (RMSECV) as the number of co-added scans was increased from 10 to 80 scans [6]. This effectively reduces the effects of random noise sources and improves measurement precision [6]. Decreasing the resolution of Fourier-transform spectra provides the same improvement in spectral reproducibility and SNR [6, 7]. In Raman spectroscopy, optimization of measurement parameters includes laser power, integration time, spot size, sample positioning, and replicate scan averaging. For quantitative applications, the Raman signal response must be calibrated. Calibration is obtained by determining a Raman-specific response function value using a standard reference material (SRM). The relative intensity correction function is created by dividing the proper spectral shape of the SRM by its uncorrected spectrum [8, 9]. All future Raman intensities are corrected using the response function determined for each instrument and probe combination.

Establishing a robust sample presentation procedure will help to maintain the integrity of the spectral data collected over time. Data integrity refers to the assurance that spectral information remains accurate and consistent over the entire life cycle of the method. Sample presentation problems and other anomalies during sample introduction may compromise the quality of the spectrum. These anomalies include a dirty measurement window, bad reference spectrum, excessive noise or vibration, poor contact with the probe or window, light-scattering artifacts such as air, bubbles, or particles, foam formation, stray light, light scattering, and careless execution.

During development, the slit width and spot size should be optimized to obtain good-quality spectra with the best signal strength. In ultraviolet, visible, and near-infrared techniques, narrow bands relative to the instrument's bandwidth will lead to deviations in Beer's law because a proportion of light bypasses the sample and reaches the detector [10]. In Raman spectroscopy, a spacer may be needed for surface measurements, whereas proper vial positioning is needed to ensure the laser focuses inside the sample container [11].

In the early stages of development, it may also be necessary to review and sometimes improve the primary method procedure used to assign a concentration or property of interest to each calibration spectrum. The measurement error in the primary method can affect the goodness of fit and prediction error obtained in spectroscopy. To minimize the propagation of error, the accuracy in the primary methods should be established and optimized if necessary to ensure the final procedure meets the business goals for the analysis. The typical range is between 1 and 2%.

3.3.2 Sample Accessories

Many different types of measurement cells are available for liquid measurements. Glass is a common choice and is much cheaper than spectroscopic-grade windows and quartz cuvettes. It is ideally used at wavelengths above 350 nm and begins to absorb strongly near 290 nm. Quartz cells are manufactured with different grades of silica; standard Infrasil® (65% transmission at 250 nm), infrared-grade Herasil® (85% transmission at 200 nm), and Suprasil® (90% transmission at 200 nm). Their cost is much higher than glass, and thus, quartz cells are only used when it is necessary to observe in the ultraviolet region below 350 nm. Polystyrene cells are significantly cheaper, and light transmission is unobstructed to about 340 nm [12].

Liquid cells for stationary atline and offline measurements are available for all spectroscopy techniques, and many are interchangeable. A standard cuvette typically has a 10 mm pathlength and holds about 3.5 ml of liquid. Semi-micro and micro-cuvettes range from 300 µl to 1.4 ml, respectively. Long- and short-pathlength cells range from 1 mm up to 100 mm in diameter. Cylindrical cells are used for gas monitoring, and flow and stirrer cells are used for reaction monitoring. Disposables made from polystyrene are an excellent option for measurements in the 340 nm–900 nm range, and acrylic (PMMA) in the 300–900 nm range. Scintillation glass and dram vials are often used in near-infrared and Raman measurements owing to their low cost and ease of cleanup. For process applications, an interactance immersion probe with a variable pathlength is used in the near-infrared and ultraviolet–visible regions for clear and translucent aqueous samples. A top-load attenuated total reflectance (ATR) accessory is the accessory of choice for mid-infrared quantitative measurement of liquids due to the ease of sample introduction and clean-up between measurements. Salt (typically NaCl) plates and a disposable card with a polyethylene window that is IR transparent below 2849 cm^{-1} are used for qualitative determinations. Still, they are generally not used for quantitative transmission owing to the difficulty in controlling sample thickness.

Certain types of glass or plastic containers can become a significant source of measurement error and chemical instability [13]. For example, pharmaceutical products are often measured directly through a certified damage-resistant glass vial. Verify the consistency, set specifications, or build robustness to deviations within the training set by using multiple lots. Watari et al. demonstrated measurable differences in the pathlength and peak intensity of bands near 7130 cm^{-1} (1402 nm) and 4515 cm^{-1} (2214 nm), attributed to the variances in hydroxyl molecules in different lots of disposable vials [14].

Standard glass vials can also degrade chemical ingredients over time due to the reaction with leachable sodium [13]. Specialty-treated high-quality glass vials are available that prevent leachable ions from interacting with samples [13].

In transmission spectroscopy, the volume of a liquid sample measured must be optimized when bulk measurements are desired to avoid detector saturation. When measuring liquids in sealed containers, air bubbles in samples will cause light scatter and should be removed or coaxed toward the surface by gently tapping. Liquids with particulates may need to be filtered if transmission mode is used. Preparing samples for ultraviolet, visible, or fluorescence spectroscopy procedures may involve additives or derivatization to enhance color, absorption, or emission response [15].

For solid samples, various accessories are available for each spectroscopy technique. An integrating sphere accessory is often used in ultraviolet, visible, and near-infrared measurements. ATR in the mid-infrared region can be used for quantitative measurements of highly viscous materials. However, it is more challenging to use for quantitative analysis of solids because constant back pressure is required to ensure good sample contact with the sensor surface. Diffuse reflectance infrared spectroscopy can be used in laboratory procedures to determine materials in powdered form quantitatively. The classical approach requires careful and elaborate sample preparation using potassium bromide (KBr) to dilute the powdered material [16]. Reflectance measurements are transformed into Kubelka–Munk units to establish a linear relationship between spectral intensity and concentration. The theory applies to measurements of molecules in a non-absorbing matrix, with a constant scattering coefficient and infinite sample dilution [17].

Most solids can be measured "as is" in the near-infrared region without detector saturation. In near-infrared measurements of samples, the spectrum may include the spectral fingerprint of a plastic container or bag, but high-grade glass or spectroscopic-grade windows can be purchased that are optically transparent in the near-infrared region [18]. For near-infrared measurements, a top-load diffuse reflectance stationary accessory or probe can be used for a wide range of solids, powders, slurries, grains, and animal or biological samples. Many vendors offer rotating cups for measuring heterogeneous samples atline or offline using laboratory instruments. The sample is placed into a beaker with a spectroscopic-grade quartz or infrared-grade glass window, which rotates between measurement scans. This enables signal averaging across a wider portion of the sample. When measuring powders or grains, removing foreign material and large particles with a sieve is beneficial especially when they are not representative of the bulk material [19]. This approach helps to improve the reproducibility of material chemistry, packing uniformity, and light-scattering variance.

In process applications of spectroscopy, the type of fiber and the number of fibers packed in a bundle are important considerations. Ultraviolet, visible, near-infrared, and Raman can be configured for process analysis using silver-lined fiber cores. Silver halide fibers have shown significantly better mechanical properties for the mid-infrared region than the often-used brittle chalcogenide glass fibers. In process Raman, fused silica core fibers have gained popularity because they are low-cost, stable, and robust for most applications [20]. Generally, the more return fibers used in a bundle, the better the collection efficiency, but it will drive the cost higher [21]. Minimizing the fiber optic length will reduce light leakage and improve collection efficiency. Specialized fibers for measuring over long distances can be used. For example, Mitsubishi Diaguide®, a germanium-doped SiO_2 fiber core, was designed to reject background emissions in Raman spectra collected over hundreds of meters [22].

3.3.3 Instrument-Specific Considerations

Figure 3.1 illustrates a setup for measuring a heterogeneous mixture of polypropylene and polyethylene polymer pellets by Raman and near-infrared spectroscopy. One of the major differences between these techniques is the sampling area. Owing to the small spot size of the laser, the Raman sampling area at the laser focus point is about 100 μm [23]. Thus, it is much more difficult to obtain a representative measurement of heterogeneous samples using Raman instruments [23]. One approach would be to collect several spectra replicates across a wide sample area to obtain a more representative measurement. On the other hand, most near-infrared instruments measure a large spot size, typically from 5 to 17 mm in diameter, and some accessories can rotate or move the sample between scans.

(a) (b)

Figure 3.1 (a) Measurement of heterogeneous polymer pellets by Raman with a fiber optic probe. *Source*: Deborah A. Peru. (b) Measurement of heterogeneous polymer pellets by Near-infrared on top-load diffuse reflectance accessory. *Source*: Deborah A. Peru.

Raman spectroscopy has excellent molecular selectivity and can be used for non-contact analysis through glass or plastic. Raman is generally free of water interference and thus is ideal for measuring aqueous solutions [24]. However, the highly energetic laser may decompose sensitive samples, particularly biologicals [24]. Other ingredients, matrix components, or trace impurities can generate a photoluminescence (fluorescence) response during Raman measurements. This response can overwhelm the detection of Raman scattering bands from the analyte of interest. This is because photoluminescence has a much higher probability of occurring than Raman scattering [25]. In these instances, photobleaching or using a different Raman laser line can help minimize this interfering response [25]. Certain preprocessing techniques will reduce fluorescence interference but can weaken or distort the remaining Raman band signal. The use of the 1064 nm near-infrared laser band will help reduce unwanted fluorescence, but integration times may need to increase to get a better SNR. To substantially enhance Raman scattering, a gold or silver substrate can improve sensitivity orders of magnitude by generating a local plasmic field [26].

Conversely, near-infrared spectroscopy does not suffer from the fluorescence problem and can also measure samples through plastic or glass containers. The limiting factor in near-infrared spectroscopy is the lack of specificity. The near-infrared spectrum is complex because it contains many overlapping vibrational overtones and combination bands that require multivariate techniques to separate closely related materials and develop predictive models [19, 27, 28]. Lyophilized products can form an unevenly dispersed freeze-dried pellet [29]. For direct measurement of residual moisture by near-infrared through glass, it may be necessary to optimize how and where the sample is analyzed within the container to obtain the most relevant information [30].

Temperature control of the samples and instrument is important for all spectroscopy methods because of the effects on band shape, position, and intensity [10, 19, 27, 31, 32]. In some applications, when product temperature cannot be adequately controlled, it is considered a process variable. As such, the temperature of the calibration samples can vary in temperature to build robustness to changes over time [27]. In process applications, temperature control using heat exchangers and temperature-controlled flow cells will help to minimize spectral shifts and structural changes in hydrogen bonds due to temperature

[28]. Including temperature as a variable (within a modest range) will ensure the effect on the spectrum is factored into the regression vectors determined when using principal component (PC) based regression techniques such as partial least squares (PLS) regression.

3.4 Data Analytics

Procedure design incorporates learning about the relationship within and between variables, applying concepts in correlation and covariance, and statistical detection of unusual spectral behavior. Proper utilization of data analytics during model building enhances the prediction outcome. Various data analytic tools and techniques are used to better understand the information in both the [X] block and [Y] block variables on a statistical basis.

Data analytics is useful for designing, evaluating, and optimizing the training set. This book uses the term "training set" synonymously with calibration standards. The choice of standards directly affects the procedure's overall accuracy, precision, and long-term robustness. The distribution (frequency and concentration of standards), range, and number of standards affect the calibration outcome. Multivariate methods can benefit from the inclusion of expected sources of random error that become embedded in the spectrum over time. Including variables unrelated to chemistry, such as product temperature and particle size, over the ranges expected will also improve long-term robustness. Other variances, such as lot, date, time, and instrument hardware, must be considered. The decision on how to handle these types of variances will depend on the application and circumstances. In general, as the number of variances included in the "training set" increases, the specificity and accuracy of the method may begin to decline, and thus, it is necessary to seek a balance between robustness and accuracy [33].

Chapters 4–6 contain an in-depth discussion of other essential topics in data analytics, such as the types of regression algorithms, wavelength selection strategies, and when and how to use spectral preprocessing techniques to improve precision and remove unwanted measurement artifacts.

3.5 Data Set Organization

The spectral data is known as the [X] block of variables. These include the absorbance, intensity, reflectance, or transformed reflectance (e.g. Kubelka Munk) values for each wavelength measured with the spectrophotometer. The [Y] block contains the dependent variable(s) assigned to each spectrum in the training set.

Data matrices may need to be oriented either column-wise or row-wise depending on the type of quantitative algorithm being used and the software requirements [34]. When exporting data collected with the instrument software, it may be necessary to transpose the spectra for post-analysis row-wise, as shown in Figure 3.2. Because the data is organized separately, we can also apply data analytics and statistical diagnostics to evaluate each matrix independently before creating a correlation between them. A thorough understanding of the data sets helps to create a more intuitive (commonsense) approach when modeling begins.

$$\begin{array}{ccccc} x_{11} & A_{12} & A_{13} & \cdots\cdots & A_{1s} \\ x_{21} & A_{22} & A_{23} & \cdots\cdots & A_{2s} \\ x_{31} & A_{32} & A_{33} & \cdots\cdots & A_{3s} \\ x_{41} & A_{42} & A_{43} & \cdots\cdots & A_{4s} \\ x_{n1} & A_{n2} & A_{n4} & \cdots\cdots & A_{ns} \end{array} \qquad \begin{array}{ccc} C_{11} & \cdots\cdots & C_{1p} \\ C_{21} & \cdots\cdots & C_{2p} \\ C_{31} & \cdots\cdots & C_{3p} \\ C_{41} & \cdots\cdots & C_{4p} \\ C_{n1} & \cdots\cdots & C_{np} \end{array}$$

Figure 3.2 Example setup of the [X] and [Y] block orientations. Depending on the algorithm and software requirements, the data set is organized row-wise or column-wise for post-analysis. *Source*: Deborah A. Peru.

3.6 Training Set Designs

The samples used to calibrate an instrument to predict concentration or property of interest in a future unknown sample are known as the "training set" (a.k.a calibration standards). Most spectroscopy procedures are considered secondary methods because the spectra are assigned a concentration based on a primary analysis method, such as high-performance liquid chromatography (HPLC). The main issue regarding the training set is where the samples will come from, how they are prepared and stored before analysis, and if they capture the full gamut of spectral variance expected in future samples. Sample size planning (SSP) is an important stage of method development. Building a model with acceptable performance requires knowledge of the minimal number of samples needed to reach a statistically meaningful result [8]. This becomes an even more important issue for process analytical applications because the data generated is time-consuming and costly [35]. To decrease the workload, pioneers in near-infrared proposed a method for selecting the optimal number of training set samples and wavelengths to develop a multilinear regression based on the error tolerance and the number of sample constituents measured. These researchers concluded that incorporating too many wavelengths in the model often degraded the prediction performance [36].

Pure materials can be mixed together to create calibration and validation standards for univariate and simple multicomponent applications. In process applications, environmental, forensic, or medical applications where many other factors can influence spectral signatures, standards should be collected from a similar process, and the concentration should be determined by a reference (or compendial) method after spectral data collection. Some developers use the rule of thumb that the concentration range of the analyte of interest should be at least five times the standard deviation (SD) or reproducibility of the reference method [37].

In complex applications, it is helpful to identify important factors that contribute to spectral variance, such as chemical composition, physical texture, bulk density, temperature, particle size, color, etc. If the number (n) of key variances is known, the minimum number of standards should be at least $3n$, and ideally, $10n$ or more should be used. Methods developed using near-infrared instrumentation may require a more extensive training set because of the broad overlapping absorbance bands in the near-infrared region, particularly in mixtures. For complex systems, 30–50 standards should be selected to establish feasibility [38]. As will be discussed further in this chapter, design of experiment (DOE) is a statistical approach often used in training set design that will ensure the main effects (and interaction effects) caused by key variables are captured within the training set with a minimum number of standards.

Some regression software may automatically divide the dataset into a training set and validation subset based on variability in the spectrum [X] and prediction variable [Y] response [39]. The training set is used for model training, while the validation test set evaluates and optimizes model performance [40, 41]. The user may have options for selecting the percentage or number of standards to keep in the training set, and the software will parse the data based on statistical requirements, e.g. Euclidean distance or Mahalanobis distance (MD) determined in PC scores space.

3.6.1 Univariate Regression Methods

Univariate procedures can be used whenever the analyte of interest has a unique spectral response, and the spectral response is free from spectral interferences by other components. Most univariate methods use far fewer calibration standards than multivariate calibrations. However. frequent re-calibration may also be necessary since many sources of chemical and physical variances in the samples cannot be corrected in these simple regression equations. The calibration standards should be prepared to cover the full range of concentrations expected to be measured during validation and in future unknown samples. A pure material in the form of a liquid, powder, or solid can be used to prepare standards gravimetrically using the same sample matrix or diluent used when measuring unknown materials. When diluting samples with volatile organic liquids care must be taken to avoid evaporation losses and to account for density differences between organic liquids. The reference materials can be stored in a regulated environment, desiccant, or dark room to avoid chemical degradation over time. The concentration can be calculated using the volume or mass added, assuming no material loss, or by the primary method. PH, temperature, buffer capacity, and ionic strength can significantly change the spectral response. Thus, these variances are usually fixed or adjusted within a narrow range of tolerance. When dissolving solid standard materials, vortexing or sonication may be used to ensure dissolution is complete. A series of working standards may be prepared for instrument calibration by diluting with a solvent and maintaining the same pH or buffering range.

For solids applications, a set of working standards can be prepared similarly using a pure material substance measured by mass and mixed with an inert material. The powders can be uniformly mixed using a Wig-L-Bug® mixer, amalgamator, or other mixing device and then packed into the appropriate sample container.

Raman data can be modeled using univariate and multivariate regression methods. A univariate approach typically extracts quantitative information from key Raman bands using the peak area or band intensity [42]. Correlations to properties of interest (form or composition) are created using band ratios. Although univariate data analysis is straightforward, it requires that the components of interest have distinguishing and unique Raman bands. When important bands overlap in biological tissue, pharmaceutical formulations, or bioprocesses, multivariate data analysis techniques often improve understanding at the molecular level and for predictive modeling [43]. Multivariate data analysis (also referred to as chemometrics) is widely used in biomedical Raman spectroscopy owing to the complexity of these systems.

3.6.2 Multivariate Regression Methods

Multivariate modeling refers to the analysis of one or more analytes using multiple variables measured with an instrument. Each component in a mixture has a unique spectral fingerprint and may participate in interactions resulting in new peaks or peak shapes. While material properties in univariate methods are often fully investigated, the behaviors and distributions of the individual components in multivariate systems are sometimes not well understood [10]. When working with multivariate systems, it is desirable to design a "training set" in a multivariate way that spans the full range of concentration for each major ingredient in the unknown sample. Variables affecting the spectral signature should be included in the "training set," and modeling parameters should be carefully selected to account for interferences and offsets due to measurement conditions. Selecting samples or standards for use should include all the types of variations expected in any future prediction. The inclusion of systematic and random errors in the training set is sometimes referred to as the "bias-variance trade-off" [44]. If there are too many variables, the model is too complex and variance dominates. Whereas, if the training set only captures a few variances, the regression is not complex enough, and bias dominates the uncertainty in the regression model [44]. In complex formulas, there are a number of interactions that can occur between ingredients that involve hydrogen bonding and covalent bonding. These molecular interactions are influenced by factors such as pH, and ionic conditions, all of which, can change the spectral response. Thus, standards should mimic the chemical and physical characteristics of future unknown samples and contain enough samples to statistically define relationships between component concentrations and spectral variables [44]. If done properly, multivariate methods can last for years before a complete recalibration becomes necessary, but model maintenance may be needed as described in Chapter 7.

Several approaches are used for selecting a representative training set. The right approach for each application can differ based on available resources and access to individual ingredients. One important criterion is to span a wider range of concentrations in anticipation of validation requirements and possible future out-of-specification samples. Multivariate methods designed poorly are not good at extrapolating beyond the design space. Standards should be measured over multiple days and shifts to build robustness to subtle changes in processing and measurement conditions.

The one-component-at-a-time approach varies one ingredient at a time while the others are kept fixed [44]. This conventional approach usually requires a large number of samples and is prone to overfitting thus reducing prediction accuracy [34, 44]. A major drawback is that while standards may span the full range of each component concentration, the approach does not capture the full range of possible combinations of component concentrations and measurement variances [34].

Another approach is the random selection approach using samples taken from the process or source of future unknown samples. The problem is that while the technique is sound for rapidly analyzing a broad range of materials, it suffers from the potential for chance correlation between components [34]. This design is prone to having more samples near the central region of the design space where processes tend to operate, yielding better accuracy in predicting in-specification samples within this narrow region [34]. However, the determination of the accuracy parameter at ±30% of the target concentration and analysis of out-of-specification unknowns often yield inaccurate results because of a lack of representation at the extreme range.

Lastly, DoE is a systematic and efficient approach for studying the relationship between multiple independent variables and key response (dependent) variables. This structured approach is commonly used for product and process optimization in a laboratory or pilot plant. During each experimental run, the spectrum for each sample can also be collected for calibration modeling building and prediction purposes. These spectra become the "training set" and contain valuable information for discovering relationships between chemical and process variables. Striving for mutual independence of variables is an essential goal in building quantitative spectroscopy methods, and DoE helps remove collinearity among spectral component variables.

3.7 Evaluation of the [Y] Block

The integrity of the [Y] variable matrix should be evaluated for quality using statistical analysis tools before regression development begins [45]. Build a summary of the data that includes measures of central tendency and dispersion, such as the mean and standard deviation. A distribution frequency plot describes the number of standards at each concentration level. Boxplots, histograms, and density curves also help graphically explain the distribution of the values.

Visualization of the spread associated with the dependent variables is an essential first step in creating a calibration curve. It is particularly important when random samples are provided for method development. One common problem in quantitative spectroscopy is that the training set is often obtained directly from an in-specification process and thus has a narrow distribution in concentration values. As illustrated in Figure 3.3, when the spread of [Y] values assigned to multivariate samples is too narrow, the R^2 for the regression equation will not be as good as a calibration curve developed by HPLC when calibration standards are prepared by serial dilution and measured. The coefficient of determination (R^2) is generally regarded as a range-sensitive statistic. Thus, the training set should have a concentration range of at least 10 times the standard deviation in the primary method measurement. Saunders reminds us that R^2 statistics are often misleading. The precision depends on sample size and should be reported with the associated confidence intervals for each set of measurements [46].

Calculate the range and plot the distribution of the concentrations assigned to each standard in the training set. A distribution plot can be created using Excel and sometimes the instrument calibration software will automatically generate a frequency distribution plot during sample selection. This is a good way to

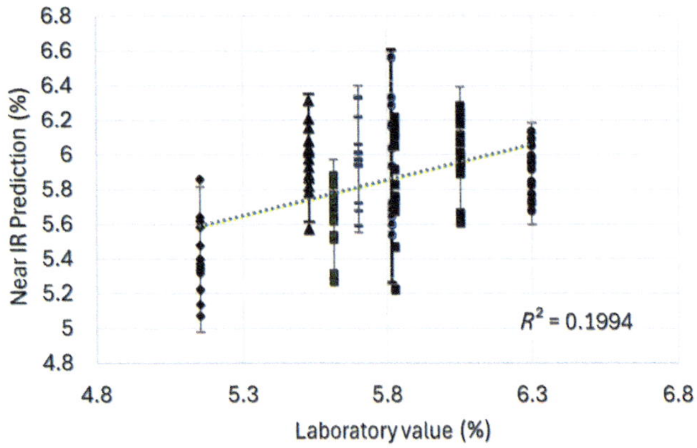

Figure 3.3 PLS moisture prediction over a 1.2% concentration range. $R^2 = 0.20$, SEC $\pm$ 0.32. *Source*: Deborah A. Peru.

visualize the spread and number of samples at each concentration level. Verify the number of standards at each concentration is consistent with your goals. Too many standards at one concentration can skew accuracy toward that level. Suppose you are developing a calibration curve to measure an FDA-regulated active material. In that case, having a set of standards at and near the target concentration range is beneficial. Additional standards should be distributed around this critical concentration and extend far enough to allow accurate validation predictions (typically 30% of the target). The bar plot shown in Figure 3.4 provides an example of a poor distribution around the target value of 4.4%. Most of the standards are at the lower level and not evenly distributed around this critical concentration.

3.7.1 Multivariate Considerations

Spectral data are inherently collinear, meaning that in a regression model, some variables are linear combinations of other variables. In regression analysis, we want to isolate the influence of each independent variable on the dependent variable. If there is a lack of independence between the independent (spectral) variables chosen and the dependent variables, it can cause instability in calculating the regression coefficients and inaccurate predictions. In Chapter 4, the author discusses how the effect of collinear variables can be overcome to some degree using principal component-based regression modeling algorithms.

Yet, the linear relationship between the analyte of interest and other ingredients and the physical attributes of the sample must be considered during model building. Establishing a correlation between the spectral response and a concentration does not necessarily mean the procedure is specific or statistically sound. Correlations may exist without causation when the variable of interest is confounded with another variable and when the spectra respond to a variable interaction [10]. When collinearity exists between the analyte of interest and other excipient components, a more robust training set design structure is required.

Suppose the concentrations of the major components within the calibration training set are all known. In that case, the Pearson correlation coefficients can be calculated for each pair of ingredients within the "training set" as shown in Figure 3.5. Components with correlations greater than 0.7, are considered collinear and may weaken the specificity of a multivariate method based on the degree of similarity in spectral response. Restructuring the training set by including samples that disrupt collinear relationships between variables is helpful. This can often be accomplished in combination with surrogate standards.

The condition number (η) is a useful statistical parameter that describes the degree of collinearity within any matrix of variables [47]. It is calculated using certain statistical software packages and pertains to the sensitivity of a matrix to inversion. Data sets considered ill-conditioned will have a large condition

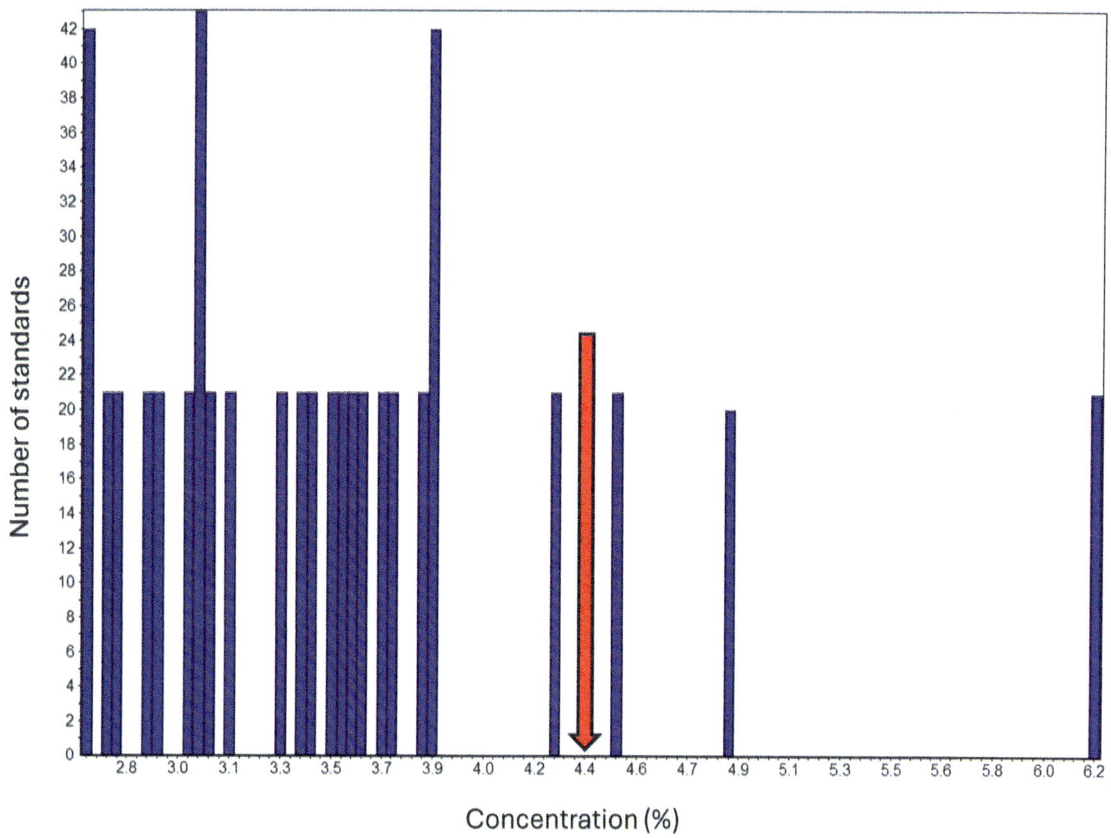

Figure 3.4 Bar plot illustration of a poorly distributed training set for measurement of an analyte target concentration of 4.4%. *Source*: Deborah A. Peru.

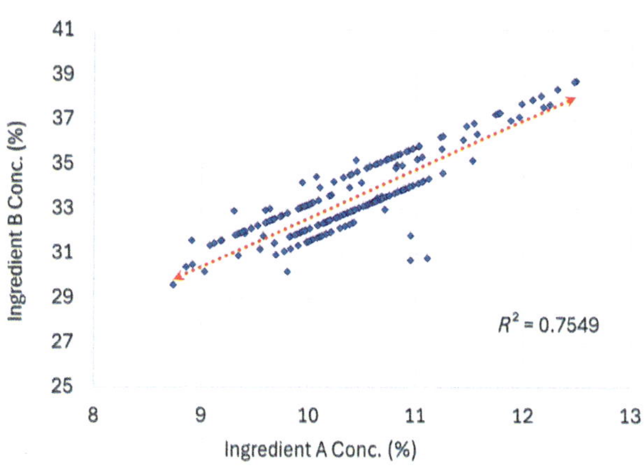

Figure 3.5 Example of a cross-correlation chart for main ingredients in a mixture. *Source*: Deborah A. Peru.

number (typically > 30). In practicality, a matrix with a high condition number is almost singular, and the computation of its inverse, or solution of a linear system of equations, is prone to large numerical errors. A matrix that is not invertible has a condition number equal to infinity. The condition number, η (Eq. (3.1)) is defined as the ratio of the largest and smallest eigenvalues calculated after the data is decomposed using linear math, typically singular value decomposition.

$$\eta = \mu_{max} / \mu_{min} \tag{3.1}$$

Most spectral software packages do not include this type of calculation. However, Matlab and most traditional statistical software packages will calculate η on either the [X] block or [Y] block of data. The DoE produces training sets with an ideal condition number of close to 1.

In the next chapter, you will learn about direct and inverse modeling methods. Direct modeling is used when the concentration for all components is known and included in the [Y] matrix. Because inversion of the [Y] matrix is needed to solve for the regression vector, k, collinearity among these variables cannot exist:

$$[X] = k * [Y] \tag{3.2}$$

where; $k = (Y^T Y)^{-1} Y^T$.

When inverse regression methods are used, the [X] matrix or wavelengths chosen cannot be linearly correlated. When a set of [X] variables has a high degree of covariance with the same y variable, the inversion of the [X] matrix is prone to error or cannot be determined.

$$[Y] = b * [X] \tag{3.3}$$

where; $b = (X^T X)^{-1} X^T [y]$.

Near-infrared spectra consist of broad overtones and combination bands. The vibrational signals from each component are mostly overlapped, making extracting analyte-specific information in a multi-component system a more complex task [48]. Preprocessing techniques and multivariate modeling methods can mitigate some of the effects of interfering components. A well-designed calibration set is an equally important factor in deriving models with the specificity and accuracy required to determine future unknown samples [48].

The primary purpose of PLS regression is to create a set of successive orthogonal factors (a.k.a principal components) from the [X] matrix to maximize the covariance with the reference attributes [49]. Principal component-based methods, such as PLS, attempt to mitigate the collinearity among the [X] variables. Yet, when the [Y] dependent variables are highly collinear, the specificity of a regression method can still become compromised and cause inaccurate predictions because of a poorly defined regression coefficient.

Understanding these basic principles can help design a "training set" that fully expands the design space for your system in a multivariate way and enhances specificity when needed. In summary, collinearity within the data set is one of many important sources of errors in regression development. Training set designs that minimize collinearity among ingredients, excipients, fillers, and other variables will provide more accurate and stable results.

3.8 Design of Experiment

Design of experiments (DoE) is a statistics-based experimental tool used to identify the critical factors and optimal conditions for creating new products or processes by performing a minimum number of experiments. DoE is harmonious with the Quality by Design (QbD) concepts used for product and process development and recommended by the FDA [50]. Building experimental certainty and reproducibility into a process provides confidence that formulations are robust and quality is built into the final product.

Using the QbD approach for development and testing reduces the time needed to establish compliance criteria and gets products to market faster [50, 51].

DoE is also considered a multivariate tool that provides guidance when creating a training set for multivariate quantitation. This design technique can be integrated when calibration standards are prepared in a laboratory or pilot plant. The advantages gained by using DoE for training set development include:

- Maximizes performance of the model using a minimum number of standards and resources to provide accurate and efficient information
- Interactions between components can be determined and modeled
- The significance of each variable can be characterized
- The relationships between conditions and response variables can be identified
- An acceptable design space and range of component concentrations can be established
- Outlier or anomalous data can be more easily recognized
- A more robust calibration and validation process is achieved
- Training set design with an even distribution across the range of expected variance is established.

A key goal of regression analysis is to create a relationship between the independent and dependent variables. When dependent variables are highly correlated (collinear), it becomes difficult for the model to estimate a true relationship between the [X] and [Y] variables. Thus, a designed set of calibration standards often provides the best opportunity for removing collinearity among ingredients to generate improved regression coefficients even when chemometric methods are used. In many instances, DoE can result in a more efficient calibration process that uses a much smaller number of standards and improves prediction performance [48]. In other situations, applying concepts of DoE when designing sampling strategies in harsh and restrictive environments is a more efficient and less costly approach for developing a multivariate calibration using process analytical technology (PAT) instruments without sacrificing predictive capability [44]. In near-infrared applications, particle property variations in powders, pastes, and slurries are often observed in the spectrum [19, 51]. Thus, applying concepts of DoE can help to mitigate their effect. Similarly, Sparen et al. showed how differences appear in the transmission Raman spectra of tablets due to particle-size variations not accounted for during calibration design [52].

In the pharmaceutical industry, the effect of a DoE modeling building approach on the prediction performance of a quantitative near-infrared procedure was investigated using a five-component tablet formulation [48]. These researchers concluded that using a statistically designed calibration set during multivariate calibration is critical, and optimum performance was achieved more efficiently using a well-designed training set [48]

The application of a DoE strategy is perfectly adapted when product or process experiments are conducted in a laboratory and when scale-up experiments are scheduled in the pilot plant. The execution of a DoE process is a very efficient way to establish the extreme range of concentrations for each ingredient within the formulation design space, which can then be supplemented with true in-process samples. At this stage of development, looking forward to validation requirements to ensure the range of concentrations is covered to support validation parameter testing is helpful. In some industries, this may be ±50% of the target concentration.

The use of DoE helps to create structured variation within the training set. The designs often used include mixture designs, two-level or multi-level full and fractional factorial, and central composite designs. When resources are limited, and the available number of samples is fewer than required by traditional designs, D and L optimal designs are used [44]. The number of standards necessary for a two-level full factorial design is 2 k, where k is the number of factors. As the number of factors in a two-level factorial design increases, the number of standards necessary to build a full factorial design increases quickly. For example, a two-level full factorial design with six factors requires 64 standards or formulation experiments. Identify the high and low

Table 3.1 Example of a full-factorial design with three (3) main ingredients.

Ingredient A	Ingredient B	Ingredient C
1	1	1
−1	1	1
1	−1	1
−1	−1	1
1	1	−1
−1	1	−1
1	−1	−1
−1	−1	−1
0	0	0
0	0	0
0	0	0

−1 = low concentration, 1 = highest concentration, 0 = mid-concentration.
Source: Deborah A. Peru.

concentrations that define the product design space. Then, a formulation grid alternating the high and low concentrations, represented as +1 and −1, is created in a pattern format, as shown in Table 3.1. Including the midpoint (0) for each major ingredient further describes the center of the design, often the target analyte concentration, and replication is helpful to describe the intrinsic variability in the process (i.e. making, measurement, and modeling variances). For a fractional factorial design, only half the number of standards (experiments) are prepared, and interactions are not fully resolved from the main constituents. The primary analysis method most often determines the dependent values for the analyte of interest. Alternatively, a concentration may be assigned by the mass or volume used to create the mixed standards. Many developers will supplement a designed training set with samples obtained from the source or process when future unknown samples are analyzed. The unknown samples are analyzed by the spectrophotometer first, followed by the primary method, and the data is included in the regression equation.

A D-optimal design is used when constraints prevent all experiments from being executed due to either resource constraints, disallowed ingredient combinations, or problems at extreme levels. A mixture design is another design sometimes used in multivariate calibration projects. The design involves varying a major ingredient within the design space and using a filler to sum the total percentage of the formulation to 100%. The filler combines base ingredients, usually inactive materials or solvents that may comprise most of the formula.

Statistical software packages (e.g. Sartorius Umetrics® Suite and Aspen Tech Unscrambler) contain a full array of data analytic tools, including DoE and multivariate analysis (PCA, PLS, etc.). Once the experiments are completed, the spectra and other data collected can be analyzed using analysis of variance (ANOVA) and other statistical methods to generate contour plots and to graphically represent the relationship between the critical factors and the critical responses. The data set can be used to identify the optimal process conditions and for developing multivariate calibrations for future predictions of unknown samples measured with the instrument. Ideally, these experiments should be randomized to avoid systematic bias related to the sequence. Figure 3.6 graphically depicts a training set design created using two three-variable full-factorial designs. The calibration region is expanded to support validation requirements of ±30% and a narrower

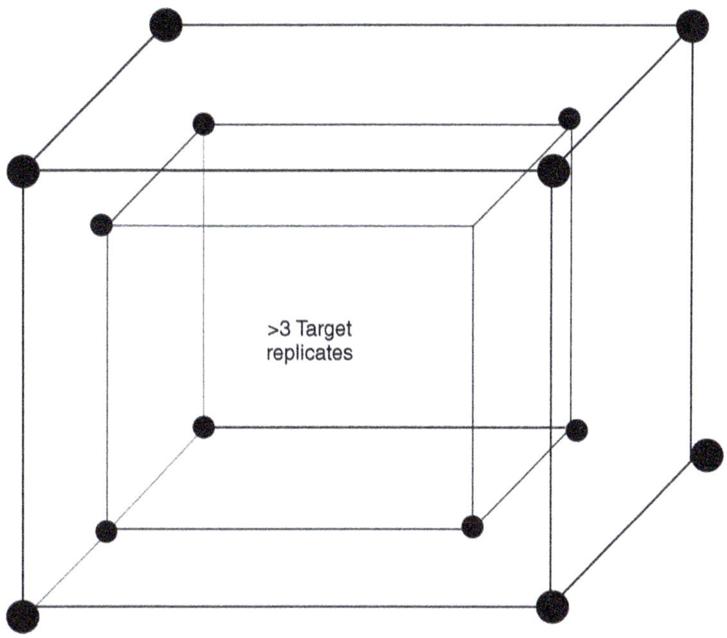

Figure 3.6 Application of DoE for "training set" development shows two full-factorial designs and in-specification (*target samples in replicate*) samples pulled from the process. *Source*: Deborah A. Peru.

design encompassing the typical manufacturing design space. Additional in-specification samples (*in replicate*) are represented in the center of the design to enhance the model's robustness for target samples.

3.9 Primary Method Characterization

Characterizing the errors associated with the primary method is essential to the method development process. Review the validation report associated with the primary method to determine if its degree of accuracy and precision meets the performance goals. Sometimes, the goal may be to develop a spectroscopy procedure that performs better than the primary method because it can analyze samples "as is" without introducing sample preparation or analyte recovery errors. This often can be accomplished provided the spectroscopy procedure does not propagate the same errors associated with the primary method when assigning analyte values.

One way to enhance the new procedure is to optimize the performance of the primary method before spectroscopy calibration begins. Take the necessary time to identify all of the sources of error. Work toward minimizing sampling errors and any systematic errors during development. Minimize the time delay between spectral data collection and the primary method analysis. Be sure both techniques measure the same representative portion of the substance. And use optimal storage conditions and minimize storage time to prevent changes in chemistry.

Most compendial procedures require a smaller weight or volume of sample than what is often measured by spectroscopy instruments. The sample size requirement for analysis is prescribed in the standard operating procedure (SOP) to yield a specific range of final concentrations. This often eliminates the problem of excessive titrant and eluent and avoids column saturation in chromatography. Intermediate precision can be reduced using control standards at prescribed measurement intervals [53]. Measuring more than one replicate may be economically unfeasible due to the reagent cost and expertise needed for some analytical procedures. Because the goal should be to create a spectroscopy method that is virtually independent of sample handling and measurement errors, consider what type of modification(s) may be necessary to avoid the propagation of errors due to the primary analysis method.

3.10 Evaluation of the [X] Block

3.10.1 Introduction to Chemometrics

Svante Wold first proposed the term chemometrics in the early 1970s as a class of statistical tools that utilize linear algebra to solve complex chemistry problems. Chemo, derived from the Latin word chemi, refers to chemical or chemical reaction. Metrics, derived from metrica in Latin, denotes the science of measuring. Chemometrics (a.k.a. principal component analysis) has become a branch of analytical chemistry used to solve complex problems in instrumental data analysis and calibration. In spectroscopy, chemometrics functions very much like mathematical chromatography. When applied properly, it provides information by dissecting a complex spectrum and presenting the extracted information in various ways to enhance the chemical understanding of the system. Chemometric algorithms are often essential in quantitative applications. Spectral data is inherently collinear, and thus, modeling using orthogonalized principal component vectors has important advantages compared with univariate modeling methods. Often referred to as the "multivariate advantage" it includes:

- Precision improves by $\sqrt{n}$, where n = number of wavelengths in model
- More preprocessing tools are available for removing baseline and other artifacts unrelated to chemistry
- More statistical tools are available for outlier detection based on spectral fingerprints
- The relationship between wavelengths provides new information for understanding
- Interferences can be modeled to minimize their effect
- Latent variables and intermediates are discoverable.

The PCA forms the basis for many chemometric methods used to analyze spectroscopic data for exploratory variable relationships and for qualitative and quantitative model building. In this chapter, we begin our discussion of chemometrics because it is a data analytics tool often used to evaluate the [X] data block for variable relationships when designing a new spectroscopy procedure. Chapters 4 and 5 will continue the discussion of chemometrics, and several excellent books on chemometrics written for spectroscopists are available [34, 54, 55].

The PCA transforms many variables in column and row space to form linear combinations of factors known as "principal components (PCs)." The algorithm reduces the dimensionality of the spectral data set because the relevant information is extracted and compressed into fewer variables than in the original data set. Each PC axis comprises a linear combination of orthogonal scores and loadings, often called factors.

$$[X] = S * L \tag{3.4}$$

Each row vector of the loading matrix [L] consists of a single loading value for each spectral wavelength. The magnitude of a particular loading indicates how much information that wavelength contributes to the overall variance within the data set. The scores [S] define how much a particular sample contributes to each principal component. Each PC is normalized and orthogonal to the next, thus eliminating the problems associated with modeling collinear x variables. The maximum number of possible principal components is equal to the number of wavelengths in the spectrum or one less than the number of samples in the [X] matrix.

The principal component vectors are ranked by decreasing order of variance. The first principal component explains the most significant variation in the data set. What is not explained in the first principal component is explained by the second, and so on. Since most spectral variances are expressed in the first few PCs, the data set can be visualized in two- or three-dimensional plots.

One of the most essential advantages of using principal components for data analytics is the ability to plot and visualize all the data in two- or three-dimensional plots. Exploratory analysis using PCA is often performed before regression modeling begins and is an excellent way to identify outliers and inconsistencies, such as clustering, within the data set. The PCA loadings help select regions to include in the quantitative model and provide information for understanding material chemistry. The PCA scores help detect outliers, systematic variances, and in the evaluation of preprocessing math on spectral reproducibility.

3.10.2 PCA Scores

The PCA scores can be plotted in multiple ways to provide unique information about the relationships between samples and within replicates. The scores also contain information about the intrinsic relationship between the spectral fingerprint and the dependent variable(s) of interest. Figure 3.7 shows the relationship between the first principal component sample scores and the moisture concentration assigned to samples measured by diffuse reflectance near-infrared. The samples were measured in five different laboratories using the same model instrument. A unique color was assigned to samples measured in each laboratory (categorical variable). The plot shows a general trend with concentration, but the data source highlighted in orange deviates notably from the trend line. A deep dive investigation revealed a critical deviation in the data collection method (DCM) used by this particular laboratory. The DCM used for spectral measurement was established without reference standardization. Reference standardization uses a National Institute of Standards and Technology (NIST)-traceable 99% reflectance standard to linearize the absorbance scale axis for more consistent near-infrared quantitative measurements. This procedure is done after lamp changes to normalize for the slight differences in lamp intensity. Spectral data measured without reference standardization have the potential for developing bias between instruments within the same instrument over time owing to minor changes in the absorbance scale.

When spectral data sets are complex and variance relates to multiple sources, the first principal component may not fully describe a change in analyte concentration. In this case, several principal component vectors

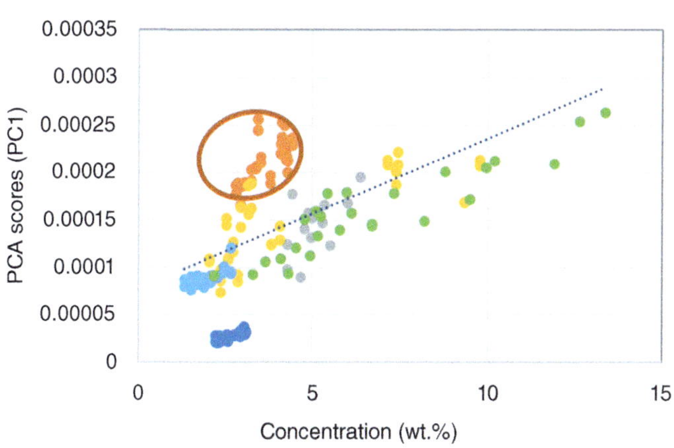

Figure 3.7 Plot of PC 1 scores vs. concentration for near-infrared data set from six different laboratories. *Source:* Deborah A. Peru.

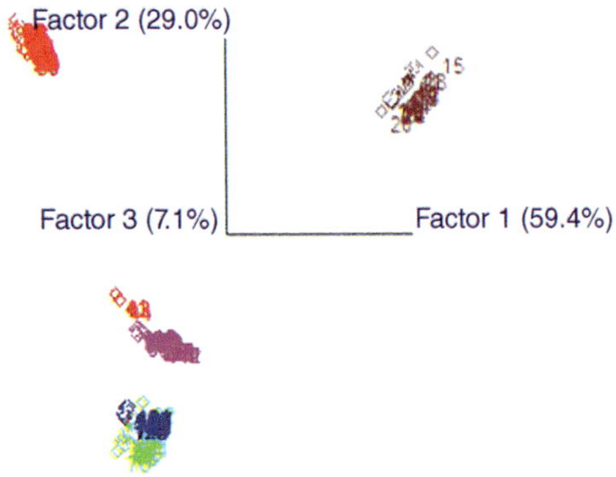

Figure 3.8 3D scores plot of PCA scores for factors 1, 2, and 3 describing 95% of the variance in the data set. Spectra from five (5) different laboratories are plotted by color. *Source*: Deborah A. Peru.

will be required to explain the analyte of interest fully. Similarly, larger regions of the spectrum are often required to fully develop indirect relationships for latent variables with other, measurable properties [56].

The scores can also be plotted in two or three dimensions, which is an excellent way to make class assignments and detect outliers. Figure 3.8 illustrates the clustering of spectral data using a three-dimensional PCA scores plot. There are many reasons why spectra will cluster or group together during modeling. Differences in the sample preparation or manufacturing environment, lot-to-lot variations, production batch variations, equipment differences, and day or time of sampling and analysis are among the standard explanations. If clustering amongst a subset of samples is found, this information provides valuable insight that furthers the understanding of the chemical system under investigation and helps build better models.

3.10.3 Principal Component Analysis Loadings

The loadings generated for each principal component provide valuable information for understanding the chemical system and detecting nuances and behaviors that are not easily found when overlaying spectra.

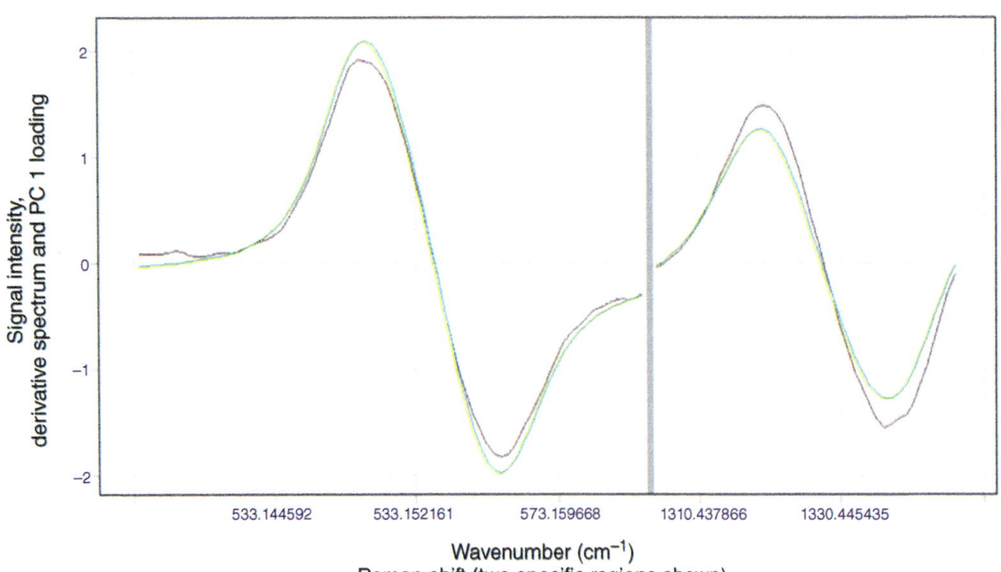

Figure 3.9 Comparison of the PLS PC 1 loading (green trace) and 1st derivative spectrum of caffeine (black trace) measured by Raman spectroscopy scaled using standard normal variate (SNV) for the two modeling regions. *Source*: Deborah A. Peru.

The loading vectors for each principal component can be plotted to visualize important spectral features and information regarding the analyte of interest. They may reveal new bands, such as the formation of intermediates during a reaction, peak shifting, or an interaction band.

The raw material spectral fingerprint for the analyte of interest often provides a useful interpretation of the loading vectors. As shown in Figure 3.9, the first PC loading vector obtained during PLS regression is nearly identical to the first derivative Raman spectrum of pure caffeine. In this example, a PLS prediction model for caffeine was developed using the first derivative of the Raman spectra collected for the training set. The two prominent regions for Raman scattering of caffeine are near the 556 cm^{-1} band resulting from the C=O–N deformation in the purine backbone and the C–N stretching band at 1332 cm^{-1}. The fact that the first PC loading is similar to the derivatized spectrum of pure caffeine indicates that the model is specific for this ingredient and that the largest variance within the training set is due to the variance in caffeine concentration.

The loadings are a mathematical expression for the chemical changes occurring within the spectral data at the molecular level. In Figure 3.10, a set of ultraviolet data is shown for a series of standards containing various concentrations of histidine. Histidine is an essential amino acid not made by the human body but is naturally found in many food sources. The loadings for PC 1 and PC 2 depict a significant band position variance as the concentration changes from high to low; hydrogen bonding and other ionic effects have a much stronger influence at lower concentration levels.

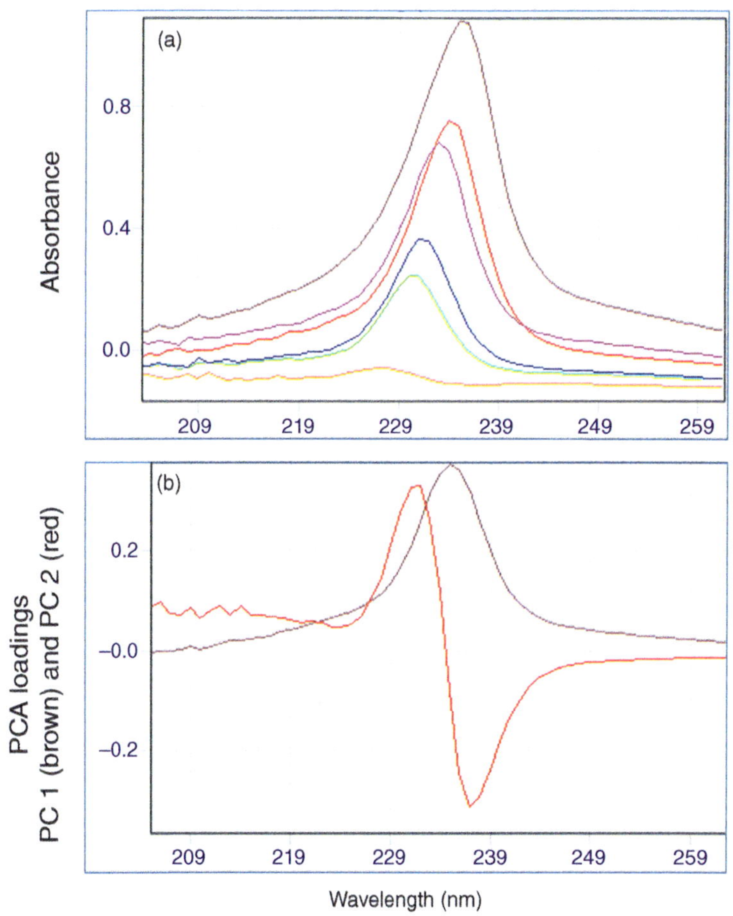

Figure 3.10 Comparison of PC1 and PC2 Loadings with the original spectra. *Source*: Deborah A. Peru.

For other chemical systems, the loadings may review the formation of a new chemical band due to intermediate formation or ingredient interaction bands. The loadings interpret the spectral behavior not easily observed in the original spectra, particularly in big data analysis, when overlays become a web of confusion. Because PCA methods will attempt to model systematic noise, it is valuable to eliminate as many errors as possible during data collection and before modeling begins.

3.11 Outlier Detection Tools

It is useful to become familiar with the various outlier detection tools available when using multivariate modeling techniques. When analyzing the data sets, you will encounter three basic types of outliers: spectral outliers, concentration outliers, and samples that exhibit both types of outliers. Outlier detection is important during model building and when using the model to predict unknown samples. During regression model building, samples with high leverage can pull the model in their direction, depending on their relative size and the size of the training set. Thus, statistical metrics are used to better understand how much samples deviate and why.

3.11.1 Spectral Outliers

A "spectral outlier" is a spectrum that significantly differs from the other spectra in the training set. Spectral outliers are due to measurement errors caused by a sudden vibration, bad reference scan, poor sample contact, deviation in temperature, or physical change in the sample, causing the spectral fingerprint to be outside the norm. Detection of spectral outliers can be done using several approaches. Visual inspection of the original spectra can help identify spectral artifacts when data sets are small (<25), but is not the most efficient or accurate way when the data set is large.

Inspection of spectral residuals is a better approach for identifying samples and variables that are poorly fit. Once the model is created, many software applications calculate a spectral residual, the difference between the original and predicted spectrum. Numerically, an individual spectral residual can be compared to the average training set residual to verify whether an individual spectral residual is of reasonable magnitude. Most software applications include statistical diagnostics that involve the use of spectral residuals. For example, the *Q*-residuals express the magnitude of the variation remaining in each sample that the model does not explain during a prediction using the final model [57]. A 95% confidence interval is determined based on the training set, and these limits are used to flag future samples as potential outliers. Likewise, an *F*-value determines the significance of the residual compared to the average residual in the training set. From this, a probability (*P*) of belonging is determined at the 95% confidence interval level [57].

The classical outlier detection approach uses the Mahalanobis squared distance to tag outliers above a predefined chi-squared (χ^2) distribution for p degrees of freedom [58]. Depending on the sensitivity required and spread in the training set, some developers may set outlier detection to above 10. MD is a multidimensional parameter assigned to each sample typically determined using the PCA scores for n factors selected in the model. The MD describes the sample's distance from the training set mean in PC scores space. A plot of the spectral residuals vs. the MD is shown in Figure 3.11 for a three-factor PCA model. The green lines show the 95% probability threshold limits (set by the chemometrics software [57]). Five percent of the samples would be expected to fall outside that cutoff [57]. Samples falling outside one or both of the thresholds are potential outliers. In this plot, the group in brown has an average MD more than four standard deviations away from the data set mean. Still, the sample residuals (y-axis) are all within the 95% boundary limit set for the training set. None of the spectra are considered outliers in this particular example of PCA used for exploratory purposes.

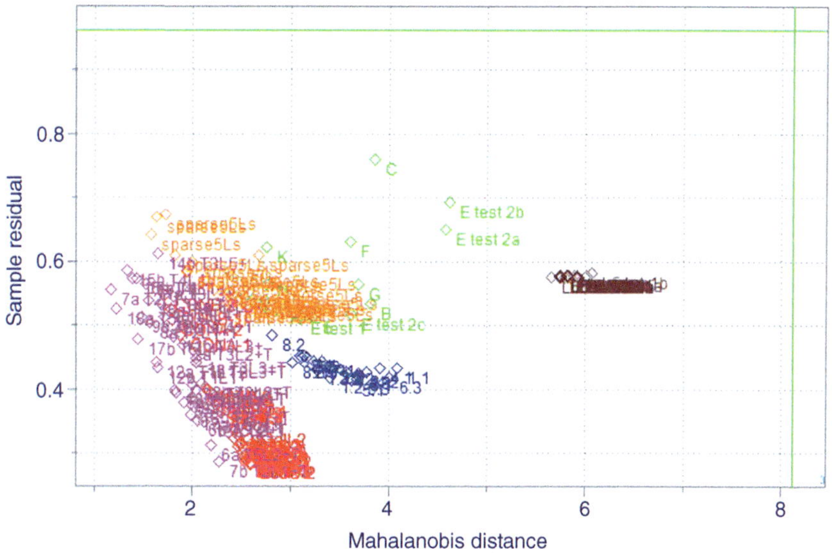

Figure 3.11 Plot of Mahalanobis distance (*x*-axis) vs. sample residuals (*y*-axis) used for outlier detection. The same data set is used in Figure 3.8. *Source*: Deborah A. Peru.

Some chemometric software packages will determine Hotelling's *t*-squared statistic in either PCA scores or wavelength space to determine the significance of each spectrum compared to the training set mean. It is a multivariate version of Student's *t*-test for evaluating the equivalence of spectral means. Hotelling's *t*-squared values use MD to measure each sample spectrum's distance from the model's center. Following, an MD of zero (0) means a perfect fit. Based on the training set, a 95% confidence interval is created to identify future outliers. However, a data set with a tight variance will have very small boundaries, and the potential to remove a sample that may explain a future unknown sample dramatically increases.

3.11.2 Concentration Outliers

Concentration outliers are errors found in the [Y] block variables. These types of errors are typically caused by human error, either because a wrong concentration value was assigned to a training sample or because of a measurement error during the primary method of analysis. Concentration outliers occasionally occur because the sample is unique and beyond the calibration limits. In these instances, the sample may be very important for improving the method's robustness since it may describe an essential feature in a future sample. The use of a residual plot is a powerful way to identify outliers. Residual plots help to detect remaining nonlinearity in the regression model because they visually enhance patterns in the residuals. Generally, one or two samples with an unusual residual are not indicative of a modeling problem but may be more indicative of a sampling or measurement problem. If patterns or trends in the residual plots are detected, a closer look at the data will be needed to resolve the problem.

During prediction, models that make erroneous prediction errors can increase the risk of releasing an out-of-specification finished product. Type II errors are samples with high prediction errors that are not detected as spectral outliers. One approach proposed for catching Type II errors in spectroscopy is based on a cross-correlation comparison between sample spectra measured over a specific region. The correlation values are converted using Fisher's *z*-transform, while a *z*-test of the transformed values is performed to screen out the outlier spectra [59].

3.12 Third-Party Software

In addition to the software provided with your instruments, several external software platforms can analyze your data sets. Many companies require the final procedure to be developed using vendor software, especially when the software is CFR 21 part 11 compliant. However, many of the concepts in data analytics discussed herein may not be available when some vendor software packages are used for development. Therefore, other platforms may help gain knowledge that can be used to build a final procedure within the vendor software platform.

3.13 Analytical Controls

The last important element of procedure design is implementing an analytical control strategy. An analytical control strategy includes instructions and procedures for statistical sampling, replication of measurements, and instrument and method verification testing. The procedure must specify how many lots or samples must be measured and the number of replicates per sample to improve precision. The location for obtaining samples, storage conditions, and how often measurements are to be made must also be considered. The analytical control strategy should include system suitability testing using reference standards to ensure the analytical procedure performs as expected during routine use throughout its lifecycle. Performance verification testing of the instrument and procedure should be conducted at regular intervals. Validation is further discussed in Chapter 7. On-going risk assessment and model maintenance should be in place to minimize misdiagnosis of a future sample being out-of-specification [2].

USP general chapter <1220> on validation procedures discusses the need for including analytical controls. The analytical control strategy plays a pivotal role in lifecycle management to ensure analytical testing procedures deliver accurate and reliable information that meets specification requirements [60]. Having a sampling strategy and replication strategy in place will help improve the consistency of results over the lifetime of the procedure, particularly when measuring heterogeneous materials.

3.13.1 Sampling Strategy

Designing and optimizing a sampling procedure may require consultation with a statistician. Review current practices and procedures in place for the compendial method of analysis. Chapter 2 reviews various approaches for representative sampling. In general, a larger sample size (volume or weight) will be more representative of the true population. Larger sample portions can be ground, mixed, sonicated, or vortexed to ensure a more representative aliquot is measured. Optimize the spot size or instrument slit width to ensure a more representative portion of the sample is measured. Try to normalize the sample size measured by the primary and secondary (spectroscopy) methods when differences in sample volume or weight exist, particularly during the calibration and validation stages.

3.13.2 Replication Strategy

The methodology used to calibrate and measure unknown samples for a meaningful scientific analysis should include a replication strategy to reduce measurement uncertainty. The degree of uncertainty is justified by precision requirements to ensure the risk of making inappropriate quality decisions is minimized [53]. Different control strategies may be required in different labs and when instrumentation is used for different purposes (e.g. release testing vs. process monitoring). During method development, strategies should be embraced to improve measurement uncertainty and meet validation requirements of robustness (interday precision) and repeatability (intraday precision).

Factors contributing to the measurement variability must be identified and well understood. A good replication strategy will reduce measurement errors that give rise to most of the variation [53]. Signal averaging, either by co-adding scans, decreasing resolution, or using derivatives with applied smoothing functions, can reduce measurement errors caused by random noise sources. The use of replication (multiple analyses of the same sample) should be included in both the primary method and the spectroscopy procedures. Many laboratories measure three aliquots taken from the same mixed sample portion and each aliquot three times.

When developing a regression model, spectral replicates are often included as individual spectra, each assigned the same concentration or property value determined by the primary method. This approach eliminates distortion of spectral averages by including a potential spectral outlier. On the other hand, when multiple replicate determinations are made by the primary method, the average is calculated and assigned to each spectral replicate. If there is uncertainty regarding the validity of one of the replicate measurements, statistical guidance using the Z score (equation below) can be used where X_1 is the individual replicated in questions, μ is the mean for all replicates, and σ is the standard deviation for all replicates.

$$Z = \left(\frac{X_1 - \mu}{\sigma}\right) \tag{3.5}$$

Any replicate with a Z score more significant than three (3) is generally considered an outlier. As discussed above, for spectral outliers, model-based statistics and diagnostic tools, e.g. spectral residual and MD in PCA scores space, are often used to find outliers in the spectral data.

3.13.3 Verification Strategy

Once validation is completed and the quantitative method is implemented, an effective analytical control strategy will include ongoing instrument and method performance tests to ensure optimal performance during its lifetime. An instrument performance verification test ensures the instrument performs well each time a sample is measured. A series of instrument performance tests that review instrument noise, SNR, peak positions, and peak height of an SRM is typically included within the vendor software. For process systems and Raman instruments, various voltages, gains, and laser powers may be monitored and tracked using control charting methods.

If you follow the ICH or USP guidelines, a NIST-traceable reference standard for wavelength certification is required. The standard is purchased with a certificate of known wavelength positions and performed regularly, such as once a month. The photometric (y-axis) response may be tested to ensure long-term stability using a series of photometric standards with different transmission or reflectance responses to monitor the linearity in the detector response.

With some instruments, these same SRMs are used to adjust the wavelength and photometric response to correct for instrumental drift that may occur over time. Reference standardization and intensity corrections are performed on near-infrared and Raman instruments, respectively, to ensure the y-axis is normalized when new light sources are installed. Chapter 7 further describes the purpose and importance of instrument qualification procedures.

3.14 Calibration Protocol and Report

In cGMP environments, and especially in the pharmaceutical industry, a calibration protocol or written document may be required, especially for any procedure intended to replace a compendial method [61]. The protocol must be approved before execution and contains the scope of work, timeline, and experimental details for calibrating the instrument. This written document includes the calibration set design, the concentration range, a description of how the calibration standards will be prepared or where in the process they will be obtained, the storage conditions, replication and numbering systems, and the primary method SOP used to assign a concentration or property of interest. Information regarding instrument software versions and other statistical software licenses must be documented. The protocol may also contain the routine analysis procedure for spectral data collection, as well as the instrument installation, operation, and performance qualification reports. In a cGMP laboratory (e.g. pharmaceutical industry), the performance qualification report will need to be approved to change the instrument's status from "not in use" to "in use" before calibration data is collected. Lastly, the calibration report should include analyst training records for those collecting calibration and validation data. Once approved, the calibration protocol can be executed, and a report will need to be written summarizing the data and results and then approved. The same type of documentation will be needed to validate a new procedure.

3.15 Conclusion

Evaluate the dependent and independent variables separately to verify the integrity of the data. Univariate statistical tools help to understand the distribution of values for one variable at a time. In contrast, multivariate analysis tools provide a deeper understanding of the relationship between many variables and samples. Visualize the relationships within and between spectra and their assigned quantitative values, and use a variety of plots to represent the data in different ways to provide interpretation, new insights, and understanding.

A training set designed to incorporate the range of variables and response space for a given application is paramount for complex chemistries involving multiple sources of interferences and spectral effects. Training sets incorporating orthogonal array designs that strive to establish mutual independence of key variables will most often outperform ill-conditioned matrixes.

An analytical control strategy should be incorporated into the procedure to minimize the propagation of errors. Outlier detection tools and residual plots identify anomalous behavior and model inaccuracies. In the following two chapters, we plan to discuss quantitative model building using chemometrics and other statistical techniques for optimizing calibration models during development.

References

1 ICH Q2 (R2) (2023). *Validation of Analytical Procedures*. Switzerland, International Council for Harmonisation of Technical Requirements for Pharmaceuticals for Human Use. ICH. https://www.ich.org/page/quality-guidelines (accessed 30 Nov. 2023).

2 USP <1220> (2022). *Analytical Procedure Lifecyle*. Maryland: United States Pharmacopeia. USP–NF. https://www.uspnf.com/ (accessed 11 Dec. 2023).

3 ICH Q14 (2023). *Analytical Procedure Development*. International Council for Harmonization of Technical Requirements for Pharmaceuticals for Human Use. ICH. https://www.ich.org/page/quality-guidelines (accessed 30 Nov. 2023).

4 Guiraldelli, A. (2022). Introduction to new general chapter <1220> analytical procedure life cycle. https://www.usp.org/sites/default/files/usp/document/user-forums/04-introduction-to-new-gc-1220-analytical-procedure-life-cycle.pdf (accessed 29 November 2023).

5 Chaursiya, A. and Dumpala, R. (2020). A review on revision of ICH Q2 (R1) and new ICH Q14 guidance. *Global Journal for Pharma and Allied Sciences* 1 (6): 1–6.

6 Barra, I., Khiari, L., Haefele, S.M. et al. (2021). Optimizing setup of scan number in FTIR spectroscopy using the moment distance index and PLS regression: Application to soil spectroscopy. *Scientific Reports* 11: 13358.

7 Chen, R., Heinz, R. (2024). FTIR microscopy: How experimental decisions affect the signal-to-noise ratio. Thermo Fisher Scientific Technical Note 54706. https://assets.thermofisher.com/TFS-Assets/CAD/Technical-Notes/rapttir-snr-technote-tn54706-en.pdf (accessed 17 April 2024).

8 Guo, S. and Popp, J. (2023). Key steps in the workflow to analyze Raman spectra. *Spectroscopy* 38 (10): 30–33.

9 B&W Tek Technical Note (2021). The importance of relative intensity correction of Raman data and how to utilize it for i-Raman series portable Raman instruments in BWSpec software. Metrohm (accessed 20 April 2024).

10 Mark, H. and Workman, J. (2003). *Statistics in Spectroscopy*. San Diego, CA: Elsevier Academic Press.

11 Seasholtz, M.B., Archibald, D.D., Lorber, A., and Kowalski, B.R. (1989). Quantitative analysis of liquid fuel mixtures with the use of Fourier transform near-IR Raman spectroscopy. *Applied Spectroscopy* 43 (6): 1067–1072.

12 The basics of UV-Vis spectrophotometry: A primer. https://www.agilent.com/cs/library/primers/public/primer-uv-vis-basics-5980-1397en-agilent.pdf (accessed 21 April 2024).

13 Berthelette, K.D., Xu, Y., Zhang, Z., et al. (2023). Effect of sample vials on analyte stability: Mitigating in-vial degradation of pharmaceuticals during LC analysis. Waters Corporation Application Note 135108411. The Effect of Sample Vials on Analyte Stability (waters.com) (accessed 21 April 2024).

14 Watari, M., Du, Y., and Ozaki, Y. (2007). Variations in predicted values from near-infrared spectra of samples in vials by using a calibration model developed from spectra of samples in vials: Causes of the variations and compensation methods. *Applied Spectroscopy* 61 (4): 397–405.

15 Adegoke, O. (2012). Chemical derivatization methodologies for UV-visible spectrophotometric determination of pharmaceuticals. *International Journal of Pharmaceutical Sciences Review and Research* 14 (2): 6–24.

16 Fahelelbom, K.M., Mustafa, K., Saleh, A. et al. (2022). Recent applications of quantitative analytical FTIR spectroscopy in pharmaceutical, biomedical, and clinical fields: A brief review. *Reviews in Analytical Chemistry* 41 (1): 21–33.

17 Thermo Fischer Scientific (2022). FT-IR – Introduction to attenuated total reflection (ATR) spectroscopy. https://knowledge1.thermofisher.com/Molecular_Spectroscopy/Fourier_Transform_(FT)_Infrared_(IR)_and_Near_Infrared_(NIR)/FT-IR_and_FT-NIR_Operator_Manuals/FT-IR_-_Introduction_to_Attenuated_Total_Reflection_(ATR)_Spectroscopy (accessed 19 April 2024).

18 Near-infrared region measurement and related considerations Part 1 near-infrared: Part 1 (Shimadzu Corporation) (accessed 21 April 2024).

19 Williams, P. and Norris, K. (2001). *Near-Infrared Technology: In the Agricultural and Food Industries*, 2e. Cereals & Grains Assn.

20 Huang, M., Sun, S., Saini, T.S. et al. (2023). Raman amplification at 2.2 μm in silicon core fibers with prospects for extended mid-infrared source generation. *Light: Science & Applications* 12: 209.

21 Process Insights Technical Bulletin (2024). How to Select the Best Fiber Optic Cable For High-Performance Spectroscopic Analysis. https://www.process-insights.com/products-3/nir-uv-vis-process-and-lab-analyzers/fiber-optic-cable-selection/ (accessed 21 April 2024).

22 Myrick, M.L. and Angel, S.M. (1990). Elimination of background in fiber-optic Raman measurements. *Applied Spectroscopy* 44 (4): 565–570.

23 Zhao, J., Bakeev, K.A., and Zhou, Z.J. (2019). Advancements in Raman technology for identifying real-world samples. *Proceedings of SPIE* 10983: 109830L.

24 Buckley, K. and Ryder, A.G. (2017). Applications of Raman spectroscopy in biopharmaceutical manufacturing: A short review. *Applied Spectroscopy* 71 (6): 1085–1116.

25 Tuschel, D. (2016). Selecting an excitation wavelength for Raman spectroscopy. *Spectroscopy* 31 (3): 14–23.

26 Inscore, F., Shende, C., Sengupta, A. et al. (2011). Detection of drugs of abuse in saliva by surface-enhanced Raman spectroscopy (SERS). *Applied Spectroscopy* 65 (9): 1004–1008.

27 Burns, D.A. and Ciurczak, E.W. (ed.) (2007). *Handbook of Near-Infrared Analysis*, 3e. CRC Press. https://doi.org/10.1201/9781420007374.

28 Sun, L., Hsiung, C., Pederson, C.G. et al. (2016). Pharmaceutical raw material identification using miniature near-infrared (MicroNIR) spectroscopy and supervised pattern recognition using support vector machine. *Applied Spectroscopy* 70 (5): 816–825.

29 Hirsch, J. (2007). Near-infrared analysis of critical parameters in lyophilized materials. Application Note: 50911 Thermo Fisher Scientific, Madison, WI, USA.

30 Duncan, D., Veale, J.R., Victor, K., and de Goeij, A.H. (2023). Laser-based headspace moisture analysis for rapid nondestructive moisture determination of lyophilized products. In: *Principles and Practices of Lyophilization in Product Development and Manufacturing*, AAPS Advances in the Pharmaceutical Sciences Series, vol. 59. Cham: Springer.

31 Jensen, P.S., Bak, J., and Andersson-Engels, S. (2003). Influence of temperature on water and aqueous glucose absorption spectra in the near- and mid-infrared regions at physiologically relevant temperatures. *Applied Spectroscopy* 57 (1): 28–36.

32 Beć, K.B., Grabska, J., and Huck, C.W. (2020). Near-infrared spectroscopy in bio-applications. *Molecules* 25 (12): 2948.

33 Bian, X. (2022). Methods for improving prediction ability of model. In: *Chemometric Methods in Analytical Spectroscopy Technology*. Singapore: Springer. https://doi.org/10.1007/978-981-19-1625-0_14.

34 Kramer, R. (1998). *Chemometric Techniques for Quantitative Analysis*. New York: Marcel Dekker Inc.

35 Ali, N., Girnus, S., Rösch, P. et al. (2018). Sample-size planning for multivariate data: A Raman-spectroscopy-based example. *Analytical Chemistry* 90 (21): 12485–12492.

36 Honigs, D.E., Hieftje, G.M., and Hirschfeld, T. (1984). Number of samples and wavelengths required for the training set in near-infrared reflectance spectroscopy. *Applied Spectroscopy* 38 (6): 844–847.

37 Metrohm USA Application Note (2018). Near-Infrared Spectroscopy: Quantitative analysis according to ASTM E1655. https://www.metrohm.com/en_us/applications/whitepaper/wp-029.html (accessed 6 Dec. 2023).

38 ASTM E1655-05 (2012). *Standard Practices for Infrared Multivariate Quantitative Analysis*, ASTM International. West Conshohocken, PA: ASTM International – Standards Worldwide.

39 Galvão, R.K.H., Araujo, M.C.U., José, G.E. et al. (2005). A method for calibration and validation subset partitioning. *Talanta* 67 (4): 736–740.

40 Guo, Z., Chen, Q., Chen, L. et al. (2011). Optimization of informative spectral variables for the quantification of EGCG in green tea using Fourier transform near-infrared (FT-NIR) spectroscopy and multivariate calibration. *Applied Spectroscopy* 65 (9): 1062–1067.

41 Mattu, M.J., Small, G.W., and Arnold, M.A. (1997). Application of multivariate calibration techniques to quantitative analysis of bandpass-filtered Fourier transform infrared interferogram data. *Applied Spectroscopy* 51 (9): 1369–1376.

42 Skoulika, S.G. and Georgiou, C.A. (2000). Univariate and multivariate calibration for the quantitative determination of methyl-parathion in pesticide formulations by FT-Raman spectroscopy. *Applied Spectroscopy* 54 (5): 747–752.

43 Esmonde-White, K.A., Cuellar, M., Uerpmann, C. et al. (2017). Raman spectroscopy as a process analytical technology for pharmaceutical manufacturing and bioprocessing. *Analytical and Bioanalytical Chemistry* 409 (3): 637–649.

44 Sadergaski, L.R., Toney, G.K., Delmau, L.H., and Myhre, K.G. (2021). Chemometrics and experimental design for the quantification of nitrate salts in nitric acid: Near-infrared spectroscopy absorption analysis. *Applied Spectroscopy* 75 (9): 1155–1167.

45 Oliveri, P., and Forina, M. (2012). Chapter 2: Data analysis and chemometrics. *Chemical Analysis of Food: Techniques and Applications*. Elsevier Inc. https://www.academia.edu/16829504/Data_Analysis_and_Chemometrics (accessed 19 Aug. 2023).

46 Saunders, L.J., Russell, R.A., and Crabb, D.P. (2012). The coefficient of determination: What determines a useful R-2 statistic? *Investigative Ophthalmology & Visual Science* 53: 6830–6832.

47 Battermanc, S. (1992). The use of singular value variance-decomposition proportions in FT-IR analysis of gases and vapors. *Applied Spectroscopy* 46 (5): 800–806.

48 Bondi, R.W. Jr., Igne, B., Drennen, J.K., and Anderson, C.A. (2021). Effect of experimental design on the prediction performance of calibration models based on near-infrared spectroscopy for pharmaceutical applications. *Applied Spectroscopy* 66 (12): 1442–1453.

49 Xu, Z., Chen, X., Meng, L. et al. (2019). Sample consensus model and unsupervised variable consensus model for improving the accuracy of a calibration model. *Applied Spectroscopy* 73 (7): 747–758.

50 FDA (2004). PAT—A framework for innovative pharmaceutical development, manufacturing, and quality assurance. Guidance for Industry PAT (fda.gov) (accessed 24 April 2024).

51 Kim, E.J., Kim, J.H., Kim, M.S. et al. (2021). Process analytical technology tools for monitoring pharmaceutical unit operations: A control strategy for continuous process verification. *Pharmaceutics* 13 (6): 919.

52 Sparén, A., Hartman, M., Fransson, M. et al. (2015). Matrix effects in quantitative assessment of pharmaceutical tablets using transmission Raman and near-infrared (NIR) spectroscopy. *Applied Spectroscopy* 69 (5): 580–589.

53 Borman, P., Schofield, T., and Lansky, D. (2021). Reducing uncertainty of an analytical method through efficient use of replication. *Pharmaceutical Technology* 45: 48–56.

54 Mark, H. and Workman, J. (2007). *Chemometrics in Spectroscopy*. Academic Press.

55 Beebe, K., Pell, R., and Seasholtz, M.B. (1998). *Chemometrics: A Practical Guide*. New York: John Wiley & Sons, Inc.

56 Brown, S.D. (1995). Chemical systems under indirect observation: Latent properties and chemometrics. *Applied Spectroscopy* 49 (12): 14A–31A.

57 Infometrix, Inc. (2014). *Pirouette Multivariate Data Analysis Software Version 4.5*. Infometrix, Inc.

58 Ghorbani, H. (2019). Mahalanobis distance and its application for detecting multivariate outliers. *Facta Universitatis* 34: 583.

59 Abookasis, D. and Workman, J.J. (2011). Application of spectra cross-correlation for Type II outliers screening during multivariate near-infrared spectroscopic analysis of whole blood. *Chemometrics and Intelligent Laboratory Systems* 107 (2): 303–311.

60 Kovacs, E., Ermer, J., McGregor, P.L. et al. (2016). Analytical control strategy. *Pharmacopeial Forum* 42 (5): 1–29.

61 FDA Document Number ORA-LAB.5.5. Laboratory manual volume II. Food and Drug Administration Office of Regulatory Affairs ORA. https://www.fda.gov/media/73942/download (accessed 3 March 2024).

4

Types of Quantitative Methods

Charles E. Miller

ChEM Analytics LLC, Lansdale, PA, USA

4.1 Introduction

At a high level, quantitative data analysis involves the generation of numerical outputs from one or more input values, using a model or equation.

Quantitative data analysis has always been a cornerstone of analytical chemistry. This critical symbiotic relationship stems from the simple fact that no analytical measurement result is 100% specific. Put another way, no analytical measurement system *directly* produces the property of interest from the measurement. Even the earliest measurement systems, including thermometers and balances, relied on a calibration that relates the raw measurement system output(s) to a property of interest, and this calibration must be generated through quantitative analysis.

In recent years, the importance and diversity of quantitative analysis methods being applied to analytical measurement systems have received increased attention for two main reasons: (1) the increased complexity of the measurement systems that need to be calibrated, and (2) the increased business demands on the timeliness, versatility, accuracy, and specificity of the calibrated systems. This chapter serves as an introduction to these methods and how they add value to analytical systems.

4.2 Uses of Quantitative Analysis

Based on the discussion above, a key use of quantitative analysis is to calibrate an analytical measurement system. This applies to both the simpler systems that require only a calibration to a primary standard (e.g., balance, pH electrode, thermocouple) and more complex multivariate measurement systems like near-infrared (NIR) and Raman spectrometers. In the latter case, the objective is to create a more convenient and relevant spectroscopic method to serve as a surrogate for a more labor- and resource-intensive reference analytical method. For this case, a calibration model generated through quantitative analysis is a critical component of the surrogate method, because this model is required to impart analytical specificity to the method. Consequently, the key quality attributes of such calibration models are focused on metrics like method accuracy, precision, specificity, and robustness.

In more recent years, the ever-increasing volume of data being generated and historized in today's operations is driving an additional use case: Exploratory Analysis. For this case, quantitative analysis is used as an efficient means to extract information from a very large dataset. The model generated by quantitative analysis is not meant to be put into online service but provides a means to reveal previously hidden correlations

Practical Quantitative Vibrational and Electronic Spectroscopy: A Guide for Developing, Optimizing, and Validating Procedures, First Edition. Edited by Deborah A. Peru.

Companion website: www.wiley.com/go/peruelectronicspectroscopy

and relationships within the data that could be used to improve understanding or optimize processes or products. In this case, the key quality attributes of the model are its transparency or ease of interpretation. As shown below, some of the more complex quantitative methods generate a plethora of model parameters, and efficient perusal and plotting of all these parameters to find value-adding information is a common workflow.

4.3 Nomenclature and Convention

To efficiently navigate the diverse landscape of quantitative methods, one must familiarize themselves with the common terminology and conventions used in the field. There are many references available for understanding these methods, with my favorites cited here [1–3].

4.3.1 Variables and Observations

In quantitative analysis, the data used for analysis are arranged by two attributes: Variables and Observations. Each variable denotes a different measured property, and each observation denotes a single instance of observing each property. For most data sets used in quantitative analysis, the convention is to express the data in a two-dimensional array, where the columns are arranged by variable, and the rows are arranged by observation. On occasion, problems arise that are best accommodated by expressing the data in three- or higher-dimensional arrays. One example of such problems is batch process data analysis, where the data are most conveniently contained in a three-dimensional array based on three attributes: variables, observations, and batches.

4.3.2 Dependent and Independent Variables

In quantitative analysis, not all variables are treated the same. The variables used as inputs to the quantitative method are called *independent* variables, or "X-variables," as the letter "X" is typically used to denote such variables in mathematical equations. The variables used as outputs from the method are called *dependent* variables, or "Y-variables," which are typically denoted using the letter "Y." It is important to note that, for any given application, the variables assigned to X (independent) and Y (dependent) depend on how the analyst chooses to formulate the problem, and this, in turn, depends on the objective of the analysis. For example, in the case of a simple sensor calibration, like that of a pH electrode, the independent variable (X) is typically assigned to the sensor response (e.g., measured potential, in mV), and the dependent variable (Y) is assigned to the property that the sensor is designed to measure (e.g., pH value).

4.3.3 Univariate and Multivariate

In the context of quantitative analysis, *univariate* refers to a case where a single independent X-variable is used as an input and a single dependent Y-variable is used as an output. In contrast, *multivariate* refers to a case where multiple independent or multiple dependent variables are used in the same problem. In univariate cases, the scope of quantitative methods is relatively limited, with *linear regression* (LR) being the default method. Not surprisingly, the scope of possible methods expands considerably for multivariate cases due to the larger number of variables being considered and the different ways to handle possible correlations within the sets of dependent or independent variables.

4.3.4 Regression

Regression is defined as a statistical method to calculate the relationship between independent and dependent variables. Also referred to as Least Squares Regression, this method calculates a relationship (model) between the X and Y variables such that the sum of squared differences between the observed data and the model-estimated data is minimized. For a given set of X and Y data that is sufficiently conditioned, the regression solution consists of a set of model parameters, one for each independent variable, and these are calculated via a closed-form mathematical equation, as discussed later.

4.3.5 Parametric

Within the context of quantitative analysis for analytical chemistry, this term refers to any mathematical transform of raw analytical instrument outputs (X-variables) to produce a new parameter that better reflects the property of interest. A common example of this is calculating a spectroscopic or chromatographic peak area from the raw spectral or chromatographic profile generated by the instrument. Provided that the peak being integrated is baseline resolved and sufficiently isolated from interfering peaks, this peak area parameter can represent a more sensitive and robust dependent variable for determining component concentration.

4.3.6 Overdetermined and Underdetermined

As briefly alluded to above, datasets must satisfy some prerequisites for regression analysis to even be possible. A good general rule is that the number of independent observations must equal or exceed the number of dependent variables to enable regression analysis. This refers to the *overdetermined* case, where sufficient independent observations are available to estimate the model parameters by regression. In contrast, an *underdetermined* case occurs when the number of independent observations is insufficient to enable the estimation of the model parameters via regression.

4.3.7 Scalars, Vectors, and Matrices

A scalar contains a single value expressed as a lower-case character (x) in mathematical equations. A vector contains a series of multiple values in a one-dimensional array and is expressed as a lower-case bold ($\mathbf{x}$) character. A matrix contains a series of multiple values organized in a two-dimensional array and is expressed as an upper-case bold ($\mathbf{X}$) character. For this chapter, the rows in a matrix refer to observations, and the columns refer to variables, following the convention discussed above. Finally, the presence of a caret ("^") symbol above a character denotes a model-estimated value, as opposed to a measured or observed value.

4.3.8 Model

A model is an object produced from quantitative analysis. Once generated, the model can either be interpreted to obtain an improved understanding of the system that generated the data, or it can be deployed to generate predicted outputs that can be used for monitoring, control, and other purposes. Drilling down a bit deeper, a model can be thought of as consisting of two key components: the ***model form*** and the ***model parameters***. For example, an LR model has the model *form* $\mathbf{y} = a\mathbf{x} + b$, where a and b are the model *parameters* that are to be determined using least squares regression.

4.4 Key Properties of Quantitative Methods

This section defines the key attributes of a quantitative method, which can be useful for understanding the differences between methods. These attributes can be defined as follows:

- **Empirical vs. Mechanistic:** An empirical method relies only on observations (data) collected from the system to determine the model, whereas a mechanistic method relies only on a theoretical understanding of the system. In most practical cases, quantitative methods contain both empirical and mechanistic elements and vary by the relative extent to which they rely on empirical fit vs. mechanistic knowledge. For example, a calibration scheme based on the LR method uses empirical least squares fitting to determine model parameters, but it could also use mechanistic knowledge of the system to identify *which* specific dependent variable (X-variable) is to be used in the regression.
- **Direct vs. Inverse:** A direct model uses a model form that follows an expected causal relationship, whereas an inverse model uses a form that is inverted from this relationship. A good example of a direct model form in analytical chemistry is the Beer's Law model: Absorbance = Coefficient × Concentration. This is the case because changes in the independent variable (analyte concentration) are deemed to cause changes in the dependent variable (absorbance). If the model form were to be flipped so that the dependent variable is concentration (e.g., Concentration = Coefficient × Absorbance), then the model form would be considered an inverse model, because spectral absorption changes do not actually cause concentration changes.
- **Univariate vs. Multivariate:** Univariate methods use only one independent variable, whereas multivariate methods use multiple independent variables.
- **Linear vs. Non-linear:** Linear methods presume a linear relationship between X- and Y-variables, utilizing, for example, the model equation $y = ax + b$ for a univariate case. In contrast, non-linear methods presume a non-linear relationship between X- and Y- variables, for example, $y = ax + bx^2 + cx^3 + d$, or $y = ax + e^{-x/b} + c$. Note that non-linear models often require additional terms to fit the relationships between X- and Y-variables.
- **Simple vs. Complex:** This refers to the level of complication of the model form, and consequently to the complexity of the workflow and algorithm required to determine the model parameters. On one extreme, the LR method uses a simple model form ($y = ax + $b) and the least squares method of parameter estimation, which are easily understood. On the other hand, the Artificial Neural Network (ANN) method has higher complexity, as the model form can be very complex (containing multiple layers of nodes and operations between the nodes, see Figure 4.9 below), and consequently parameter estimation involves a complex series of operations as well, most commonly utilizing the back-propagation algorithm [4].
- **Transparency:** This property refers to the ability to "look inside" or interpret a model once it is calculated. This isn't necessarily a function of the method itself but rather how it is implemented in a commercial or customized model-building solution. On one extreme are the "black box" solutions, where one has limited or no access to the model information: neither the algorithms used to generate the models nor the values of parameters that are generated. On the other extreme are solutions that provide all the information regarding how models are generated and the parameters for a given model generation case. In the latter case, many commercial solutions even provide pre-configured plot templates for the model parameters, which enable efficient model review and interpretation. Model transparency is a key requirement for the exploratory analysis use case, and is gaining appreciation for the calibration use case, especially for applications that have regulatory oversight.
- **Deterministic vs. Stochastic:** In the context of quantitative analysis, a deterministic method refers to a method that uses an algorithm with a fixed sequence of steps based on a closed-form mathematical expression, which thus generates the same model result when applied to the same set of model training

data. In contrast, a stochastic method includes an element of randomness to the calculation of the model, thus generating numerically different results when applied to the same population of model training data multiple times.

4.5 Quantitative Analysis Workflow

When comparing quantitative analysis methods, it is also helpful to understand under what context they are used. Figure 4.1 shows a typical high-level workflow for quantitative analysis. This workflow comprises five key objects and five key actions, as described below [5, 6].

4.5.1 The Five Key Workflow Objects

Training Data: These are the data that are used to identify or build the model. A key requirement of this data is that it adequately represents the states that are relevant to the modeling objective. This is a key requirement because the scope of relevance of the resulting model is defined by the population of objects in this dataset, regardless of how the subsequent workflow actions are executed. With this in mind, it is advised that the objects in this data set not only span the range of states required to support the modeling objective but are also reasonably well distributed across these states if possible.

Subject Matter Knowledge: This is the prior information that can be used to direct the model-building process. For example, one can use pure component analytical profiles (e.g., spectra) either explicitly to generate synthetic objects within the training data or implicitly to guide the optimal selection of X-variables to be used in a regression model. Another good example is using prior knowledge of the sampling and measurement systems to converge upon an optimal X-data preprocessing scheme for spectroscopic and other analytical instrument data.

Unvalidated Model: This is the result of executing the quantitative analysis method, as well as any required preprocessing methods, on the training data. For the surrogate method calibration use case, the model object is deemed to be optimized but has not yet been validated for its intended use. For the exploratory analysis use case, such unvalidated models are generally acceptable for use, as they are only used to understand the correlation structure within the training data and thus do not require formal validation.

Test Data: These are the data used to validate the model after it is already optimized and built. In stark contrast to the Training Data discussed earlier, these data are used solely to assess the quality of the model,

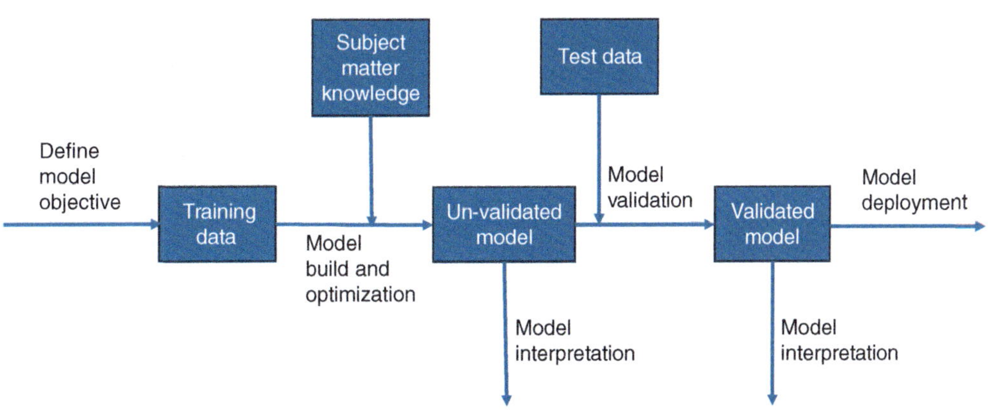

Figure 4.1 Typical workflow used for quantitative analysis. *Source*: Charles Miller.

and they define the range of states for which the model can be compliantly applied for the stated objective. A good rule of thumb is that the Training Data establishes the range of conditions where the model **could** be applied, whereas the Test Data defines the range of conditions where the model **shall** be applied.

Validated Model: This is simply an elevated form of the Unvalidated Model discussed earlier, which has passed formal validation. For the surrogate method calibration use case, this validated model is deemed suitable for deployment for its intended use. Accompanying this validated model are the key validation metrics that were used to evaluate the model, such as precision, accuracy, linearity, and reproducibility [7, 8].

4.5.2 The Five Key Workflow Actions

Define Modeling Objective: This is by far the most critical action in the quantitative analysis workflow, as it establishes the connection of the model to its intended use, and it influences all subsequent actions in the workflow. It can also be the most challenging action, as it is typically impacted by a diverse array of interests and stakeholders: including manufacturing operations, lab operations, product QA, process engineering, and even IT/automation. For regulated use cases, it is typical that a formal risk assessment (RA) is done with the input of all stakeholders to establish and define the set of validation metrics to be used, as well as to specify the method's performance requirements to support its intended use. To underscore the criticality of this item for life science use cases, several regulatory guidance documents [9] address the key parts of this item, including the selection of appropriate validation elements for different use cases.

Model Build and Optimization: Constructing the model from the training data is typically an open-ended non-linear exercise involving multiple iterations of many sub-tasks, including sample selection, variable selection, X-data preprocessing, and model generation. It is important to note that there is no firm "recipe" for this task: the model builder is free to explore different pathways using various tools (algorithms) at their disposal. However, regardless of the *pathway* taken, the final *objective* of this task is very clear: to produce a model that is optimized for the objective, and which would pass the subsequent model validation step. With this in mind, a key tool that is utilized during this step is called cross-validation [1, 5, 6], which is simply a "simulated validation" process that involves partitioning the objects in the training data to provide an early and approximate assessment of subsequent model validation outcomes.

Model Interpretation: Once a model is built, the parameters of the generated model can be viewed to reveal its inner workings, as well as key trends and structures within the training data. Some good examples of this task include viewing the values of regression coefficients to assess the importance of different X-variables for predicting Y, viewing scores, and loadings from a latent variable method such as partial least squares (PLS) to explore structures in the training data, and viewing residuals of the model to assess the quality of fit and unmodeled features in the X-data and Y-data. This is, of course, a key task for the Exploratory Analysis use case. Still, it can also be very relevant for the analytical method calibration use case, especially those that require regulatory oversight, where method transparency is important.

Model Validation: This involves challenging a model using a carefully designed test set of objects that were *not* used to train the model. A true validation exercise involves completely isolated model build and model validation procedures. Furthermore, for highly regulated use cases, a detailed validation protocol, which includes the design of the test set objects and detailed success criteria, must be documented and approved *before* validation is executed. If the outcome of the validation is successful, in that all previously stated success criteria are met, then the model is deemed to be validated and fit for use [7–9].

Model Deployment: As a final step for the analytical calibration use case, the validated model must be deployed onto the platform that is used to execute the model. Such deployment platforms vary in complexity, digital architecture, and error-checking features. However, in all cases, it is critical that the model is correctly configured on the platform so that it generates outputs as expected when it is in service.

For highly regulated use cases, it is also important that the platform supports both user access controls and transparent change management policies to properly address security, data integrity, and traceability requirements [10].

4.5.3 The Five Potential Pitfalls

While executing the quantitative analysis workflow, one must be careful to address several risks that could derail the process and lead to sub-optimal outcomes. The most commonly-encountered issues are described below:

Wrong Model Objective: This pitfall can be very problematic, and could result in considerable unproductive effort. It is also very people-oriented: The best way to avoid it is to practice inclusive principles, ensuring that all relevant stakeholders are involved in defining the objective. In some cases, persistence can also be valuable, as many stakeholders might not readily share their thoughts on important method requirements, such as accuracy, precision, and scope.

Insufficient Data: This applies to both training and validation data and is particularly important for empirical methods, which rely heavily on data to train and validate the model. It is also important to note that "sufficiency" in this context applies to multiple properties of the data, notably:

- The *number* of objects
- The *distribution* of these objects within the range of states that are required for training and validation, and
- The *quality* of the data.

Fortunately, several statistical and exploratory analysis tools, including principal components analysis (PCA) [1, 3, 6], can be utilized to efficiently assess potential data insufficiencies.

Overfitting: For empirical modeling methods, the risk of overfitting a model is significant, as there are typically no strict controls in the model build software that limit model complexity. A more complex overfit model might provide more visually appealing fit results when applied to training data, but will likely not provide a commensurate improvement in model performance when it is put into service. A good philosophical rule of thumb is that the model should fit the *underlying structure* in the training data but not *the data itself*! Carefully designed test set validation and cross-validation experiments can be executed to determine the optimal model complexity for a given use case, thus avoiding the overfitting pitfall [1, 5, 6].

Outliers: The presence of outliers in the training data can bias the identification of the model and can also lead to model overfitting. However, on the flip side, one needs to be careful about the overzealous removal of potential outliers from the training data, which can unwittingly eliminate useful objects that span the required observation space. Whether to include or exclude detected outliers in the training data depends on the model objective, and it should lean heavily on prior knowledge of the system. For multivariate data sets, the standard multivariate analysis (MVA) methods of PCA and PLS regression can be very useful for efficiently detecting and identifying outliers [5, 6, 11].

Model Deployment Errors: Model deployment platforms vary widely in their workflows and error-checking capabilities, but there is always a risk of the model not being accurately installed on the platform. This can lead to either failure of the platform to generate model outputs at all, or (worse yet) generation of incorrect model outputs on the platform. Some deployment platforms apply a useful "simulation" feature, which can be instrumental in detecting such errors. Specifically, this feature can be used to apply a known test set object (with expected outputs) to an installed model object on the platform. Verification of correct deployment can then be done by comparing the platform-generated outputs to the expected outputs.

4.6 Quantitative Methods

In this section, several commonly used quantitative methods will be described. For each of these methods, Table 4.1 provides values of the key method attributes that were discussed earlier. This table can be considered a general reference for comparing the properties of these methods, and thus their suitability for addressing specific modeling objectives.

To illustrate the performances of different quantitative calibration methods, a dataset containing 70 NIR transmission spectra of styrene–butadiene copolymers of varying composition will be used [12]. These samples vary in the relative amounts of four known chemical components: cis-butadiene, trans-butadiene, 1,2-butadiene, and styrene, which were all determined using a reference analytical method. The NIR spectra of these copolymer samples over the spectral range of 1570–1850 nm are shown in Figure 4.2.

4.6.1 Linear Regression

Linear regression is the simplest quantitative method to be discussed and is expressed by the following model equation [1, 2]:

$$y = b \cdot x + b_0 + e \tag{4.1}$$

where y is the dependent variable, x is the independent variable, b, and b_0 are model parameters, and e represents model error. In most analytical use cases, x represents a measurement value, and y represents a key sample property to be determined. When LR is used in this way, the parameters b and b_0 are often referred to as "span" and "zero" coefficients, respectively. Most typical applications of LR use it as an inverse method, where the y variable represents the property of interest (e.g., concentration) and the x variable represents the measurement response.

Given a suitable set of n calibration objects, resulting in a series of n paired (x, y) data points producing arrays $\mathbf{x}$ and $\mathbf{y}$, the optimal values of the parameters b and b_0 are determined using the least squares regression algorithm:

$$\hat{\mathbf{b}} = \left(\mathbf{X}_{\text{aug}}{}^{t}\mathbf{X}_{\text{aug}}\right)^{-1}\mathbf{X}_{\text{aug}}{}^{t}\mathbf{y} \tag{4.2}$$

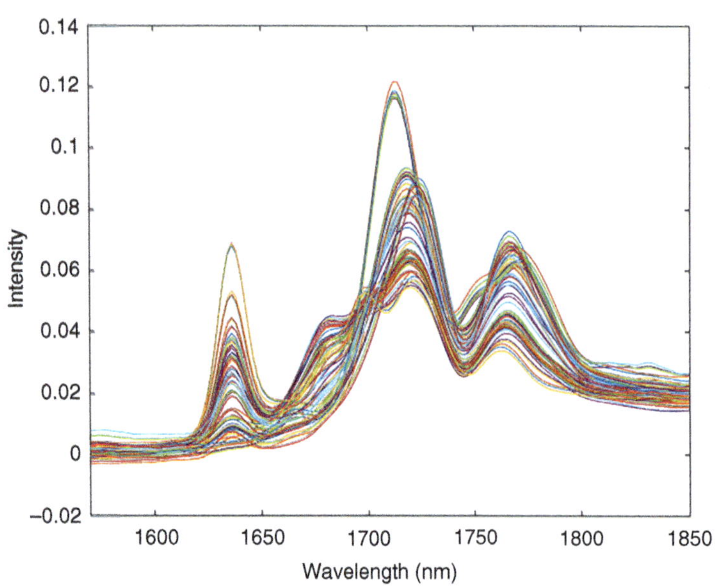

Figure 4.2 NIR spectra of 70 styrene–butadiene copolymers, used to demonstrate quantitative calibration methods. *Source*: Charles Miller.

Table 4.1 Summary of key attributes of quantitative methods discussed in this chapter.

Attribute	Linear regression (LR)	Parametric methods	Multiple linear regression (MLR)	Principal components regression (PCR)	Partial least squares (PLS) regression	Classical least squares (CLS)	Extended CLS	Multivariate curve resolution (MCR)	Advanced methods (ANN, SVM, LWR, CART)
Empirical or mechanistic focus	Both	Mechanistic	Both	Empirical	Empirical	Mechanistic	Both	Empirical	Empirical
Direct or inverse model, for typical use cases?	Inverse	Inverse	Inverse	Inverse	Inverse	Direct	Direct	Direct	Inverse
Univariate-multivariate	Univariate	Can be either	Multivariate	Multivariate	Multivariate	Multivariate	Multivariate	Multivariate	Multivariate
Linearity of model form	Linear	Can be linear or non-linear	Linear	Linear	Linear	Linear	Can be linear or non-linear	Linear	Can be linear or non-linear
Simplicity	Simple	Simple	Medium	Medium	Medium	Medium	Complex	Medium	Complex
Interpretability	High	High	High	Medium	Medium	High	High	Medium	Low
Deterministic or stochastic	Deterministic	Deterministic	Deterministic	Deterministic	Deterministic	Deterministic	Deterministic	Can be deterministic or stochastic	Can be deterministic or stochastic

where the $\mathbf{X}_{aug}$ matrix contains the x values augmented with a column of ones, and the $\hat{\mathbf{b}}$ vector contains the estimated coefficients b and b_0 from Eq. (4.1). The LR solution calculated in Eq. (4.2) represents a line in $<x, y>$ space (the two-dimensional space defined by the x and y variables) that minimizes the error (e), or the sums of squared deviations of the individual data points from the line. This is also referred to as the "Line of Best Fit". Once the LR model is fitted, the model error can be estimated as the difference between the measured y values and the y values estimated by the model ($\hat{\mathbf{y}}$):

$$\hat{\mathbf{e}} = (\mathbf{y} - \hat{\mathbf{y}}) = (\mathbf{y} - \mathbf{X}_{aug}\hat{\mathbf{b}}) \tag{4.3}$$

From this, various figures of merit of the LR model can be calculated, including the root mean squared error of estimate (RMSEE).

$$\text{RMSEE} = \sqrt{\frac{\sum_{i=1}^{N} (y_i - \hat{y}_i)^2}{N - 1}} \tag{4.4}$$

While LR is a commonly used method for quantitative analysis, many users are not aware of the underlying assumptions associated with the method, listed below:

- Model errors (**e**) are assumed to be normally distributed
- The underlying relationship between x and y is assumed to be linear, and
- All of the model error is assumed to be associated with the independent variable (y).

Although it can be argued that none of these assumptions are *perfectly* true for all *practical* use cases, this has not precluded the effective use of LR as a quantitative analysis method in analytical chemistry. With that said, cases of strong deviations from these assumptions have led to the development of alternative quantitative methods, some of which are discussed below.

For the styrene–butadiene NIR dataset, the only spectral band that appears to be "pure" with respect to composition is the characteristic 1,2-vinyl unsaturation band at approximately 1640 nm. If the spectral intensities at this wavelength (after a simple baseline offset correction using 1570 nm) are plotted against the known 1,2-butadiene concentration, the result is obtained in Figure 4.3. The LR solution line is also provided in Figure 4.3, along with the RMSEE fit error. Results indicate a somewhat good fit of the LR model, with a few instances of training objects deviating from the line of best fit.

4.6.2 Parametric Methods

Many of the earliest quantitative methods in analytical chemistry involved the conversion of raw analyzer signals into values that are more relevant for quantitation. These methods utilize a mechanistic understanding of sample chemistry and the measurement system to define specific analytical peaks or instrument responses to be used for analysis. Once the analytical peaks are identified, a specific metric based on the peak profiles, such as peak height, peak area, or peak ratio, establishes a relationship between chemistry and the measurement system. Peak fitting, another method commonly applied in spectroscopy and chromatography [13], determines the locations and intensities of overlapping peaks in a profile based on the assumed shape of each of the peaks. Once the parameterization scheme is defined and the metrics are calculated for a population of training set standards, the values of this metric are used as the X-variable in a regression against the known property values (Y-variable) of the same standards to define the calibration model.

Although the earliest usage of parametric methods was partly driven by the limited computing capabilities of the early generations of instrument workstations and software packages, these methods still have relevance today due to their simplicity and transparency. They are less susceptible to overfitting, less demanding

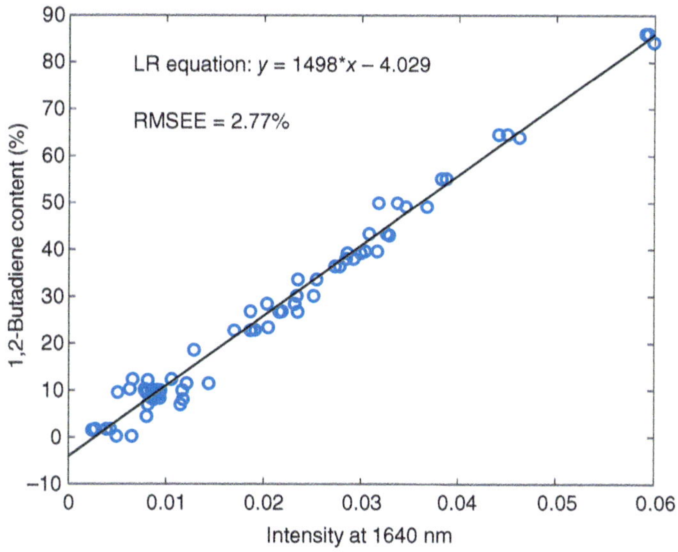

Figure 4.3 Scatter plot of the intensity of the 1,2-vinyl 1640 nm NIR band and the known 1,2-butadiene content of the styrene–butadiene copolymers, after simple baseline offset correction using the intensity at 1570 nm. The linear regression solution line is overlaid, and the fit error (RMSEE) is displayed. *Source*: Charles Miller.

with respect to the number of calibration standards required, and easier to deploy in the field. There are many recent examples of such methods in the literature, including peak areas in UV–vis–NIR spectra for color measurement [14], peak ratio of Fourier transform infrared (FTIR) bands for crystallinity index of bone [15], standard deviation of absorbance ratios from FTIR to assess epoxy resin dispersion [16], curve fitting of the Amide I FTIR band region for analysis of protein mixtures [17], and UV spectral peak ratios for DNA and RNA analysis [18].

In the styrene–butadiene NIR dataset case, one could attempt an integration of the 1,2-vinyl band to correlate to the 1,2-butadiene concentration. The results of LR of the integrated 1640 nm peak area across the range 1620–1656 nm (after a simple baseline offset correction using 1570 nm) vs. the 1,2-butadiene concentration is shown in Figure 4.4. Note that this result appears to show a poorer fit than the earlier LR method that uses only the single NIR intensity at 1640 nm, suggesting that the 1,2-vinyl band at 1640 nm (defined from 1620 to 1656 nm) might not be "pure," as initially postulated.

4.6.3 Multiple Linear Regression

Multiple Linear Regression (MLR) [1, 2, 6] is simply an extension of LR, where multiple independent X-variables are used in the regression. It can be used in cases where a single X-variable cannot provide sufficient analytical specificity for the model objective. In spectroscopy and chromatography, this can occur in cases where the peak for the analyte or property of interest is not "pure," in that it has significant overlap with peaks of other components in the material being analyzed. The model equation for MLR is given below:

$$\mathbf{y} = \mathbf{bX} + \mathbf{e} \tag{4.5}$$

where $\mathbf{y}$ contains the values of the dependent variable for the calibration standards, $\mathbf{X}$ contains the values of each of the m dependent variables for the calibration standards, $\mathbf{b}$ contains the MLR regression coefficients, and $\mathbf{e}$ contains the model errors. Given a suitable set of n calibration objects, resulting in a series of

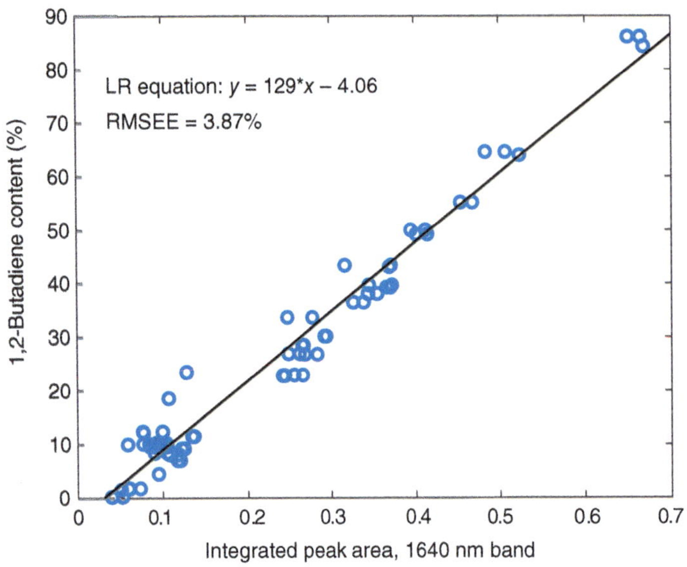

Figure 4.4 Results of linear regression of the integrated 1640 nm peak area from 1620 to 1656 nm (after baseline offset correction using 1570 nm) vs. the 1,2-butadiene concentration. The linear regression solution line is overlaid, and the fit error (RMSEE) is displayed. *Source*: Charles Miller.

n paired $(x_1, x_2, x_3, \ldots, x_m, y)$ data points producing arrays **X** and **y**, the optimal values of the parameters **b** are determined using least squares regression:

$$\widehat{\mathbf{b}_{\mathbf{MLR}}} = (\mathbf{X}^t\mathbf{X})^{-1}\mathbf{X}^t\mathbf{y} \tag{4.6}$$

Once the MLR model is fitted, the Y-value associated with a new array of X-data ($\mathbf{x_p}$) can be estimated using Eq. (4.7):

$$\widehat{y_{\mathrm{mlr}}} = \widehat{\mathbf{b}_{\mathbf{MLR}}} * \mathbf{x}_{\mathrm{p}} \tag{4.7}$$

Equation (4.6), and specifically the $(\mathbf{X}^t\mathbf{X})^{-1}$ term in the equation, can be used to highlight the pitfalls of MLR. First of all, $\mathbf{X}^t\mathbf{X}$ cannot be inverted in an underdetermined case: where there are more variables (m) than objects (n) in X. This is unfortunately the case in most spectroscopy applications, where the number of spectral wavelength variables is in the hundreds, or even the thousands. Furthermore, even in an over-determined case ($n > m$), $\mathbf{X}^t\mathbf{X}$ cannot be inverted in cases of singularity: when any two columns (or variables) in X are correlated to one another. Similarly, if any two columns in X are *almost* correlated (nominally correlated but not numerically correlated due to noise in the data), the inversion of $\mathbf{X}^t\mathbf{X}$ becomes unstable. Again, this is rather common in spectroscopy, where adjacent wavelengths in a spectrum often have a high degree of correlation. Finally, even in cases where the previous two conditions are not present, MLR is still susceptible to overfitting, through the usage of too many X-variables in the regression.

The above issues underscore the importance of X-variable selection in MLR, both with respect to the number of variables and the properties of these variables. To address this, some software packages utilize empirical-based variable searching algorithms, including step-wise search and all-possible-combination search, and utilize cross-validation to avoid overfitting. Of course, such selection could also be influenced by prior knowledge of the chemistry and spectroscopy of the application to more efficiently converge upon specific X-variables that are known to be relevant to the modeling objective. Once developed and validated, an MLR model is simple and thus easy to deploy and communicate to key stakeholders.

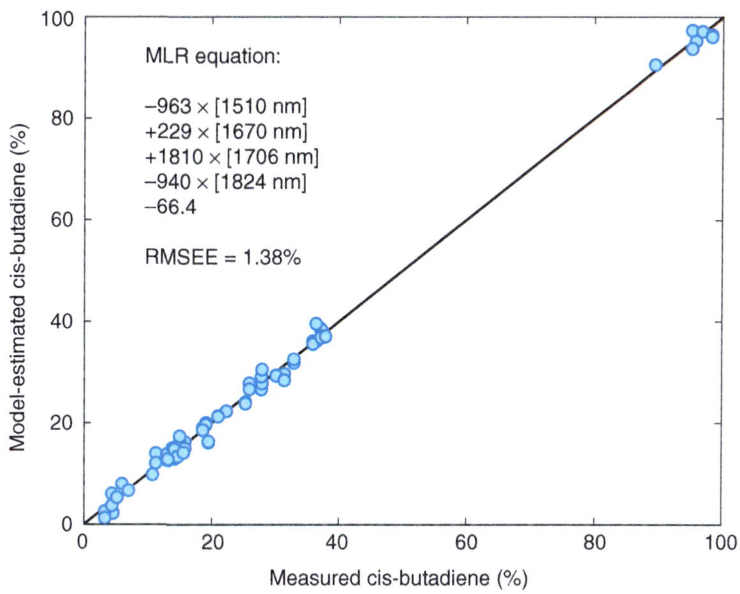

Figure 4.5 Results for MLR regression to cis-butadiene content in the copolymers, using the step-wise MLR variable method to select four NIR wavelengths: 1706, 1824, 1670, and 1570 nm. *Source*: Charles Miller.

To illustrate the MLR method on the styrene–butadiene dataset, the step-wise MLR variable selection method was used to build an MLR model for the cis-butadiene content in the copolymers. Four variables were selected based on the prior knowledge that four major chemical components vary independently in the calibration samples. The four X-variables chosen by this method are 1706, 1824, 1670, and 1570 nm, in that order. These four selected variables were used to build an MLR regression model for cis-butadiene content, the results of which are shown in Figure 4.5. Note the improved (reduced) RMSEP for the MLR case when compared to the LR and univariate parametric cases (Figures 4.3 and 4.4) indicating the improved specificity of the MLR method that utilizes multiple wavelengths in the model.

4.6.4 Latent Variable Methods

Although the pitfalls of MLR can be partially addressed through careful selection of X-variables, as discussed above, latent variable methods provide an alternative means to address these issues [1, 3, 6]. Instead of addressing the underdetermined state and X variable collinearity by selecting individual X variables to use in an MLR, these methods compress the original m X-variables into a much smaller number of k latent variables, each of which is simply a linear combination of the original m variables. The coefficients of these linear combinations are determined based on the covariance structure in the training data to be used in the regression.

Although it is not a regression method, PCA provides a useful background for the family of latent variable regression methods used in quantitative analysis. Given an $n \times m$ training set of X data, PCA proceeds to decompose these X data into a set of k principal components per the model equation below:

$$\mathbf{X} = \mathbf{T}\mathbf{P}^{t} + \mathbf{E} \tag{4.8}$$

where $\mathbf{T}$ is an $n \times k$ matrix of PCA scores, $\mathbf{P}$ is a $k \times m$ matrix of PCA loadings, and $\mathbf{E}$ is a $n \times m$ residual matrix. Each sequential latent variable (or Principal Component) 1,2, … , k is determined such that the maximum (remaining) variance in the X data is explained, thus minimizing the residual $\mathbf{E}$. Although one, in principle, could calculate as many principal components as the minimum of n and m, in most practical cases a much lower number of PCs ($k \ll \min(n,m)$) are needed to capture the majority of structured variance in **X**. Once calculated for a given set of training data **X**, the PCs can be considered a newer, simpler coordinate system for explaining the spectra in **X**, where the loadings **P** are essentially the "translator" between the original m variables and the new k-dimensional coordinate system. A key property of these new k variables expressed in **P** is that they are orthogonal to one another, thus providing a more efficient means for exploratory analysis of **X** as well as eliminating the singularity risks associated with MLR. The key tradeoff of this orthogonality relates to model interpretability: The loading profiles in **P** rarely correspond to pure component spectral profiles and more commonly relate to *differences* in pure component spectra or pure spectral effects [19, 20].

4.6.4.1 Principal Component Regression

With a simpler set of orthogonal k PCs generated by PCA, one can proceed with a regression of these PCs against Y without concern about underdetermined status or singularity. This regression constitutes Principal Component Regression (PCR) [1, 6], which for a single Y variable utilizes the model in Eq. (4.9) below:

$$\mathbf{y} = \mathbf{Tq} + \mathbf{f} \tag{4.9}$$

where **y** contains the values of the dependent variable for the training set standards, **T** contains the PCA scores from Eq. (4.8), **q** contains the Y loadings for the PCR model, and **f** contains the Y-residuals. Given the PCA scores **T** generated from the training set X-data and corresponding Y-data from the training set (**y**), the Y loadings for PCR ($\widehat{\mathbf{q}_{\mathbf{pcr}}}$) are estimated using the least squares algorithm:

$$\widehat{\mathbf{q}_{\mathbf{pcr}}} = (\mathbf{T}^t\mathbf{T})^{-1}\mathbf{T}^t\mathbf{y} \tag{4.10}$$

Compared to the matrix inversion for MLR (Eq. (4.6)) above, the inversion of $\mathbf{T}^t\mathbf{T}$ is very stable, due to the aforementioned orthogonality of the PCs. PCR regression coefficients using Eq. (4.11) are then calculated from the estimated Y-loadings and X-loadings:

$$\widehat{\mathbf{b}_{\mathbf{pcr}}} = \hat{\mathbf{P}}\widehat{\mathbf{q}_{\mathbf{pcr}}} \tag{4.11}$$

Once the PCR model is fitted, the Y-value associated with a new array of X-data ($\mathbf{x_p}$) can be calculated using Eq. (4.12):

$$\widehat{y_{\mathrm{pcr}}} = \widehat{\mathbf{b}_{\mathbf{pcr}}}\mathbf{x}_{\mathrm{p}} \tag{4.12}$$

The advantage of PCR over MLR is that one can utilize multiple spectral (X) variables in the regression model to impart improved specificity of the method without concerns about singularity. Overfitting is still a concern for PCR, and can occur when one uses too many PCs in the model. This can be mitigated through well-designed cross-validation experiments to help guide the user to an optimal number of PCs. However, one remaining disadvantage of PCR is that the covariance of X and Y is not utilized during the PCA decomposition of X, and in practice, this can lead to requiring more PCs than desired to get a sufficient fit to Y, especially in cases where the relevant variance in X is "buried" under other stronger variance components in X that are not relevant for predicting Y. A good example of such a case in spectroscopy would be one where the analyte of interest has a rather weak absorption band that is overlapped with several other stronger absorption bands from interfering components.

4.6.4.2 Partial Least Squares

To address the issue discussed above for PCR, one could explicitly utilize the Y data during the decomposition of the X data, in order to generate a different set of PCs (called *latent variables*) that not only explain the variance in X but also explain the covariance between X and Y. Like PCR, PLS utilizes the same model equations for both X and Y (Eqs. (4.8) and (4.9), respectively), but utilizes different algorithms than PCR to address this dual objective. There are several commonly used algorithms for PLS, which are a bit more complex than the algorithm for PCR, and the reader is directed to other references for details [1, 21–23]. The main outputs from a PLS regression include the X loadings **P**, the X scores **T**, and Y loadings **q**, as well as the residuals **E** and **f**, just as in PCR. However, PLS also generates a matrix of loading weights, **W**, which has the same dimensionality as the X loadings **P**, but differ in that they are used to calculate scores directly from the spectra ($\mathbf{T} = \mathbf{XW}$), instead of calculating the spectra from the scores ($\mathbf{X} = \mathbf{TP}^t$) as in Eq. (4.8). The PLS regression coefficients can then be calculated using the PLS analog to the PCR regression coefficient equation in Eq. (4.11), namely:

$$\widehat{\mathbf{b}_{\mathbf{pls}}} = \mathbf{W}(\mathbf{PW})^{-1}\mathbf{q}_{\mathbf{pls}} \tag{4.13}$$

Once the PLS model is fitted, the Y-value associated with a new array of X-data ($\mathbf{x}_\mathbf{p}$) can be calculated using Eq. (4.14):

$$\widehat{y_{\text{pls}}} = \widehat{\mathbf{b}_{\mathbf{pls}}}\mathbf{x}_{\text{p}} \tag{4.14}$$

As in PCA and PCR, the loadings **P** from PLS are orthogonal and thus have limited interpretability.

As both PCR and PLS are latent variable regression methods, they perform similarly for most use cases. However, the fact that PLS considers both X and Y variance when determining the latent variables means that this method can lead to a more efficient explanation of Y using fewer latent variables, and it is thus less susceptible to overfitting. This could be particularly noticeable in cases where the Y-relevant variance in **X** is small in magnitude relative to the interfering variance in **X**, for example, where the analyte peak of interest is very weak compared to interfering spectral peaks.

To provide an illustration of the difference between PCR and PLS, Figure 4.6 provides the explained X and Y variance for PCR and PLS regressions to cis-butadiene concentration of styrene–butadiene copolymers using the same data set. Note that in the PLS case, more of the Y variance is explained, and less of the X variance is explained, after only 2 PLS latent variables. In fact, the variance curves in Figure 4.6 suggest that one might only need to utilize 2 factors (latent variables) in the PLS model, whereas 3 or 4 factors (principal components) might be needed to obtain a similar Y prediction performance of the PCR model. This result underscores the increased efficiency of PLS for determining factors that are relevant for predicting Y.

Finally, it should be noted that PLS can be extended to the case where multiple *y* variables are contained in the **Y** matrix, and this is commonly called "PLS-2" regression. PLS-2 operates on the same principle as PLS, where the objective is to construct latent variables that explain the most variance in both the X and Y data. However, PLS-2 must also explain the covariance between the multiple Y variables. PLS-2 has not seen commonplace usage in quantitative analysis but is quite popular in exploratory analysis to efficiently interpret common and unique variance components for two discrete blocks of multivariate data [1].

4.6.5 Direct Calibration Methods

4.6.5.1 Classical Least Squares

As stated previously, a direct calibration method uses a model form that follows an expected causal relationship. Classical least squares (CLS) is a method that utilizes the classical Beer's Law model for spectroscopic data, which is based on the causal relationship of component concentrations in a mixture producing well-defined spectral profiles, using the model equation:

$$\mathbf{X} = \mathbf{CS}^t + \mathbf{E} \tag{4.15}$$

where **C** contains the concentrations of *p* chemical components, **S** contains the spectra of the *p* pure components, and **E** contains the model error. This model simply says that any single spectrum in **X** is the result of a linear combination of the *p* pure component spectra in **S** [1, 24].

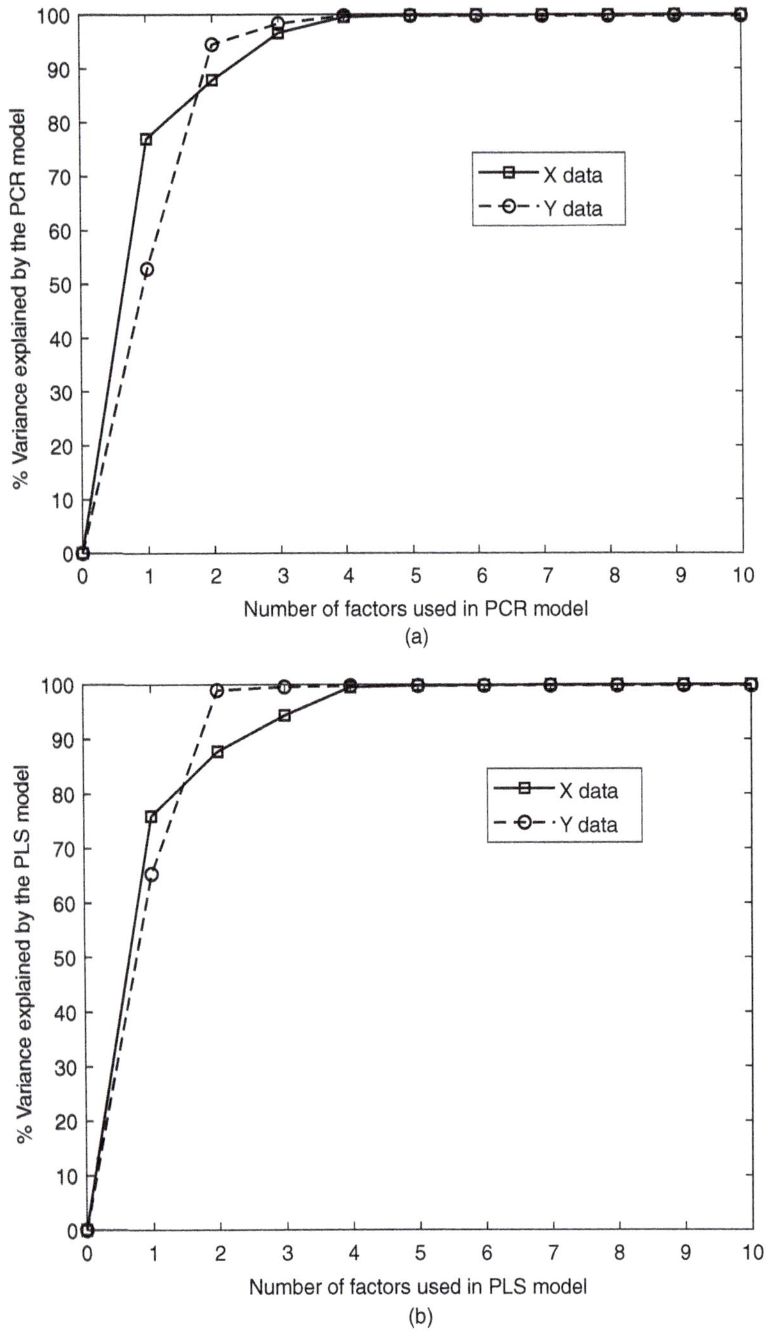

Figure 4.6 Comparison of percentage explained X and Y variance for PCR (a) and PLS (b) regressions to cis-butadiene concentration of styrene–butadiene copolymers. *Source*: Charles Miller.

To use the CLS method for quantitative analysis, one must first know the value of p: the number of spectroscopically active components in the mixtures. Once this is known, or estimated, calibration of a CLS model consists of obtaining the pure component spectra in **S**. This can be done in one of two ways:

1) Experimentally, through measurement of isolated pure components on the instrument, or
2) Inferentially, through least squares analysis of a data set where the concentrations of the pure components in all samples (**C**) are known.

Option 1 is viable in cases where all components can be chemically isolated. Option 2 can be used if **C** is known for all of the standards in **X**, and utilizes the least squares regression method to estimate **S** from the known concentrations and the mixture spectra **X**:

$$\hat{\mathbf{S}} = (\mathbf{C}^t\mathbf{C})^{-1}\mathbf{C}^t\mathbf{X} \tag{4.16}$$

Once **S** is calculated, the concentrations of all of the components in an unknown sample ($\mathbf{c_p}$) can be calculated from the spectrum of the unknown ($\mathbf{x_p}$) using least squares:

$$\widehat{\mathbf{c_p}} = \mathbf{x_p}\mathbf{S}(\mathbf{S}\mathbf{S}^t)^{-1} \tag{4.17}$$

In cases where it can be used, CLS has several appealing features. As a quantitative analysis method, a single model can cover all of the p components in the mixtures, and closure constraints on the concentrations **C** can be imposed to enable efficient multiplicative (pathlength) corrections of the spectra. It is also a very intuitive and transparent model in that the model parameters **S** contain pure component spectra, and are not constrained to be orthogonal like the loadings (**P**) in PCR and PLS.

Unfortunately, CLS is not always a viable regression option due to insufficient prior knowledge or inaccuracies of the linear additive assumption inherent in the method. Furthermore, it cannot be used in cases where the properties of interest are not concentrations. Finally, when estimating **S** from pure component concentrations **C**, one must avoid cases of highly correlated concentration profiles, in which case the inversion of $\mathbf{C}^t\mathbf{C}$ in Eq. (4.16) can be unstable. However, the CLS method can be extended to provide some flexibility in cases where not all relevant pure components and spectral effects are known *a priori*, and this is the subject of the next section.

For the styrene–butadiene case, where all of the four major components in all 70 samples are known via a reference analytical method, the option (2) method mentioned above can be used to estimate the four pure component spectra **S**. Figure 4.7 shows the CLS-estimated pure component spectra of the four components. To provide an interesting comparison, the four X variables selected in the stepwise MLR method for cis-butadiene (1706, 1824, 1670, and 1570 nm) are denoted. Note that the first variable selected, 1706 nm, clearly coincides with a strong cis-butadiene peak, which is unfortunately overlapped with bands from the other components. As a result, it appears that the other three MLR-selected variables account for either interfering components (1824 nm, 1670 nm) or spectral baseline offset variations (1570 nm).

4.6.5.2 Extended Classical Least Squares

The CLS model in Eq. (4.15) is quite rigid. It assumes linear additivity of all of the pure components in the mixture and cannot accommodate the presence of additional unknown components or other relevant spectral effects. Fortunately, the CLS model can be extended [25, 26] to include these effects, per Eq. (4.18).

$$\mathbf{X} = \mathbf{C}\mathbf{S}^t + \mathbf{D}\mathbf{S_2}^t + \mathbf{E} \tag{4.18}$$

where the term $\mathbf{D}\mathbf{S_2}^t$ captures these additional components or spectral effects. Like **S**, $\mathbf{S_2}$ contains basis spectra, or profiles, corresponding to interfering components or effects, and **D** contains the "intensities" of

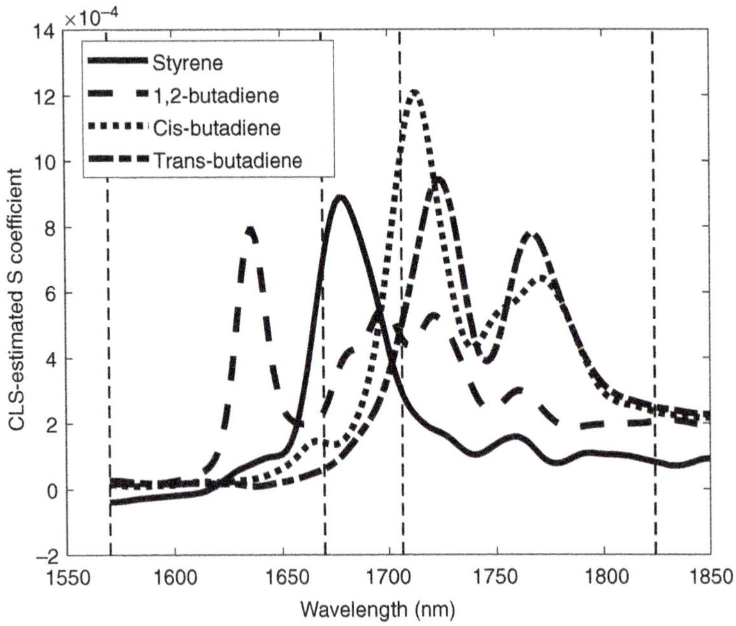

Figure 4.7 CLS-based estimate of the four pure component spectra. For reference, the four wavelengths selected for the stepwise MLR regression to the cis-butadiene concentration are denoted by vertical dashed lines. *Source*: Charles Miller.

these effects. Such interfering components in $\mathbf{S}_2$ could include baseline functions, pure spectra of additional chemical components that are not represented in **S**, or other profiles that account for non-idealities in the CLS model.

Unlike CLS, the basis vectors in $\mathbf{S}_2$ cannot typically be calculated using least squares, as their "concentrations" or intensities in the training set standards are not typically known in advance. Instead, they can either be arbitrarily assigned to specific functions, such as a curved or quadratic baseline function, or calculated from an independent set of experiments focused on assessing interfering components in the system. For example, they could be assigned as loadings from a PCA analysis of a set of spectra obtained from replicate sample analyses, or from a PCA analysis of a set of spectra of a single standard where various sample conditions are modified. In this way, extended CLS can have the appealing property of explicitly accounting for interfering effects that are not, or cannot, be present in the training data. One specific method that utilizes this principle is Prediction Augmented CLS (PA-CLS), which uses results from actual predictions of the CLS model on field samples to determine more relevant interference basis profiles [27].

The concept of model overfitting is just as relevant for extended CLS as it is for other methods and is caused by using too many basis vectors in **S** and $\mathbf{S}_2$. As for any of the quantitative methods discussed in this chapter, proper fit-for-purpose model validation is critical for any to avoid overfitting and to effectively realize the value of these model extensions.

4.6.5.3 Multivariate Curve Resolution

Multivariate curve resolution (MCR) [28–31] uses the same classical Beer's Law model (Eq. (4.15)) as the CLS method discussed earlier. However, the estimates of the pure component concentrations (**C**) and pure component spectra (**S**) are made using an iterative algorithm that exploits prior knowledge-based constraints regarding the nature of the profiles in **C** and **S**. Commonly applied constraints in **C** include closure and non-negativity, and non-negativity is also commonly imposed on **S**. One can also impose additional constraints on **C** and **S** based on prior knowledge of the spectroscopy or chemistry of the system. The usage of such constraints is critical for MCR, as there are two inherent ambiguities associated with the model equation (Eq. (4.15)) in cases where neither **C** nor **S** are known in advance:

- **Rotational Ambiguity:** Different linear combinations (rotations) of the pure component spectra in **S** will result in the same model fit error (**E**), provided that the pure concentration matrix **C** is correspondingly counter-rotated, and
- **Multiplicative Ambiguity:** The same model fit error can be obtained by both upscaling **S** and, correspondingly, downscaling **C**, or *vice versa*.

However, by imposing sufficient and accurate constraints on the MCR solution, one can address these ambiguities and obtain highly relevant estimates of **S** and **C**.

Multivariate curve resolution is typically used as an exploratory method but can also be utilized in quantitative analysis. For example, estimates of **S** from an MCR analysis could be used in a CLS or extended CLS model to calculate concentrations (**C**) per Eq. (4.17) or (4.18).

The results of applying the MCR method to the styrene–butadiene spectral data are shown in Figure 4.8. In this case, 4 components were specified, and non-negativity constraints were applied to both **C** and **S**. These results somewhat resemble the CLS-estimated profiles (Figure 4.7) but show some clear distinctions. First, note that the components are referred to by component number, and not by chemical species, as in the CLS case. Secondly, note that the Y-axis scales are orders of magnitude different: in the CLS case, the values refer to true absorptivities because the concentrations **C** were known in advance, whereas the MCR result reflects the multiplicative ambiguity mentioned earlier. Finally, note that the shapes of the profiles from MCR do not exactly match the shapes of the profiles from CLS, reflecting the rotational ambiguity mentioned earlier. This result underscores the importance of utilizing prior knowledge of the chemistry and spectroscopy of the system under study to produce more intuitively meaningful results.

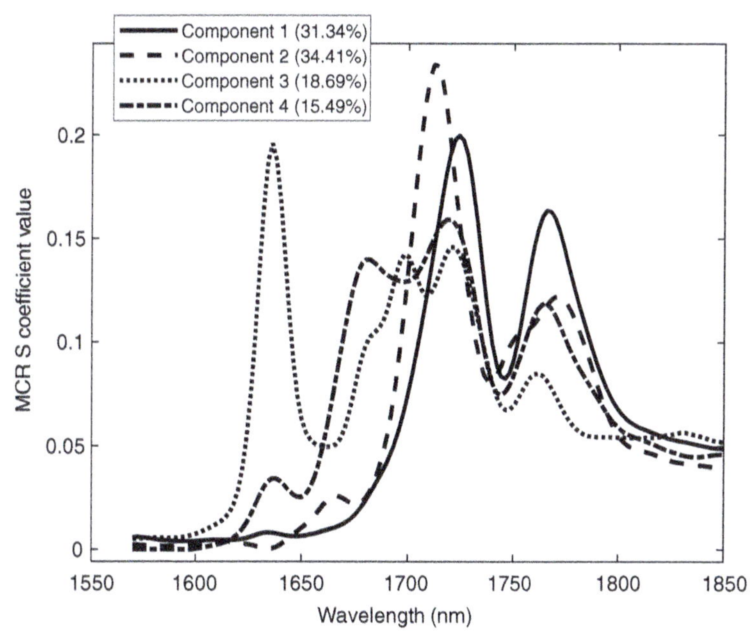

Figure 4.8 Result of applying MCR to the X data from the styrene–butadiene dataset. 4 components were specified, and non-negativity constraints were applied to **C** and **S**. *Source:* Charles Miller.

4.6.6 Advanced Methods

Most of the quantitative methods discussed above are inherently linear, in that the model equations assume linear relationships between the X variables and between the X and Y variables. This does not necessarily mean that they cannot be effective in cases where these relationships are non-linear. In fact, it is common to utilize increased model complexity with such methods to effectively address deviations from perfect linearity, for example, approximating an inherent non-linear relationship using two or more linear principal components, latent variables, or extended CLS basis spectra. However, even these adaptations can lead to poor results in systems with strong inherent non-linearities.

The first step toward addressing non-linear relationships is to detect their presence, and this is most effectively accomplished through exploratory analysis methods like PCA. The presence of non-linear or curved trajectories in PCA scores readily indicates non-linearities between the PCA input variables, which can include X and Y variables. Once detected, many non-linearities can be sufficiently addressed through minor extensions of the methods already discussed above, for example, through non-linear parameterization (section 4.5.2), non-linear transformation of individual X variables, or the addition of non-linear X terms in the MLR or PLS model equations (Eqs. (4.5), (4.8), and (4.9)). In cases where these methods are insufficient, other methods can be utilized, which are reviewed below.

4.6.6.1 Artificial Neural Network

As the name suggests, this method utilizes a model form that resembles neurons in the brain [32]. Although several types of ANNs are used for quantitative analysis, this chapter will focus on the basic feed-forward neural network, a schematic of which is provided in Figure 4.9. Like the methods discussed earlier, the ANN method has input values (nodes I1–I4 in the schematic), one or more output values (node O1), and model parameters ($W_{i,j}$ and $WO_{i,j}$). However, there are three ways in which ANNs differ from the previously discussed methods:

1) The ANN model has an additional set of "hidden nodes," denoted by H1–H3 and H1′–H3′ in Figure 4.9. As a result, the model structure is best expressed using a map (like Figure 4.9), rather than a simple mathematical expression.

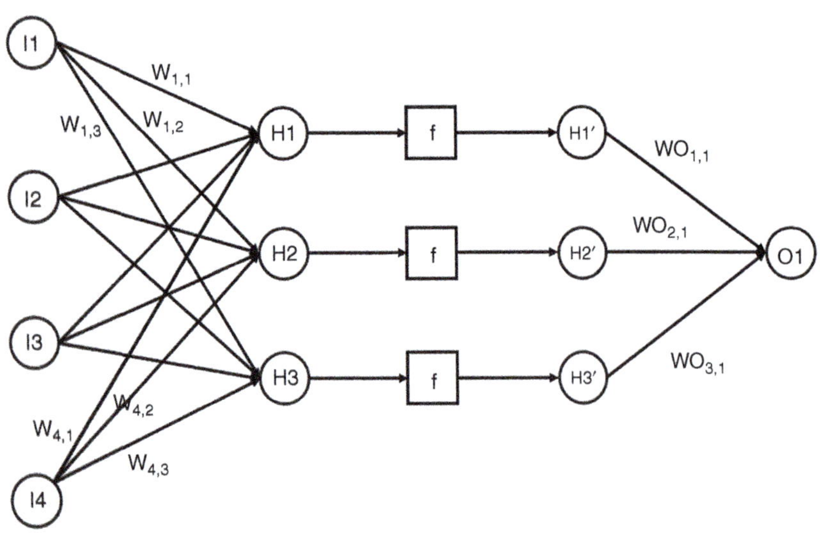

Figure 4.9 Example of a simple Feed Forward ANN model structure, containing one hidden layer that contains three hidden nodes. *Source*: Charles Miller.

Table 4.2 Fit results of six ANN models using 1–6 hidden nodes for determining cis-butadiene concentration in the styrene–butadiene samples.

Number of ANN hidden layer nodes	RMSEE for cis-butadiene (%)
1	2.70
2	1.52
3	1.13
4	1.05
5	0.96
6	1.07

2) The model parameters are not determined by a simple algorithm like least squares regression that uses all of the training objects simultaneously, but rather by an optimization algorithm, such as back propagation [4], which starts with a random selection of parameters and proceeds to adjust these through sequential introduction of each individual training object to the network.
3) It allows for non-linear modeling using non-linear transfer functions (f) in the model structure.

Despite these differences, it is worth noting that the PCR method mentioned earlier can be considered a special case of the ANN model shown in Figure 4.9, where the input weights of the ANN ($W_{i,j}$) are the PCR loadings (**P**), the output weights of the ANN ($WO_{i,j}$) are the Y loadings (**q**), and the transfer functions in the ANN hidden layer are simple linear functions.

The features described above make ANNs effective in cases where there are unknown non-linear relationships that are not easily mitigated using one of the strategies mentioned earlier. However, ANNs can be susceptible to overfitting from the use of multiple hidden layers and many nodes within these layers, and thus, model validation is particularly important.

Table 4.2 provides the RMSEE fit results obtained from generating a series of ANN models, using the structure in Figure 4.9, to the styrene–butadiene data for determination of cis-butadiene concentration. These models differ by the number of nodes that are in a single hidden layer of the ANN, ranging from one to six hidden nodes. Note that an ANN structure using three hidden nodes appears to be optimal for this case, which is consistent with the earlier PCR and PLS results that suggested the use of 2–4 factors in the model. This is also consistent with the knowledge that four known components in the samples are bound by closure, thus leading to the expectation of 3 independent components in the data.

4.6.6.2 Support Vector Machine Regression

The Support Vector Machine (SVM), and its regression analog Support Vector Machine Regression (SVM-R) were first introduced as machine learning tools decades ago, and since then, have gained considerable traction in analytical chemistry. One key distinction between SVM-R and other methods discussed above is that it can address strong non-linearities in the X data using non-linear transfer functions, known as kernels. Common non-linear kernels in SVM-R include Polynomial, Gaussian Radial Basis Function (RBF), and Sigmoid, and these can be used to re-map the original X variables into a higher-dimensional space from which one can obtain an improved linear correlation to Y. A second unique property of SVM-R is that the algorithm does not necessarily utilize all training objects equally but instead imposes a "cost penalty" on objects that are poorly predicted. As a result, SVM-R, in essence, has a built-in robust component that tolerates outliers in the training data set to some extent. A final distinction of SVM-R is that the resulting model is not expressed as a set of regression coefficients but rather as a subset of the training set objects, called "support vectors," which are most relevant for identifying the model.

The SVM-R method requires prior specification of some key parameters, including the type of kernel, the C parameter, which defines the cost penalty to be assessed for poorly predicted training objects, and γ, which defines how far the influence of each individual training object reaches in the space. It is common to use an empirical grid search method to find the optimal values of C and γ, and trial and error to determine the optimal kernel. More information on SVM can be found in several references [33].

4.6.6.3 Locally Weighted Regression

Locally weighted regression (LWR) is designed to address cases where the X–Y relationship varies across different regions of the training dataset, where a single global model can have limited effectiveness. Like SVM, LWR does not utilize all training set objects equally: the objects closest to the sample to be predicted are weighted the highest. Several variants of LWR have been in practice for many years, and perhaps one of the simplest variants of these involves a three-stage process:

1) identify a subset of N objects in the training set that are closest to the prediction object, using a global PCA model based on the X variables, followed by
2) construct a local regression model to Y using only this subset of N objects, then
3) use this model to predict Y for the prediction object.

Locally weighted regression involves both a PCA projection and a local regression for each analysis cycle, as the subset population of N objects will depend on the properties of the unknown's X variable profile. The key method parameters in LWR are the maximum distance that the local objects can have from the prediction object for step 1 and the regression model type to use in step 2.

Locally weighted regression can be very effective in cases where there is a general but non-linear trend between X and Y. It is also generally the case that the local regression models are much less complex than any attempted global regression model that utilizes the entire training dataset. However, LWR requires collecting and maintaining a sufficiently large set of training objects that adequately span the entire range of conditions that are expected to be encountered. It also requires a deployment software platform that supports the more complex workflow of LWR prediction and involves a longer prediction cycle than other regression methods. Details on LWR can be found in reference [34].

4.6.6.4 Classification and Regression Trees

The classification and regression trees (CART) method uses decision trees to build classifiers and predictors that can accommodate highly non-linear data structures. Decision trees are structures that resemble a flowchart, consisting of the following:

- a single **root node** that is the starting point: where the initial decision is made,
- **internal nodes** that represent decisions,
- **branches**, which represent the outcomes of these decisions, and
- **terminal nodes**, or **leaf nodes**, that each correspond to a final result.

To illustrate CART for a very simple case, consider the data shown in the scatter plot in Figure 4.10a, displaying 20 training objects for a single input variable X and a single output variable Y. It is very clear from Figure 4.10a that the overall relationship between X and Y is very non-linear, but still follows a clear trend. Figure 4.10b shows one possible regression tree that corresponds to the scatter plot in Figure 4.10a. In this specific case, the root node is based on an X value of 1.0, which then splits the 20 training objects into two internal nodes containing subsets of 8 and 12 objects. Each of these subsets is then split again based on other values of X, and the process is repeated until a minimum number of objects

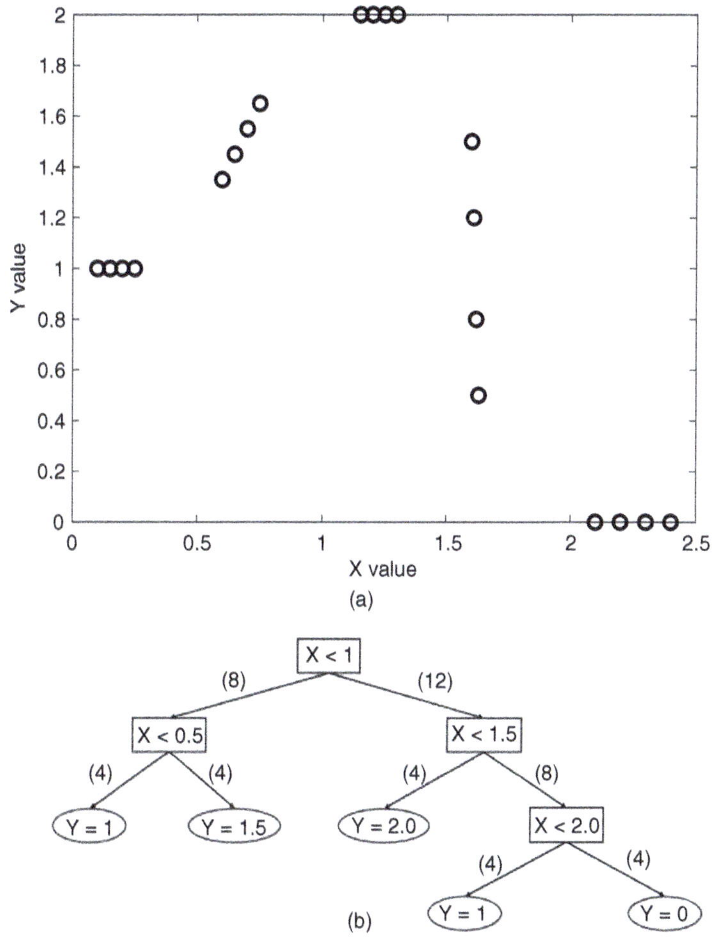

Figure 4.10 (a) Scatter plot of a training data set of objects with a single X variable and a single Y variable, showing a non-linear X/Y relationship. (b) A regression tree that captures the non-linear X/Y structure, with the minimum number of objects per node set at 4. Root nodes and internal nodes are denoted as rectangles and terminal nodes as ovals. The number of objects proceeding along each branch is provided in parentheses. *Source*: Charles Miller.

per node is reached. In the case of the tree shown in Figure 4.10b, this minimum number is 4 objects, which results in 5 terminal nodes.

Decision trees, like flow charts, are quite intuitive in their operation, but their usage in quantitative regression problems is somewhat less intuitive. Unlike most of the other regression methods discussed above, where a predicted Y value is generated from a closed-form mathematical expression, regression trees generate a discrete outcome for a predicted Y value for a given prediction object, where this value is based on the specific terminal node that is encountered for the prediction object.

The algorithm for developing the optimal decision values and tree structures involves iterative evaluations of different dataset-splitting schemes based on minimization of Y prediction residuals. Like all other regression methods, trees can be overfitted, and this can occur if one specifies a minimum number of objects per node that is too low. For example, in the case illustrated here in Figure 4.10b, one could have specified a minimum number of objects per node of 2 instead of 4, which would result in further splits in the tree structure. This would improve the overall fit of the method to the training data but might result in more significant errors when the method is applied to new data. As a final note, regression trees provide a useful introduction to several other popular methods that are in use today, including Random Forest and X-Gradient Boosted (XG-Boost) regression, which improve on the CART principle by constructing and evaluating multiple regression trees for any given problem. Details on CART and similar methods can be found in reference [35].

4.7 Conclusion

This chapter provided the background for quantitative analysis methods in analytical chemistry, along with the key properties and considerations associated with their use. A wide range of commonly used methods was discussed in detail, with key distinctions between these illustrated using various examples. Although the methodologies are numerous and continually increasing, the principles regarding their usage remain the same: persistent focus on the method objective, avoiding needless model complexity, ensuring appropriate model validation, and seeking sufficient model transparency.

References

1 Martens, H. and Næs, T. (1989). *Multivariate Calibration*. Chichester: Wiley.

2 Box, G.E.P., Hunter, W.G., and Hunter, J.S. (1978). *Statistics for Experimenters: An Introduction to Design, Data Analysis and Model Building*. New York: Wiley.

3 Mardia, K.V., Kent, J.T., and Bibby, J.M. (1979). *Multivariate Analysis*. London: Academic Press.

4 Rumelhart, D., Hinton, G., and Williams, R. (1986). Learning representations by back-propagating errors. *Nature* 323: 533–536.

5 Miller, C.E. (2018). The "how" of multivariate analysis (MVA) in the pharmaceutical industry: a holistic approach. In: *Multivariate Analysis in the Pharmaceutical Industry* (ed. A.P. Ferreira, J.C. Menezes, and M. Tobyn), 93–124. San Diego: Academic Press.

6 Miller, C.E. (2010). Chemometrics in process analytical technology (PAT). In: *Process Analytical Technology*, 2e (ed. K. Bakeev), 353–438. Chichester, UK: Wiley.

7 U.S. Department of Health and Human Services, Food and Drug Administration, Center for Drug Evaluation and Research (CDER), Pharmaceutical Quality/CMC (2021). Development and submission of near infrared analytical procedures. Guidance for Industry. https://www.fda.gov/regulatory-information/search-fda-guidance-documents/development-and-submission-near-infrared-analytical-procedures (accessed 29 January 2025).

8 European Medicines Agency, Committee for Human Medicinal Products (CHMP), Committee for Veterinary Medicinal Products (CVMP) (2014). Guideline on the use of near infrared spectroscopy by the pharmaceutical industry and the data requirements for new submissions and variations (accessed 10 January 2025).

9 U.S. Department of Health and Human Services Food and Drug Administration, Center for Drug Evaluation and Research (CDER), Center for Biologics Evaluation and Research (CBER). (2021). ICH Q12. Technical and Regulatory Considerations for Pharmaceutical Product Lifecycle Management, Guidance for Industry (accessed 10 January 2025).

10 U.S. Department of Health and Human Services Food and Drug Administration, Center for Drug Evaluation and Research (CDER), Center for Biologics Evaluation and Research (CBER), Center for Veterinary

Medicine (CVM), Pharmaceutical Quality/Manufacturing Standards (CGMP) (2016). Data integrity and compliance with CGMP guidance for industry (accessed 5 January 2025).
11 Miller, C.E. (1995). The use of chemometric techniques in process analytical method development and operation. *Chemometrics and Intelligent Laboratory Systems* 30 (1): 11–22.
12 Miller, C.E., Eichinger, B.E., Gurley, T.W., and Hermiller, J.G. (1990). Determination of microstructure and composition in butadiene and styrene-butadiene polymers by near-infrared spectroscopy. *Analytical Chemistry* 62: 1778–1785.
13 Vandeginste, B.G.M. and De Galen, L. (1975). Critical evaluation of curve fitting in infrared spectrometry. *Analytical Chemistry* 47 (13): 2124.
14 Peru, D., Panariello, S., and Cox, R. (2025) Colorimetric determination of tea-stained seashells using the B&W tek quest X UV-vis-NIR spectrometer. Spectroscopy Online. https://doi.org/10.56530/spectroscopy.zt2966i4 (accessed 3 April 2025).
15 Alvarez-Lloret, P., Rodriguez-Navarro, A., Romanek, C.S. et al. (2006). Quantitative analysis of bone mineral using FTIR. *Macla* 6: 45–47.
16 Yoshida, S. (2014). Quantitative evaluation of an epoxy resin dispersion by infrared spectroscopy. *Polymer Journal* 46: 430–434.
17 Maurizio Baldassarre, M., Li, C., Eremina, N. et al. (2015). Simultaneous fitting of absorption spectra and their second derivatives for an improved analysis of protein infrared spectra. *Molecules* 20: 12599–12622.
18 Tom, J. (2023). UV-Vis spectroscopy: principle, strengths and limitations and applications. https://www.technologynetworks.com/analysis/articles/uv-vis-spectroscopy-principle-strengths-and-limitations-and-applications-349865 (accessed 10 February 2025).
19 Miller, C.E. and Eichinger, B.E. (1991). Analysis of RIM polyurethanes by near infrared diffuse reflectance spectroscopy. *Journal of Applied Polymer Science* 42 (8): 2169.
20 Miller, C.E. (1993). Use of near-infrared spectroscopy to determine the composition of high-density/low-density polyethylene blend films. *Applied Spectroscopy* 47 (2): 222.
21 Pell, R., Ramos, S., and Manne, R. (2007). The model space in partial least squares regression. *Journal of Chemometrics* 21: 165–172.
22 Bro, R. and Elden, L. (2008). PLS works. *Journal of Chemometrics* 23: 69–71.
23 Wold, S., Hoy, M., Martens, H. et al. (2009). The PLS model space revisited. *Journal of Chemometrics* 23 (2): 67–68.
24 Brown, C.W. and Obremski, C.W. (1984). Multicomponent quantitative analysis. *Applied Spectroscopy Reviews* 20: 373–418.
25 Gallagher, N.B., Goyetche, R., Amigo, J.M., and Kucheryavskiy, S. (2024). Extended least squares (ELS) and generalized least squares (GLS) for clutter suppression in hyperspectral images: a theoretical description. *Chemometrics and Intelligent Laboratory Systems* 244: 105032.
26 Gallagher, N.B. (2019). Chapter 2.9 – Classical least squares for detection and classification. In: *Data Handling in Science and Technology*, vol. 32 (ed. J.M. Amigo), 231–246. Elsevier.
27 Haaland, D.M. and Melgaard, D.K. (2000). New prediction-augmented classical least squares (PACLS) methods: application to unmodeled interferents. *Applied Spectroscopy* 54: 1303.
28 Gemperline, P.J. and Cash, E. (2003). Advantages of soft versus hard constraints in self-modeling curve resolution problems. Alternating least squares with penalty functions. *Analytical Chemistry* 75: 4236.
29 de Juan, A. and Tauler, R. (2006). Multivariate curve resolution (MCR) from 2000: progress in concepts and applications. *Critical Reviews in Analytical Chemistry* 36: 163–176.
30 Lawton, W.H. and Sylvestre, E.A. (1971). Self modeling curve resolution. *Technometrics* 13: 617–633.

31 Tauler, R., Kowalski, B., and Fleming, S. (1993). Multivariate curve resolution applied to spectral data from multiple runs of an industrial process. *Analytical Chemistry* 65: 2040–2047.
32 Næs, T., Kvaal, K., Isaksson, T., and Miller, C. (1993). Artificial neural networks in multivariate calibration. *Journal of Near Infrared Spectroscopy* 1: 1–11.
33 Vapnik, V.N. (1995). *The Nature of Statistical Learning Theory*. New York: Springer.
34 Naes, T., Isaksson, T., and Kowalski, B. (1990). Locally weighted regression and scatter correction for near infrared reflectance data. *Analytical Chemistry* 62 (7): 664–673.
35 Loh, W.-Y. (2014). Fifty years of classification and regression trees. *International Statistical Review* 82 (3): 329–348.

5

Variable Selection Strategies

Charles E. Miller

ChEM Analytics LLC, Lansdale, PA, USA

5.1 Introduction to Variable Selection

Variable Selection, also known as "Feature Selection," is a very important part of the multivariate calibration development workflow [1]. This is particularly the case for methods developed using training data with many independent (X) variables, where many of the variables could be irrelevant for obtaining the objective of the method. It is important to note that, in this context, the term "irrelevant" can refer to variables exhibiting random variance (e.g., "noise") or variables exhibiting systematic variance that is unrelated to the method objective. The presence of many such irrelevant variables in a training data set is not only detrimental to the performance of the resulting method, but it can also mask the relevant variance in the data during model development, thus preventing one from converging upon an effective method. With the above in mind, variable selection can be thought of as an extension of the analytical specificity mission, which was discussed previously in Chapter 4 for the analytical method calibration use case.

Just like quantitative methods, discussed in Section 4.4, variable selection methods span a wide range of properties, including complexity of use, their underlying assumptions, their dependency on empirical vs. mechanistic workflows, and the amount of data and computing power that is required to use them. As we proceed through these methods, please keep in mind that empirically driven selection methods are closely tied to the specific sample population that is fed to the algorithm, which is typically the model training data set. Therefore, any future changes in key training data set attributes, including sample population, outliers, initial spectral range, spectral preprocessing scheme, or changes to the data source (driven by, for example, changes in instrument configuration, sampling interface, or sampling scheme), can render a previously generated variable selection outcome irrelevant. This has two key implications:

- Changes in *any* of these key data set attributes should prompt a reassessment of the optimal variable subset, regardless of the level of rigor that was used for the original variable subset selection.
- These key dataset attributes should be considered "design variables" when executing variable selection. For example, if a developer executes separate variable selections for different relevant combinations of initial spectral range, preprocessing scheme, and sampling interface (as in a Design of Experiments, or DOE), they would likely obtain different optimal selections for each case.

Another key consideration for variable selection is the specific criterion used to optimize the selection. For mechanistic selection, this criterion is simple: variables are selected based on domain knowledge of the system, choosing variables that are expected to be relevant for the method objective. In the case of spectroscopy methods, these are typically variables corresponding to known absorption bands of the

Practical Quantitative Vibrational and Electronic Spectroscopy: A Guide for Developing, Optimizing, and Validating Procedures, First Edition. Edited by Deborah A. Peru.

Companion website: www.wiley.com/go/peruelectronicspectroscopy

analyte or property of interest. For empirical selection, the criterion is typically the optimization of a model performance metric, such as the model fit, cross-validation error, or independent test set validation error obtained after building a method using the specified variable subset. In this case, linking the criterion to the method's objective can be done through careful validation design. For example, if the objective is to build a robust predictor for analyte "A," the validation design should be set up to split the training and test sets in a way that sufficiently challenges the model's ability to predict analyte "A."

A final consideration for empirical variable selection is the possibility of overfitting the selection, especially for cases where there are many more variables than samples in the dataset. As the number of candidate variables to select from increases, it becomes more plausible that random noise in some of these variables across a smaller population of samples has a spurious correlation to the property of interest. A common overfitting pitfall that many method developers fall into is using the same sample population for both variable selection and subsequent model evaluation. To avoid this situation, it is critical to *design the variable selection and model evaluation procedures as separate tasks that are independent of one another*. For example, one could select two different independent test set sample subpopulations for variable selection and model evaluation to avoid getting overly optimistic model evaluations that are not reflective of future performance. Useful discussions of this pitfall and how to address it are provided in several references [2, 3].

5.2 Why Variable Selection?

In the early days of quantitative analysis, variable selection was the norm, driven mainly by limitations in computing resources. Once multivariate analysis (MVA) methods were made accessible to method developers, *via* moderately improved computing resources, their inherent capability for feature selection was immediately made apparent. In fact, many early MVA users felt that an additional variable selection workflow would not be necessary due to this inherent selection capability of MVA. Subsequent years of deploying MVA models in the field, however, exposed the bitter irony: that MVA methods by themselves leave a lot of potential method specificity on the table, and that variable selection can still be used to realize significant performance improvement. The reasons for this improvement can be summarized below:

1) Variable selection can be used to enhance the analytical specificity of a method beyond that obtained from the quantitative method alone, even for MVA methods.
2) From an operational perspective, variable selection reduces the burden on the quantitative method to explain variance components in X that are irrelevant to the method objective, thus allowing more efficient modeling of the relevant variance in X.
3) This in turn allows one to build more parsimonious models, which are easier to manage, less likely to encounter outlier conditions when in service, and less likely to require updating once they are put into service.

With this in mind, variable selection can improve the resiliency and effectiveness of a method for its specific objective.

5.3 Key Attributes of Variable Selection Methods

The variable selection methods discussed below differ in respect to three key attributes. Note that these attributes are also discussed in the context of quantitative methods (Section 4.4), although their meaning is slightly different in the context of variable selection.

- **Empirical vs. Mechanistic:** Just like model-building algorithms, variable selection algorithms can be positioned along a continuum from purely empirical to purely mechanistic. Empirical methods are driven

Table 5.1 Summary of the key attributes of the variable selection methods discussed in this chapter.

Attribute	Manual selection	Empirical MLR	S/N method	MVA model parameters	Jackknifing	Interval methods	Genetic algorithm
Empirical vs. mechanistic focus	Mechanistic	Empirical	Empirical	Empirical	Empirical	Empirical	Empirical
Univariate vs. multivariate	Univariate	Univariate	Univariate	Multivariate	Multivariate	Multivariate	Multivariate
Complexity	Low	Low	Medium	Medium	High	High	High

entirely by data measured on the system, whereas mechanistic methods are driven entirely by prior mechanistic knowledge of the chemistry, physics, and spectroscopy of the system. Most methods lie somewhere in between these extremes.

- **Univariate vs. Multivariate:** In the context of variable selection, univariate methods consider each variable candidate as an independent variable, whereas multivariate methods account for the possibility that variables can have differing degrees of correlation to one another.
- **Simple vs. Complex:** This property refers to both the number of adjustable parameters in the variable selection algorithm and the degree of user interaction that is required to run the algorithm. Simple methods have a clear workflow requiring one or two parameters, and complex methods tend to have many parameters or require frequent user interaction.

Table 5.1 provides a high-level summary of the variable selection methods discussed below, using the attributes discussed above.

5.4 Constraints of Commercial Solutions

In a perfect world, one would have easy access to all of the selection methods discussed below, but alas, we are often constrained by the commercial solutions provided to us. Recognizing this practical constraint, along with the Parsimony Principle discussed earlier in Section 5.2, it is generally recommended to first utilize simpler methods and resort to more complex methods if the outcome is not sufficient. Simpler methods are also more amenable to coding in third-party scripting languages such as Python or M-code, which are now supported in several commercial solutions. Of course, one can resort to more complex selection algorithms if they are available, and if the simpler methods do not provide acceptable outcomes.

It is important to note that, from a method development workflow perspective (see Figure 4.1), variable selection is exclusively in the realm of model development and optimization and not an active part of model deployment. As a result, such selection can be made using any available third-party software, followed by the manual specification of the optimal variable subset in the primary model-building solution (i.e., the model-building software that is used for generating the model object to be deployed). However, when using such a "manual transfer" approach, it is critical that one accurately import the training data into the third-party solution, and then accurately specify the selected variable indices in the primary solution, to avoid unexpected errors. Finally, for regulated use cases, one might also be required to provide traceability and transparency to the variable selection process, requiring sufficient documentation of this process from the third-party solution.

5.5 Manual Selection

This is the simplest method of variable selection, which requires mechanistic knowledge of the chemistry and measurement science (e.g., spectroscopy or chromatography) of the application. In the case of spectroscopy, prior knowledge of the key functional groups or chemical species associated with the application, along with their characteristic spectroscopic signatures, can be utilized to specify variable ranges that are more specific to the modeling objective, or to exclude ranges that are known to be irrelevant to the objective. In the case of the styrene butadiene polymer example used in Chapter 4, such knowledge was used to select a characteristic 1,2-vinyl band at 1640 nm to be used for a simple multiple linear regression (MLR) calibration to 1,2-butadiene (see Figure 4.3).

It should also be mentioned that such manual selection can be combined with empirical selection methods (discussed below) to determine an optimal variable subset. For example, mechanistic knowledge can be used first to exclude variable regions that are known to be irrelevant, with empirical methods then being applied to only the resulting variable subset. Such a hybrid approach often presents a good balance between effective utilization of prior knowledge and ability to accommodate unknown factors through the empirical approach.

Regardless of the approach taken, there are several key considerations associated with manual feature selection:

- **Depth of Prior Knowledge:** The potential effectiveness of the manual approach can be enhanced through deeper understanding of the chemistry and spectroscopy of the application. For example, this knowledge might identify the key functional groups and their characteristic spectral bands and could also include information on the higher-order interaction effects of these groups in the spectroscopy. Such higher-order effects could include hydrogen bonding, other intermolecular interactions, and tertiary structures of the compounds in the application.
- **Accuracy of Prior Knowledge:** Of course, the prior knowledge used for manual selection might not be accurate, in which case the subsequent modeling results could be poorer than expected. In such cases, one can either interpret the residuals, coefficients, and loadings from the poorly performing model to reveal potential root causes of the poor outcome, or simply revert to a purely empirical variable selection approach.
- **Inherent Analytical Specificity:** Some analytical methods produce profiles with well-isolated features, while others produce highly overlapped features. The former case better allows one to identify sufficiently isolated relevant features for manual selection, whereas for the latter case one is almost certain to encounter manually selected features that are not isolated (or pure), thus requiring one to resort to empirical selection methods.
- **Non-chemical Effects:** These effects, including baseline shifts and multiplicative effects, cannot generally be assigned to isolated variables in the X data, and their effects are more often spread out across multiple X variables. Careful filtering of these effects out of the X data, using for example Multiplicative Signal Correction (MSC) Standard Normal Variate (SNV) or Savistky–Golay derivatives, and can render profiles more conducive to manual selection.

5.6 Empirical Methods

In cases where manual selection based on mechanistic knowledge is of limited or no value, one has access to a wide range of empirically driven feature selection methods to improve their method. These empirical methods vary in complexity, underlying assumptions, computing requirements, and workflow.

5.6.1 Empirical Multiple Linear Regression Methods

In the early days of quantitative NIR spectroscopy during the 1980s, when computing resources were relatively limited, many of the instrument supplier's software packages supported the development and deployment of empirical multiple linear regression (MLR) methods [4, 5]. These solutions typically utilized either "Stepwise MLR" or "All Possible Combinations MLR" algorithms, as described below:

- **Stepwise MLR:** Starting with the complete set of m variables in the dataset ($X_1, X_2, \ldots, X_m$), sequentially select analytical wavelengths one at a time to generate a linear MLR model $Y = a_1X_{s1} + a_2X_{s2} + \cdots a_iX_{si}$, where the optimal selection at each step is based on the empirical linear regression (LR) or MLR fit results to the Y variable. One can continue to select X variables to add to the MLR model until optimal model performance is achieved.
- **All Possible Combinations (APC) MLR:** First, specify the number of X variables to use in a linear MLR model (i), then assess the MLR model fits using all possible combinations of the full set of m X variables in the training set. Select the specific subset of i variables that provides the best model performance.

The Stepwise MLR method is simpler and less demanding computationally, but can converge upon local optima. The APC method will find the optimal MLR selection for a given set of m X variables, at the cost of more computational demand. Early versions of these methods used MLR model fit as the optimization criterion, which posed the risk of model overfitting through the use of too many selected X variables, but current versions avoid this risk and exploit today's much higher computing power by using cross-validation error or test set validation error as the criterion. In addition, more recent versions utilize the residuals from the MLR regressions, along with significance testing, to enhance the variable selection criterion.

Although the discussions above assume a linear MLR model, as shown previously in Eq. (4.5), some versions of these selection methods support non-linear terms in the MLR equation. For example, if one needs to correct for multiplicative variance in the spectral data, which is common in near-infrared (NIR) diffuse reflectance spectroscopy, some solutions allow for the selection of an X variable to be used as a denominator term in the MLR equation (e.g., $Y = a_1(X_1/X_d)$).

5.6.2 Signal-to-Noise Method

This method, which was published in 1992 [6], calculates a "signal-to-noise" (S/N) statistic for each of the m X variables in the data, derived from m linear regressions each X variable's intensity to the Y variable, per the following equation:

$$\mathbf{y} = b_1\mathbf{x}_k + b_0\mathbf{1} + \mathbf{e}_{y,k} \tag{5.1}$$

where $\mathbf{y}$ contains the values of the Y variable for the n objects in the training set, $\mathbf{x}_k$ contains the intensities of the kth X variable ($k = 1$ to m) for the n objects in the training set, $\mathbf{1}$ is a vector of ones (to account for offset), and $\mathbf{e}_{y,k}$ is the residual. For this selection method, the slope coefficient b_1 represents the "Signal," as this reflects the degree to which this specific X variable can predict Y. Conversely, the residual term $\mathbf{e}_{y,k}$ represents the "Noise," as this reflects the overall error of prediction for this specific X variable.

For each X variable, linear regression is used to produce estimates of the coefficients, b_1 and b_0, from which the estimated residuals $\mathbf{e}_{y,k}$ are calculated. The "S/N" metric for each variable is then calculated by the following equation:

$$S/N_k = \hat{b}_1 / (\widehat{\mathbf{e}_{y,k}} * \widehat{\mathbf{e}_{y,k}}) \tag{5.2}$$

Like the newer versions of empirical MLR selection methods, this method utilizes both the "signal" and "noise" when assessing variables. Unlike these methods, it provides a complete profile of S/N values

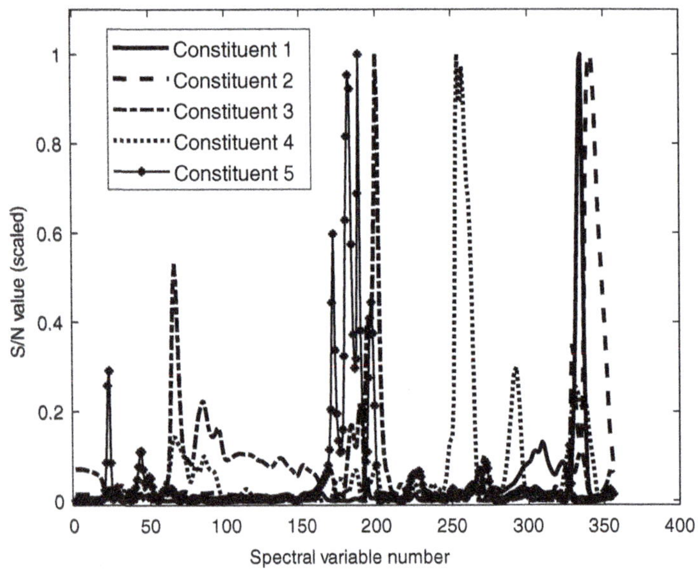

Figure 5.1 Results of applying the S/N method to a dataset containing 355 candidate X variables, for 5 different analytes (Y variables). S/N values are scaled for clarity. *Source*: Charles Miller.

across the spectral range, from which one can select variable subsets using mechanistic knowledge and statistical significance metrics. It is also important to note that the "noise" term in this method captures all deviation from ideal univariate fit for each X variable: both random noise and systematic deviations (e.g., non-linearities). Finally, this method treats each X variable independently, not explicitly accounting for covariance in the X variables. As a result, this method is computationally efficient, but does not necessarily select the optimal variable subset when X and Y covariance are considered.

An example of the use of the S/N method for an NIR process analytical dataset with 355 candidate variables and 5 different Y-variables is provided in a previous reference [7]. This involved five separate executions of the algorithm, one for each of the Y variables. The S/N profiles, shown in Figure 5.1, suggest different variable subsets for the five Y variables. From these S/N profiles, one could either select the variable subset manually based on known spectroscopic bands in the profiles, or by using a statistically derived S/N threshold value. In this specific example, note that this method is generally able to detect independent variable subregions for the five different analytes.

5.6.3 Multivariate Analysis Model Parameters

The multivariate latent variable methods discussed in the previous chapter (Section 4.5.4) [8–11] have the inherent ability to assess the multivariate relevance of X variables for predicting Y. For example, inspection of the regression coefficients and loadings obtained from a partial least squares (PLS) regression or principal component regression (PCR) that uses all of the X variable candidates can be used to assess the relative importance of the X variables. In general, the magnitudes of the regression coefficients (positive or negative values) indicate the importance of the X variables for predicting Y in the presence of interfering effects in the training data set. In addition, for PLS regression, the loading profiles for the first few latent variables in the model typically reveal the specific variance components in X that are most relevant for predicting Y.

Although this method accounts for multivariate covariance in X and Y and utilizes metrics that are already calculated when generating a latent variable model, there is significant subjectivity regarding the manual selection of variables from these metrics. One must also consider the number of latent variables reflected in the regression coefficients, as well as which latent variables are most relevant for predicting Y when using the loading profiles. One way to address this subjectivity issue is to calculate a Very Important Variable (or VIP) metric from the model parameters. Several such VIP metrics have been defined for this purpose, one of which is described in [12]. However, one must still consider the model rank (number of latent variables) when using such metrics to assess variable importance, as the VIP profile can change significantly when changing the model rank.

Figure 5.2 shows the regression coefficients and first two latent variable loadings from a PLS regression to cis-butadiene content for the styrene butadiene dataset that was presented earlier in Chapter 4. Note that the strong positive band at approximately 1705–1710 nm in both the regression coefficient and in the loadings for latent variables 1 and 2 corresponds to the first variable selected in stepwise MLR, at 1706 nm (Section 4.5.3), which is a known band for cis-butadiene. However, note also the other strong positive and negative bands in these profiles, which likely reflect the fact that interfering components also have absorptions in this region.

Model parameters obtained from other exploratory multivariate methods like principal component analysis (PCA), classical least squares (CLS), and multivariate curve resolution (MCR) (see Sections 4.5.4, 4.5.5.1, and 4.5.5.3) can also be used to determine the variables associated with pure components and pure effects in the training data set. Such insight can be useful for selecting variables for non-quantitative methods (e.g., classification), but they can be of limited use in selecting variables for regression problems, as neither of these methods explicitly use the Y data when building the model, and thus some level of subjectivity would be required to enable variable selection for these cases.

5.6.4 Jackknifing

The cross-validation process described in the previous chapter (Chapter 4) is typically used only to provide a conservative estimate of model prediction error. However, for multivariate methods like PLS regression, the process of executing cross-validation, namely the multiple partitioning of the training dataset into simulated training and test sets followed by model build and test prediction for each partitioning case, conveniently lends itself to calculating uncertainty estimates of key model parameters, such as the scores, loadings, and regression coefficients [13]. Specifically, the uncertainty estimates of the regression coefficients can be used to assess the significance of each variable for prediction: any variable for which the uncertainty range spans zero is deemed insignificant for prediction.

Although this jackknifing method is computationally demanding for large datasets, it has several key advantages vs. the methods described thus far. When it is applied to multivariate modeling methods, like PLS regression, PCA, and PLS-DA classification, it explicitly accounts for covariance in X and Y. Furthermore, the cross-validation dataset partitioning scheme can be customized to address specific performance attributes of the resulting model, which is commonly referred to as "Fit For Purpose Validation." For example, partitioning based on sampling location assesses the ability of the model to predict well across different sampling locations. This means that the selected variables are associated not only with the specific training data set, but also with this specific performance attribute. As a result, a different cross-validation partitioning scheme that targets a different attribute might result in a different set of selected variables.

Figure 5.3 below shows the results of applying this jackknifing method to the styrene butadiene dataset, where the Y variable once again is the cis-butadiene concentration. In this case, cross-validation was done using 20 randomly selected object subsets, and the optimal model was obtained at 3 PLS latent variables. A subset of 114 variables were selected from the original set of 141 variables, as indicated in Figure 5.3. Comparing this result to the stepwise MLR regression result obtained in the previous chapter (Section 4.5.3) reveals good agreement: of the selected MLR wavelengths (1706, 1824, 1670, and 1570 nm) all but 1670 nm are included in the subset of wavelengths selected by this method, and even in this case an adjacent wavelength of 1666 nm is still included in the jackknife results. This result underscores the similar outcomes of

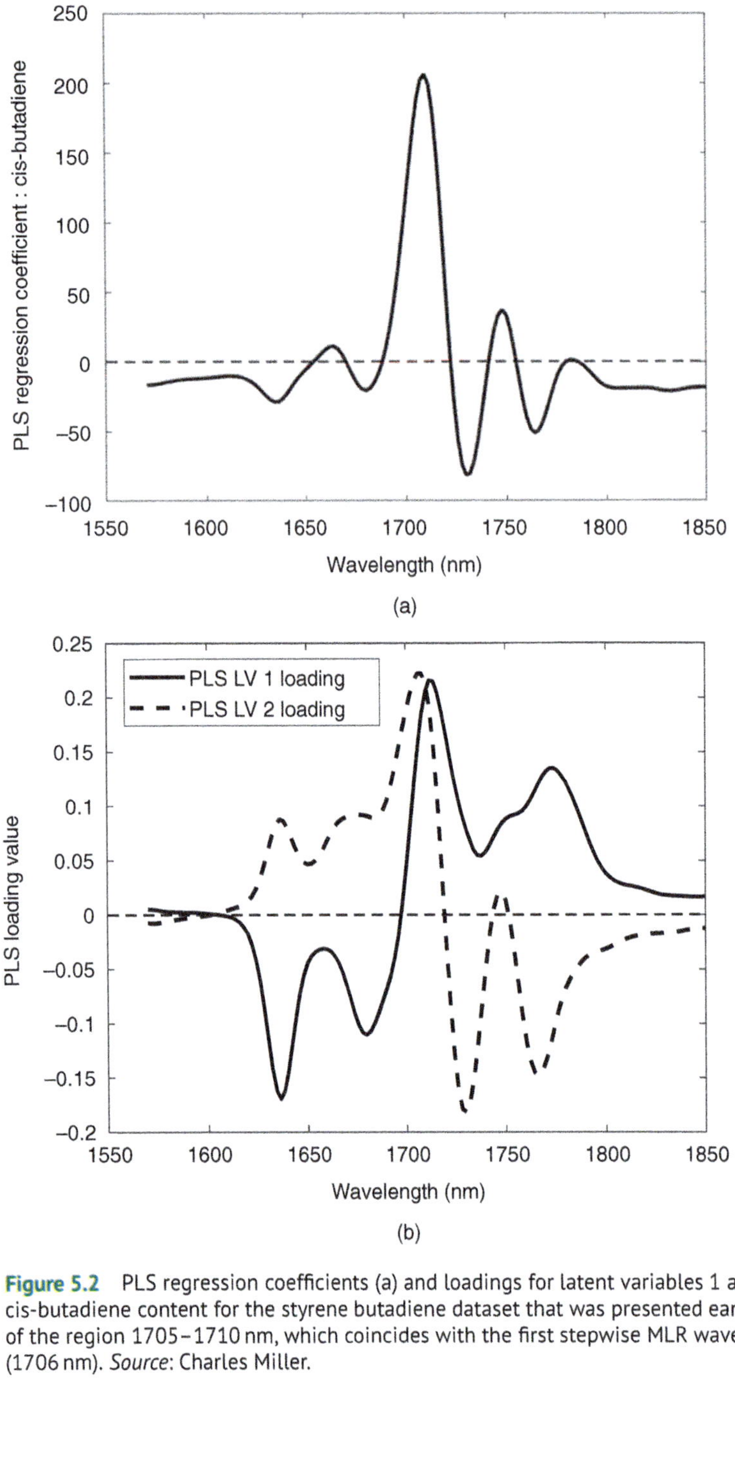

Figure 5.2 PLS regression coefficients (a) and loadings for latent variables 1 and 2 (b), from a PLS regression of the cis-butadiene content for the styrene butadiene dataset that was presented earlier in Chapter 4. Note the prominence of the region 1705–1710 nm, which coincides with the first stepwise MLR wavelength chosen for cis-butadiene (1706 nm). *Source*: Charles Miller.

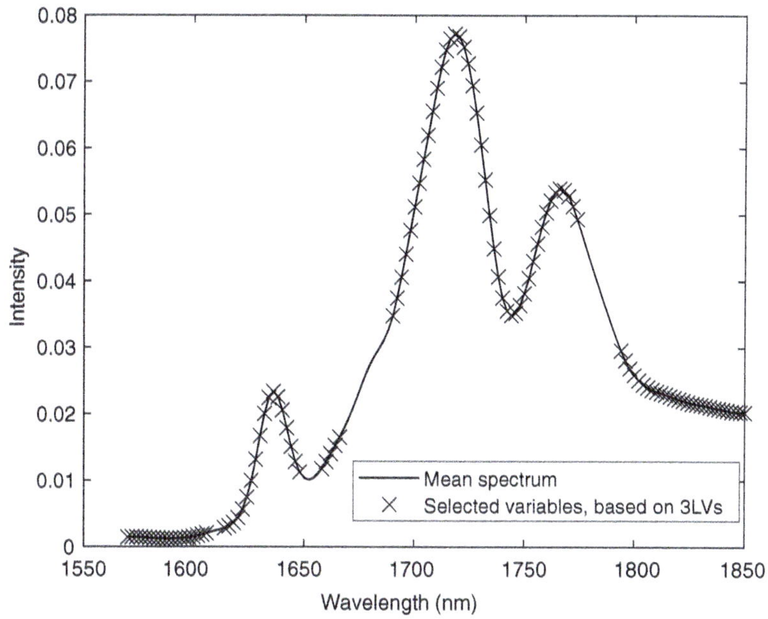

Figure 5.3 Result of applying the jackknifing method to the styrene butadiene dataset, where Y is the cis-butadiene concentration, and cross-validation was based on 20 randomly selected subset, and the optimal model rank is 3 latent variables. 114 of the 141 variables were selected using this method. *Source*: Charles Miller.

the two methods, while also demonstrating their key differences: as the jackknifing method accounts for covariance in the X variables, it is better suited for cases when one ultimately intends to use multivariate calibration methods for model build.

5.6.5 Interval Methods

These methods involve the sequential addition or exclusion of *intervals* of contiguous variables, where selection is based on the model performance improvement that is obtained through the addition or exclusion of the interval to the variable set [14]. The model performance is typically specified as the cross-validation error of a model built using the specified interval set, at the optimal number of latent variables. There are two distinct operating modes for interval selection methods:

- **Forward Mode:** Starting with no selected variables, the first cycle involves the selection of the interval that results in the best model performance. Subsequent cycles involve the selection of an additional interval which, when combined with the already selected interval(s), results in the best model performance. Cycles continue until either a prespecified number of intervals are selected, or when model performance does not improve.
- **Reverse Mode:** Starting with the complete set of variables, the first cycle involves the *exclusion* of the interval that results in the best model performance. Subsequent cycles involve the *exclusion* of an additional interval such that the remaining set of intervals provides the best model performance. Cycles continue until either a prespecified number of intervals are excluded, or when model performance does not improve.

There are two key method parameters that need to be specified for interval methods: the interval size (or number of variables per interval) and the stoppage criterion. The latter can be based on either a fixed number of intervals, or the number of intervals at which no further model improvement is obtained. Interval methods of selection must be used for data where the X variables reflect a continuous property, such as wavelength or wavenumber (for spectroscopic data) or time (for time-series data), in which case each interval likely contains highly correlated X variables. Furthermore, this selection method can be applied to both regression methods like PLS and PCR, or to classification methods like PLS Discriminant Analysis (PLS-DA).

Note that the forward mode of selection resembles stepwise MLR, which was discussed previously (Sections 4.5.3 and 5.6.1), which makes this method rather intuitive and easy to understand. Key differences from stepwise MLR are that *intervals* of variables can be selected instead of individual variables, and that the covariance of X variables can be accounted for through the use of latent variable methods like PLS and PCR. Still, the stepwise nature of the selection presents the potential to select local rather than global optima.

Figure 5.4 shows the result of forward iPLS selection using the styrene butadiene dataset, where the Y variable is cis-butadiene, interval size is 10 variables, and the stopping criterion is the number of intervals at which there is no model improvement. In this case, three intervals (30 variables) are selected from the complete set of 141 variables, and this resulted in a less complex PLS model (3 LVs) with a similar cross-validation error vs. a more complex model (using 4 LVs) where all 141 variables were used. Note that the variable set selected by this method includes only one of the 4 variables that were selected by stepwise MLR (1706 nm), and this discrepancy is likely driven by the high degree of overlap of the interfering components, the high covariance of the X variables, and the sub-optimal searching properties of the step-wise MLR method.

5.6.6 Genetic Algorithm

As mentioned previously, stepwise variable selection methods like Stepwise MLR and Interval methods are relatively intuitive and computationally efficient but are prone to selecting local optima. On the flip side, an exhaustive search for the single global optimum variable set for a given problem can be computationally prohibitive: even for a relatively small set of 100 X variables, there are $10 \times e^{100}$ possible combinations that would need to be tested! With this in mind, one could do a "smarter" and more efficient search of this vast feature space to execute a selection that balances these considerations.

One such searching method, called Genetic Algorithm (GA), has seen considerable usage in multivariate calibration. The reader is referred to several publications [15–17] for details, but at a high level GA operates in the following manner:

1) A total number of variable subset candidates (or intervals) k is specified, along with several other GA parameters, such as number of latent variables, cross-validation scheme, number of variable subsets selected in each "generation," and "mutation" rate.
2) An initial population of k variable subset candidates is randomly generated.
3) For each variable subset, a regression model is generated.
4) Each of the regression models is evaluated for prediction ability, typically using cross-validation.
5) Different pairs of variable subsets are "mated" together, based on their prediction abilities assessed in (4) above. Those variable subsets that result in the best prediction abilities are most likely to "mate" with one another, generating a new variable subset that uses some of the selected variables from one mate and some from the other. When this is done for all possible pairs of variable subsets, this process results in the calculation of a new population (new "generation") of variable subset candidates.
6) Steps (3)–(5) are repeated, until a termination condition is met.

Termination of the algorithm can be based on several conditions, such as a prespecified number of iterations, or a prespecified level of agreement between the k variable subset candidates.

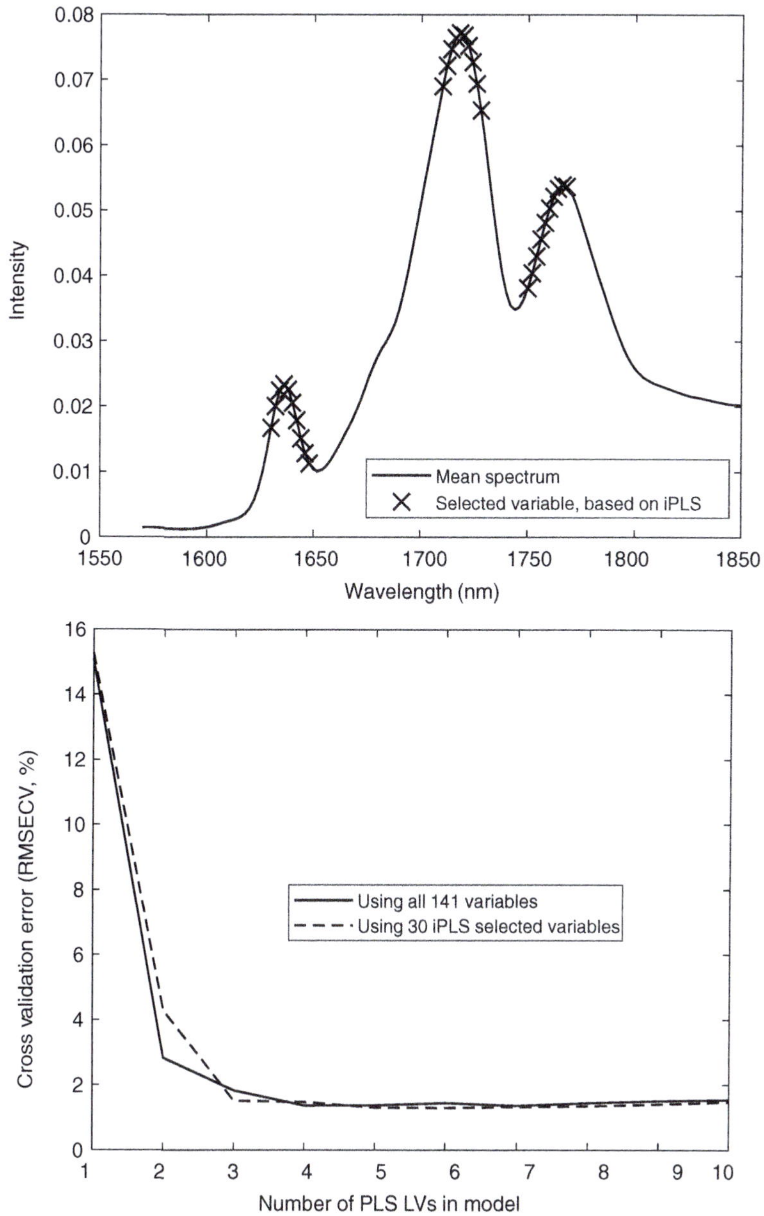

Figure 5.4 Result of applying interval PLS (i-PLS) in forward mode to the styrene butadiene dataset, where Y is the cis-butadiene concentration. Interval size is 10 variables and stoppage was based on lack of model improvement. 30 of the 141 variables were selected using this method. *Source:* Charles Miller.

Termination of the GA method results in the identification of one or more variable subsets, along with the "quality," typically the cross-validation error, associated with each subset. From this, one could arrive at the final variable subset in several ways: such as selecting the union or the intersection of the variables that appear in all of these subsets, or simply selecting the subset that generated the lowest cross-validation error. One could also use prior knowledge regarding the stability or reliability of certain variables to make the final

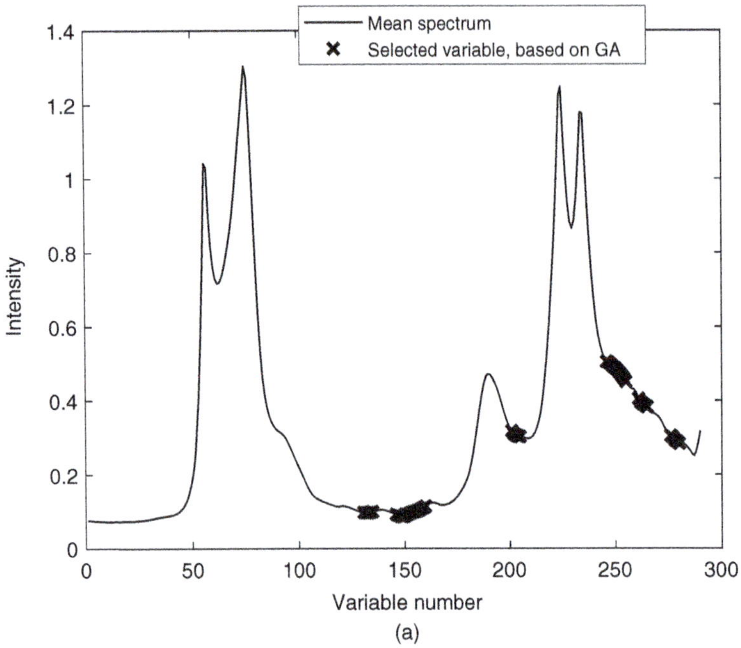

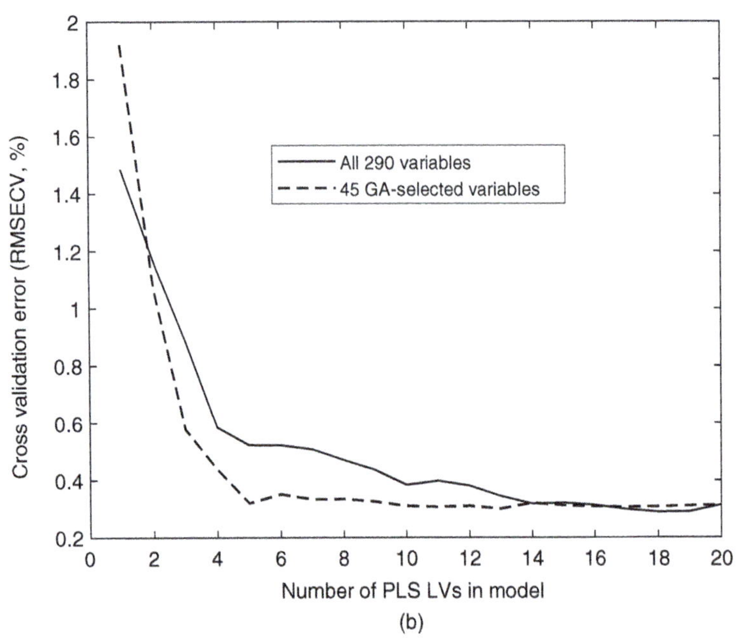

Figure 5.5 Results of GA-based variable selection for a PLS regression case discussed in [7, 11]. (a) GA-selected variables indicated by "x" labels. (b) Cross-validation error at different numbers of latent variables for the full set of 290 variables and the GA-selected subset of 45 variables. *Source*: Charles Miller.

selection. Finally, one could use GA to identify variables to *exclude* from the full X variable set, much like the "Reverse Mode" of the Interval-PLS method discussed earlier.

As the GA method starts with a randomly selected set of variable subsets, one will generate different results when applying it to the same X and Y data multiple times. As a result, it is often useful to run the GA method several times on the same data, to obtain a consensus on the selection of useful variables.

Although the GA method can do a more thorough search of the variable subset solution space than the other methods discussed above, it requires the specification of several parameters, thus requiring sufficient understanding of its operation. In some cases, the higher computational burden of GA can be partially offset through sample selection, blocking or averaging of adjacent variables (for spectroscopy data), or prior exclusion of irrelevant X variables based on subject matter expertise.

Figure 5.5 below shows the results of GA variable selection for a process analytical application that has been discussed in several references [7, 11]. The plot on the left shows the subset of 45 variables selected by the GA method, and the plot on the right shows the cross-validation error as a function of the number of latent variables for the two dataset cases: all 290 variables vs. the 45 GA-selected variables. Note that the GA-based selection in this case not only results in a lower cross-validation error, but it also results in a much less complex PLS regression model: requiring only 5 LVs instead of 10 or 13 LVs. The non-contiguous, "fractionated" nature of the selected variable subset (see Figure 5.5a) can be somewhat problematic when transferring such a model for online use, but this can be mitigated through the use of wider wavelength intervals in the GA, or through manual re-inclusion of "gap" variables into the GA solution, thus resulting in fewer non-contiguous regions.

5.6.7 Other Methods

The variable selection methods discussed above cover most of the methods that are easily accessible in most commercial model-building packages. However, there are other notable selection methods that are worth mentioning. One example, Interactive Variable Selection PLS (IVS-PLS) is a modification of the classical PLS algorithm that involves dimension-wise re-weighting of the PLS weights to reduce the impact of non-influential variables on a per-factor (per-latent variable) basis [18]. Another method utilizes the combination of GA selection with wavelet transform data compression [19].

5.7 Preprocessing and Variable Selection

Most methods that utilize analytical profile data, such as chromatography and spectroscopy, require preprocessing to reduce the impact of irrelevant variance in the analytical data. There are many different preprocessing methods that can be utilized for this purpose, many of which are summarized in Chapter 6. In cases where both preprocessing and variable selection are required, it is very important to understand the possible interactions between these operations. Specifically, many preprocessing methods are dependent on the variable range that is specified, in that different variable subsets on the same dataset will generate different preprocessing outcomes. Table 5.2 below lists several common preprocessing methods, and whether their outcomes are impacted by the variable subset.

There are several recommendations that can be made regarding cases where variable subset-dependent preprocessing methods are being used:

- Proactively exclude variables that are known to be of no value in the analysis (e.g., noisy variables, or variables with instrumental artifacts), as these can adversely affect the preprocessing.

Table 5.2 List of common preprocessing methods, and whether their outcome depends on the variable subset.

Preprocessing method	Does the preprocessing outcome depend on the variable subset?
Mean centering	No
Autoscaling	No
Derivatives (Savitsky Golay)	No
Normalization	Yes
Baseline correction	Yes
SNV	Yes
MSC	Yes
Extended multiplicative signal correction (EMSC)	Yes

- Consider whether preprocessing based on the entire variable range might produce better results than preprocessing based on a variable subset. For example, baseline correction methods can benefit from usage of variables that describe the spectral baseline (for example, regions of no spectral absorption bands), which are typically not deemed to be useful for prediction.
- For variable selection methods that involve empirical searching of the feature space (e.g., interval methods and GA), consider whether to apply preprocessing one time to the data before running the algorithm, or whether to apply preprocessing to each candidate variable subset that is evaluated within the algorithm.
- For all of the above, leverage your prior knowledge of the chemistry and spectroscopy of the application to guide your decision.

5.8 Conclusion

This chapter provided the background and rationale for variable selection methods in method development, along with the key attributes of these methods. Although most of the methods discussed are empirical in nature, it was shown that prior mechanistic knowledge of the chemistry and spectroscopy of the application can be leveraged to realize more efficient and optimal results. Improved access to computational resources has facilitated the use of more complex searching algorithms, but it is important to note that any changes in the sample set population, or changes in the system used to generate the sample data, can prompt the need to reassess the optimal variable subset. Finally, the potential for overfitting by using the same sample population for both variable selection and subsequent model evaluation was discussed.

References

1 Balabin, R.M. and Smirnov, S.V. (2011). Variable selection in near-infrared spectroscopy: benchmarking of feature selection methods on biodiesel data. *Analytica Chimica Acta* 692 (1–2): 63–72.

2 Anderssen, E., Dyrstad, K., Westad, F., and Martens, H. (2006). Reducing over-optimism in variable selection by cross-model validation. *Chemometrics and Intelligent Laboratory Systems* 84: 69–74.

3 Gidskehaug, L., Anderssen, E., and Alsberg, B.K. (2008). Cross model validation and optimisation of bilinear regression models. *Chemometrics and Intelligent Laboratory Systems* 93: 1–10.

4 Hocking, R.R. (1976). The analysis and selection of variables in linear regression. *Biometrics* 32: 1–49.

5 Draper, N. and Smith, H. (1981). *Applied Regression Analysis*, 2e. New York: Wiley.

6 Brown, P.J. (1991). Wavelength selection in multicomponent near-infrared calibration. *Journal of Chemometrics* 6: 151–161.

7 Miller, C.E. (1995). The use of chemometric techniques in process analytical method development and operation. *Chemometrics and Intelligent Laboratory Systems* 30 (1): 11–22.

8 Martens, H. and Næs, T. (1989). *Multivariate Calibration*. Chichester: Wiley.

9 Mardia, K.V., Kent, J.T., and Bibby, J.M. (1979). *Multivariate Analysis*. London: Academic Press.

10 Miller, C.E. (2018). The "how" of multivariate analysis (MVA) in the pharmaceutical industry: a holistic approach. In: *Multivariate Analysis in the Pharmaceutical Industry* (ed. A.P. Ferreira, J.C. Menezes, and M. Tobyn), 93–124. San Diego: Academic Press.

11 Miller, C.E. (2010). Chemometrics in process analytical technology (PAT). In: *Process Analytical Technology*, 2e (ed. K. Bakeev), 353–438. Chichester, UK: Wiley.

12 Galindo-Prieto, B., Trygg, J., and Geladi, P. (2017). A new approach for variable influence on projection (VIP) in O2PLS models. *Chemometrics and Intelligent Laboratory Systems* 160: 110–124.

13 Westad, F., Byström, M., and Martens, H. (1999). Modified Jack-knifing in multivariate regression for variable selection and model stability, NIR-99. *Proceedings, International Symposium on NIR Spectroscopy*, Verona, (13–18 June 1999).

14 Nergaard, L., Saudland, A., Wagner, J. et al. (2000). Interval partial least-squares regression (iPLS): a comparative chemometric study with an example from near-infrared spectroscopy. *Applied Spectroscopy* 54 (3): 413–419.

15 Leardi, R. (2001). Genetic algorithms in chemometrics and chemistry: a review. *Journal of Chemometrics* 15: 559–569.

16 Jouan-Rimbaud, D., Massart, D.-L., De Leardi, R., and Noord, O.E. (1995). Genetic algorithms as a tool for wavelength selection in multivariate calibration. *Analytical Chemistry* 67: 4295–4301.

17 Leardi, R. and Nørgaard, L. (2004). Sequential application of backward interval partial least squares and genetic algorithms for the selection of relevant spectral regions. *Journal of Chemometrics* 18: 486.

18 Lindgren, F., Geladi, P., Berglund, A. et al. (1995). Interactive variable selection (IVS) for PLS. Part II: chemical applications. *Journal of Chemometrics* 9: 331–342.

19 Depczynski, U., Jetter, K., Molt, K., and Niemoller, A. (1999). Quantitative analysis of near infrared spectra by wavelet coefficient regression using a genetic algorithm. *Chemometrics and Intelligent Laboratory Systems* 47: 179–187.

6

Preprocessing of Spectral Data

Alexis R. Weber

PerkinElmer, Albany, NY, USA

6.1 Introduction

In the field of analytical chemistry, vibrational spectroscopy is utilized to collect multivariate data, providing both qualitative and quantitative results for scientific inquiries [1–3]. The most common vibrational spectroscopy techniques are Raman, mid-infrared (Mid-IR), and near-infrared (NIR) spectroscopy. These methods have variants, such as surface-enhanced Raman spectroscopy (SERS) and attenuated total reflectance (ATR) Fourier transform infrared (FTIR) spectroscopy. These techniques produce spectra that can be analyzed using chemometric methods to create statistical models. However, before model creation, multivariate data typically needs to undergo preprocessing.

Chemometric preprocessing is a collective set of mathematical and statistical operations in the chemical data science pipeline aimed at reducing unwanted information and highlighting desired information without manipulating the true nature of the raw data. Preprocessing techniques play a crucial role in data analysis, specifically when the spectra are used for classification and calibration techniques [4, 5]. The primary objective is to meticulously prepare the data for analysis by employing various mathematical processes. This preparation is essential for enhancing the overall quality of the dataset and addressing inherent variations unrelated to the chemistry of interest. Data is organized and refined through preprocessing, paving the way for more accurate and meaningful analysis in subsequent stages.

In chemometric preprocessing, the focus is on preparing and enhancing data specifically for robust chemical analysis and interpretation. These techniques ensure the data used in models is optimized, standardized, and conducive to robust and reliable results [4, 6]. This involves using systematic and tailored techniques to address the unique challenges posed by chemical and physical variants in the spectral data, such as noise, baseline and peak variations, and interferences. While this chapter emphasizes preparing the spectral data for chemometric modeling methods, many of the same preprocessing techniques can be used for other quantitative approaches, such as linear regression and peak area analysis.

There are many types of mathematical challenges posed by artifacts in the spectra. These include baseline fluctuations, missing data, noise, variability on the Y-scale, random artifacts, and interferences. Preparing spectral data for modeling often requires using one or more mathematical procedures to normalize, filter, and correct physical artifacts. Techniques discussed in this chapter include normalization, baseline correction, smoothing, derivatives, standard normal variate (SNV), multiplicative scatter correction (MSC), and instrument-specific preprocessing tools like fluorescence removal functions for Raman spectra. This chapter provides several practical applications for each preprocessing technique and guidelines for combining multiple preprocessing steps for optimal data quality and robustness. While preprocessing techniques often

Practical Quantitative Vibrational and Electronic Spectroscopy: A Guide for Developing, Optimizing, and Validating Procedures, First Edition. Edited by Deborah A. Peru.

Companion website: www.wiley.com/go/peruelectronicspectroscopy

improve modeling outcomes, they also carry the risk of removing relevant chemical information, which can negatively affect the information gained in multivariate modeling [4, 7–9]. Therefore, it is essential to use preprocessing techniques cautiously and avoid misuse.

6.2 Artifacts in Spectra

There are many reasons why variations within spectral data exist that need to be corrected before quantitative methods can be developed. The most common variances are due to noise, baseline changes, missing data, variability on the Y-scale, random artifacts, and interferences [7]. Each artifact has its underlying cause(s), which, in the case of spectroscopy, can range from experimental error during measurements to complex interactions of light with the sample's physical structure. There are many reasons why artifacts can be present within spectra. While this chapter will not cover every possible case, the goal is to discuss the most frequently encountered artifacts during spectroscopic analysis.

6.2.1 Noise

Noise is the most common spectral artifact resulting from unwanted fluctuations in signal. Spectral noise can obscure important chemical information or interfere with accurate measurements of the desired features. This is because excessive noise makes it difficult to distinguish the true signal from the background and subtle peak shifts or intensity changes that would be otherwise detectable from a noise-free signal [7]. Many factors contribute to noise within a spectrum, including instrumental noise, which can be seen at the extreme values in Figure 6.1.

One source of instrumental noise is from the detector. Thermal noise is caused by the random motion of electrons in the detector material. Dark current noise is the signal generated by the detector in the absence

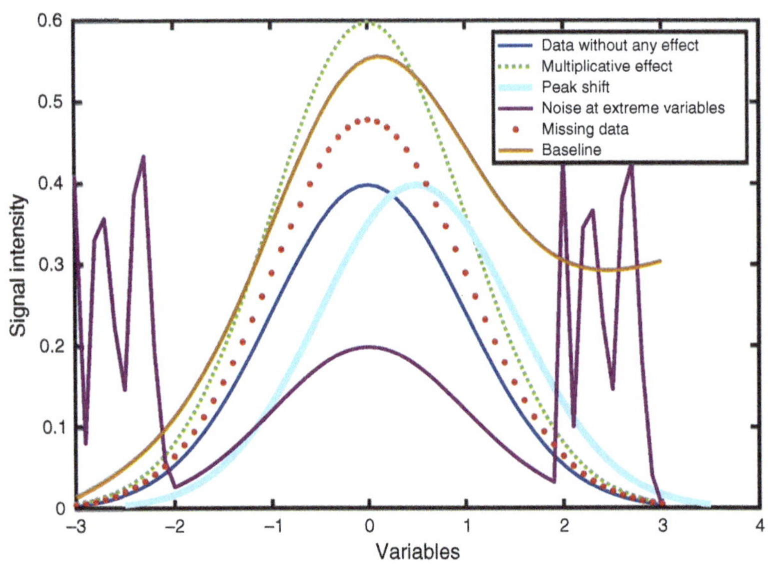

Figure 6.1 Example of several simulated artifacts for visible and near-infrared data. Data with no unwanted effect (solid blue line), multiplicative effects (dashed green line), peak shifts (solid cyan line), noise at extreme bands (solid purple line) and missing data (dotted red line), and baseline shift (solid brown line). *Source*: Reproduced from Mishra et al. [4] / with permission from Elsevier.

of light. Readout noise is introduced during the electronic readout process of the detector and often arises due to fluctuations in the readout electronics. Additionally, noise from the light source, such as fluctuations in the laser intensity or wavelength position, can introduce noise within the measurement [10, 11].

Another type of noise is environmental noise, which originates from ambient light and adds to the signal detected by the spectrometer. Sample-related noise includes fluorescence, where specific samples emit a broad background signal when excited by a light source, interfering with the detection of specific spectral features, and sample inhomogeneity, where variations in the sample's composition or physical structure result in inconsistent signals [11]. Minimizing noise in spectroscopy involves careful instrument design, proper calibration, and the use of advanced data processing techniques to filter out or correct for noise. All types of noise affect the signal-to-noise ratio (SNR) in chemical data. The SNR is the ratio of the strength of the signal to the strength of the noise [12]. A higher SNR indicates a better-quality signal, as the desired signal is stronger relative to the background noise, allowing for improved sample analysis. Techniques to improve the SNR include baseline correction, normalization, and smoothing, which reduces unwanted variations and artifacts in the data. Additionally, as discussed later in this chapter, data scaling methods may be applied to ensure that variables are comparable and contribute equally to the analysis, preventing certain variables from dominating the model.

6.2.2 Baseline Effects

Baseline changes are structured background effects characterized by continuous and low-frequency variations, with their occurrence generally dependent on the specific measurement technique used. Baseline effects can cause signals to exhibit a vertical offset or a linear or non-linear slope, as shown in Figure 6.1 [4]. Like noise, the background responsible for these baseline offsets or slopes varies depending on the analytical technique [7]. In Raman spectroscopy, fluorescence is a common cause of an elevated and sloping baseline. Biological fluids commonly possess an elevated fluorescence background and require baseline correction before chemometric analysis [13]. In NIR spectroscopy, when samples are measured in reflectance mode, the baseline can change due to stray light interacting with solid particles, powders, or droplets, leading to light scattering [4]. The multiplicative effect in NIR is caused by variations in absorptivity at longer wavelengths, leading to a sloping baseline. At these longer wavelengths, absorption is stronger, limiting light penetration to 1 mm or less. In contrast, light penetrates deeper at shorter wavelengths (closer to the visible region) due to the lower absorptivity. These factors influence the NIR baseline and should be corrected before developing quantitative methods [14]. Other factors that can cause the baseline to fluctuate include minor shifting of the sampling accessory, such as in ATR-FTIR spectroscopy, environmental changes in a laboratory, or sample preparation inconsistencies like particle-size effects or air bubbles. Typically, baseline correction involves estimating the baseline through a mathematical approach and subtracting this estimate from the measured spectrum.

6.2.3 Missing Data

Missing data refers to the absence of one or more variables within a spectrum that is expected to be present, as shown in Figure 6.1. This phenomenon can occur for several reasons, including the planned values falling outside the instrument range, instrument malfunction, communication failures between the instrument and the digital controller, the use of multiple sensors operating at different sampling rates, signal intensity saturation, and instrumental errors during data acquisition [4]. Despite the cause of these errors, the repetition of specific measurements may be too expensive, too complex, or impossible to repeat. Thus, mathematical techniques that can handle missing data, outliers, and feature selection will help improve chemometric models to focus on the more relevant chemical information [4, 8].

6.2.4 Variability of Spectral Y-Scale Axis

Variability on the Y-scale of spectra can arise from multiple phenomena, which are artifacts whose influence depends on the intensity of the original signal. In light-based techniques, these are called multiplicative effects and occur when light particles (photons) are forced to change direction due to irregularities in the sample, as demonstrated in Figure 6.1. These irregularities cause the photons to deviate from their straight path, affecting the measurement [4, 7]. In Mid-IR and Raman spectroscopy, this occurs when measuring samples that are not uniform at the point of collection. Changes in the thickness or smoothness of the sample cause the light to change direction, impacting the spectrum upon collection. Another cause for variation along the Y-scale occurs when spectra are collected from different instruments, such as two different FTIR spectrometers. This can happen when a laboratory changes its instrument but wants to compare older data to newly collected samples or when collaborating with another laboratory that uses a different instrument. The inconsistent Y-scale should be handled with instrument standardization procedures in such cases. Additionally, the infrared spectrum of a sample obtained using an ATR accessory differs from that obtained through transmission. The ATR technique causes relative shifts in band intensity and absolute shifts in frequency. The relative intensity shift is well known and can be easily corrected using an ATR-correction algorithm). However, the influence of the accessory used on the overall spectrum shape and band size is a variable that must be carefully controlled once a procedure is established.

6.2.5 Peak Shifting

Peak shifting is when signals that should ideally line up horizontally on (x-axis) do not align. This means the main bands, representing the maximum peak heights, are not in the same positions, as depicted in Figure 6.1. In spectroscopy, these peaks should appear at the same frequencies, and when they do not, are called spectral shifts, which relate to differences in frequency [4]. Causes of spectral shifts can be due to variations in temperature, pressure, viscosity, and pH of a sample [15]. In some applications, shifts in band or peak locations may be due to chemical changes within the data set that provide important information. In other applications, band shifting might occur due to instrumental reasons such as incorrect or infrequent calibration, mechanical drift, electronic drift, problems with fluctuations in the light source, and detector aging [7, 16]. To improve the procedure's robustness, x-axis and y-axis traceable standards must be implemented during instrument maintenance and qualification testing to correct and eliminate peak shifting over time.

6.2.6 Random Artifacts and Interferences

Random artifacts can arise from various sources, leading to noise and erroneous signals in the spectral data. Thermal noise, also known as Johnson–Nyquist noise, is caused by the random motion of electrons within the detector and electronic components due to thermal agitation and can be reduced by cooling the detector [15]. High-energy cosmic rays can interact with a charge-coupled device (CCD) detector, creating sharp spikes in the spectral data at random pixels, known as cosmic-ray interference, which can be identified and removed during data processing. Photon noise, or Poisson noise, arises from the statistical nature of photon emission and detection processes, affecting the SNR, especially at low light levels [15]. Environmental interferences, such as fluctuations in ambient light, vibrations, and acoustic noise, introduce variability into measurements and can be mitigated by shielding the instrument and conducting measurements in a controlled environment. These random artifacts and interferences can be minimized through proper instrument design, regular maintenance, calibration, and the use of advanced data processing techniques to filter out or correct for noise [17–19].

6.3 Preprocessing Techniques

Preprocessing tools enhance the quality of analytical spectra obtained from experimental samples used in statistical models. A spectrum is made up of hundreds to thousands of unique data points, which means that the computers and the computation algorithms used must be capable of processing large amounts of data are necessary. This type of analysis is referred to as "multidimensional statistical preprocessing." Preprocessing techniques can be row-based, performed on a single spectrum at a time, or column-based, where the mathematical correction is applied column-wise to each variable in the data set. The next few sections review common preprocessing methods applied both row-wise and column-wise.

6.4 The Role of Normalization in Quantitative Applications

Normalization, a fundamental data preprocessing technique, is widely used in various domains, such as statistics, machine learning, and chemometrics. Its role in addressing issues related to variable scales between spectra is particularly crucial in the realm of model building, as it removes baseline offset or *y*-axis variability between measurements.

In the intricate landscape of chemical data analysis, normalization serves as a vital tool to overcome challenges arising from the disparate scales and combining several chemical measurements with different magnitudes. In spectroscopy, normalization is applied when signal intensity exhibits substantial variations across different wavelengths or samples within the same class. The application of normalization ensures that each variable contributes proportionally to the analysis, irrespective of its original scale. This allows for proper quantitative classification of samples and is commonly used for NIR modeling of samples as a form of scatter correction to handle variations in particle size [3, 12]. Similarly, it is used in Mid-IR and Raman spectroscopy, where it is vital to normalize the spectra to account for factors like the thickness of the sample, particularly in biological tissue and bone analysis. Figure 6.2 shows the raw and normalized Mid-IR spectra of fatty acid blends collected with a 9-reflection ATR accessory. In Raman spectroscopy, normalization helps to correct baseline offsets caused by sample fluorescence [20, 21]. This enhances the interpretability of results and guards against the biased influence of a single measurement/spectrum solely due to its scale, which can cause skewed or biased results if not corrected adequately before modeling (chemometric or statistical).

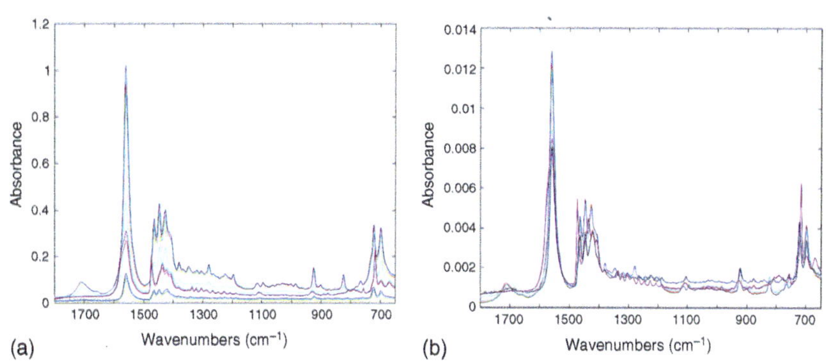

Figure 6.2 (a) RawMid-IR spectra of fatty acid blends collected with a 9-reflection ATR probe, 256 scans, 4 cm^{-1} resolution. (b) The corresponding normalized spectra. *Source*: Deborah A. Peru provided spectral data with permission. Alexis Weber.

These techniques help data analysts and model developers enhance important chemical patterns and relationships within the data set. They strengthen the foundation for informed decision-making in chemometrics and statistical modeling. Proper normalization techniques will ultimately affect the quality and reliability of future predictions made on unknown samples.

6.5 Generalized Mathematical Approaches for Normalization

Many algorithms are used for normalization, each tailored to suit specific data characteristics and analysis requirements. The commonly employed normalization techniques utilize total area, min–max scaling, or Z-score as the core mathematical function within the algorithm.

6.5.1 Normalization by Total Area

This normalization method is particularly relevant when dealing with data where the total sum or area under a curve is significant, such as vibrational spectral data. The goal is to ensure that each of the thousands of data points contributes proportionally to the overall sum. In chemical analysis, especially in techniques like spectroscopy, measuring signals, or absorbance at different wavelengths is common. The total area under the curve represents the cumulative intensity of these signals over the entire spectrum. Normalization by total area involves adjusting the values at each data point so that the sum of all values equals a predefined constant value, typically one [6]. The normalization process involves dividing each data point (intensity) by the total sum of all data points, ensuring that the resulting normalized values represent the proportion of the total area contributed by each data point. This is one of the most common forms of normalization used in spectroscopic preprocessing to overlay and compare samples with markedly different concentrations or scales [22, 23]. In quantitative analysis, normalization ensures that all samples contribute equally to the final model. Without normalization, some samples can have such severe multiplicative scaling effects that they will not be significant contributors to the variance and, as a result, will not be considered important by many multivariate techniques [24].

6.5.2 Normalization by Min–Max Scaling

This method transforms the data to a predefined range, often between 0 and 1. This method of normalization is commonly utilized for qualitative analysis, such as in clustering or classification models. By compressing the data within this standardized interval, min–max scaling ensures uniformity and facilitates fair comparisons between variables with distinct scales [25]. When using this on spectral data, it is common to choose a single point on your spectrum that is the maximum or minimum on your y-axis [15]. In some applications, the use of a reference band to correct for light-scattering differences or differences in sample thickness is a form of normalization often used in biological tissue applications [26].

6.5.3 Normalization by Z-Score Normalization

Z-score normalization, also known as standardization, adjusts the data set to have a mean of 0 and a standard deviation of 1. This differs from SNV as it is applied across the entire dataset or as a specific feature for general statistical standardization. On the other hand, SNV is applied to individual spectra to correct for scatter effects, with a focus on removing multiplicative noise. This method is particularly effective when the distribution of the data is approximately Gaussian, allowing for a consistent representation of the data across different scales [27].

6.6 Choosing a Normalization Method

Choosing the proper normalization method for chemometric analysis in vibrational spectroscopy depends on several factors, including the nature of the data, the specific application, and the desired outcomes. The nature of the data is crucial; for instance, if your data has significant variations in signal intensity due to undesired differences in sample concentration or pathlength, methods like vector normalization or area normalization can be effective. If your spectra are prone to baseline drift, SNV, and mean centering (MC) can help correct these issues, which will be discussed later. The specific application also plays a significant role. For quantitative analysis, where consistent peak areas are crucial, area normalization is often preferred [28, 29]. When comparing spectra from different samples or conditions, vector normalization ensures that the overall intensity is comparable across spectra.

In statistical preprocessing, there are two main ways that data can be affected. Row-wise (sample-wise) and column-wise (variable-wise) computations serve different purposes and impact the data in various ways. Row-wise preprocessing techniques are performed across each sample (row) independently, often affecting each sample as a whole [30]. Row-wise computations can alter the intrinsic relationships within each sample. For instance, normalizing each row can make the overall magnitude of variables irrelevant, focusing instead on the relative values within each observation. In contrast, column-wise preprocessing of the data is applied to each variable (column) across all samples and affects the data set as a whole [4]. This standardizes spectral features across the samples, allowing for a more unbiased data comparison. This is especially useful when some spectra have slightly different Y-scale intensity values, such as when data is collected with the same instrument on different days. In general, row-wise preprocessing methods focus on individual samples and can alter relationships within the sample, often enhancing the interpretability of internal patterns. While column-wise methods standardize features across samples, ensuring consistent scaling and facilitating the application of statistical models [31].

The desired outcome may guide the choice of normalization method. If your data contains significant outliers, robust normalization methods like SNV can help minimize their impact. This is because normalization techniques like SNV affect your data along rows, and therefore, the presence of outliers would not affect the entire set of preprocessed data. These considerations help select the most appropriate normalization technique to ensure accurate and reliable chemometric analysis. There are several ways to determine if you have applied the correct preprocessing method(s). One of the most common is the trial-and-error approach to select the best method or combination of methods based on performance metrics such as standard error of calibration (SEC), standard error of prediction (SEP), and even SNR can be used. An advanced approach would involve the use of an automated algorithm to determine the optimal combination of preprocessing steps [7]. The primary goal for preprocessing is to ensure that the resulting output still correctly represents the original data.

6.7 Standard Normal Variate

Standard normal variate (SNV) is a prominent preprocessing technique in the field of spectroscopy to correct for scattering effects and is commonly applied when analyzing NIR or Mid-IR spectroscopic data [3, 12]. Its primary objective is to elevate spectral data quality by effectively addressing multiplicative interferences, including baseline variations and intensity differences, in one operation. SNV provides these benefits without significantly distorting the spectral shape or compromising information related to the chemical composition of samples [12]. This technique is considered "set independent" as it does not rely on an ideal reference spectrum, unlike MSC, which will be discussed in the next section. It operates independently on each spectrum by subtracting the mean of the spectrum and then dividing by its standard deviation [32]. This approach is efficient for removing constant baseline effects and scaling differences from spectra, provided that the original scale of the spectra is not of interest and allows the remaining spectra to be comparable [2, 3]. This method is similar to z-score normalization but is applied row-wise instead of column-wise.

The benefits of using SNV include (1) correcting for baseline variations, (2) adding robustness in intensity variations due to measurement error, and (3) preserving spectral information. These benefits are discussed further below:

1) **Baseline Correction:** SNV excels at effectively removing baseline variations and offsets. This is particularly valuable in scenarios where the baseline's shape interferes with accurately identifying spectral features. By standardizing the data, SNV helps achieve more consistent and reliable results across different samples.
2) **Maintains Robustness:** SNV effectively normalizes spectral intensity levels and provides robustness to variations in overall intensity differences across different samples. This ensures that spectral patterns are not unduly influenced by differences in signal strength caused by measurement variation, thus reducing the multiplicative effects and enhancing the changes due to variations in chemistry [33].
3) **Preservation of Spectral Information:** Unlike other preprocessing techniques, SNV is designed to correct multiplicative effects while preserving the essential spectral information relevant to the chemical composition of samples. Therefore, it improves the performance of multivariate statistical techniques such as principal component analysis (PCA), principal component regression (PCR), andpartial least squares (PLS) regression by reducing the influence of scatter and other unwanted variations [34].

6.7.1 The Mathematics of Standard Normal Variate

An SNV preprocessing step is conducted independently on each spectrum and involves three steps:

1) **Calculate Mean and Standard Deviation:** Calculate the mean intensity of all wavelengths and standard deviation within each spectrum.
2) **Standardize the Spectrum:** Subtract the mean intensity from each variable across the entire spectrum.
3) **Divide by the Standard Deviation:** Divide each mean-subtracted intensity by the standard deviation to complete the normalization. The result is an adjusted spectrum with a mean intensity of zero and a standard deviation of one.

Standard normal variate is often sensitive to the chosen wavelength range for preprocessing. Ideally, the SNV calculations should be performed over a larger wavelength region than the calibration range. If the software provides an option for selecting the SNV endpoints, the left and right endpoint wavelengths should cover a wider range of wavelengths than the calibration region. In addition, the endpoints should ideally fall on a flat region of the spectrum to ensure reproducibility while capturing the range of desired artifacts to remove. Determining the correct endpoints may take some trial and error to determine the ideal range. Other modeling software packages don't allow users to set the SNV range and utilize the same region used for calibration development.

6.7.2 Application of Standard Normal Variate in Multivariate Analysis

Standard normal variate is frequently employed as a preprocessing step before utilizing multivariate analysis techniques, such as PCA or PLS. This application enhances the performance of these methods performed on spectroscopic data. SNV has been used to improve the capabilities of food composition analysis and by being paired with hyperspectral imaging to identify amounts of talcum [35, 36]. SNV is extensively used in spectroscopy, especially in NIR, as depicted in Figure 6.3 [37]. Commonly, SNV is used after calculating the

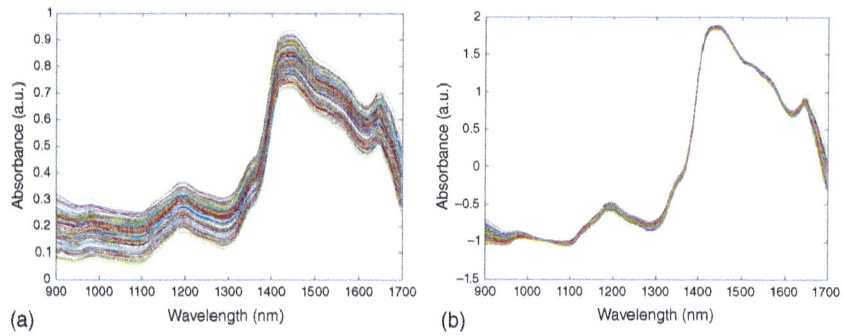

Figure 6.3 Near-infrared (NIR) raw spectra and standard normalized variable (SNV) pretreated spectra of whole pine nuts, (a) the raw spectra, (b) the SNV pretreated spectra. *Source*: Li et al. [37] / MDPI / CC BY 4.0.

derivative of the spectrum, but different combinations can be evaluated and optimized based on performance statistics [12].

The use of SNV has many practical applications when used properly. Consider a large dataset of Mid-IR spectra acquired from different samples over time. The use of SNV before modeling ensures that baseline variations, often caused by instrumental artifacts, are effectively corrected [1, 3]. SNV normalizes the spectra, allowing for robust comparisons when intensity levels vary between samples. Under these conditions, no additional normalization methods would be required [38].

An SNV's ability to correct multiplicative interferences and its emphasis on preserving essential spectral information help ensure accurate and meaningful analyses across various analytical techniques and applications. However, it is always appropriate to approach any preprocessing method with caution. While SNV is effective for many spectral data types, it may not always be the best choice for data with strong baseline variations or significant noise. Other preprocessing methods like MSC or other baseline correction techniques might be more appropriate. Likewise, SNV is not valid when baseline variations are helpful for the model, such as when particle-size differences are of interest.

6.8 Multiplicative Scatter Correction

Multiplicative scatter correction (MSC) is one of the cornerstones in chemometric preprocessing, tailored explicitly for vibrational spectroscopic data. Its primary objective is the correction of unwanted variations arising from light scatter effects, which often stem from physical phenomena like particle-size variations, inhomogeneity, or alterations in sample presentation. This preprocessing method is ideal for measuring absorption data collected in reflectance mode. MSC overcomes scattering effects that can cause multiplicative artifacts in the spectra, ensuring that the variations that are present in the spectral data reflect chemical changes and not physical inconsistencies between samples [33]. Thus, it allows for a better comparison between spectra collected from different samples, resulting in more reliable spectral data necessary for chemometric modeling.

6.8.1 The Fundamental Concept of Multiplicative Scatter Correction

Multiplicative scatter correction works by applying a correction factor to each spectrum that fully describes the reflectance behavior of a stable "ideal" sample unaffected by particle-scattering phenomenon. The "ideal" spectrum is often the mean spectrum, but an external reference spectrum may be more appropriate depending on the data. This reference spectrum serves as a baseline representation of the scatter presented in the dataset. To create this reference spectrum, an average spectrum is calculated from the calibration/training dataset [39]. Next, linear regression is applied between the sample and the "ideal or average" spectrum. This produces a slope and an intercept used for correction for each sample spectrum [40]. The final step is to create the MSC corrected spectrum; this is done by subtracting the intercept from the spectrum and dividing the wavelength by slope. Because the reference spectrum is created by averaging the spectra in a dataset, this technique is considered "set dependent" and must be calculated from scratch for each unique dataset [14]. Therefore, this technique may be computationally intensive for larger data sets and may require much processing power and time. There is an additional assumption that the "ideal" spectrum applies to all future spectra measured.

6.8.2 Normalizing Spectra Before Multiplicative Scatter Correction

Since this technique is set-dependent, it may be necessary to normalize the individual spectra in the dataset, such as by dividing each spectrum by a reference spectrum before applying MSC. Normalizing in this manner (often referred to as y-axis normalization) can help stabilize the variance across spectra, especially when there are significant variations in intensity. Providing a more consistent starting point can make the MSC process more effective [14]. However, if normalization significantly changes the baseline or slope, it could interfere with the scatter correction process that MSC is designed to handle.

Extended-MSC (EMSC) can enhance the performance by incorporating more complex models (e.g., polynomial fitting) to correct for both multiplicative and additive effects in the spectra [41]. In cases where simple filtering methods modify the spectral baseline or slope, E-MSC might be more appropriate as it can handle such variations more flexibly, mitigating potential complications arising from pre-normalization.

6.8.3 Important Advantages of Multiplicative Scatter Correction

There are two main advantages of using MSC. First, MSC can reduce unwanted scatter effects without compromising the true chemical variations inherent in the data. This preservation is vital for obtaining accurate and reliable chemical information from spectral datasets [14, 42]. The second advantage is it enhances the performance of multivariate analysis techniques, such as PCA or PLS, by providing a more accurate representation of the underlying chemical structure [14].

An MSC is a sophisticated and powerful technique to preprocess Raman and NIR spectra. In Raman spectroscopy, MSC is often used to remove sloped baselines frequently caused by the underlying fluorescence of matrix components, as shown in Figure 6.4 [43]. The application of MSC with different chemometric methods for the identification and quantification of different food constituents in samples has been widely used [44, 45].

6.9 Baseline Correction

Baseline correction is a fundamental preprocessing step in spectroscopic data analysis aimed at removing baseline drift and background noise from the spectra. In analytical measurements, the baseline signifies the signal level observed when no analyte or target substance is present, serving as a reference point against which changes due to the presence of the analyte are measured [46]. However, various factors often compromise this pristine baseline, leading to fluctuations that can obscure the true analytical signal. Both matrix components and instrumental variations, such as detector noise, drift, and temperature fluctuations, can all cause baseline changes.

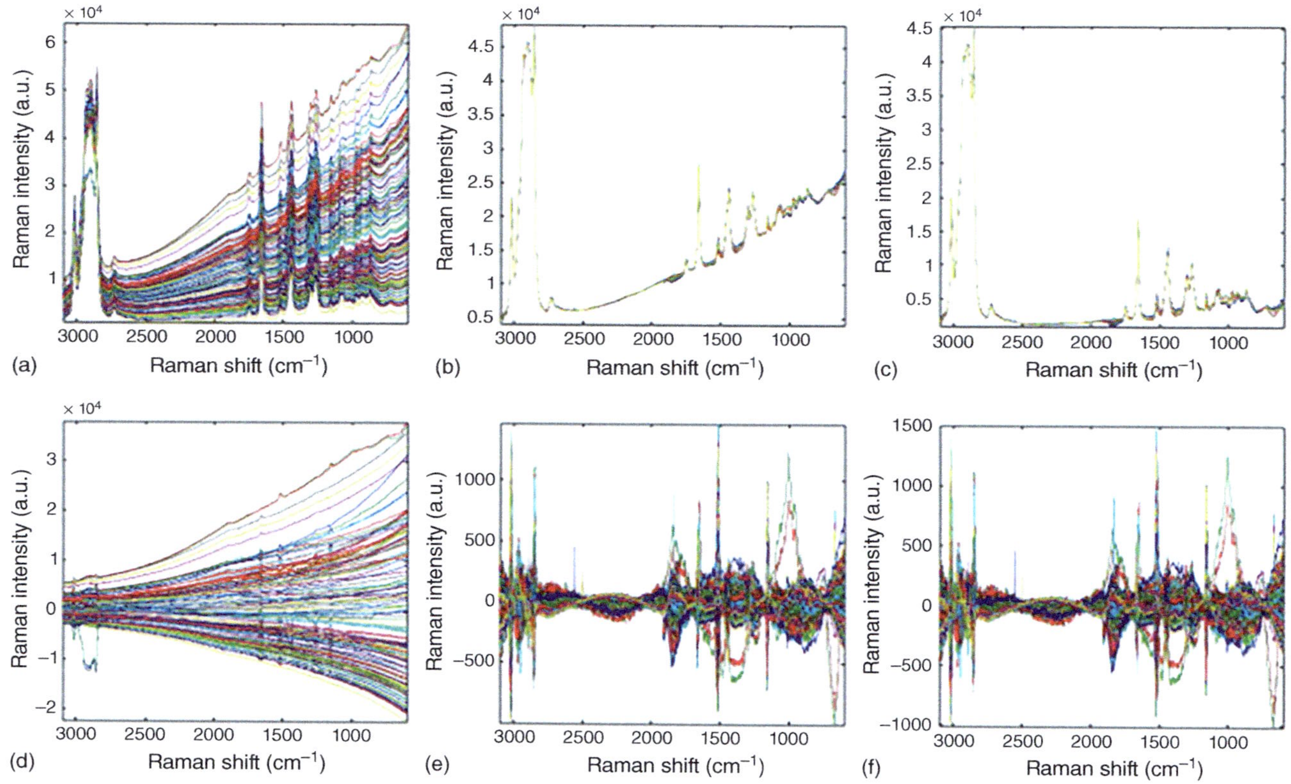

Figure 6.4 Raw Raman spectra of salmon oils (a), and corresponding spectra subjected to basic EMSC (b) and basic EMSC using a reference spectrum with a reduced baseline slope (c). Corresponding mean-centered data sets are provided in the lower rows (d–f), respectively. *Source*: Reproduced from Afseth and Kohler [43] / with permission from Elsevier.

These baseline variations can significantly impact the SNR, making it difficult to discern the presence of the analyte. A lower SNR indicates higher noise levels relative to the signal, resulting in a noisier baseline that can obscure genuine features within the data [4, 7]. Baseline correction techniques designed to improve the stability and accuracy of the baseline are crucial for reliable data interpretation, better SNR, and accurate representation of genuine features within the data.

The main objective of a baseline correction is to apply a filter that removes unwanted variations caused by instrumental and sample-related factors, such as varying sample thickness and fluorescent noise in Raman spectroscopy [13, 46].

6.9.1 Methods of Baseline Correction

A plethora of methods exist for baseline correction, each catering to the specific characteristics of the data. Various software packages provide tools for baseline correction, including MATLAB, R, Python (SciPy, NumPy), and specialized spectroscopy software like OPUS, OMNIC, and OriginLab. Some standard methods for baseline correction are described here.

1) **Single-Point Correction:** This straightforward technique selects a single point in the spectrum, typically a region free of analyte peaks. The value of the spectrum at this reference point is then subtracted from all other points, shifting the entire spectrum vertically so that the baseline at the reference point is set to zero. The main advantages of single-point baseline correction are its simplicity and efficiency, making it a quick solution that requires minimal computational resources and can be applied rapidly to large datasets. However, it is most effective for spectra with a constant baseline offset [12]. The method has limitations when dealing with spectra that have complex or sloped baseline variations or drifts, as it assumes a uniform baseline shift. Additionally, the correction accuracy depends heavily on correctly identifying a true baseline point [46].
2) **Multipoint Correction:** Multipoint baseline correction is an advanced technique used in spectroscopic data analysis to remove baseline variations and enhance the clarity of spectral features. Unlike single-point baseline correction, multipoint baseline correction uses multiple points across the spectrum to define and subtract the baseline [15, 47]. The process begins with selecting multiple reference points from the spectrum, ideally in regions free of analyte peaks, which are assumed to represent the true baseline. A model of the baseline is then constructed using these points through methods such as polynomial fitting or asymmetric least squares (ALS), described below. This modeled baseline is subtracted from the original spectrum, resulting in a corrected spectrum with minimized baseline variations [25, 48]. Multipoint baseline correction is advantageous due to its flexibility in handling complex baseline variations and its improved accuracy from using multiple reference points. These methods can work on sloped or variable baselines, making them suitable for most spectroscopic techniques, including Mid-IR and Raman spectroscopy.
3) **Polynomial Fitting:** In this method, a mathematical function, typically a polynomial, is fitted to the baseline region of the data. Subsequently, this function is subtracted from the entire spectrum or chromatogram, nullifying baseline drift or offset. Different orders of polynomials are used depending on the shape of the baseline. Polynomial fitting is effective when a polynomial function can adequately approximate the baseline, as demonstrated in the analysis of narcotics in solid mixtures [49]. Figure 6.5 shows the effect of polynomial baseline correction of biological samples measured with Raman spectroscopy. The analyst must choose the correct polynomial function to avoid overcorrecting the baseline, which can be seen when the spectral data points appear below 0 on the y-axis.

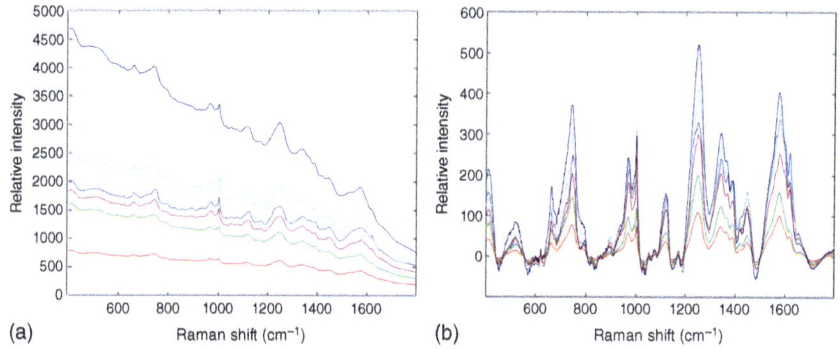

Figure 6.5 The Raman spectra of blood samples. (a) Before pre-processing. (b) After 5th-order polynomial baseline correction. *Source*: Alexis Weber.

4) **Asymmetric Least Squares (ALS):** ALS is an iterative algorithm that minimizes the differences between the observed data and the fitted baseline while allowing for asymmetry [7]. ALS automatically identifies and corrects baseline variations by optimizing a cost function that considers both the fitting error and a penalty term for asymmetry. It is robust and suitable for correcting baselines with asymmetric features.

Other methods used for baseline correction that are not as common include wavelet transformation, top-hat transformation, iterative curve fitting, and morphological operations [50]. The choice of a specific baseline correction method depends on the characteristics of the data, the nature of the baseline distortions, and the desired outcome of the analysis. Analysts often experiment with different techniques to determine the most effective approach for a particular dataset. Mitigating unwanted variations in the baseline sets the stage for a more precise and meaningful interpretation of analytical measurements, especially when identifying or quantifying analytes of low concentrations within a complex matrix [3]. As technology advances and analytical methodologies evolve, the mastery of baseline correction remains a critical skill for analysts and researchers alike, contributing significantly to the robustness and accuracy of analytical processes.

6.10 Spectral Derivatives

Derivatives, specifically the first and second derivatives, represent versatile tools in chemometric preprocessing. They offer a nuanced approach to analyzing spectral data derived from techniques such as NIR, Mid-IR, and Raman spectroscopy and non-vibrational spectroscopic techniques. This method removes subtle variations in the baseline and slope. It plays a dual role in extracting additional information from spectra while enhancing features that might be obscured by noise or overlapping signals [2, 48]. The first and second derivatives remove first-order baseline effects and create a zero crossing Y-axis, though this process differs from the previously discussed baseline correction methods [51]. The fundamental concepts of derivatives and their applications in chemometric analysis are discussed below.

Choosing to use the first or second derivative depends upon the unique characteristics of the spectral data and the specific objectives of the analysis. While derivatives effectively enhance spectral features, a cautious approach is imperative to avoid introducing noise and artifacts or removing information during smoothing, often applied when derivatives are used. Striking a balance between noise reduction and preserving crucial information is paramount. Therefore, it is necessary to tailor your preprocessing methods to the intricacies of the specific data for a reliable and insightful chemometric analysis.

Savitsky Golay derivatives are by far the most universally applied derivatives in spectral preprocessing. However, the Norris gap derivative (named after Karl Norris, a pioneer in NIR) is still available in many NIR spectral software packages. It is equally applicable when processing NIR spectral data sets. In this type of derivative, the chosen segment is calculated as the difference between two data points, while the gap size, usually set to 0, is the distance between the segments. These two points are computed as averages over the segment for a smoothing effect. For example, if the segment is set to 10, five absorbances to the right of the gap are averaged and five absorbances to the left of the gap are averaged, and these points are subtracted to yield the smoothed derivative point at the gap location.

6.10.1 First Derivative Preprocessing

The first derivative, a fundamental mathematical concept, measures the slope of the curve at each spectral point. The resulting spectra have the spectral range along the X-axis vs. the slope of the curve on the Y-axis. This technique is primarily used in preprocessing NIR, Mid-IR, and Raman spectra. This derivative serves as a dynamic lens, emphasizing variations and edges within the spectrum, making it invaluable for discerning peaks and resolving overlapping or not well-resolved bands [3]. Applying the first derivative can effectively remove first-order baseline drifts and offsets, as using the derivative tends to cancel out constant or slowly varying background signals. This results in a spectrum that is more focused on the actual peaks and less affected by baseline variations [3, 4]. Post derivative preprocessing, the spectrum will have emphasized inflection points corresponding to the peak positions.

There are many situations when first derivatives become the preprocessing technique of choice for post-analysis. One application is for “Peak Detection.” In this scenario, when spectral peaks are broad or when there is significant baseline drift, the first derivative becomes a reliable tool for detecting peaks with enhanced sensitivity [12]. Another example is in “Edge Detection.” The derivative’s capacity to accentuate changes in intensity aids in the identification of edges and abrupt transitions within the spectrum.

In Raman [52] and Mid-IR [48] spectroscopy, the first derivative can be used to resolve overlapping peaks in the fingerprint region, where many molecular vibrations occur [53]. For instance, applying the first derivative to a Raman spectrum of a biological sample can help distinguish between closely spaced peaks corresponding to different biochemical components, such as proteins, lipids, and nucleic acids.

As with all preprocessing methods, it is important to be aware of the techniques’ limitations. Derivatives can amplify high-frequency noise present in the data, which can sometimes necessitate prior smoothing of the data. Additionally, while the first derivative narrows the band of interest and enhances the identification of peak positions, it does not retain the original intensity or band position information, which can be a limitation in the interpretation of certain quantitative applications.

6.10.2 Second Derivative Preprocessing

Building upon the foundation discussed for the first derivative, the second derivative provides a deeper understanding of changes in the slope of the spectrum. This application is incredibly useful for resolving closely spaced peaks and refining subtle spectral features [54]. Compared to the first derivative, the second derivative will produce peaks that are sharper and more defined and are more capable of removing broad and complex/variable background signals [15]. In NIR spectroscopy, the second derivative is almost always used to remove second-order or multiplicative effects that occur in diffuse reflectance analysis of solids. The multiplicative scattering effect causes a broad-sloping baseline to occur near the combination band region that is easily corrected. The second derivative is more commonly used when trying to determine the components in

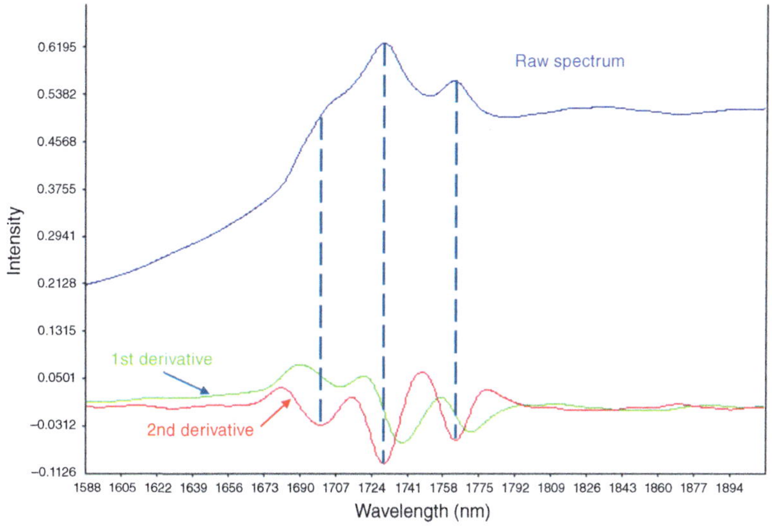

Figure 6.6 Comparison of the original raw NIR spectrum and after first derivative and second derivative preprocessing. *Source*: Courtesy of Deborah A. Peru, Editor with Permission.

a complex matrix [2, 55]. The interpretative of the second derivative is slightly easier since the on-peak band position in the original spectrum is 180 inverted in the negative direction. Figure 6.6 shows a comparison between NIR spectrum of animal feed after being processed with the first and second derivatives [56].

6.10.3 Benefits of Using Derivatized Spectra

The two most important reasons for using derivatives in spectral preprocessing is the ability to improve the resolution of nearby peaks and for baseline correction. The second derivative distinguishes individual components and enhances peak resolution in complex spectra with closely spaced peaks at the price of enhancing spectral noise. For baseline correction, the second derivative effectively mitigates the impact of baseline drift and sloped baselines and begins the deconvolution process of resolving overlapping bands, thus contributing to a more precise representation of the spectral features [57].

6.10.4 Integration with Other Preprocessing Techniques

Derivatives are a powerful function that often finds synergy with other preprocessing techniques, amplifying their impact on spectral data quality:

1) **Smoothing:** When coupled with smoothing techniques, derivatives contribute to an enhanced SNR. Smoothing aids in reducing noise, allowing for a more precise representation of spectral features when chemometric analysis methods are used [58]. Figure 6.7 shows the integration of smoothing with the gap derivative. Figure 6.7b uses 1-segment smoothing that does not remove the additional noise introduced during derivatization. Figure 6.7c uses a 10-segment smoothing function, considered optimal for this data type and resolution. Figure 6.7d illustrates the results using a 20-segment gap derivative, resulting in an over-smoothed spectrum with many spectral features removed.
2) **Scatter Correction:** Combining scatter correction with derivatives is essential for improving spectral data quality. Scatter correction reduces variations caused by physical sample properties, while derivatives enhance the chemical information by highlighting peaks and transitions, making the overall analysis more specific, robust, and accurate [12].

3) **Normalization:** After normalization, derivatives (first or second) are applied to enhance the differences between spectra and remove baseline shifts. Derivatives emphasize the peaks and transitions, making the features of interest more prominent [12].

6.11 Smoothing Functions

Smoothing functions enhance the interpretability of complex datasets by improving the SNR and removing unwanted fluctuations. This technique finds relevance in chemometrics as researchers grapple with intricate spectroscopic data. These datasets often harbor inherent noise, challenging the extraction of meaningful information [7]. Smoothing enhances the performance of derivatives by removing added noise. It also provides additional protection and improves the resiliency from long-term shifting in the x-wavelengths, thus improving the robustness of the chemometric method [5].

Achieving a delicate balance when applying smoothing methods is essential. Excessive smoothing can remove critical information and introduce artifacts, which negates the critical work done during the preprocessing stages [15]. In cases where analytes are present at low concentrations, resulting in small peaks,

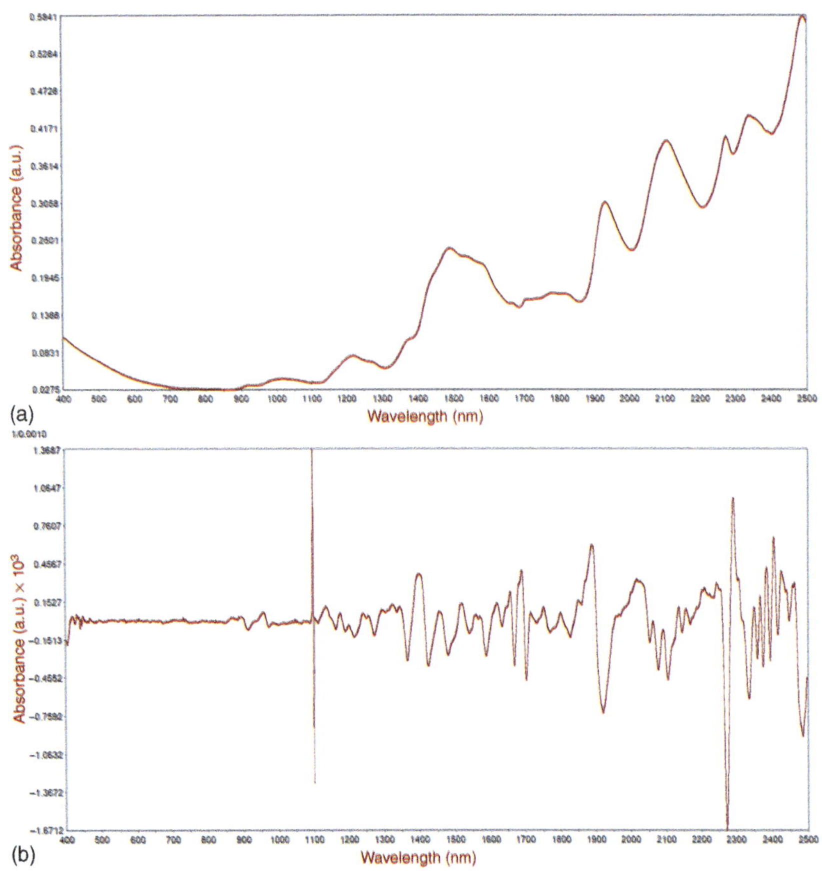

Figure 6.7 Raw spectrum (a), gap second derivative 1 segment smoothing (b), gap second derivative with 10 segment smoothing (c), gap second derivative 20 segment smoothing (d). *Source*: Courtesy of Deborah A. Peru, Editor with Permission.

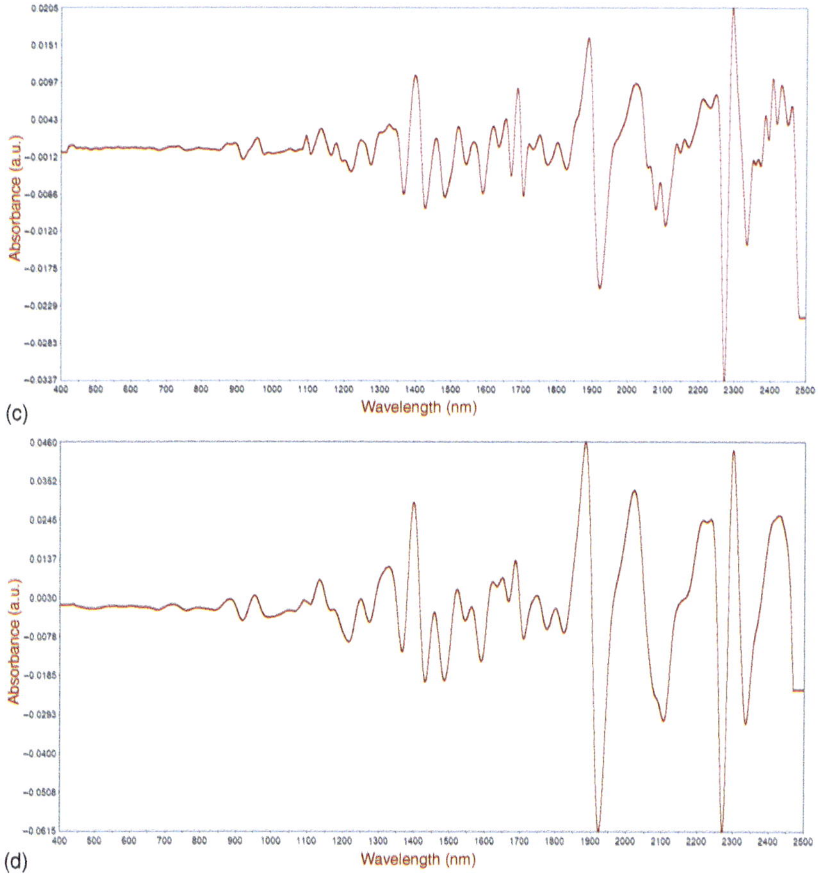

Figure 6.7 (*Continued*)

overuse of smoothing techniques can mask this data, causing a loss of crucial information. Thus, the selection of smoothing methods and parameters requires careful consideration based on the data's characteristics and the specific analysis objectives [3].

6.11.1 Common Smoothing Methods

The smoothing process typically involves the application of mathematical functions or algorithms that suppress high-frequency noise while preserving broader features of interest. Several smoothing methods are employed in vibrational spectroscopy data analysis, each chosen based on the data's characteristics and analysis goals:

1) **Moving Average Smoothing:** A window or kernel traverses the individual data points, averaging values within a user-selected range to generate a smoothed point [59]. It then calculates the average intensity within a moving window along the spectrum, reducing high-frequency noise while preserving the overall shape of spectral bands [60]. This method is particularly suitable for data with relatively low noise levels and where preserving the shape of peaks is essential. Its simplicity and effectiveness make it ideal for minor noise reduction. However, if the window size is too large, it can lead to peak broadening and the loss of fine spectral details, which can diminish the accuracy of the spectral analysis.

2) **Savitzky–Golay Smoothing:** This technique fits successive subsets of the data to a low-degree polynomial using linear least squares [61]. The central point of each subset is then replaced with the value of the polynomial at that point. This method is widely used in spectroscopy due to its ability to preserve peak shapes while effectively reducing noise. It maintains the integrity of sharp spectral features better than simple moving average smoothing, making it highly advantageous for detailed spectral analysis [62]. However, the technique requires careful selection of the polynomial order and window size to avoid over-smoothing, which can otherwise diminish the quality of the spectral data.
3) **Gaussian Smoothing:** Gaussian smoothing is a technique that applies a Gaussian function to weigh data points within a specified window, giving more importance to points closer to the center of the window. This method effectively reduces noise while preserving the general shape of spectral features, making it particularly suitable for datasets characterized by Gaussian noise [63]. The technique works by convolving the data with a Gaussian kernel, where the width of the kernel (standard deviation) determines the extent of smoothing [64]. Gaussian smoothing is widely used in image processing, signal processing, and various spectroscopic applications due to its ability to enhance the SNR without significantly distorting the underlying data. However, the computational demand is higher compared to simpler methods like moving average smoothing, due to the need for convolution operations.

6.11.2 Generalized Application of Data Smoothing

In chemometrics, smoothing is applied to diverse data types, including spectra and chromatograms. In chromatography, it aids in baseline correction, contributing to the precision of peak identification. In spectroscopy, smoothing enhances the SNR, facilitating peak identification and spectral feature analysis when applied correctly [61]. Understanding the complexities of smoothing in chemometrics and spectroscopy is essential for researchers and analysts seeking reliable insights from complex datasets.

Knowing when data has been smoothed too much is a delicate balance for maintaining data quality and relevant spectral information while eliminating noise. Over-smoothing can be identified by the loss of sharp features, where significant peaks or valleys in the spectrum become flattened or broadened, leading to the loss of critical information. The distortion of spectral shapes makes it challenging to identify and quantify features accurately, another sign that the data is over-smoothed. Likewise, a reduced SNR, where the signal's amplitude is diminished more than the noise, can indicate over-smoothing [48]. Visual inspection by plotting smoothed data alongside raw data can reveal excessive smoothing if there are significant deviations at known feature points. Furthermore, if analytical results from smoothed data are inconsistent or unexpected, the data may have been overly smoothed.

Several methods can assess the degree of smoothing. Cross-validation techniques can compare the performance of models built with smoothed vs. unsmoothed data, with poorer performance indicating over-smoothing. Residual analysis, which examines differences between raw and smoothed data, can highlight significant data loss at essential features. Carefully tuning the parameters of the smoothing algorithm, such as window size and polynomial order, and evaluating their impact on the data can help find an optimal balance. Finally, using synthetic data or reference spectra with known characteristics to test the smoothing technique in a controlled environment can also help understand when over-smoothing occurs. By monitoring these indicators and methods, one can apply the right amount of smoothing to retain essential spectral features while reducing noise, ensuring high-quality data for analysis.

6.12 Mean Centering

Mean centering (MC) is a popular data preprocessing technique that finds applications across diverse fields, including chemometrics and statistics. This fundamental step involves shifting the values of each variable in a dataset independently to ensure that the mean (average) of each variable becomes zero [65]. The primary objective is to eliminate the overall offset or baseline from the data, thereby facilitating a more focused examination of variations around the mean and improving the interpretation of the spectra. Mean centering is almost always applied to the training/calibration dataset when creating PLS or PCR models [5].

Without delving into the mathematical equations, mean centering operates by independently adjusting the values of each variable in a dataset [66]. The adjustment is performed in such a way that the average value of each variable is centered at zero. It is important to note that mean centering is a technique that is performed on both the [X] and [Y] blocks of data. The calculation involves scaling the data around zero by calculating the average spectrum and average concentration for the training set and then subtracting the average from each sample spectrum and concentration [65]. Post mean centering, the differences between spectra are enhanced, both in terms of concentration and spectral response [67].

6.12.1 Practical Benefits of Mean Centering

In practical terms, consider a dataset with multiple variables, each representing different features or measurements. This approach is particularly advantageous in scenarios where the absolute magnitudes of variables are less critical than their relative variations. While the resulting spectra may seem more visually complex, the use of mean centering does allow for better chemometric analysis when creating statistical models. Mean centering is essential for PCA, as it ensures that the principal components are oriented along directions of maximum variance relative to the mean [5]. On the other hand, with PLS, mean centering helps model the relationship between predictor and response variables by removing the overall mean. Figure 6.8 shows an example of mean-centered spectra collected with a UV–Vis instrument.

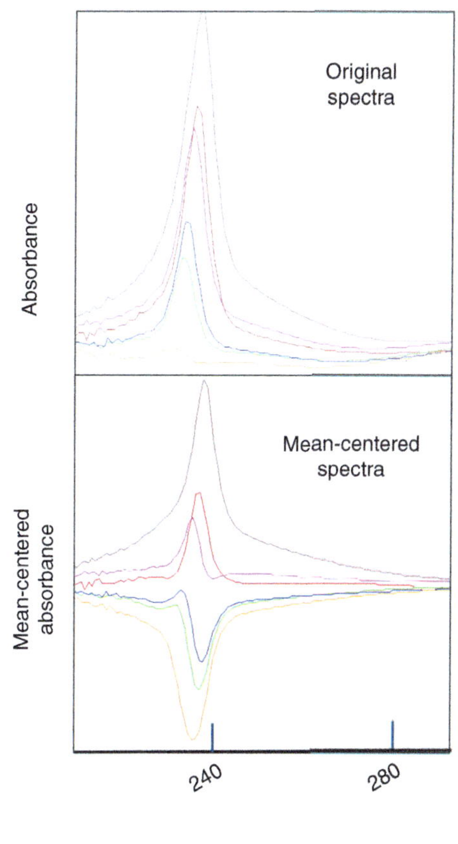

Figure 6.8 Comparison of ultraviolet spectroscopy spectral data set before and after mean centering applied. *Source*: Courtesy of Deborah A. Peru, Editor with Permission.

One key benefit of MC is that it removes the overall offset or shift of peaks present in the data. This shift occurs when the individual variables or features in a dataset have different mean values, which can dominate the variation in the data [68]. Without mean centering, these mean differences can obscure the true relationships between variables and introduce bias into the statistical analysis. By subtracting the mean of each variable, mean centering eliminates this irrelevant offset, allowing the underlying patterns and variations to emerge. This step simplifies data interpretation and prevents the overall shift from affecting multivariate techniques, which rely on understanding relationships between variables rather than their absolute values.

Secondly, MC will enhance the interpretability of the data set. The technique enhances the interpretability of multivariate analyses [68]. By focusing on relative variations, mean centering simplifies the interpretation of analysis results. This allows analysts to identify small differences between samples more easily and makes interpreting loadings and scores plots straighter forward when trying to identify minute differences between samples [69].

6.13 Autoscaling

Autoscaling, also called feature scaling or standardization in some fields, is a preprocessing technique widely employed in chemometrics and machine learning. The fundamental aim of autoscaling is to establish comparable scales for all variables, features, or measurements within a dataset [66]. Autoscaling is used when data from multiple sources with different Y scales are combined for modeling, a process known as data fusion. Autoscaling is commonly used in process applications when several instruments and devices are used in line for product and process optimization.

Mathematically, autoscaling is a column-wise scaling method that transforms variables to possess a mean of zero (a.k.a. mean-centering) and a unit variance [66]. Autoscaling does not address individual baseline variations and thus is often combined with other processing tools. For example, SNV is a row-wise preprocessing technique that would first be applied to take care of baseline effects, followed by autoscaling to scale the Y variables.

The math involved with autoscaling data involves two basic steps:

Step 1: Determine the mean and standard deviation for each band intensity across the selected spectral range and column-wise across the entire [X] data set. For quantitative applications, the mean and standard deviation are also determined for each dependent variable in the [Y] matrix.

Step 2: Transform each variable in [X] and [Y] by subtracting the mean and then dividing the mean-centered variable by the standard deviation.

6.13.1 Benefits of Autoscaling Data

Autoscaling offers a significant advantage, mainly when processing data from multiple sources or datasets with variables on different scales. This normalization process is essential when dealing with variables with vastly different ranges on the Y-scale, as it prevents variables with larger magnitudes from dominating the analysis.

Autoscaling is typically applied before employing multivariate modeling techniques such as PCA, PLS, or regression when equal weighting of all variables is beneficial. Autoscaling ensures variables with disparate scales are equally weighted. Without autoscaling, variables with larger numerical ranges may skew the results, leading to biased models that favor these variables. Autoscaling enhances model performance and the expression of more meaningful patterns from complex datasets [70]. This is particularly crucial in algorithms using data from multiple devices, distance measures, and optimization techniques, where variable scale can impact results [68]. Autoscaling facilitates faster convergence when optimization algorithms are used on multi-source data sets.

In quantitative regression modeling applications, autoscaling creates a consistent scale for regression coefficients and principal component loadings, thus enhancing their interpretability. This ensures that the importance of each variable is comprehensible and comparable.

6.13.2 Applications of Autoscaling

In practical terms, autoscaling is most often applied when datasets contain univariate and multivariable input variables ranging from discrete temperature and pressure values to spectra, each measured in different units. Variables with larger scales may disproportionately influence the analysis, potentially masking the true relationships within the data [71].

Autoscaling can be applied when spectral data is collected over long periods of time in the same laboratory or when spectral data sets are collected with different instruments. In Raman spectroscopy, the intensity of Raman bands may be consistent between spectra collected on the same instrument. However, when a new laser is installed or data is collected on another instrument, there may be differences in the Y-scale intensity. A PCA model might interpret these two data sets differently without autoscaling due to the inconsistency of the Y-axis (intensity scale).

The application of autoscaling addresses challenges posed by varying scales among variables, promoting fairness in comparisons and preventing undue influence on analysis outcomes. As an integral part of the preprocessing toolbox, autoscaling lays the foundation for robust and meaningful modeling of diverse variables in many fields of science and statistics.

6.14 Cosmic-Ray Removal in Raman Spectra

Cosmic rays are high-energy particles, primarily protons and atomic nuclei, originating from outside the Earth's atmosphere. The impact of cosmic rays on spectroscopy is particularly noticeable in high-sensitivity instruments and measurements over extended periods. This is encountered mainly in Raman spectroscopy using integrated circuit detectors such as a CCD. When it comes to spectroscopy, cosmic rays can sometimes interfere with measurements and result in unwanted variable information such as sharp spikes in the collected spectra [72]. These artifacts are often seen as intermittent spikes or peaks in the spectrum that introduce noise and can distort preprocessing tasks and chemometric modeling of the data.

The energy from cosmic rays can ionize molecules or cause electronic excitation, creating spurious signals in the detector, as shown in Figure 6.9 [48]. Since these events are random and unrelated to the analyzed sample, they introduce variability and inaccuracies into the recorded spectra [73].

Researchers typically employ various techniques to mitigate the effects of cosmic rays in spectroscopy. These may include visual inspection, data interpolation, smoothing, threshold designation, or specialized algorithms designed to detect and remove cosmic-ray artifacts [72–74]. Specialized software can replace the cosmic-ray intensity value using an interpolation method to estimate the baseline intensity value at the spike position based on the data surrounding the area. Another approach is to remove cosmic rays manually and to replace the cosmic-ray intensity with the average or median values of nearby non-affected points [48].

Careful experimental procedures, including minimizing the exposure time during data acquisition or incorporating shielding hardware, can help reduce the impact of cosmic rays on spectral measurements. It's essential to be aware of the potential presence of cosmic-ray artifacts in spectroscopic data and to take appropriate steps to address them during data analysis and interpretation. However, removing cosmic rays should be done using a context-specific approach to not alter the data present in the spectrum. There is no

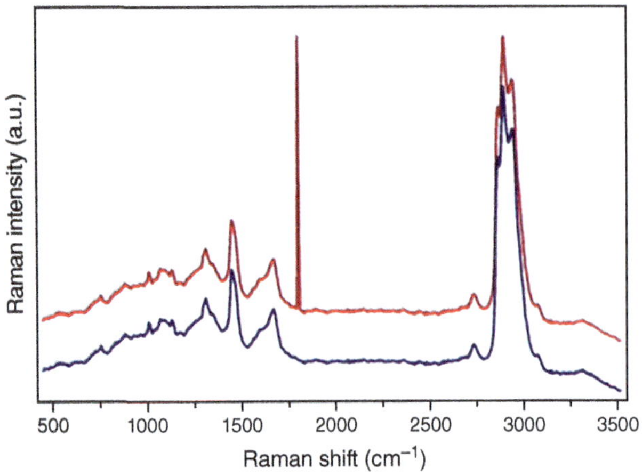

Figure 6.9 Example of cosmic-ray correction of Raman spectra. Red trace – Raman spectrum obtained from the cerebellum of a Syrian hamster showing a strong cosmic-ray feature around 1795 cm^{-1}. Blue: the corrected Raman spectrum (spectra are shifted along the *y*-axis for clarity). *Source*: Reproduced from Lasch [48] / with permission from Elsevier.

optimal way to remove cosmic rays. The best approach is based on the specific characteristics of the instrument. Adopt an interactive process. Once the first cosmic band is removed, assess the quality of the spectrum. If necessary, the removal algorithm can be refined to improve the quality of the resulting spectrum.

6.15 Spectral Filtering

Filtering is a general term used to describe several preprocessing techniques commonly used in various fields, including signal processing and spectroscopy, to enhance the quality of data by selectively modifying or removing certain frequencies or components [75, 76]. The goal is to improve the SNR, highlight relevant features, and address unwanted variations in the data. Preprocessing typically uses several important filtering methods, including low-pass filtering, high-pass filtering, band-pass filtering, and smoothing. The following sections explain each approach.

6.15.1 Low-Pass Filtering

Low-pass filters are designed to allow low-frequency components to pass through while decreasing high-frequency noise. These filters are particularly useful for smoothing spectra and reducing high-frequency noise that may arise from electronic sources or environmental interferences. For instance, when dealing with spectra that have sharp noise spikes, applying a low-pass filter helps to smooth out these spikes while retaining the broader spectral features, thereby improving the overall quality of the spectral data and is commonly used when working with Mid-IR and Raman data [48]. Baseline variations, such as drift or offsets in the intensity, can be treated as low-frequency components. Applying a low-pass filter selectively removes these baseline variations, allowing for a clearer view of the spectral features [11].

6.15.2 High-Pass Filtering

Unlike low-pass filters, high-pass filters are designed to remove low-frequency components, such as baseline drifts, while retaining high-frequency details. These filters are effective for enhancing sharp spectral features and eliminating baseline drifts. For example, in Raman spectroscopy, where baseline drift or noise obscures important spectral details, applying a high-pass filter helps to isolate these details by removing the slowly varying baseline, thereby improving the clarity and accuracy of the spectral analysis [48]. The derivatives used in spectral preprocessing are another example of high-band-pass filtering.

6.15.3 Band-Pass Filtering

Band-pass filters are combinations of high- and low-pass filters designed to allow frequencies within a specific range to pass through while decreasing frequencies outside this range, making them ideal for isolating particular spectral features of interest. For example, when analyzing a specific vibrational mode that falls within a known frequency range, a band-pass filter can be used to focus on this range and remove irrelevant spectral components, thereby enhancing the analysis of the desired spectral features [48]. A good example of a band-pass filter is the Fourier self-deconvolution filter, which can enhance resolution and reduce noise simultaneously.

6.15.4 Savitzky–Golay Filtering

Savitzky–Golay filtering uses a polynomial smoothing function to fit successive subsets of data points with a low-degree polynomial. It is beneficial for preserving the shape of spectral peaks while reducing noise. For example, in Mid-IR spectroscopy, where it is crucial to maintain peak shapes for accurate analysis, the Savitzky–Golay filter can smooth the data while retaining the integrity of the peaks, ensuring that critical spectral features are not distorted [48].

The choice of filter type and parameters depends on the quality and type of data and the objectives of the analysis. It's important to carefully assess the impact of filtering on the data, considering potential trade-offs between noise reduction and preservation of relevant information. Additionally, transparent reporting of the filtering methods used is crucial for reproducibility and interpretation of results.

6.16 Fluorescence Removal Functions for Raman Spectroscopy

Fluorescence removal functions are preprocessing techniques designed to mitigate or eliminate the impact of fluorescence interference in Raman spectra. In Raman spectroscopy, fluorescence signals can overlap with or mask the informative bands, making it challenging to analyze the underlying chemical information accurately. Fluorescence removal functions aim to enhance the accuracy of the analysis by addressing this interference. Commonly used fluorescence removal functions include baseline correction, spectral subtraction, MSC, and fluorescence modeling and removal algorithms. The choice of fluorescence removal function depends on the nature of the fluorescence interference, the available reference information, and the specific characteristics of the data. It's important to carefully validate the effectiveness by reviewing the impact on the accuracy and reliability of subsequent analyses. Transparent reporting of the fluorescence removal techniques is essential for reproducibility and interpretation of spectroscopic results.

6.16.1 Baseline Correction

Baseline correction is a crucial preprocessing technique in spectroscopy, particularly when addressing fluorescence interference. Fluorescence signals often create a broad, sloping background in the spectrum that can obscure important chemical information. By effectively modeling and subtracting this baseline, baseline correction allows for the recovery of the pure spectral signal and enhances the reliability of spectral interpretation and more accurate downstream quantitative analysis.

Weber et al. highlight that several mathematical approaches can be used to model the baseline, with polynomial fitting being one of the most widely adopted methods. In this technique, a polynomial of appropriate degree is fitted to the non-informative portion of the spectrum (i.e., regions without peaks), modeling the baseline as a continuous, smooth curve. This polynomial is then subtracted from the original spectrum to correct for the fluorescence background. Higher-degree polynomials can be used to account for more complex baselines, though caution must be taken to avoid overfitting, which could distort the actual spectral features [13, 69, 77].

6.16.2 Spectral Subtraction

Spectral subtraction involves acquiring a reference spectrum of the fluorescence alone and subtracting it from the measured spectrum. This helps remove the fluorescence component, leaving the desired signals. However, acquiring or creating a reliable reference spectrum of fluorescence can be challenging. Two procedures are often used. Polynomial fitting (PolyFit) and modified multi-polynomial fitting (ModPoly) are effective approaches for modeling the fluorescence background. The background can subsequently be subtracted from the measured spectrum in an automated fashion, improving the clarity and accuracy of the Raman signal in unknown samples [46, 78, 79]. The method is designed to be robust and applicable across a range of biological samples, and the paper demonstrates its effectiveness through various case studies.

6.16.3 Multiplicative Scatter Correction

Multiplicative scatter correction (MSC) is a technique that aims to correct for both baseline variations and multiplicative effects, including fluorescence. It involves normalizing the data by dividing each spectrum by a reference spectrum. While MSC is not explicitly designed for fluorescence removal, it can be effective in some instances [52].

6.16.4 Fluorescence Modeling and Removal Algorithms

Some specialized algorithms are designed explicitly for fluorescence removal. These algorithms model the fluorescence component and attempt to separate it from the measured spectra [80]. Examples include polynomial-fitting-based methods and algorithms based on mathematical or statistical fluorescence models.

6.17 Combining Preprocessing Tools

Combining preprocessing techniques is common in data analysis, particularly when vibrational spectroscopy is paired with chemometrics, which is prevalent with high-dimensional and complex datasets. Different preprocessing methods address specific issues in the data, and the order in which they are applied can influence the effectiveness of each technique and the overall outcome. The choice and sequence of preprocessing steps depend on the data's specific characteristics and the analysis goals.

Identifying the best method for a specific data set will take time and should not be rushed when working with spectra [5]. A generalized seven-step guideline to follow when combining preprocessing techniques is described:

Step 1: Understand the Data: Begin by understanding the main features and characteristics of your data. Consider factors such as baseline variations, noise, outliers, and unwanted artifacts. Visualize the raw data to identify patterns and challenges that may need to be addressed in preprocessing.

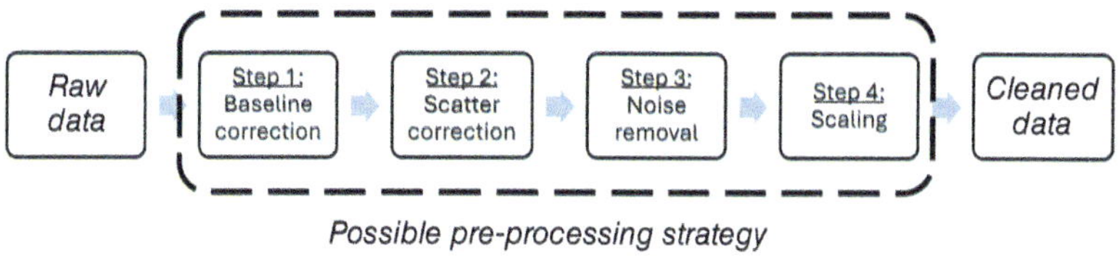

Figure 6.10 A complete pre-processing strategy for IR data. *Source*: Reproduced from Engel et al. [7] / with permission from Elsevier.

Step 2: Define Objectives: Clearly define your objectives and the challenges you aim to address through preprocessing. Common goals include removing noise, correcting baseline variations, handling outliers, and improving the SNR.

Step 3: Optimized the Sequential Order of Operation: Preprocessing techniques are often applied sequentially. Start with fundamental techniques and gradually progress to more advanced methods [7, 48]. One example of a preprocessing strategy is shown in Figure 6.10, depicting the standard steps for NIR and Mid-IR samples [7]. This proposed method also works with Raman spectroscopy data, as all these data types possess similar artifacts, baseline changes, scattering effects, and noise [7]. The order and type of preprocessing steps can be modified based on the data set and goals. The next section will discuss further details about the order of operations.

Step 4: Evaluate Intermediate Results: Assess the impact of each preprocessing step on the data. Visualization and statistical metrics (such as standard deviation) can help you understand how each technique influences the data's characteristics; basic chemometric methods, such as PCA scores plots, can assist in this process.

Step 5: Iterative Process: Preprocessing is often an iterative process. After applying initial techniques, it's common to reassess the data and consider additional steps based on the observed effects. Iterative preprocessing helps balance noise reduction and preservation of relevant information [4]. Document the preprocessing tools used, the order, and the combination of techniques during development, and run mathematical experiments to determine their effect on SNR and the statistical outcome of modeling and prediction.

Step 6: Validate Preprocessing Choices: Assess the impact of preprocessing on the performance of your chosen analysis or model. Use validation techniques, such as cross-validation, to ensure that preprocessing enhances the robustness, generalizability, and predictive performance of results.

Step 7: Document and Report: Document the preprocessing steps applied to your data. Transparent reporting ensures the reproducibility of your analysis and allows others to understand the decisions made during data preprocessing.

6.18 Commonly Applied Sequences (Order of Operations)

The unique order and combination of preprocessing techniques will vary based on the type of spectroscopy and the objectives of the analysis. The order of operations in preprocessing matters significantly, as the sequence of operations can affect the quality of the data and the results from subsequent analysis. In many instances, the order of operation is customized to achieve optimal results. When sequencing preprocessing steps, it is generally accepted to first perform operations that involve changes in the data row-wise (e.g., derivatives) before doing column-wise preprocessing steps (e.g., mean centering) [66]. Figure 6.11 provides a generalized order of operations when using multiple preprocessing techniques.

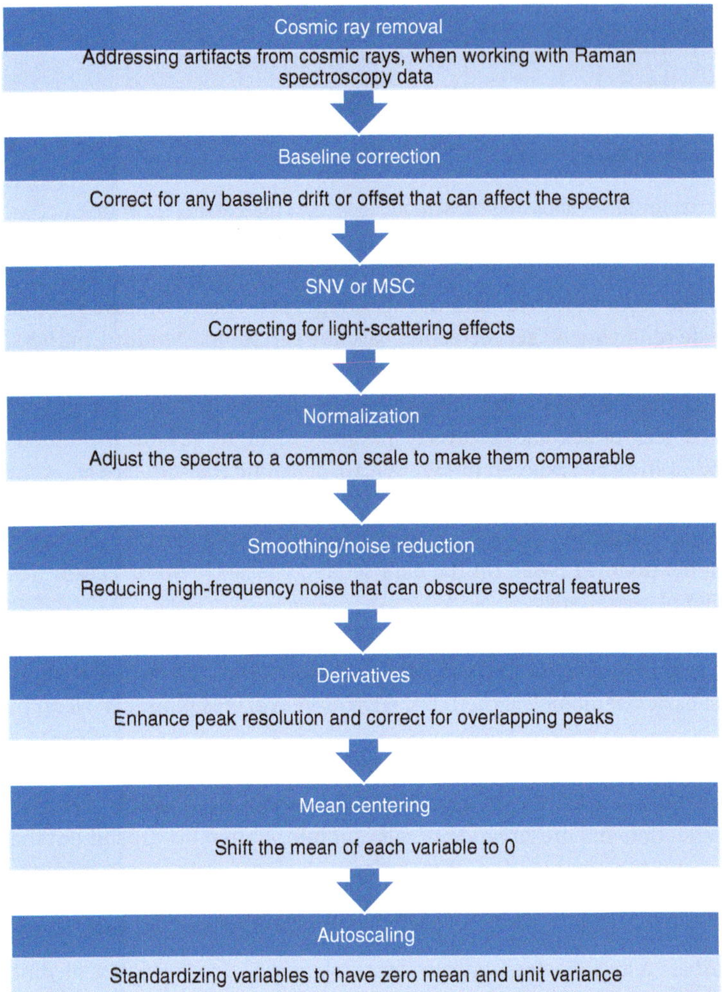

Figure 6.11 Recommended order of applying multiple combinations of preprocessing steps. *Source*: Alexis Weber.

6.18.1 Baseline Correction Before Scaling

Baseline correction is commonly performed early in the preprocessing pipeline because the operation can alter the overall scale of the data. Baseline correction before scaling will prevent the baseline correction process from affecting the mean and variance of the data.

6.18.2 Smoothing After Baseline Correction

Smoothing to reduce noise in the data is often performed after baseline correction. This helps preserve the integrity of the spectral features while minimizing the impact of baseline variations [81].

6.18.3 Derivatives After Smoothing

Derivative calculations are often applied after smoothing. Smoothing helps to reduce high-frequency noise that might be amplified during the derivative calculation [82].

6.18.4 Standard Normal Variate Before Derivative

Apply SNV before calculating the derivative ensures that the spectra are normalized, making the subsequent derivative calculation more meaningful and less affected by variations in the original intensity. Derivatives, especially the second derivative, emphasize high-frequency noise and small variations. If SNV is applied after the derivative, the noise and variations could be amplified, leading to less effective normalization [83]. However, the order may also depend on the use of smoothing with derivatives. Thus, the effect of combining these three preprocessing tools should be scrutinized during cross-validation testing and external validation set testing. In NIR applications, SNV after smoothing and derivative is often a more effective preprocessing order of operation, yielding better SEC and SEPV results.

6.18.5 Mean Centering Before Autoscaling

Mean centering is a finishing operation and is generally the last operation in a series of operations. It involves subtracting the mean from each variable to ensure that each variable is centered around zero. Scaling, such as autoscaling, can then be applied to achieve unit variance, allowing for a fair comparison of variable contributions.

6.19 Conclusion

The goal of preprocessing is to remove unwanted spectral variations not related to chemistry before modeling begins. The preprocessing tools selected for an application should be chosen carefully and with knowledge of consequences because they can potentially remove or distort the signal of interest or add other unwanted variances. It is essential to carefully consider the impact of each preprocessing step on the data and results and to validate the chosen sequence through visualization and statistical metrics. Transparent reporting of the preprocessing steps is crucial for reproducibility and the interpretation of results.

When combining several techniques, the order becomes a critical factor. Yet, once optimized, the right combination of preprocessing techniques will significantly enhance method robustness and often reduce model complexity.

Acknowledgments

I want to thank Deborah A. Peru, the editor of this book, for her helpful guidance in developing this chapter. I appreciate her support throughout the process and for providing several data sets and figures as examples of preprocessing techniques.

References

1 Barra, I., Haefele, S., Sakrabani, R. et al. (2021). Soil spectroscopy with the use of chemometrics, machine learning and pre-processing techniques in soil diagnosis: recent advances–a review. *TrAC Trends in Analytical Chemistry* 135: 116166.

2 Liland, K.H., Kohler, A., and Afseth, N.K. (2016). Model-based pre-processing in Raman spectroscopy of biological samples. *Journal of Raman Spectroscopy* 47 (6): 643–650.

3 Rinnan, Å. (2014). Pre-processing in vibrational spectroscopy–when, why and how. *Analytical Methods* 6 (18): 7124–7129.

4 Mishra, P., Biancolillo, J., Roger, M. et al. (2020). New data preprocessing trends based on ensemble of multiple preprocessing techniques. *TrAC Trends in Analytical Chemistry* 132: 116045.

5 Wise, B.M. and Roginski, R.T. (2015). A calibration model maintenance roadmap. *IFAC-PapersOnLine* 48 (8): 260–265.

6 Rácz, A., Bajusz, D., and Héberger, K. (2018). Chemometrics in analytical chemistry. In: *Applied Chemoinformatics: Achievements and Future Opportunities* (ed. T. Engel and J. Gasteiger), 471–499. Wiley.

7 Engel, J., Gerretzen, J., Szymans, E. et al. (2013). Breaking with trends in pre-processing? *TrAC Trends in Analytical Chemistry* 50: 96–106.

8 Verboven, S., Hubert, M., and Goos, P. (2012). Robust preprocessing and model selection for spectral data. *Journal of Chemometrics* 26 (6): 282–289.

9 Gerretzen, J., Szymanska, E., Jansen, J. et al. (2015). Simple and effective way for data preprocessing selection based on design of experiments. *Analytical Chemistry* 87 (24): 12096–12103.

10 Gautam, R., Vanga, S., Ariese, F. et al. (2015). Review of multidimensional data processing approaches for Raman and infrared spectroscopy. *EPJ Techniques and Instrumentation* 2 (1): 8.

11 Morais, C.L., Lima, K.M., Singh, M. et al. (2020). Tutorial: multivariate classification for vibrational spectroscopy in biological samples. *Nature Protocols* 15 (7): 2143–2162.

12 Rinnan, Å., Van Den Berg, F., and Engelsen, S.B. (2009). Review of the most common pre-processing techniques for near-infrared spectra. *TrAC Trends in Analytical Chemistry* 28 (10): 1201–1222.

13 Barber, A.P., Weber, A.R., and Lednev, I.K. (2024). Raman spectroscopy to determine the time since deposition of heated bloodstains. *Forensic Chemistry* 37: 100549.

14 Martens, H. and Stark, E. (1991). Extended multiplicative signal correction and spectral interference subtraction: new preprocessing methods for near infrared spectroscopy. *Journal of Pharmaceutical and Biomedical Analysis* 9 (8): 625–635.

15 Skoog, D.A., Holler, F.J., and Crouch, S.R. (2019). Textbook "principles of instrumental analysis." Cengage learning. https://core.ac.uk/download/pdf/232277508.pdf.

16 Jellema, R.H., Folch-Fortuny, A., and Hendriks, M.M.W.B. (2020). 3.05 – Variable shift and alignment☆. In: *Comprehensive Chemometrics*, 2e (ed. S. Brown, R. Tauler, and B. Walczak), 115–136. Oxford: Elsevier.

17 Baker, M.J., Trevisan, J., Bassan, P. et al. (2014). Using Fourier transform IR spectroscopy to analyze biological materials. *Nature Protocols* 9 (8): 1771–1791. https://doi.org/10.1038/nprot.2014.110.

18 Zancla, A., De Santis, S., Scorza, A., et al. (2020). A preliminary study on the importance of normalization methods in infrared micro-spectroscopy for biomedical applications. Proceedings of the 22nd International Workshop on ADC and DAC Modelling and Testing, Palermo, Italy.

19 Kwak, C., Choung, D., Min, S. et al. (2007). Fast determination of the ripeness stage of strawberries using infrared spectroscopy combined with principal component analysis. *Analytical Sciences* 23: 895–899.

20 Beattie, J.R., Glenn, J., Boulton, M. et al. (2009). Effect of signal intensity normalization on the multivariate analysis of spectral data in complex 'real-world' datasets. *Journal of Raman Spectroscopy* 40 (4): 429–435.

21 Kistenev, Y.V., Borisov, A.V., Samarinova, A.A. et al. (2023). A novel Raman spectroscopic method for detecting traces of blood on an interfering substrate. *Scientific Reports* 13 (1): 5384.

22 Pacheco-Londoño, L.C., Ortiz-Rivera, W., Primera-Pedrozo, O.M. et al. (2009). Vibrational spectroscopy standoff detection of explosives. *Analytical and Bioanalytical Chemistry* 395: 323–335.

23 Ralbovsky, N.M., Halamkova, L., Wall, K. et al. (2019). Screening for Alzheimer's disease using saliva: a new approach based on machine learning and raman hyperspectroscopy. *Journal of Alzheimer's Disease* 71: 1351–1359.

24 Research, E. (2021). Advanced preprocessing: sample normalization. https://wiki.eigenvector.com/index.php?title=Advanced_Preprocessing:_Sample_Normalization (cited 2024).

25 Lasch, P. and Petrich, W. (2010). Data acquisition and analysis in biomedical vibrational spectroscopy. *RSC Analytical Spectroscopy Series*, 192–225.

26 Lloret, P.Á., Navarro, A.R., and Romanek, C. (2006). Quantitative analysis of bone mineral using FTIR. *Macla: revista de la Sociedad Española de Mineralogía*. 6: 45–47.

27 Gao, Z., Ding, L., Gong, Z. et al. (2019). Image compressive sensing reconstruction based on z-score standardized group sparse representation. *IEEE Access* 7: 90640–90651.

28 Zeaiter, M., Roger, J.M., and Bellon-Maurel, V. (2005). Robustness of models developed by multivariate calibration. Part II: the influence of pre-processing methods. *TrAC Trends in Analytical Chemistry* 24 (5): 437–445.

29 Filzmoser, P. and Walczak, B. (2014). What can go wrong at the data normalization step for identification of biomarkers? *Journal of Chromatography A* 1362: 194–205.

30 Lee, L.C., Liong, C.-Y., and Jemain, A.A. (2017). The effects of column-wise manipulations on accuracy of classical classifiers with high-dimensional spectral data. *AIP Conference Proceedings*. AIP Publishing.

31 Jarvis, R.M. and Goodacre, R. (2005). Genetic algorithm optimization for pre-processing and variable selection of spectroscopic data. *Bioinformatics* 21 (7): 860–868.

32 Grisanti, E., Totska, M., Huber, S. et al. (2018). Dynamic localized SNV, peak SNV, and partial peak SNV: novel standardization methods for preprocessing of spectroscopic data used in predictive modeling. *Journal of Spectroscopy* 2018 (1): 5037572.

33 Dhanoa, M.S., Lister, S.J., Sanderson, R. et al. (1994). The link between multiplicative scatter correction (MSC) and standard normal variate (SNV) transformations of NIR spectra. *Journal of Near Infrared Spectroscopy* 2 (1): 43–47.

34 Mishra, P., Nordon, A., and Roger, J.-M. (2021). Improved prediction of tablet properties with near-infrared spectroscopy by a fusion of scatter correction techniques. *Journal of Pharmaceutical and Biomedical Analysis* 192: 113684.

35 He, H.-J., Chen, Y., Li, G. et al. (2023). Hyperspectral imaging combined with chemometrics for rapid detection of talcum powder adulterated in wheat flour. *Food Control* 144: 109378.

36 Kharbach, M., Mansouri, M.A., Taabouz, M. et al. (2023). Current application of advancing spectroscopy techniques in food analysis: data handling with chemometric approaches. *Foods* 12 (14): 2753.

37 Li, H., Jiang, D., Cao, J. et al. (2020). Near-infrared spectroscopy coupled chemometric algorithms for rapid origin identification and lipid content detection of Pinus koraiensis seeds. *Sensors* 20 (17): 4905.

38 Guo, Q., Wu, W., and Massart, D.L. (1999). The robust normal variate transform for pattern recognition with near-infrared data. *Analytica Chimica Acta* 382 (1): 87–103.

39 Rato, T.J. and Reis, M.S. (2019). SS-DAC: a systematic framework for selecting the best modeling approach and pre-processing for spectroscopic data. *Computers and Chemical Engineering* 128: 437–449.

40 Maleki, M.R., Mouazen, A.M., Ramon, H. et al. (2007). Multiplicative scatter correction during on-line measurement with near infrared spectroscopy. *Biosystems Engineering* 96 (3): 427–433.

41 dos Santos Panero, P., dos Santos Panero, F., dos Santos Panero, J. et al. (2013). Application of extended multiplicative signal correction to short-wavelength near infrared spectra of moisture in marzipan. *Journal of Data Analysis and Information Processing* 1: 30–34.

42 Carlsson, A.E. and Janné, K.L.R. (1995). Near-infrared spectroscopy as an alternative to biological testing for quality control of hyaluronan: comparison of data preprocessing methods for classification. *Applied Spectroscopy* 49 (7): 1037–1040.

43 Afseth, N.K. and Kohler, A. (2012). Extended multiplicative signal correction in vibrational spectroscopy, a tutorial. *Chemometrics and Intelligent Laboratory Systems* 117: 92–99.

44 Isaksson, T. and Næs, T. (1988). The EFFECT OF MULTIPLICATIVE SCATTER CORRECTION (MSC) and linearity improvement in NIR spectroscopy. *Applied Spectroscopy* 42 (7): 1273–1284.

45 Kanchanomai, C., Ohashi, S., Naphrom, D. et al. (2020). Non-destructive analysis of Japanese table grape qualities using near-infrared spectroscopy. *Horticulture, Environment, and Biotechnology* 61 (4): 725–733.

46 Lieber, C.A. and Mahadevan-Jansen, A. (2003). Automated method for subtraction of fluorescence from biological raman spectra. *Applied Spectroscopy* 57 (11): 1363–1367.

47 Sacré, P.Y., De Bleye, C., Chavez, P.F. et al. (2014). Data processing of vibrational chemical imaging for pharmaceutical applications. *Journal of Pharmaceutical and Biomedical Analysis* 101: 123–140.

48 Lasch, P. (2012). Spectral pre-processing for biomedical vibrational spectroscopy and microspectroscopic imaging. *Chemometrics and Intelligent Laboratory Systems* 117: 100–114.

49 Leger, M.N. and Ryder, A.G. (2006). Comparison of derivative preprocessing and automated polynomial baseline correction method for classification and quantification of narcotics in solid mixtures. *Applied Spectroscopy* 60 (2): 182–193.

50 Shao, L. and Griffiths, P.R. (2007). Automatic baseline correction by wavelet transform for quantitative open-path Fourier transform infrared spectroscopy. *Environmental Science & Technology* 41 (20): 7054–7059.

51 Li, X., Liu, Y., Jiang, X. et al. (2020). Determination and quantification of kerosene in gasoline by mid-infrared and Raman spectroscopy. *Journal of Molecular Structure* 1210: 127760.

52 Afseth, N.K., Segtnan, V.H., and Wold, J.P. (2006). Raman spectra of biological samples: a study of preprocessing methods. *Applied Spectroscopy* 60 (12): 1358–1367.

53 Oliveri, P., Malegori, C., Simonetti, R. et al. (2019). The impact of signal pre-processing on the final interpretation of analytical outcomes – a tutorial. *Analytica Chimica Acta* 1058: 9–17.

54 Wahl, J., Klint, E., Hallbeck, M. et al. (2022). Impact of preprocessing methods on the Raman spectra of brain tissue. *Biomedical Optics Express* 13 (12): 6763–6777.

55 Butler, H.J., Smith, B., Fritzsch, R. et al. (2018). Optimised spectral pre-processing for discrimination of biofluids via ATR-FTIR spectroscopy. *The Analyst* 143 (24): 6121–6134.

56 Wajizah, S. and Munawar, A.A. (2020). Near infrared spectroscopy (NIRS) data analysis for a rapid and simultaneous prediction of feed nutritive parameters. *Data in Brief* 29: 105211.

57 Liberda, D., Pieta, E., Pogoda, K. et al. (2021). The impact of preprocessing methods for a successful prostate cell lines discrimination using partial least squares regression and discriminant analysis based on Fourier transform infrared imaging. *Cells* 10 (4): 953.

58 Tafintseva, V., Shapaval, V., Blazhko, U. et al. (2021). Correcting replicate variation in spectroscopic data by machine learning and model-based pre-processing. *Chemometrics and Intelligent Laboratory Systems* 215: 104350.

59 Weakley, A.T., Griffiths, P.R., and Aston, D.E. (2012). Automatic baseline subtraction of vibrational spectra using minima identification and discrimination via adaptive, least-squares thresholding. *Applied Spectroscopy* 66 (5): 519–529.

60 Schulze, H.G., Foist, R.B., Okuda, K. et al. (2012). A small-window moving average-based fully automated baseline estimation method for Raman spectra. *Applied Spectroscopy* 66 (7): 757–764.

61 Savitzky, A. and Golay, M.J. (1964). Smoothing and differentiation of data by simplified least squares procedures. *Analytical Chemistry* 36 (8): 1627–1639.

62 Xiao, M., Chen, Y., Zheng, F. et al. (2023). Predicting the storage time of green tea by myricetin based on surface-enhanced Raman spectroscopy. *npj Science of Food* 7 (1): 28.

63 Fellows, A.P., Casford, M.T.L., and Davies, P.B. (2019). Spectral analysis and deconvolution of the amide I band of proteins presenting with high-frequency noise and baseline shifts. *Applied Spectroscopy* 74 (5): 597–615.

64 Adams, M.J. (2004). *Chemometrics in Analytical Spectroscopy*, vol. 8. 2004: Royal Society of Chemistry.

65 Sim, J., McGoverin, C., Oey, I. et al. (2024). Optimisation of vibrational spectroscopy instruments and pre-processing for classification problems across various decision parameters. *Food Innovation and Advances* 3 (1): 52–63.

66 Miller, C.E. and Igne, B. (2021). *Chapter 15 Preprocessing methodsCiurczak, E.W., Igne, B., Workman, Jr., J., & Burns, D.A.*, 289. Handbook of Near-Infrared Analysis, Boca Raton, FL: CRC Press.

67 Moros, J., Garrigues, S., and M.d.l. Guardia. (2010). Vibrational spectroscopy provides a green tool for multi-component analysis. *TrAC Trends in Analytical Chemistry* 29 (7): 578–591.

68 Peris-Díaz, M.D. and Krężel, A. (2021). A guide to good practice in chemometric methods for vibrational spectroscopy, electrochemistry, and hyphenated mass spectrometry. *TrAC Trends in Analytical Chemistry* 135: 116157.

69 Weber, A., Wojtowicz, A., Wietecha-Posłuszny, R. et al. (2024). Raman spectroscopy for the time since deposition estimation of a menstrual bloodstain. *Sensors* 24 (11): 3262.

70 Labuschagne, J.-P. and R. Pellissier. (2015). Assessing the impact of variable scaling on projection methods in the analysis of spectral data. ISI Proceedings.

71 Noda, I. (2008). Scaling techniques to enhance two-dimensional correlation spectra. *Journal of Molecular Structure* 883–884: 216–227.

72 Schulze, H.G. and Turner, R.F.B. (2013). A fast, automated, polynomial-based cosmic ray spike-removal method for the high-throughput processing of Raman spectra. *Applied Spectroscopy* 67 (4): 457–462.

73 Cappel, U.B., Bell, I.M., and Pickard, L.K. (2010). Removing cosmic ray features from Raman map data by a refined nearest neighbor comparison method as a precursor for chemometric analysis. *Applied Spectroscopy* 64 (2): 195–200.

74 Barton, S.J. and Hennelly, B.M. (2019). An Algorithm for the removal of cosmic ray artifacts in spectral data sets. *Applied Spectroscopy* 73 (8): 893–901.

75 Green, D., Camargo, F., Heisler, I.A. et al. (2018). Spectral filtering as a tool for two-dimensional spectroscopy: a theoretical model. *The Journal of Physical Chemistry* 122 (30): 6206–6213.

76 de Beer, A.G.F., Chen, Y., Scheu, R. et al. (2013). Analysis of complex spectra using fourier filtering. *The Journal of Physical Chemistry* 117 (50): 26582–26587.

77 Doty, K.C., McLaughlin, G., and Lednev, I.K. (2016). A Raman "spectroscopic clock" for bloodstain age determination: the first week after deposition. *Analytical and Bioanalytical Chemistry* 408: 3993–4001.

78 Zhao, J., Lui, H., McLean, D. et al. (2007). Automated autofluorescence background subtraction algorithm for biomedical Raman spectroscopy. *Applied Spectroscopy* 61 (11): 1225–1232.

79 Cadusch, P. et al. (2013). Improved methods for fluorescence background subtraction from Raman spectra. *Journal of Raman Spectroscopy* 44 (11): 1587–1595.

80 Cadusch, P.J., Hlaing, M.M., Wade, S.A. et al. (2013). Improved methods for fluorescence background subtraction from Raman spectra. *Journal of Raman Spectroscopy* 44 (11): 1587–1595.

81 Han, M., Dang, Y., and Han, J. (2024). Denoising and baseline correction methods for raman spectroscopy based on convolutional autoencoder: a unified solution. *Sensors* 24 (10): 3161.

82 Brown, C.D., Vega-Montoto, L., and Wentzell, P.D. (2000). Derivative preprocessing and optimal corrections for baseline drift in multivariate calibration. *Applied Spectroscopy* 54 (7): 1055–1068.

83 Huang, J., S. Romero-Torres, and M. Moshgbar (2019). Practical considerations in data pretreatment for NIR and Raman spectroscopy, American Pharmaceutical Review. Dostopno na. http://www.americanpharmaceuticalreview.com/Featured-Articles/116330-Practical-Considerations-in-Data-Pre-treatment-for-NIR-and-Raman-Spectroscopy/ [Dostop: 10-Sep].

7

Procedure Validation and Lifecycle Management

Deborah A. Peru

DP Spectroscopy and Training, LLC, Lebanon, NJ, USA

7.1 Introduction

Chapter 7 discusses the validation of spectroscopy procedures following a risk-based approach. Regardless of your industry and where in the process measurements are being made, validation of analytical testing methods is a good practice and should be performed whenever analytical methods are used for informational purposes and compliance. Be proactive! The best protection against implementing an inadequate procedure is to challenge the procedure during the development and validation stages and continue to monitor risk factor contributions that may cause the method to degrade and generate inaccurate results. Calibration models developed using multivariate math will require ongoing maintenance to preserve their integrity. Deviation in performance may occur due to instrumental drift, change in sample chemistry, or processing conditions. Instrument and method verification testing procedures help to identify when model maintenance is required.

Become familiar with validation guidelines for your industry, external guidance documents, and any company-specific guidelines developed by the quality assurance or compliance groups. Make validation a routine part of method development. Complete all instrument-related documentation, including but not limited to instrument qualification protocols and reports, validation protocols and reports, comparative studies, training, and maintenance logs, to demonstrate that your spectroscopy method meets necessary internal and external regulatory requirements.

This chapter provides a brief history and purpose of several U.S. and international agencies that develop, enhance, and promote technical quality standards for analytical testing procedures. A summary of validation parameters commonly used and related math is provided. And a general discussion of lifecycle management concerns after the spectroscopy procedures is implemented. This chapter is intended to provide a broad overview of validation strategies and current best practices as recommended by the agencies. The developer is responsible for identifying validation requirements for their particular industry to ensure the procedure is appropriately validated and "fit for purpose." While many topics will apply, this chapter does not attempt to address the validation of manufacturing processes fully. For more information on process validation refer to the FDA guidance document "Process Validation: General Principles and Practices" [1].

Practical Quantitative Vibrational and Electronic Spectroscopy: A Guide for Developing, Optimizing, and Validating Procedures, First Edition. Edited by Deborah A. Peru.

Companion website: www.wiley.com/go/peruelectronicspectroscopy

7.2 Regulatory Agencies, Guidelines, and Standard Practices

7.2.1 International Committee of Harmonization

The International Committee of Harmonization (ICH) is the youngest of the regulatory agencies. This nonprofit organization was established in 1990 and reformed in 2015 with a new focus on developing globally harmonized procedures. It is an inclusive organization consisting of both government and industrial partners. One focus is improving drug development and registration efficiencies and promoting public awareness of health-related issues [2, 3]. These guidelines embrace the current good clinical practices established by the European Union, Japan, the United States, and those of Australia, Canada, the Nordic countries, and the World Health Organization (WHO). Various appointed committees are responsible for publishing standards and guidance documents for quality assurance testing of pharmaceutical products and validation protocols for those testing procedures.

Guidelines for developing and validating analytical procedures are found in ICH Q14 and ICH Q2 (R2) [4, 5]. These guidance documents reflect the ongoing collaborative relationship between the scientific community and regulatory agencies in support of the development and lifecycle management of existing and state-of-the-art analytical procedures. ICH Q2 (R2) applies to new or revised analytical methods used for the release and stability testing of commercial drug substances and products [5]. These recommended validation protocols apply to spectroscopic procedures developed with multivariate calibration models.

The ICH Q12 on technical and regulatory considerations for pharmaceutical product lifecycle management discusses how process analytical knowledge can establish conditions for manufacturing and controlling pharmaceutical products to enhance product quality [6]. The guidance uses change control for continual improvement and includes greater flexibility in making post-approval changes.

7.2.2 Food and Drug Administration

The Food and Drug Administration (FDA) was established in 1906 with the Pure Food and Drugs Act and renamed the Food and Drug Administration in 1930. This agency reviews, approves, and regulates medical products, including pharmaceutical drugs, medical devices, and other related health products.

The FDA is responsible for publishing and enforcing the "Current Good Manufacturing Practices (cGMP)" guidelines. The guidelines in Title 21, Part 211, establish a manufacturer's minimum requirements to ensure each product batch is made consistently and high quality for its intended use [7]. Testing procedures are required to verify that pharmaceutical products for human consumption are within specification limits and that the procedures used meet the proper standards for accuracy, reliability, and suitability for the condition of use. More information can be found in the Code of Federal Regulations (CFR) Testing and Release for Distribution guidelines CFR:211.165 [8]. The FDA may periodically inspect research and manufacturing facilities under its jurisdiction. A 483 observation is a citation alerting the company of a finding [9]. The company has 15 days to fix the problem before being fined [10].

In 1997, the FDA recognized the use of electronic media and published guidelines for recording and storing electronic information. The FDA continues to work closely with pharmaceutical and related industry sponsors and review boards to establish actionable guidelines for paperless data recording and ensure electronic records' trustworthiness, reliability, and integrity [11]. Instrument vendors that offer CFR 21 Part 11 software adhere to the FDA guidance and offer security features to protect data integrity and enable electronic signatures for routine analysis and procedure modifications. The combination of the software tools provided by the vendor and the organization's record-keeping protocols enables the user organization to comply with the electronic record-keeping requirements of the FDA.

7.2.3 American Society for Testing and Materials

The American Society for Testing and Materials (ASTM) was formed in 1898 by Dr. Charles B. Dudley, a Ph.D. chemist employed by the Pennsylvania Railroad [12]. The Society changed its name to ASTM International in 2001. ASTM Committees are made up of industrial volunteers, consumers, and other groups from governments or academia. Anyone interested can participate in a technical committee once you are a member. ASTM develops and publishes standards by consensus for a wide range of materials, products, systems, and services. ASTM standards typically apply to industries pertaining to metals, construction, petroleum, and some consumer products. ASTM D standards primarily focus on materials-related testing, whereas ASTM E standards are primarily concerned with testing methods and practices. Table 7.1 (see Chapter 7 Companion Website) provides a snapshot of current ASTM Standard Practices that apply specifically to spectroscopy instrument testing and procedure development.

7.2.4 International Organization for Standardization

International Organization for Standardization (ISO) is an independent, non-governmental international organization with over 150 participating countries. The committees are responsible for developing consensus standards for sampling and testing methods concerning environmental interests. They have developed hundreds of international standards for monitoring the quality of air, water, and soil, as well as noise and radiation, and for controlling the transport of dangerous goods. ISO provides technical knowledge on environmental regulations in several countries [13]. For example, ISO 19702 pertains to the sampling and analysis of gases and vapors in fire effluents using open-path Fourier transform infrared (FTIR) spectroscopy. Other specific ISO standards have been developed using a variety of spectroscopy-based instrumentation.

7.2.5 United States Pharmacopeia

The United States Pharmacopeia (USP) was initially formed by a group of physicians in 1820 [14]. Their main goal was to begin standardizing the practice of medicine by publishing recipes for pharmaceutical preparations. The first Pharmacopoeia of the United States of America was published by the end of that year. It consisted of five sections, including a historical introduction and a section called *materia medica*, which lists 221 drugs with known efficacy and a secondary list of 71 drugs and substances of "doubtful efficacy." This first publication also contained a section on weights and measures and a section with over three hundred medicinal preparation procedures and compositions [15]. In 1888, the American Pharmacists Association published the First National Formulary (NF). In 1938, the USP and NF standards for strength, quality, purity, packaging, and labeling were established and recognized as "official," and enforcement by the FDA began [16].

The USP is no longer involved in the science of medicine but is actively involved with publishing written quality standards and testing procedures to characterize the chemical and physical properties of drugs, ingredients, and dietary supplements. The USP standards are recognized in various provisions of the U.S. Federal Food, Drug, and Cosmetic Act (FDCA) and laws, regulations, and policies the States recognize [17]. USP General Chapters below 1000 provide the minimum standards for compliance for use in a monograph, whereas General Chapters 1000 and above provide theory, guidance, and recommendations for best analytical practices.

Several guidance documents, validation standards, and qualification standards developed by the USP are often used when developing spectroscopy methods. USP Chapter <1225> is followed by the pharmaceutical and related industries and describes the validation tests and specifications for compendial methods [18]. The chapter aligns with ICH Q2(R1), "Validation of Analytical Procedures: Text and Methodology" [19].

When developing spectroscopy methods to replace compendial methods, the submission should include a description of the new procedure, why it may be preferable, and advantages and performance characteristics consistent with a full validation report.

The USP <1220> addresses the entire scope of lifecycle management of analytical procedures and refers to validation as an ongoing continuum of development, validation, and monitoring activities [20]. The process begins by establishing an Analytical Target Profile (ATP), which defines the performance requirements of the analytical procedure based on external requirements and risk assessment. Stage 1 of procedure design and development focuses on knowledge gathering, risk assessment, and the design and optimization of analytical procedure control strategies and parameters. Stage 2 involves the execution of performance qualification (PQ) procedures. Stage 3 discusses the importance of continued performance verification of procedures using routine monitoring and managing procedural changes as the product and manufacturing process evolves.

The USP <1058> contains general terminology, qualification definitions, requirements, and change control procedures for analytical instruments [21]. The General Chapters further describe specific parameters and tests pertaining to each analytical technique. USP <856> contains specific qualification tests and acceptance criteria for near-infrared (NIR) instruments with associated details found in USP <1856> [22]. Likewise, USP <858> and <1858> are dedicated to Raman instruments and demonstrate Raman's growing importance in the pharmaceutical industry [23]. USP <854> and <1854> contain specific information for mid-infrared instruments, and USP <857> and <1857> are specific for ultraviolet–visible instruments [24, 25]. These guidance documents outline specific qualification tests and acceptance criteria for all three phases of qualification: Installation qualification (IQ), operational qualification (OQ), and PQ. Table 7.2 (see Chapter 7 Companion Website) contains a summary of USP guidance documents used by spectroscopists. The following section reviews the different phases of instrument qualification in more detail.

7.3 Instrument Qualification

Qualification is the process used to ensure the instrument is functioning correctly to generate trusted results. Qualification is a series of interconnected activities performed over the lifetime of the instrument [26]. USP <1058> defines a risk-based framework for instrument qualification and includes a section on roles and responsibilities. Users are ultimately responsible for specifying their instrumentation and qualification needs [26]. Bansal et al. summarize the complicated issues associated with the various stages of analytical instrument qualification [27].

Multiple agencies (e.g., USP and ASTM) offer instrument-specific guidance documents that describe recommended technique-specific calibration and testing procedures for instrumentation before use. These guidance documents are aligned to meet cGMP requirements for testing finished products.

Most vendors offer a complete validation package that satisfies the IQ and OQ requirements, which can be implemented when new instrumentation is installed or used equipment is moved and re-installed in a new location. The design qualification (DQ) is a set of documents intended to demonstrate the design of a proposed or existing instrument satisfies all the requirements defined by the end user. In a cGMP laboratory (e.g., pharma), all qualification protocols will require a dedicated report to document the execution of qualification protocols and a summary of results. These reports often require internal regulatory approval before instruments are commissioned for use.

The increasing use of automation in the laboratory has led to recent revisions in USP <1058> to include the qualification and validation of computerized systems. Collaborations are advancing to establish a harmonized approach between GAMP (Good Automated Manufacturing Practice) and USP <1058> that reduces compliance risks in laboratory computerized systems [28].

7.3.1 Design Qualification

The suitability of a specific instrument for a given procedure begins with design qualification (DQ), a collection of activities defining the instrument's functional and operational specifications for its intended purpose. Tests are tailored to validate the functional and operational specifications of the instrument for its intended purpose. The DQ most often originates from the instrument vendor for general-purpose, off-the-shelf instrumentation. However, the USP does recommend that the user laboratory specify, when possible, how the instrument should perform for a specific application. When the instrument is moved, has a major upgrade, or the application changes, the system suitability specifications in the DQ documentation should be reviewed and updated to reflect new performance requirements [21].

7.3.2 Installation Qualification

Installation qualification (IQ) is a set of activities that ensures the instrument is delivered and installed correctly. This process demonstrates that hardware and software are properly installed in the user's location and that the environment is suitable for the instrument. The IQ document outlines all the information required to assemble all components. For the analytical instrument, this means documenting the hardware and the connections made and creating a record of the instrumental set-up activities on day one of operation. It also consists of several initial diagnostic and performance tests to verify that the instrument communicates properly with the computer or network.

The IQ activities and documentation provided may include: (1) A checklist detailing the instrument and all its components, including software manual supplies and accessories and condition, upon delivery. (2) A document listing instrument model number, supplier, serial numbers, software versions, and location. (3) Verification that the location, utilities, and environment meet the vendor's specified environmental requirements for optimal performance. (4) Installation and assembly of all components by a qualified engineer, preliminary diagnostic testing, and reference standard baseline testing to ensure the instrument meets acceptable performance criteria.

7.3.3 Operational Qualification

Operational qualification (OQ) is a series of rigorous tests demonstrating that the instrument and software are functioning correctly to meet the "fit-for-purpose" criteria for the user's environment. Individual tests are performed on the hardware components to verify that the instrument works per design specifications. The test procedures often contain a mixture of manual and automated tests to check the safety, performance, and reliability of each instrument. OQ should establish suitable performance over the operational range of wavelength, absorbance, and the evaluation of stray light and resolution (spectral bandwidth). Instrument vendors often use reference materials as part of the IQ/OQ testing protocol. Certified Reference Materials (CRMs) are preferred, but sometimes laboratory-prepared materials using high-purity reagents are used.

Testing of various software functions should also be provided during the OQ to ensure the reliability of the software for data acquisition and post-acquisition calculations when installed on a site-specific computer. The USP General Chapter <1058> recognizes the need for an integrated approach that includes both hardware and software validation and qualification. For Client–Server operations, the IQ would also test and document software functionality over a network.

7.3.4 Performance Qualification

Performance qualification (PQ) is a series of ongoing tests that verify that the instrument and method are performing as defined by the user organization. This often follows a user-defined protocol that defines a collection of measurement tests to evaluate and document the instrument and method continue to function correctly. The evaluation includes verification of both the wavelength (x-axis) and photometric (y-axis) for accuracy, precision, and resolution, testing the sample preparation procedure, and calibration for generating quantitative results.

Ideally, PQ is performed daily or whenever the instrument is used to establish its continued suitability for the intended purpose. The user must define the PQ plan, including the testing procedures, acceptance criteria, and frequency.

7.4 Certified Reference Materials

The terms, Certified Reference Material (CRM) and Standard Reference Material (SRM), are often used synonymously. A CRM is a specific material, tested and certified to have known composition and properties used for performance testing of spectroscopy instruments. The CRM must be kept clean and free from contamination, and recertification should be performed periodically to maintain the validity of the certification.

Most often, these are purchased from an accredited laboratory such as NIST (the National Institute for Science and Technology) or can be purchased from a vendor or prepared by the user in accordance with agency guidance standard procedures. Some SRMs are prepared with the same characteristics as the samples being measured for ongoing PQ (verification) testing.

The use of reference materials to align wavelength/wavenumber x-axis and y-axis in optical spectroscopy is essential for both quantitative and qualitative methods [29]. A calibrated y-axis provides an improved ability to compare spectra and relate amplitude to concentration for improved calibration performance and transfer of methods between instruments [30]. On-going standardization is required regularly to ensure instrumental drift does not occur and that the x-axis and y-axis responses are calibrated properly to ensure the ongoing trustworthiness of the results after a method is transferred for use.

The following instrument-specific sections describe some of the commonly used reference materials used for qualification, standardization, and ongoing verification of the instruments.

7.4.1 Ultraviolet–Visible

Verification of performance in accordance with USP <857> includes monitoring the wavelength accuracy, absorbance (photometric) accuracy, stray light, and resolution. The NIST provides a range of solid and liquid standards for determining wavelength accuracy, photometric accuracy, and stray light [31]. Many vendors use bands from a deuterium discharge lamp or mercury vapor arc for wavelength calibration and accuracy testing [25, 32]. The NIST-traceable SRM 2034 is a holmium oxide solution (or glass standard about 2–4 mm thick) used to verify wavelength accuracy in the 240–650 nm region. A mixture of neodymium and praseodymium can be used for the region between 730 and 870 nm.

The photometric scale (absorbance y-axis) calibration and verification testing is performed using metal on quartz (SRM 2031), potassium dichromate solution (SRM 935), or nicotinic acid solutions can be used. NIST SRM 930 are neutral-density glass filters used for photometric testing in the visible region. The accuracy is evaluated by comparing the absorbance values measured with the certified reference values provided by NIST or an equivalent agency. Instrumental drift is defined as the change of baseline absorbance with time and is measured in absorbance units (AU). Most off-the-shelf instruments require photometric stability (i.e., drift) to be <0.0004 AU/hour and <0.01 AU/day.

Because stray light can be a significant source of error in spectrophotometric measurements most vendors determine and remove stray light effects with filters at the factory. During installation of the instrument, a stray light test may be included to verify operation. Stray light is detected at a given wavelength using solutions of potassium chloride, or potassium iodide (SRM 2032) [33].

The resolution power is verified with solutions of 0.02% toluene in hexane. The ratio of the maximum absorbance @269 nm to the minimum absorbance @266 nm must be above the specification limit.

7.4.2 Fluorescence

The key parameters measured to verify the performance of a fluorescence spectrometer are the wavelength accuracy and the signal to noise. To verify the wavelength accuracy of the excitation and emission lines spectroscopy instruments, replicate measurements of a doped polymethylmethacrylate polymer (typically 35 μm thick) are used [34, 35]. Alternatively, a mercury or xenon emission line or holmium oxide, a rare-earth oxide solution, is used. A calibration photodiode is also used to verify spectral shape over the region between 200 and 800 nm.

For signal-to-noise testing, water in purist form (18-Mohm) emits a low-signal Raman band that provides a good test for evaluating optical and electronic sensitivity. Using an excitation line at 350 nm, the Raman emission at 397 nm is measured. Either the area under the curve (from 365 to 450 nm) or the 397 nm band to baseline noise level is used for the signal-to-noise measurement.

Photometric linearity testing is determined using a series of potassium dichromate samples or quinine sulfate dehydrate in sulfuric acid [36].

Spectral correction of fluorescence emission can be performed using SRMs NIST 2940, 2942 (Ce-ion-doped glass), and 2943 (Cu-ion-doped glass). The emission-response spectrum (i.e., shape and relative intensity of band) is unique for each fluorometer. For quantitative assays, Gardecki and Maroncelli demonstrated the benefits of using a set of emission standards to create an emission correction file based on the measured agreement with NIST-traceable reference spectra [37]. The correction file is subsequently used to improve accuracy in quantitative prediction applications.

7.4.3 Near-Infrared

Performance is typically characterized by measuring signal to noise, low and high flux noise, wavelength accuracy, photometric linearity, and stability. The NIST-traceable SRMs used for verification are also used for wavelength correction and photometric adjustments to correct for drift or subtle changes when hardware maintenance is performed or new lamps are installed.

Photometric linearity is typically performed by measuring four CRMs (80%, 40%, 20%, 10%) in reflectance mode and at least three glass filters in transmission mode. A certified white reflective Spectralon® standard (typically 80–99%) is used for (y-axis) reference standardization in diffuse reflectance mode. Photometric noise is determined at low- and high-light throughput, typically using a 99% reflectance standard for the high flux measurement and a 10% reflectance standard for low flux measurement.

Wavelength uncertainty testing is completed using a NIST 1920a SRM or NIST SRM 2065 glass standard consisting of a combination of rare-earth oxides. The band frequencies are listed in the certificate of analysis and recorded during the OQ testing for direct ongoing comparison and adjustment. Alternatively, the water vapor band @ 7306.74 cm^{-1} (1368.6 nm) or NIST SRM 2035, a glass transmission standard containing rare-earth oxides, can be used for wavelength uncertainty testing.

7.4.4 Mid-Infrared

The factors that impact the measurement performance of a mid-infrared instrument are spectral resolution, wavenumber accuracy, photometric accuracy, sensitivity, and absorbance scale linearity. USP <854> guideline describes qualification testing for accuracy, resolution, wavelength repeatability, and absorbance (transmittance) repeatability. Wavelength accuracy is determined using a polystyrene sample (about

3 μm thick) and measuring the wavenumber of several specific peaks. It passes the test when all peak wavenumbers are within the allowable range. Many mid-infrared vendors sell an external NIST-traceable polystyrene sample for ongoing qualification testing of the instrument.

Likewise, instrument resolution is based on the absorption spectrum of a NIST-certified polystyrene film (matte finish). The difference between absorbance at the minimum vertical axis value near 2870 cm^{-1} and the maximum value near 2849 cm^{-1} should be greater than 0.33 AU. And, the difference between the absorbance at the minimum vertical axis value near 1589 cm^{-1} and at the maximum near 1583 cm^{-1} should be greater than 0.08 AU. It should be mentioned that resolution using mid-infrared instruments depends on whether the instrument is a dispersive or Fourier Transform (FT) design. For FT instruments the resolution is affected by line shape and can be further modified based on the type of apodization factor used [38].

Repeatability involves measuring the internal polystyrene sample multiple times to determine the variance in % T (*y*-axis) and wavenumber (*x*-axis) performance.

7.4.5 Raman

The requirements for operational and PQ of Raman instruments depend on the type of hardware and are often prescribed by the vendor for the application. USP <858> provides general guidelines and acceptance criteria for wavelength accuracy and photometric precision. The vendor software may include other diagnostic parameters and verification tests, such as detector noise, dark current, and peak width. Reference standards such as polystyrene, cyclohexane, or acetaminophen are most often used for Raman shift accuracy testing. A variety of materials used for calibrating or verifying the intensity scale for Raman spectrometers are also available from NIST [30]. Various bands are measured across the full range of the Raman spectrum. Photometric precision is set relatively loosely with a tolerance of 10% from at least two measurements made with a CRM, typically polystyrene [23].

When collecting Raman data, it is important to correct the intensity response of each instrument [39]. The unique combination of hardware components and laser lines generates a unique set of spectral characteristics that affect the peak intensity. These variances are caused by differences in optical components, quantum efficiency, and detector response. To account for these variances, the relative peak intensity can be corrected across the spectrum to ensure more comparable spectral signatures and quantitative results. The NIST SRMs 2241, 2242, and 2243 are certified spectroscopic standards made of optical glass that emit a broadband luminescence spectrum when illuminated with the Raman excitation laser. The shape of the luminescence spectrum is described by a polynomial expression that relates the relative spectral intensity to the Raman shift in units of wavenumber. The NIST-generated polynomial, together with a measurement of the luminescence spectrum of the standard, is used to determine the spectral intensity-response correction that is unique to each Raman system. A relative intensity-response correction factor is applied to each Raman spectra obtained to correct the instrument-dependent artifacts [40].

7.5 Validation of Spectroscopy Procedures

Validation is a critical part of the method development lifecycle. Validation criteria and statistical requirements should be defined at the beginning stage of procedure development during the background investigation phase. Validation of an analytical procedure is a process that involves a series of experiments designed to verify that the performance of the procedure meets the statistical requirements for the intended analytical use. Furthermore, the validation protocol should demonstrate that the method is suitable for its intended purpose and meets specific performance requirements [5].

The acceptance criteria may vary depending on the industry, point of testing, and potential business and quality risks involved with an out-of-specification finding. If a regulatory agency such as ASTM, USP, ISO, or ICH governs your particular industry, review the validation guidance documents for the validation requirements and discuss validation requirements with your regulatory partners. Validation of new processes and testing methods is part of our job as scientists and engineers and should always be completed before implementing quality assurance procedures.

In a cGMP environment, and especially in the pharmaceutical industry, the U.S. and international agencies will require evidence in the form of a validation report that the analytical testing method used for product release can produce valid results. A cGMP manufacturing environment will require a validation protocol that describes in detail the testing procedures and statistical requirements used to verify the testing procedure is suitable and can produce accurate results for product release purposes. Garcia et al. detail the strategies and steps involved in analytical method validation [41].

The next section reviews the validation tests and statistical parameters most commonly included in a validation protocol. For some techniques, the scope of validation may fall short or extend beyond the generalized characteristics described in this chapter. Specific validation requirements or modifications may be required for certain techniques and applications. For example, visible or Raman instruments used for particle size analysis may not require linearity or precision testing. Still, the validation protocol may require testing different particle size ranges to ensure results are accurate, reproducible, and comparable between laboratories and studies [42].

7.6 Validation Tests and Statistical Metrics

7.6.1 Accuracy and Recovery

According to USP <1225>, the accuracy of an analytical testing procedure is defined as the closeness of the test result to truth or a reference value. Validation samples may be prepared by spiking known amounts of the analyte of interest, or the amount can be determined by an independent, well-characterized laboratory procedure. The accuracy is calculated as the percentage of recovery from the amount added or by the difference between the determined values and the accepted actual value. Assessment is determined across the specific range defined by the method. The ICH Q2(R2) validation guidelines recommend accuracy be assessed under regular test conditions using the analytical procedure as written. A minimum of nine determinations are obtained over three concentration levels (or more), covering a specified range of concentrations. ICH Q2 (R2) further describes the validation of FDA-regulated drug substances in finished products. Accuracy should be established across a specified range, typically ±30% of the target value. However, wider or smaller concentration ranges can be used depending on the application of use, calibration range, and availability of validation samples. Three (3) replicates are typically measured at each level. This test verifies the recovery of active ingredients when spiked into a placebo matrix and the agreement between the spectroscopy method and the primary analysis method. A minimum of three (3) concentrations are evaluated (e.g., 70%, 100%, and 130% of the target value) for a total of nine (9) determinations.

The recommended data in the validation report includes the mean, standard deviation (SD) of replicates, coefficient of variation (CV), and 95% confidence interval (CI) at each level. Figure 7.1 provides an example of validation data obtained for a near-infrared method developed for caffeine determination. The validation statistics are shown in Table 7.3. For NIR, the ICH Q2(R2) guideline also suggests reporting the standard error of prediction (SEP) or root mean square error of prediction (RMSEP) determined by comparison of the predicted and reference (true) values.

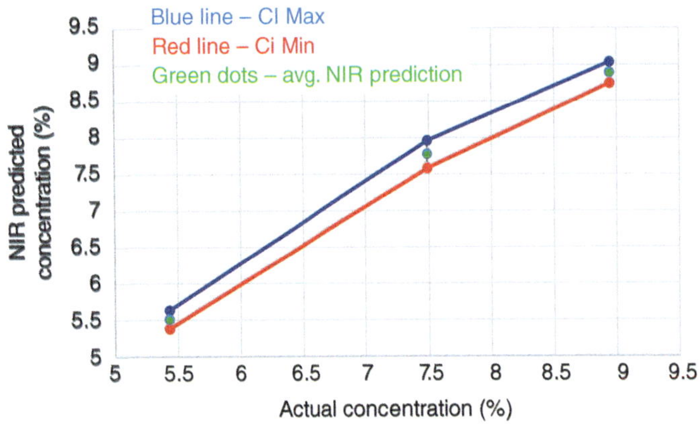

Figure 7.1 Accuracy testing at three levels with confidence intervals. *Source*: Deborah A. Peru.

Table 7.3 Accuracy and recovery validation testing results for Caffeine determination by near-infrared (NIR) Spectroscopy.

	Accuracy and recover testing results						
	Actual (%)	**NIR mean (%)**	**S.D. (±%)**	**CV (%)**	**CI (±%)**	**CI max (%)**	**CI min (%)**
Level 1 Recovery level 1 = 101.5%	5.43	5.51	0.05	0.90	0.12	5.64	5.39
Level 2 Recovery level 2 = 103.9%	7.49	7.78	0.08	0.99	0.19	7.97	7.58
Level 3 Recovery Level 3 = 99.6%	8.94	8.90	0.06	0.66	0.15	9.04	8.75
R^2	0.99						
Slope	0.97						
Bias	0.3						

Source Deborah A. Peru.

7.6.2 Specificity

The goal of specificity testing is to demonstrate interferences due to impurities, degradation products, the matrix (other ingredients), and/or process variances do not interfere with the analyte determination. The validation report should include evidence that the spectral regions relate to the analyte of interest. One way to do this is to plot the comparison or overlay of the pure component spectrum and the placebo matrix for the spectral region or wavelength(s) used in the calibration equation, as demonstrated in Figure 7.2.

For multivariate methods, the final regression vector or principal component (PC) loadings can be overlayed with the pure component fingerprint as shown in Figure 7.3. The procedure can include a qualification method to reject samples, for example, that fail to contain any analyte or when the analyte concentration is too high or low [43]. The calibration standards, spectral bands, preprocessing tools, and regression model parameters should be strategically chosen to ensure the sample matrix and other

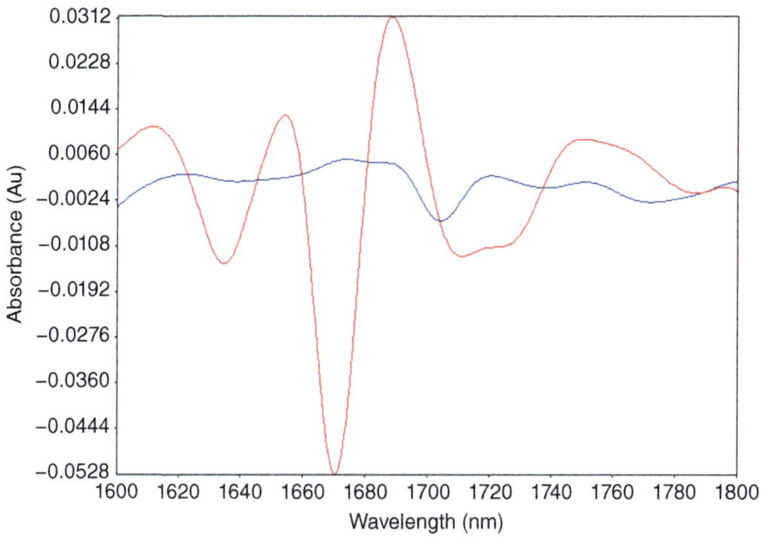

Figure 7.2 Method specificity: comparison of the 2nd derivative spectrum of the pure analyte of interest (red) and placebo matrix (blue) in the spectral region used in the calibration model. *Source*: Deborah A. Peru.

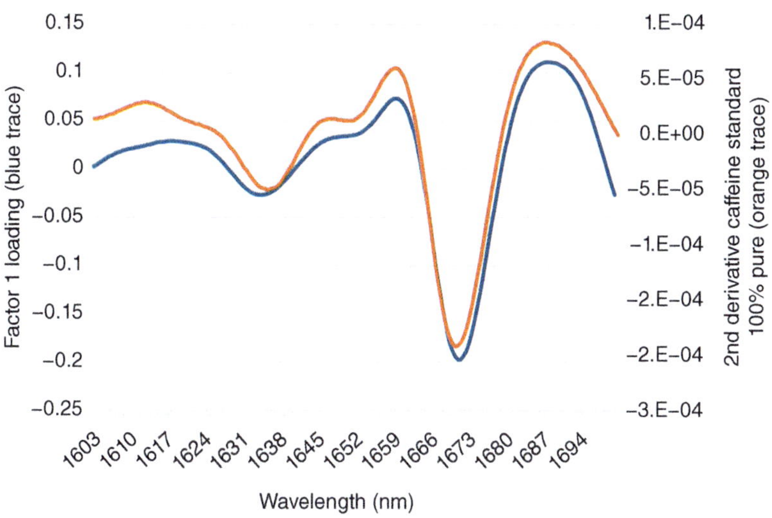

Figure 7.3 The specificity of a method is demonstrated by comparing the first PC loading (blue trace) and the spectral fingerprint of the analyte of interest with the 2nd derivative applied (orange trace). *Source*: Deborah A. Peru.

measurement variances have no remaining influence on the determination of interest. Supporting evidence of specificity and rejection of outliers containing contaminants may be provided using outlier detection statistics (Q-Residual) using spectral residuals, concentration residuals, or multivariate score statistics (Mahalanobis distance (MD)). An outlier is a spectrum that does not have the same general spectral size and shape as the other spectra. An outlier will have a large Q-residual value when compared to the corresponding residuals of other spectra.

For a pharmaceutical assay, ICH Q2(R2) recommends a "stress test" to determine the robustness of the prediction while changing the sample conditions, such as a change in the matrix or the addition of an impurity over the expected range. Depending on the application, the protocol for testing specificity may also require tests demonstrating the lack of bias under stress conditions.

7.6.3 Linearity and Range

The linearity of an analytical procedure is the ability to measure a result that is directly proportional to the "true" value over the range specified for the method (USP <1225>). The range of the procedure may be defined as the difference between the highest and lowest standard concentrations used during instrument calibration that has demonstrated a suitable level of precision, accuracy, and linearity using the procedure as written. For multivariate expressions, this definition may expand to include the range of concentrations used in the set of calibration standards.

Linearity testing is accomplished by measuring at least five (5) samples that span the operational range of the method or calibration curve. Each validation sample is measured at least three (3) times. The validation report should include a plot of the predicted concentration vs. actual concentration for the property of interest and the linear relationship between the determined value and truth (R^2, slope, and intercept) as demonstrated in Figure 7.4. The validation statistics used may include the mean, SD, coefficient of variance (CV), and CI for each concentration level of validation samples evaluated, and the overall SEP.

A concentration residual plot is often used to demonstrate the lack of bias and concentration outliers [5, 18]. The residual plot provides information on how the method error (accuracy) changes across the concentration range. A residual plot is defined as a graph of concentration residuals (y-axis) vs. the actual concentration (x-axis). This type of plot will enhance the degree of scatter in replicate determinations and will visually enhance nonlinearity that may remain in a linear model [44]. Ideally, the residual plot should show no discernable trends or patterns with respect to concentration.

Figure 7.5 shows a concentration residual plot that accentuates the prediction errors obtained using the calibration model. Several important features of the residual plot include a number of samples with errors much greater than the SEP (±0.14%) and the largest CI (±0.19%) defined in the accompanying accuracy test. Table 7.4 contains the outlier diagnostics associated with the NIR predictions for the validation set used in linearity testing. The spectra contain some minor deviations in chemistry and spectral signature for illustration purposes. The F Ratio is the ratio of the sample spectral residual vector to the average model residual. The probability value of a sample is determined from the F Ratio. When the probability exceeds

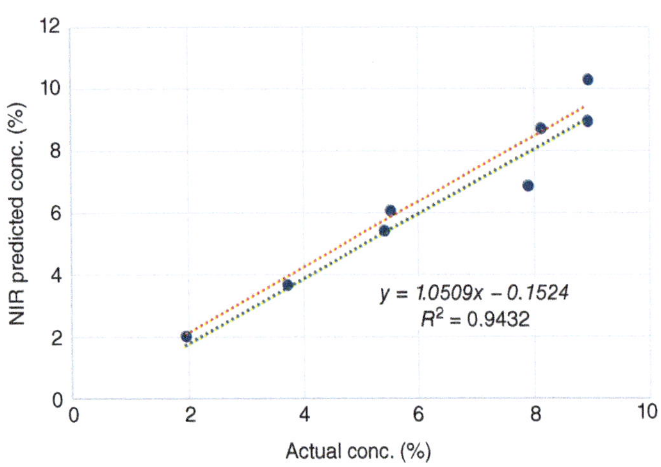

Figure 7.4 Linearity testing results during validation of Near-IR regression model. Red and blue dotted lines represent ±1 SD for replicates at each level. *Source*: Deborah A. Peru.

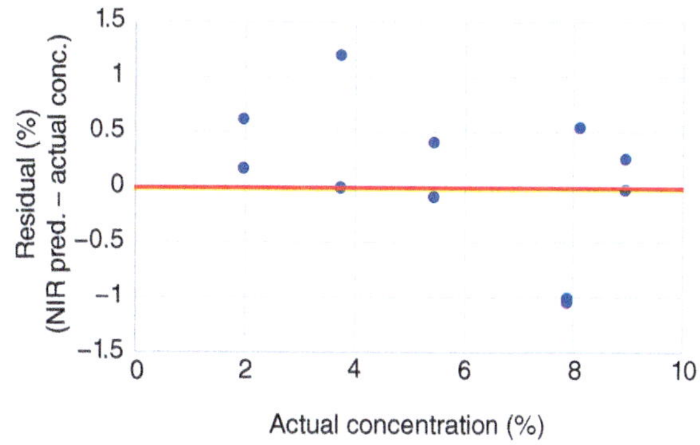

Figure 7.5 NIR prediction concentration residuals (% predicted minus % actual) vs. actual concentration obtained in linearity testing at five levels. *Source*: Deborah A. Peru.

Table 7.4 Linearity testing results using a near-infrared PLS regression model developed for caffeine (for illustration purposes only).

Sample	MD	F ratio	Probability	Leverage	Caffeine Actual (%)	Caffeine predicted (%)	Residual (%)	Recovery (%)
C-04 Rep 1	1.78	1.70	0.77	0.20	1.96	2.15	0.19	109.64
C-04 Rep 2	1.77	1.83	0.79	0.20	1.96	2.15	0.20	109.98
C-04 Rep 3	1.39	**36.06**	**1.00**	0.15	**1.96**	**2.53**	0.57	**129.20**
C-07 Rep 1	0.50	0.54	0.52	0.06	3.72	3.69	−0.03	99.25
C-07 Rep 2	0.49	0.49	0.50	0.05	3.72	3.71	−0.01	99.71
C-07 Rep 3	0.02	**114.12**	**1.00**	0.00	**3.72**	**5.12**	1.40	**137.66**
C-11 Rep 1	0.00	0.86	0.62	0.00	5.43	5.31	−0.12	97.82
C-11 Rep2	0.00	0.77	0.59	0.00	5.43	5.30	−0.13	97.61
C-11 Rep 3	0.03	66.04	**1.00**	0.00	5.43	5.91	0.48	108.75
C-16 Rep1	0.31	0.43	0.47	0.03	7.90	6.84	−1.06	86.63
C-16 Rep2	0.31	0.44	0.47	0.03	7.90	6.84	−1.06	86.60
C-16 Rep3	0.25	17.31	**1.00**	0.03	7.90	6.69	−1.21	84.71
C-18 Rep 1	1.92	0.53	0.51	0.21	8.94	8.90	−0.05	99.48
C-18 Rep2	1.95	0.48	0.49	0.22	8.94	8.92	−0.03	99.72
C-18 Rep3	2.52	**70.07**	**1.00**	0.28	8.94	9.39	0.45	**105.01**
						Average bias	**−0.03**	

Bold represent samples which are outliers or unusual compared to other validation samples.
Source: Deborah A. Peru.

the 95% level established for the training set, the sample is considered to be different from the population of samples used in calibration. As probability increases toward 1, the chance of becoming an outlier increases. Leverage represents the sample's distance to the centroid of the calibration set. MD is determined in either score's space or in wavelength space and is similar in concept to an SD. Large numbers (>3) would indicate dissimilarity between the spectra of the validation samples as compared with the calibration model samples.

In the petroleum industry, the validation of at-line or on-line multivariate analyzers will closely follow ASTM Standard Practice D 6122 and D 6708 when used to determine a physical, chemical, or quality attribute. Statistical correlation between the two test methods is assessed, and the study must show that precision and bias are within acceptable limits.

7.6.4 Precision (Repeatability)

Repeatability is often referred to as intraday error or intraoperator error. This test measures the degree of scatter in replicate measurements made on the same homogeneous sample within a short timeframe. The test evaluates the ability to replicate results using the same analyst in the same laboratory with the same instrument. It captures the repeatability of the sample preparation steps and introduction of the sample into the spectrophotometer.

The general chapter <1225> recommends assessing precision by measuring the concentration of six (6) independently prepared samples of the same lot near the target concentration level as shown in Table 7.5 and Figure 7.6. Samples should be obtained from the target source and prepared for analysis using the same analytical procedure used in routine testing. Alternatively, nine (9) determinations over the range for the procedure may be used (e.g., three (3) replicates at three (3) different concentrations [5]). For example, three (3) replicates at three (3) different concentrations spanning the range of the concentrations used during instrument calibration. If the sample chemistry changes quickly, the test may need to be modified to avoid the propagation of error caused by chemical changes. For process analyzers, the analysis may include six (6) consecutive measurements from the same process. The validation report typically reports the SD, CI, and CV commonly referred to as the relative standard deviation (RSD) for the replicate measurements.

Table 7.5 Reproducibility test of precision in Raman prediction.

Caffeine reproducibility	Raman pred (%)	Lab (%)	Residual (%)
C-18_100LP_1Sec_Rep1	8.28	8.94	−0.67
C-18_100LP_1Sec_Rep 2	8.68	8.94	−0.26
C-18_100LP_1Sec_Rep 3	8.75	8.94	−0.19
C-18_100LP_1Sec_Rep 4	9.04	8.94	0.09
C-18_100LP_1Sec_Rep 5	8.94	8.94	0.00
C-18_100LP_1Sec_Rep 6	9.01	8.94	0.06
Mean (%)	8.78		
SD (±%)	0.29		
CV (%)	3.25		
CI (±%)	0.30		

Source: Deborah A. Peru.

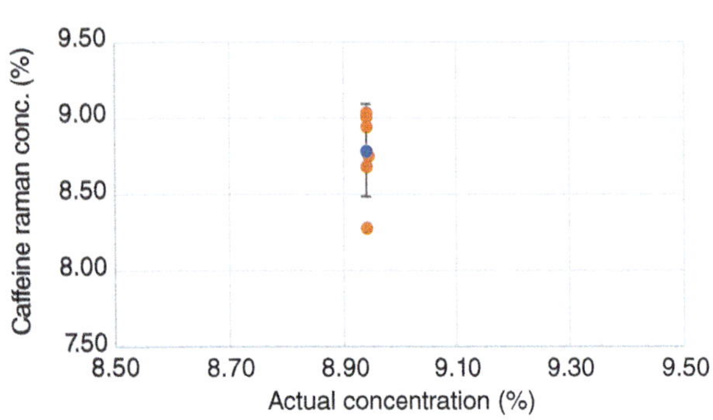

Figure 7.6 Intraday (reproducibility) precision test. *Source*: Deborah A. Peru.

Table 7.6 Intermediate precision data and statistical results.

Sample	NIR prediction (%)	Sample	NIR prediction (%)
Day 1 rep 1	2.65	Day 2 rep 1	2.42
Day 1 rep 2	2.51	Day 2 rep 2	2.25
Day 1 rep 3	2.37	Day 2 rep 3	2.52
Day 1 rep 4	2.40	Day 2 rep 4	2.54
Day 1 rep 5	2.31	Day 2 rep 5	2.77
Mean	2.45	Mean	2.50
SD (±%)	0.13	SD (±%)	0.19
CI (±%)	0.17	CI (±%)	0.24

Source: Deborah A. Peru.

7.6.5 Precision (Intermediate Precision)

Intermediate precision evaluates inter-day error within the same laboratory. This precision test evaluates the procedure's ruggedness and includes additional random errors that occur over consecutive days, possibly caused by different operators. In this regard, it is an initial estimate of long-term variability. Typical variations that are studied include different days, environmental conditions, analysts, and processing equipment. Interlaboratory variations using different analytical instruments in other laboratories are captured separately in a collaborative study [18]. The extent to which intermediate precision is evaluated depends on the circumstances and variances introduced when using the procedure [5]. Ideally, the variables included in this precision test should be chosen based on understanding and risk assessment [4, 5]. Studying these effects using a design of experiment approach vs. one at a time is recommended [5].

Data presented for intermediate precision, shown in Table 7.6 and Figure 7.7, is obtained by repeating the entire analytical measurement procedure on five (5) identical samples measured over two (2) days. Because intermediate precision is intended to capture long-term reproducibility, the variables come from executing the procedure using separate aliquots of the same homogeneous sample. Independent sample preparation of each test sample should be made daily, and different analysts should possibly be used if that is the norm. Modifications may be necessary in instances when the sample chemistry changes quickly. It is recommended that replicates measurements of the same material be measured daily. The validation report should include a comparison of the mean, SD, CV, and CI for each day testing is completed. Precision criteria are based on the statistical requirements defined for the application.

7.6.6 Robustness

Robustness is tested by deliberate variations of the analytical procedure parameters. For example, an NIR spectrum when small changes in the chemical and physical nature of the sample are introduced. The effect of small variations in particle size, moisture content, and sources of API and other formulation ingredients should be studied during the development of the procedure [4, 5]. In some instances, the robustness of multivariate calibration models can be improved by selecting a training set that includes normal parameter variations that can cause spectral signature variability. However, keep in mind that as the robustness of the model increases, the model's specificity and prediction accuracy will begin to degrade [45].

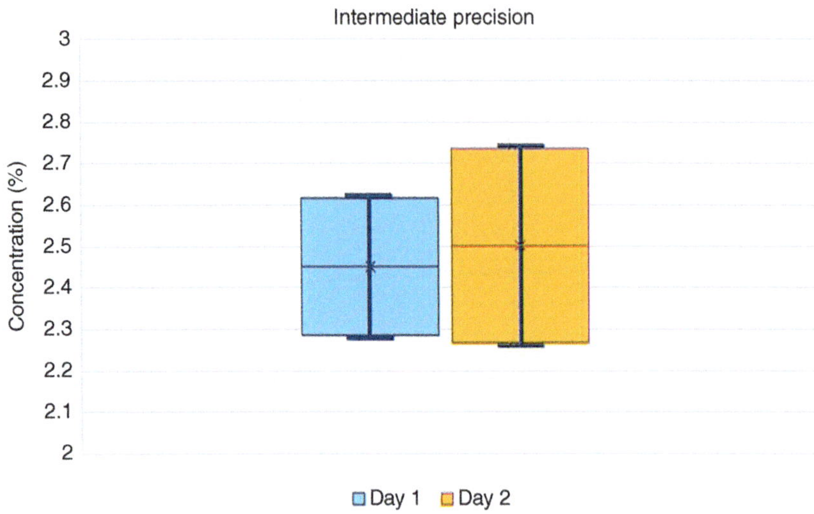

Figure 7.7 Plot of intermediate precision data; mean (*x*) and confidence intervals determined for each day of analysis. *Source*: Deborah A. Peru.

7.6.7 Detection Limit and Limit of Quantitation

The limit of detection is sometimes described as the lowest concentration that can be detected by the procedure [5]. ICH guidelines (Q2(R2) and Q2(R1)) outline three methods for determining a method detection limit (DL) based on either (a) visual evaluation, (b) signal to noise, or (c) SD of the method [19]. The most commonly used method is based on the prediction variance or SD in replicate determinations of a placebo sample containing zero amount (or near zero) of the analyte of interest. Alternatively, the DL can be determined using the standard error of calibration (SEC) from the equation (see Table 7.8 below). The calculated DL is defined as 3.3 times the SD of the average blank sample determination or three times the SEC.

The limit of quantitation (LOD) is the minimum concentration that yields a quantitative measurement near the target concentration with acceptable precision and accuracy [19]. In most applications, the LOD is determined as 10 times the regression equation SEC or the SD in replicate determinations of a blank sample.

Table 7.8 in the Companion Website further describes the equations and associated math for the statistical parameters commonly used when validation methods.

7.7 Lifecycle Management

7.7.1 Real-World Implementation

During the implementation phase, it is advisable to begin using the new spectroscopy method alongside the compendial method for a period of time. This real-world comparative testing period will help build confidence in using the new procedure and give you a chance to discover any other problems with the method during use. ASTM D 6122 for petroleum testing refers to this step as a local validation.

In the consumer products industry, the Quality and Compliance groups may require a comparative study before approval of a new release testing method. Likewise, the FDA will require a comparative study when new alternative methods replace compendial methods for pharmaceutical testing [46]. The ASTM E2935 also describes a similar procedure for equivalency testing when a new instrument or apparatus is implemented in the laboratory for routine determinations [47]. In larger companies, transferring procedures to other laboratories and for finished product quality testing is also very common. The USP Chapter <1224> describes the process for qualification of a validated analytical testing procedure that is transferred for use in another laboratory than it originated [48].

The method comparison report will contain various statistical parameters and plots. For example, use a paired *t*-test, scatter plot of concentration residuals, and the averaged bias between the methods calculated from the differences between each measurement pair. Samples should also be obtained over multiple shifts, days, and possibly weeks. The Bland–Altman analysis is another approach to evaluate if a bias exists between the mean differences and to estimate the 95% agreement interval [49].

Another task to perform during technology transfer is to ensure that proper training of all personnel using the procedure is completed. ISO 9001 is a globally recognized standard for quality management systems. One of the mandatory requirements is maintaining training records and qualification documentation for analysts. Training should include proper use of the analytical procedure, instrument maintenance, and education on the calibration's vulnerability to unexpected process upsets.

7.7.2 Risk Factor Assessment

A predictive model is valid only within the defined range of variances described in the calibration model. As discussed in Chapter 3, it is important to leverage the entire design space, both process and ingredient characteristics, during model development. Insufficient or inappropriate description of the analytical procedure or lack of analytical control strategies can lead to enhanced risk of bias, lack of precision, and out-of-compliance measurements [20]. Variables particularly troublesome for spectroscopy procedures include temperature, humidity, sampling, drift in instrument and ancillary equipment, environmental parameters, and materials characteristics. Risk assessment is partially an administrative process that begins with describing the apparatus (spectrophotometer) model, serial number, software name, and version or firmware version being used [50]. Sometimes, the system manager, location, and asset number are defined.

An analytical control strategy ensures the procedure operates as expected during routine use and throughout its lifecycle. It consists of controls derived from prior knowledge and current understanding of the analytical procedure obtained during development, risk assessment, and robustness evaluation [20]. An analytical control strategy should be defined before validation is executed and confirmed after validation has been finalized [5]. A system suitability study helps to identify potential risk factors that can cause the performance of the procedure to deviate. Specific control strategy elements that will improve a quantitative procedure's long-term robustness include PQ, model maintenance, replication strategy, environmental controls, training program, establishing procedure limitations, and cleaning protocols.

7.7.3 Performance Verification

Once the spectroscopy procedure is fully validated and implemented, it should be monitored during routine use to ensure it continually remains fit for its intended purpose. Procedure monitoring is another facet of the analytical control strategy. The standard operating procedure should include steps for ongoing instrument and method verification testing. In continuous manufacturing processes, the automated output of instrument performance verification and model performance diagnostics provide continual guidance when instrument maintenance or model updating is needed [1]. If the spectroscopy procedure does not meet the established system suitability requirements, a risk-based decision should be made for continued use. In most instances, the method is quarantined until the instrument, calibration model, and entire procedure are evaluated and appropriate actions taken to amend the problem.

- **Instrument Verification:** Establish a daily, weekly, and monthly procedure to track the instrument's ongoing performance and to ensure it performs well each time it is used to measure a sample. An instrument performance test is often included within the vendor software to assess instrumental noise when no sample is in the detector's path. This performance test typically includes a signal-to-noise evaluation, sometimes measurement of various voltages, gains, or laser powers.
  When following the ICH or USP guidelines, wavelength (*x*-axis) certification testing may be performed regularly using a NIST or equivalent traceable standard. Photometric linearity (*y*-axis) certification testing is also required. The USP general chapters for each specific type of spectroscopy provide further information on specific tests and reference standards to use. Ongoing performance verification protocols should include a preventive maintenance plan and continual documentation of repairs and change controls.
- **Method Verification:** A second level of testing pertains to the method and often involves converting the measurement of a sample into a quantitative value. Method verification is frequently done using a verification standard and a check sample. A verification standard is a sample similar in composition to the target sample population. It should have a long shelf life to enable testing daily, and the results are often evaluated using trend analysis to provide evidence the procedure is performing within the expected range of variance. A visual inspection of the trend plot quickly assesses when instrument maintenance is required.
  A check sample is performed using a typical unknown sample obtained for routine analysis. The sample is first analyzed using the spectroscopy procedure, followed by the primary method, and the results are compared. The difference should be within the CI for the average difference defined during the validation stage of implementation (Bland Altman comparative testing). The ASTM standard practice D 6122 for validation of multivariate methods for petroleum product testing also discusses the need to determine on a regular basis that the analyzer results remain in statistical agreement with the primary test method. Prediction deviations are monitored using statistical quality control charts. If deviations beyond a reasonable level are found, a decision to stop using the spectroscopy method to complete a root cause analysis may be needed. Performance verification and lifecycle maintenance of an analytical procedure are further described in USP <1220>.

7.7.4 Multivariate Calibration (Model) Maintenance

Implementation of multivariate calibration models should include a model lifecycle plan. Even the best calibration models may not capture all the variances that can happen over time, and thus, updating the model periodically will ensure the predicted results can be trusted [51]. Multivariate models may drift over time or become inaccurate for several reasons, including (a) a change in the instrument's hardware, (b) a change in an ingredient or physical characteristic of the samples measured, or (c) a change in the range or design space. In some instances, updating the model becomes unavoidable due to a significant change in the sample matrix (i.e., change in pH, particle size, raw material, interferent) or new instrumentation hardware (i.e., new fiber optic probe, light source, detector) that can change the relationship between the spectrum and concentration. The effect of some of these variations can be minimized by selecting the proper combination of calibration standards during the development of the "training set" or by use of design of experiments (DoE). Some models can be inherently robust owing to the preprocessing techniques and wavelengths chosen for calibration modeling.

A lifecycle plan will continuously evaluate the effects of the process, raw material, and equipment changes to ensure the results remain valid and relevant. The plan should include strategies for monitoring the instrument diagnostics, spectral residuals, and concentration residuals, along with an action plan in response to unplanned deviations and planned changes.

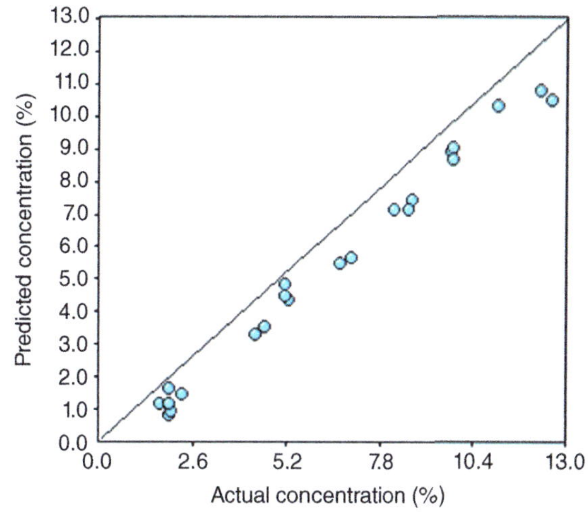

Figure 7.8 Prediction bias occurs when the spectroscopy method deviates consistently and often by the same amount from the reference method results. *Source*: Deborah A. Peru.

For example, a consistent bias (Figure 7.8) between the spectroscopy and primary methods will prompt a root cause analysis to be performed to identify the cause. A bias is often caused by changes in hardware, solvent, or analyst. In some other cases, a bias may evolve due to a change in sample temperature or where one or more raw materials are sourced. A simple shift or bias adjustment may be justifiable in a few instances. In other cases, adding new spectra to the model with their respective reference method values may correct for changes over time. Removal of older spectra that may not be representative or have spectral features that do not properly reflect the current conditions (due to hardware replacement or change in sample chemistry) may be necessary. Any change to the calibration model will require a new or partial validation study. Automated model updating is another approach evaluated in recent years and may be a good solution for critical continuous operations that rely on spectroscopy for manufacturing guidance and control. The reader can refer to Chapter 18 for more information on automated model maintenance.

7.7.5 Revalidation

Revalidation becomes necessary whenever the scope of the method (i.e., matrix or formulation) has significantly changed, and results are consistently outside the specifications defined for the procedure during validation testing. For example, when a major change occurs in the matrix (formulation), instrument, or operating conditions [41]. The model may be susceptible to more significant errors when impurities or changes in material sourcing are introduced. When the calibration equation (multivariate model) is updated with new spectral data or modifications are made to the modeling parameters, a full or partial revalidation is performed [5]. While this effort may be daunting at first, for regulated product release testing, it is easier and cheaper to be compliant rather than face the consequences of remediating noncompliance, which is the order of magnitude more costly [52].

7.7.6 Calibration Transfer of Procedures

A common occurrence in large companies is developing a new analytical procedure for technology transfer to another location, typically at a manufacturing site. The instrumentation used for spectroscopy measurements is likely the same model but with a different serial number. The manufacturing process may also differ in size and hardware components. Ultimately, the spectral fingerprint of the samples measured may be slightly different. In this case, the relationship between the unknown samples and the calibration model has changed in several ways. To handle this situation, calibration transfer algorithms are used to map the response of the measurement system in its current state back to the response during calibration [53]. A set of transfer samples are typically measured on the original and transfer instruments for this purpose. Standardization algorithms such as direct standardization (DS) are applied to create a correction factor using the data sets.

Another approach is to execute an interlaboratory validation study. USP <1224> contains guidance on performing the transfer of analytical procedures originating in another laboratory. Similarly, ASTM E2056 specifies calibration and qualification data set requirements for interlaboratory qualification procedures using surrogate mixtures [54]. The practice applies to the qualification of ultraviolet, visible, NIR, fluorescence, and Raman spectrophotometers calibrated using multivariate techniques (multiple linear regression, (MLR), principal component regression, (PCR), partial least squares (PLS) regression.

7.8 Conclusion

This chapter described the fundamentals and figures of merit for validating spectroscopy procedures. The breadth of spectroscopy applications and quantitative methods used to assess products is diverse; thus, each application should be carefully reviewed, and an individual validation protocol should be written based on a risk-based assessment of the application.

As described, part of the scope of the validation process is to confirm that the method is suited for its intended purpose and to establish the range of statistical error allowed for the predicted concentration or property of interest. Although the requirements of validation have been fully described and documented by the regulatory authorities (USP, FDA, ICH, and ASTM), the approach to validation may vary by industry and point of measurement. For example, the validation requirements may differ when procedures are in research and development compared with product release testing. USP <1210> describes an alternative statistical approach for validating bioanalytical procedures, combining accuracy and precision metrics to build accuracy profiles [55]. This approach is illustrated in Chapter 14 during the validation testing of a Raman procedure used to determine the active ingredient concentration in pharmaceutical solid dosage forms. Meanwhile, applications developed for petrochemicals, forensics, or environmental monitoring may adhere to a completely different set of validation criteria.

Lastly, the implementation plan must include an analytical controls strategy to ensure the procedure generates valid results throughout its lifetime.

References

1 FDA (2011). Process validation: general principles and practices. Guidance for Industry. US Food and Drug Administration, Maryland. https://www.fda.gov/regulatory-information/search-fda-guidance-documents/process-validation-general-principles-and-practices.

2 Overview of ICH (2023). International Council for Harmonization of Technical Requirements for Pharmaceuticals for Human Use. https://admin.ich.org/sites/default/files/2023-11/OverviewOfICH_2023_1115.pdf (accessed 13 June 2024).

3 EMA ICH E6(R2) (2016). Guideline for good clinical practice. European Medicines Agency. https://www.ema.europa.eu/en/documents/scientific-guideline/ich-guideline-good-clinical-practice-e6r2-step-5_en.pdf.

4 ICH Q14 (2023). Analytical procedure development. International Council for Harmonization of Technical Requirements for Pharmaceuticals for Human Use. https://www.ich.org/page/quality-guidelines.

5 ICH Q2 (R2) (2023). Validation of analytical procedures. International Council for Harmonisation of Technical Requirements for Pharmaceuticals for Human Use, Switzerland. https://www.ich.org/page/quality-guidelines.

6 ICH Q12 (2019). Technical and regulatory considerations for pharmaceutical product lifecycle management. International Council for Harmonisation of Technical Requirements for Pharmaceuticals for Human Use, Switzerland. https://www.ich.org/page/quality-guidelines.

7 Code of Federal Regulations (CFR), Title 21 CFR 211 (2023). Current good manufacturing practice for finished pharmaceutical products. Food and Drug Administration, Maryland. https://www.accessdata.fda.gov/scripts/cdrh/cfdocs/cfCFR/CFRSearch.cfm?CFRPart=211.

8 Code of Federal Regulations (CFR) Title 21 Chapter I Subchapter C Part 211 Subpart I 211.165 (2023). Testing and release for distribution. Food and Drug Administration, Maryland. https://www.ecfr.gov/current/title-21/chapter-I/subchapter-C/part-211/subpart-I/section-211.165.

9 FDA (2023). Inspectional observations and citations. US Food and Drug Administration, Maryland. https://www.fda.gov/inspections-compliance-enforcement-and-criminal-investigations/inspection-references/inspectional-observations-and-citations#A_Short_Description_of_Citations (accessed 17 January 2024).

10 ProPharma Group, MIS Limited (2023). 15 days of panic: you received FDA form 483 warning letter, now what? England. https://www.propharmagroup.com/thought-leadership/fda-form-483-15-days-of-panic (accessed 9 January 2024).

11 FDA (2024). Electronic systems, electronic records, and electronic signatures in clinical investigations: questions and answers. Maryland, US Food and Drug Administration. https://www.fda.gov/regulatory-information/search-fda-guidance-documents/electronic-systems-electronic-records-and-electronic-signatures-clinical-investigations-questions (accessed 8 January 2024).

12 Tschida, C. (2015). Introduction to ASTM. Textile technology source. Advanced Textiles Association, Roseville, MN. https://textiletechsource.com/2015/10/07/introduction-to-astm/#:~:text=The%20American%20Society%20for%20Testing%20and%20Materials%20was,2001%2C%20it%20changed%20its%20name%20to%20ASTM%20International (accessed 13 June 2024).

13 International Organization for Standardization (2009). Environmental management. ISO, Switzerland. ISBN 978–92-67-1050. https://www.iso.org/files/live/sites/isoorg/files/archive/pdf/en/theiso14000family_2009.pdf (accessed 8 January 2024).

14 U.S. Pharmacopeia (2020). Building trust for over 200 years: a timeline of USP. https://www.usp.org/200-anniversary/usp-timeline (accessed 9 January 2024).

15 Anderson, L. and Higby, G. (1995). *Spirit of Voluntarism: A Legacy of Commitment and Contribution – The United States Pharmacopeia 1820–1995 (USP Medical History)*. Maryland: United States Pharmacopeial Convention.

16 McCarthy, R. (2016). The history of American pharmacy. University of Connecticut School of Pharmacy. https://aihp.org/wp-content/uploads/2018/08/8-Evolution-of-the-Pharmacopoeia.pdf.

17 U.S. Pharmacopeial Convention (2018). USP and FDA. Working together to protect public health. EA772F_2018–10. https://www.usp.org/sites/default/files/usp/document/about/public-policy/USP-and-US-FDA-a-partnership.pdf.

18 USP <1225> (2017). Validation of compendial procedures. United States Pharmacopeia, Maryland.http://ftp.uspbpep.com/v29240/usp29nf24s0_c1225.html

19 ICH Q2 (R1) (2005). Validation of analytical procedures: text and methodology. International Council for Harmonisation of Technical Requirements for Pharmaceuticals for Human Use, Switzerland. https://www.ich.org/page/quality-guidelines.

20 USP <1220> (2022). Analytical procedure lifecyle. United States Pharmacopeia, Maryland. https://www.uspnf.com/.

21 USP <1058> (2017). Analytical instrument qualification. United States Pharmacopeia, Maryland. https://www.uspnf.com/.

22 USP <856> (2020). Near-infrared spectroscopy. United States Pharmacopeia, Maryland. https://www.uspnf.com/.

23 USP <1858> (2020). Raman spectroscopy. United States Pharmacopeia, Maryland. https://doi.usp.org/USPNF/USPNF_M1017_02_01.html.

24 USP <1854> (2023) Mid-infrared spectroscopy—theory and practice. United States Pharmacopeia, Maryland. https://doi.usp.org/USPNF/USPNF_M3208_04_01.html.

25 USP <857> (2020). Ultraviolet-visible spectroscopy. United States Pharmacopeia, Maryland. https://www.uspnf.com/.

26 McDowall, R.D. (2017). What's new in the new USP? *Spectroscopy* 32 (9): 24–30.

27 Bansal, S.K., Layloff, T., Bush, E.D. et al. (2004). Qualification of analytical instruments for use in the pharmaceutical industry: a scientific approach. *AAPS PharmSciTech* 5 (1): cArticle 22.

28 Vuolo-Schuessler, L., Newton, M.E., Smith, P. et al. (2014). Harmonizing USP <1058> and GAMP for analytical instrument qualification. *Pharmaceutical Engineering* 34: 46–56.

29 Mark, H. and Workman, J. (2019). Using reference materials, part 1: standards for aligning the x-axis. *Spectroscopy* 34 (2): 36–52.

30 Workman, J. and Mark, H. (2019). Using reference materials, part II: photometric standards. *Spectroscopy* 34 (10): 18–28.

31 Agilent Technical Note (1997). Operational qualification and performance verification of UV-visible spectrophotometers. Publication Number 5965-7438E. https://www.manuallib.com/download/pdf11/AGILENT-OPERATIONAL-QUALIFICATION-PERFORMANCE-VERIFICATION-OF-UV-VISIBLE-SPECTROPHOTOMETERS-MANUAL.PDF (accessed 5 January 2024).

32 American Standard for Testing and Materials. (2022). Standard practice for describing and measuring performance of ultraviolet and visible spectrophotometers (ASTM E275–08). ASTM International, Pennsylvania. https://www.astm.org/.

33 American Standard for Testing and Materials. Standard test method for estimating stray radiant power ratio of dispersive spectrophotometers by the opaque filter method (ASTM method E387–04). ASTM International, Pennsylvania.

34 USP <853> (2020). Fluorescence spectroscopy. United States Pharmacopeia, Maryland. https://doi.usp.org/USPNF/USPNF_M3207_01_01.html.

35 American Standard for Testing and Materials (2023). Standard Test Method for Wavelength Accuracy of Spectral Bandwidth Fluorescence Spectrometers (ASTM E388–04). ASTM International, Pennsylvania. https://www.astm.org/.

36 American Standard for Testing and Materials. Standard test method for linearity of fluorescence measuring systems (ASTM E578–07). ASTM International, Pennsylvania. https://www.astm.org/.

37 Gardecki, J.A. and Maroncelli, M. (1998). Set of secondary emission standards for calibration of the spectral responsivity in emission spectroscopy. *Applied Spectroscopy* 52 (9): 1179–1189.

38 Hoult, R.A., Perston, B., and Spragg.R.A. (2013). Polystyrene film as a standard for testing FT-IR spectrometers. *Spectroscopy* 28 (2): 38–43.

39 Choquette, S.J., Etz, E.S., Hurst, W.S. et al. (2007). Relative intensity correction of raman spectrometers: NIST SRMs 2241 through 2243 for 785 nm, 532 nm, and 488 nm/514.5 nm excitation. *Applied Spectroscopy* 61 (2): 117–129.

40 Metrohm USA (2021). Technical note: the importance of relative intensity correction of raman data. https://www.metrohm.com/en_us/applications/bw-tek-applikationen/410000055-a.html?referrer=https://www.bing.com/.

41 García, P., Buffoni, E., Gomes, F., and Quero, J.L. (2011). Chapter 1: Analytical method validation. In: *Wide Spectra of Quality Control* (ed. I. Akyar). InTech.

42 Brander, S.M., Renick, V.C., Foley, M.M. et al. (2020). Sampling and quality assurance and quality control: a guide for scientists investigating the occurrence of microplastics across matrices. *Applied Spectroscopy* 74 (9): 1099–1125.

43 Mark, H., Ritchie, G., Roller, R. et al. (2002). Validation of a near-infrared transmission spectroscopic procedure, Part A: validation protocols. *Journal of Pharmaceutical and Biomedical Analysis* 28: 251–260.

44 Beebe, K., Pell, R.J., and Seasholtz, M.B. (1998). *Chemometrics: A Practical Guide*. New York: Wiley.

45 Igne, B. and McGeorge, G. (2021). Chapter 23: Method development and lifecyle. In: *Handbook of Near-Infrared Analysis*, 4e (ed. E.W. Ciurczak, B. Igne, J. Workman Jr., and D.A. Burns). CRC Press.

46 FDA (2015). Analytical procedures and methods validation for drugs and biologics. U.S. Department of Health and Human Services Food and Drug Administration Center for Drug Evaluation and Research (CDER). Center for Biologics Evaluation and Research (CBER), Pharmaceutical Quality/CMC, Maryland. https://www.fda.gov/regulatory-information/search-fda-guidance-documents/analytical-procedures-and-methods-validation-drugs-and-biologics (accessed 17 January 2024).

47 American Standard for Testing and Materials. (2021). Standard practice for evaluating equivalence of two testing pocesses(ASTM Method E2935). ASTM International, Pennsylvania. https://www.astm.org/.

48 USP <1224> (2017). Transfer of analytical procedures. United States Pharmacopeia, Maryland. https://doi.usp.org/USPNF/USPNF_M5511_04_01.html.

49 Giavarina, D. (2015). Understanding Bland Altman analysis. *Biochemia Medica* 25 (2): 141–151.

50 Burgess, C. and McDowall, R.D. (2013). An integrated risk assessment for analytical instruments and computerized laboratory systems. *Spectroscopy* 28 (11): 21–26.

51 Wise, B.M. and Roginski, R.T. (2015). Model maintenance: THE unrecognized cost in PAT and QbD. *Chemistry Today* 33 (2): 38–43.

52 McDowall, R.D. (2020). Do you really understand the cost of noncompliance? *Spectroscopy* 35 (11): 13–22.

53 Wise, B.M. and Roginski, R.T. (2015). A calibration model maintenance roadmap. *IFAC-PapersOnLine* 48 (8): 260–265.

54 American Standard for Testing and Materials. (2024) Standard practice for qualifying spectrometers and spectrophotometers for use in multivariate analyses, calibrated using surrogate mixtures (ASTM Method E2056). ASTM International, Pennsylvania. https://www.astm.org/.

55 USP <1210> (2023). Statistical tools for procedure validation. United States Pharmacopeia, Maryland.

Part II

Quantitative Applications and Statistical Analysis of Spectral Data

8

Practical Considerations and Benefits of Using Vibrational Spectroscopy in Process Environments

John Wasylyk

Bristol Myers Squibb, Yardville, NJ, USA

8.1 Introduction

Process analytical technology (PAT) is a multidisciplinary subject that aims to use a variety of analytical tools to acquire a wide range of process information, usually in real time and in a fully automated fashion. Implementation usually occurs in a manufacturing environment that imposes many demanding requirements atypical of laboratory instruments. Compared to many other analytical techniques, vibrational spectroscopy techniques such as mid-infrared (MIR), near-infrared (NIR), and Raman are especially amenable to PAT implementation. They provide fast measurements (with a time resolution of seconds to minutes) and offer facile inline or online sampling probes or at-line implementation. Spectroscopy provides comprehensive compositional information for most of the major constituents without sample preparation, dilution, or other pre-treatment, as is often necessary for other analytical techniques. A seminal work, issued by the Food and Drug Administration (FDA) in September 2004, titled, "Guidance for Industry: PAT – A framework for innovative pharmaceutical development, manufacturing and quality assurance" set the framework that holds today [1].

Prior to and since the release of the FDA guidance on PAT, several well-written books and numerous review articles on the use of vibrational spectroscopy for process monitoring have been published, and many of them remain highly relevant today. For example, *Process Analytical Technology* (Ed. Katherine Bakeev) covers many in-depth details on the use of vibrational spectroscopies in a PAT environment [2]. More recent articles cover this topic as advances in instrument design, sampling, and chemometric modeling add to the growing application of PAT [3–5].

Figure 8.1 illustrates the various sampling and implementation modes of vibrational spectroscopy. The *in situ* and in-line sampling mode is the simplest conceptually but requires direct interfacing with the reactor, which may be challenging to implement and maintain or troubleshoot. Online sampling is carried out by directing a process stream (often called a slipstream) from the main reactor or pipe to the process spectrometer (also known as an analyzer) and then back to the process stream or to waste. On-line sampling requires more hardware but provides ease of operation and the possibility to isolate the analyzer from the rest of the process. The at-line (near to the sampling point) and off-line (remote laboratory) analysis is also widely carried out throughout the pharmaceutical and chemical industry, often in a quality control (QC) lab environment, and the instrumentation used is usually like those used in general laboratories.

This chapter summarizes the benefits and practical aspects of implementing vibrational spectroscopy for process monitoring and control in the chemical and pharmaceutical industries.

Practical Quantitative Vibrational and Electronic Spectroscopy: A Guide for Developing, Optimizing, and Validating Procedures, First Edition. Edited by Deborah A. Peru.

Companion website: www.wiley.com/go/peruelectronicspectroscopy

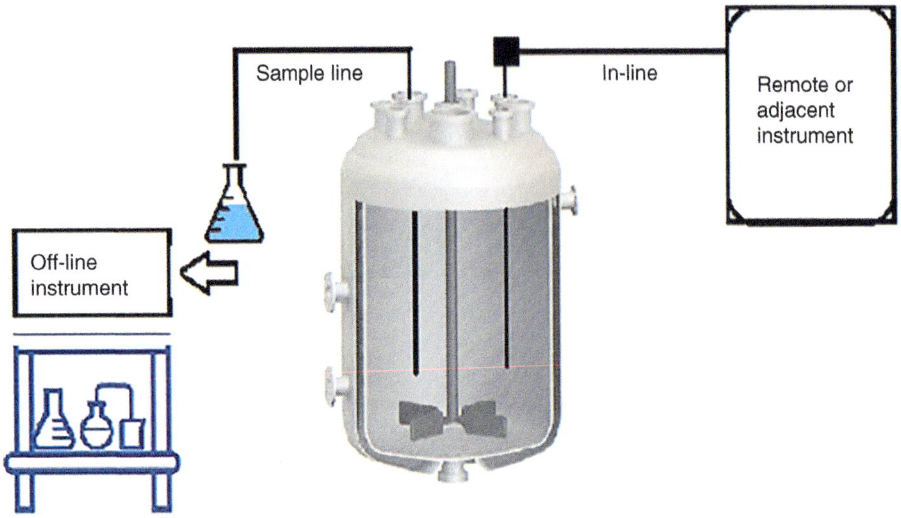

Figure 8.1 Illustration of the various sampling modes: *in situ*, in-line, on-line, at-line within a manufacturing environment. *Source*: John Wasylyk, BMS.

8.2 Process Instrumentation

While the fundamental spectroscopy principles remain the same, process instruments usually appear to be different from their laboratory counterparts due to the need for them to be ruggedized and user-friendly by typical users ranging from early-stage method development scientists to late-stage plant operators. Many of the optical components used in process instruments are engineered to sustain plant conditions, and sometimes, laboratory instruments can be placed within specialized enclosures to help the instrument survive in typical plant conditions (temperature and humidity swing, vibration, etc.). While several chapters in this book contain essential information on MIR, NIR, and Raman instrumentation, this chapter will focus on the unique aspects of their PAT implementation. Numerous factors must be considered in selecting the right tools, such as price, reliability, effectiveness, fit for purpose, local instrument troubleshooting, and technical support.

8.2.1 Mid-Infrared Spectroscopy

Almost all MIR instruments currently used in process applications are based on Fourier-transform infrared (FTIR) spectrometers, which utilize interferometers to convert time-domain signals to frequency-domain spectra. Many laboratory FTIR spectrometers are based on the Michelson interferometer configuration with a moving mirror and, therefore, cannot readily tolerate the mechanical vibration typically encountered in a manufacturing environment. Unlike Raman and near-IR spectroscopy, which can overcome this problem by using fiber-optical cables to connect sampling probes to spectrometers in a general laboratory area, fiber-optical cables with good transmission in the MIR region are currently limited to lengths of at most several meters. This necessitates the location of the MIR spectrometer next to the point of measurement. A dedicated FTIR analyzer thus often utilizes a more ruggedized interferometer configuration to minimize vibrational impact.

The most frequently used PAT sampling mode in MIR is attenuated total reflectance (ATR). The main advantage of the ATR sampling mode lies in its universal applicability. The effective pathlength is typically on the order of a few microns to tens of microns, dependent on the penetration and the number of reflective bounces in the ATR crystal. The ATR crystals are commonly made of diamond, zinc selenide, germanium, although other specialty ATR crystals are available. Such a pathlength is ideal to monitor the concentration change of major reaction species. ATR sampling can be implemented via an immersion probe or a flow cell. Commercial ATR probes are available from multiple vendors, such as Mettler Toledo and Hellma Analytics (formerly Axiom Analytical). Transmission mode is an alternative sampling mode frequently used for gas composition analysis. Care needs to be taken to ensure that the gas cell, as well as the entire transfer tubing, is properly heat-traced to avoid condensation. Under favorable circumstances, even a regular liquid flow cell can be used, though this tends to only be for special applications due to the conflicting requirements of having a sufficiently thin liquid layer (usually less than 1 mm) not to saturate the spectral region of interest, and at the same time being able to maintain a constant flow within the narrow space between two IR windows without causing too high a back pressure.

8.2.2 Near-Infrared Spectroscopy

NIR PAT instruments are either based on the Fourier-transform or dispersive design. FT-NIR instruments are often built similarly to their FTIR counterpart. NIR offers two significant advantages over MIR. First, it can use long silica-based fiber-optical cables with little attenuation, which means that the instrument can be remotely located away from the point of measurement. Second, most organic molecules have orders of magnitude weaker extinction coefficient in the NIR region than in MIR, and a pathlength on the order of mm to cm is often ideal for NIR. This means that robust transmission probes can be used. Such transmission probes tend to be less prone to error due to fouling, as a 1-μm film formation on a MIR ATR probe can render its results completely meaningless but can be tolerated well with a 1 cm transmission NIR measurement. One disadvantage of NIR compared to MIR is its lower chemical specificity in general, as some of the functional groups (carbonyl, for example) do not absorb significantly in the NIR region. However, H-containing functional groups such as OH, NH, and CH all absorb strongly in NIR, and appropriate chemometric models can be developed to correlate spectral changes to properties of interest.

8.2.3 Raman Spectroscopy

In Raman spectroscopy, samples are illuminated with a monochromatic laser beam, which interacts with molecules (solid, liquid, gas, including simple and complex mixtures) and originates a scattered light. The scattered light, having a frequency different from that of incident light (inelastic scattering), is used to construct a Raman spectrum, which arises from the inelastic collision between incident monochromatic radiation and molecules of the sample. When monochromatic radiation strikes a sample, it scatters in all directions after its interaction with sample molecules. Much of this scattered radiation has a frequency equal to the frequency of incident radiation and constitutes Rayleigh scattering. Only a small fraction of scattered radiation has a frequency different from the frequency of incident radiation and constitutes Raman scattering. As a result, sensitivity can be a concern.

Due to scattering dependence on the wavelength of incident radiation, selecting a fit-for-purpose monochromatic laser wavelength is critical to success. The most common laser is 785 nm, which, in many cases, provides sufficient molecular response without amplifying fluorescence, which can saturate the most commonly used Charge-Coupled Device (CCD) as a detector. (Listed below in Table 8.1 are typical laser wavelengths and their selection considerations [6].)

The Raman spectrum is presented as an intensity vs. wavelength shift. Raman spectra can be recorded over a range of 4000 to approximately 200 cm^{-1}, and advances in spectrometers design suppress the laser line, which provides spectral details below 0 cm^{-1} reaching into the anti-Stokes spectral region, as Coherents'

Table 8.1 Raman excitation wavelength considerations and applications.

Excitation wavelength	Spectral region	Best applications	Limitations	Comments
249, 266, 325 nm	UV	Proteins, DNA, RNA, resonance Raman, standoff explosives	Cost and performance of optics and lasers, sample damage	Useful for fluorescence prone samples
532 nm	Visible	Inorganics. Minerals, carbon materials (fibers, nanotubes)	High probability of fluorescence interferences	Excellent Raman signal and signal-to-noise ratio (SNR) if no fluorescence
638 nm	Visible	Biomedical applications, surface enhanced Raman Spectroscopy (SERS)	Fluorescence interferences	
785 nm	Near-IR	Chemicals and organic materials, SERS	Limited fluorescence interferences	Often preferred choice. Good SNR and resolution
993, 1000, 1064 nm	Near-IR	Pigment-rich samples and samples that fluoresce with 785 nm excitation	Poor Raman signal (efficiency) and poor SNR	Minimal fluorescence interferences

THz-Raman Spectroscopy Systems [7]. The spectral region in the 200 to 0 cm^{-1} range allows for quantitation of amorphous material [8].

During sampling, the incident radiation is back-scattered, the Raman signal is dispersed using the diffraction grating, and this dispersed light is directed onto a CCD. This radiation excites the pixels of the CCD, and the charge gained from the excitation is electronically read and presented as a spectrum of the collection range. Advances in grating have allowed suppression of the laser line.

Since most Raman laser excitation sources use visible or NIR light, they can use silica-based optical fibers similar to those used for NIR (though often with different sizes of silica core due to the different types of light sources utilized by these two spectroscopy). As a result, the process Raman instrument is often located away from the measurement point and sometimes can be placed in a non-hazardous (rated area) environment.

8.3 Process Sampling and Control

Sampling plays a critical role in process analytical analysis [9]. The location of the sampling point has to be chosen carefully to maximize representative sampling and minimize the potential sources of error. In general, high-flow and high-shear zones are better than stagnant zones, which is ideal for homogeneous reactions. In the case of in-line sampling with a probe, high flow/high shear will limit or prevent probe tip fouling, which can appear as a coating. The same applies to on-line sampling, while at-line requires true representative sampling.

Heterogeneous reactions follow the same pattern, but they have the added challenges of uneven settling, eddy formation in the flow stream, and phase separation. Particles can also interfere when offset sampling is used. For example, in Raman spectroscopy, the laser beam's focus may be either at the tip of the sampling probe or offset for deeper penetration into the reaction stream.

If the process stream has varying degrees of particulates, sampling will limit the collection of data on the liquid portion. In contrast, it may be critical in some applications to track and control particle size and shape [10]. In these instances, *in situ* particle analysis instruments such as the FBRM (Focused Beam Reflectance

Measurement) are often used. Off-set focal length in heterogeneous mixtures is critical in Raman as long pathlengths can prohibit collecting data on the status of the amount of uncrystallized material since most of the signal will be blocked by crystals at the collection window on the probe [11]. This is less of a concern if only crystal polymorphism is of interest. FBRM, with its fixed off-set focal length, provides chord length and shape. Probe placement should be interfaced in a highly agitated sampling point.

8.4 Area Classification and ATEX Rating System

Hazardous locations are classified according to the types of volatile compounds that may be present at any given time. When dealing with a potentially hazardous area, the worst-case scenario becomes the rule and is often not considered until later stages in the development cycle, resulting in potential delays.

There are two main systems used to classify hazardous locations, (1) the "Class" or "Division" system (Table 8.2) and (2) the "Zone-Based" system (Table 8.3). The former is primarily used in the United States and is defined by the National Fire Protection Association (NFPA) and the National Electric Code (NEC); the latter is an internationally accepted classification defined by the European Union directive 94/9/EC and the International Electrochemical Commission (IEC), also referred to as ATEX (Atmospheres Explosives). Both systems differ in assigning hazardous levels [12–16].

The hazards lie in the electrical components, which have design characteristics with the specific goal of preventing an explosion from occurring or mitigating its effects should it occur. Three elements are required for an explosion to occur: (1) a flammable substance (gas, liquid, solid) that is present in sufficient amounts to create an explosive mixture, (2) an oxidizer such as oxygen or air, and (3) an ignition source (flame, spark, high-heat surface). All three components must be present.

In the Class/Division System, the class refers to the compounds present, while the division refers to whether they are found under normal circumstances or not. Hazardous location classification may be conducted on a room-by-room basis, or at the general facility level.

Table 8.2 NFPA and NEC classes, divisions, and groups.

		Divisions	
Class	**Groups**	**Division 1**	**Division 2**
I Gases, Vapors and Liquids	A. Acetylene B. Hydrogen C. Ether D. Hydrocarbons, fuels, solvents, etc.	Explosive or ignitable gases or vapors are present under normal operating conditions	Explosive or ignitable gases or vapors NOT normally present (may exist accidentally) under normal operating conditions
II Dusts	E. Metal dusts F. Carbon dusts G. Flour, grain, wood or chemical dusts	Combustible dust in air under normal operating conditions	Dust not normally in the air at ignitable concentrations (may occur accidentally)
III Fibers and Flyings	Textiles, wood chips, etc. (Ignitable but NOT explosive)	Easily ignitable fibers and flyings are handled, manufactured or used	Easily ignitable fibers are stored, or handled

Table 8.3 ATEX rating system zones and groups.

Zone/explosive material	Zone description	Group/description
0/Gases 20/Dusts	Explosive material(s) is continuously present or for long periods of time	I/Underground and confined space use [a]II/Places with a hazardous atmosphere, excluding Group I
1/Gases 21/Dusts	Explosive material(s) is likely to occur during normal operations	
2/Gases 22/Dusts	Explosive material(s) not likely to occur during normal operation and transient if it does occur	

a) Note: Divided into subgroups according to the explosive characteristics of the location where the equipment will operate.

The Zone system is similar to the Classes found in the Class/Division system, but it replaces Divisions with Zones and uses a different nomenclature for Groups. It is the foundation for the ATEX rating system. It is important to note that specifying the Class is not mandatory in the Zone System, as opposed to the US system, whereby compounds are implicit in the zone types.

The challenge of meeting the classification requirements can be attained by several approaches based on the type of spectrometer. In the case of MIR and NIR, the instruments are typically placed within three (3) meters of the reactor in a plant setting due to the constraints on the length of optical fibers, although recent advances in fiber technology have increased the maximum length. This requires the instrument to be purged, typically with nitrogen or house air, to maintain positive pressure inside the instrument housing. A purged computer will be required unless a fiber-optic network can be utilized to allow instrument control in a non-rated or safe zone. Another consideration is that a cooling source will be required if the instrument in its enclosure generates a sizable amount of heat. A Raman spectrometer follows the same requirements, but since the Raman signal can travel over long distances of fiber, the instrument can be in a safe zone. The probes are considered conduits of the spectral beams, and the concern becomes the process compatibility with the material of construction.

8.5 Data Handling in PAT Applications

Statistics offers a wide range of helpful tools that can be used for converting data into information. Univariate methods, which consider one variable at a time, independently of the others, are less common when considering application in PAT as they supply just partial answers to the problem under study. They underutilize the potential for discovering comprehensive information embodied in the data, leaving out inter-correlation between variables, limiting the understanding of processes, and negatively impacting control. Regardless, in very simple cases, univariate analysis can be sufficient and simpler to understand and justify when it comes to filing with a regulatory agency. Since spectroscopic techniques produce profiles containing a high amount of information, they can be profitably exploited using multivariate mathematics and statistical (chemometric) techniques.

Multivariate strategies will consider underlying correlations, allowing a more complete interpretation of data [17, 18]. Principal component analysis (PCA) and partial least squares (PLS) are well-established techniques for analyzing multivariate process data. PCA interprets spectral variations along the direction representing the largest sources of variance and captures the representative properties of that data. It is an unsupervised method whereby the resulting components do not necessarily reveal patterns directly related to the classification. The first principal components represent the largest variances in the spectral dataset, which can very likely be the intragroup variations. Determining which components are directly related to the classification can be a challenge and often requires *a priori* knowledge of the process. PLS is a supervised method that mines projections so that the different groups show the best separation possible. PLS is not completely robust to the intragroup variations, and any unwanted variations on PLS may result in overfitting, leading to inaccurate predictions [19].

PAT spectral data sets can contain a tremendous amount of data. Additional chemometric statistical algorithms are often useful for processing this data and are discussed in other chapters. Classical Least Squares (CLS) Fitting, Multivariate Curve Resolution (MCR), or PCA are tools for useful for understanding the distribution of components [20]. Hierarchical Clustering Analysis (HCA) and Divisive Clustering Analysis (DCA) are helpful for classification and grouping related spectra/samples together.

8.5.1 Spectral Analysis Workflow

The early stages of the Development Cycle typically include off-line and in-line analyses. As the Development Cycle matures and gears change toward focusing on driving the product to the manufacturing environment. Quality by Design (QbD) plays a major role to improve product yield and hold production cost at bay [21]. This is where PAT becomes invaluable to the point where it impacts the manufacturing process, lends itself to process control strategy, maintenance of Critical Quality Attributes (CQA) and life cycle management [22, 23].

The successful implementation of PAT often utilizes a networked system consisting of one or more analyzers, chemometric or visualization software to convert raw data into values, database(s), and process tools to relay the information for control. Feeding PAT data back to a control system provides fine-tuning of the manufacturing process to maintain or improve product quality and includes instrument integrity (Figure 8.2). This may be through feedback and feed-forward control (Figure 8.3). One caveat to the successful implementation in a regulated environment is ensuring data integrity, maintenance, and security. FDA's CFR 21 Part 11 governs compliance in the United States and Europe by EU Annex 11.

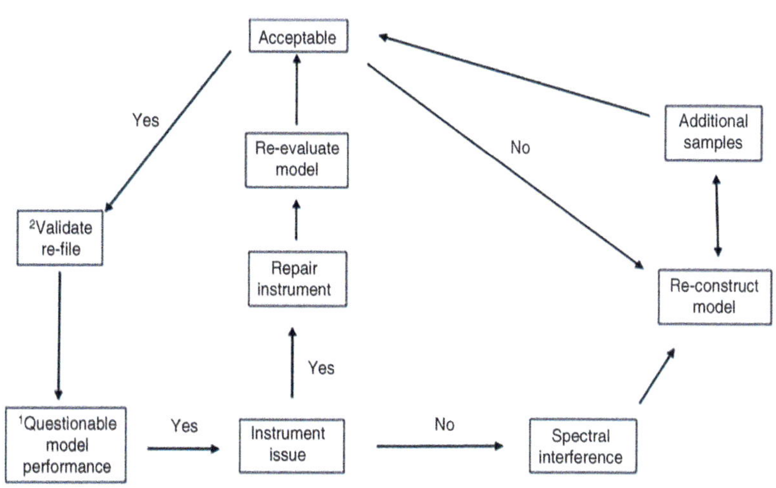

¹Frequency of outliers determined by statistical analysis is greater than the tolerable limit
²Validation and refiling with regulatory agencies depends on the root cause. Instrument issues\ would require on re-qualification when in a GMP environment

Figure 8.2 Instrument and maintenance workflow. *Source:* John Wasylyk, BMS.

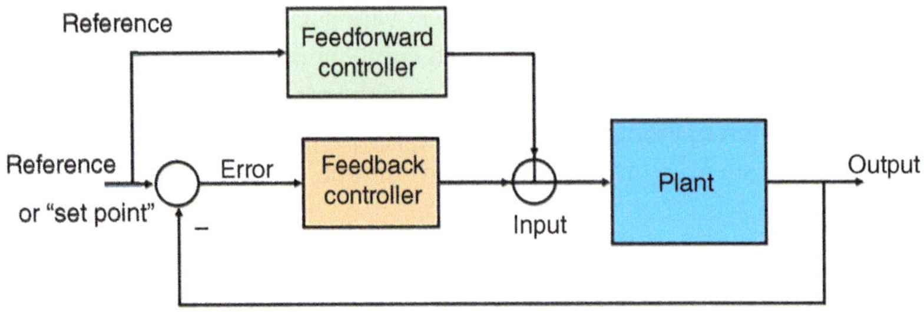

Figure 8.3 Basic diagram of feedback and feedforward with set point for reference comparison. *Source*: John Wasylyk, BMS.

Chemometric models are the enablers of spectroscopy. Spectroscopic techniques yield a wealth of information, and the application of multivariate mathematics and statistics provides the tools to exploit it. Multivariate analysis methods (PCA, PLS, etc.) are used to deconvolute complex data, but these models must be maintained during long-term use. Model maintenance must be considered for any spectroscopy-based application to preserve and/or improve models over a period. Some technical concerns with overseeing the models include spectral pre-processing, wavelength (variable) selection, data dimension reduction, specificity of a quantitative calibration, pattern recognition, calibration transfer, and multispectral data fusion.

There are several factors that may lead to the need for model maintenance, such as variability in the process equipment that affects the mixing efficiency, temperature variation during the process, and sampling challenges. Models must be re-evaluated during the scale-up of processes and when production is moved from reactor vessel to reactor vessel or site to site. The quality or variability in raw material inputs may also cause slight changes in chemistry or add impurities that affect the models. Any changes to the input or process variations can influence the reliability of a model. During the development cycle, changes to the process and raw material must be added to the model to build robustness. Instrument variations are another factor that can lead to inaccurate results. Variations may arise from the excitation source, which can change over time, spectral collection components, fouling of spectral-sampling points, and even disruption of integral components due to excessive solvent or vapor exposure, temperature fluctuations, and vibration.

If feasible, models should be evaluated on multiple identical instruments during the development cycle. More recently, applications using digitized models are being evaluated for transferability across instruments from different vendors. Periodic evaluation of the performance of a model must be made routine either by external testing or internal automated analyses. Internal techniques to track performance can be accomplished by statistical analysis of spectral residuals using Q and Hotelling T^2 to evaluate model status in real time [24]. Thorough evaluation, including automated on-the-fly analysis of outliers, allows for rapid signaling when a process is deviating. Protocols governing model control (CFR 21 Part 11 and internal controls) must be established [25].

8.6 Data Integrity and Compliance

The original "Code of Federal Regulations 21" or CFR21 is over 20 years old and was written in response to soaring costs and security concerns associated with distributing, storing, and retrieving pharmaceutical and medical manufacturing records. The code extends to those organizations involved in, at least in later stages, materials for a clinical trial, research, development, and manufacturing. These regulations also encompass contract research organizations (CROs) and contract manufacturing organizations (CMOs). The U.S. Regulation Title CFR21 Part 11 explicitly regulates electronic records and signatures so that the integrity of the data and workflow matches that of paper records [26]. Traceability of all data collected and the actions that follow is paramount to implementing a fully compliant system. Encompassing the execution of the compliant system is the establishment of Standard Operating Procedures (SOPs) and recordable training to ensure total compliance with data control.

Similar regulations are in place in other countries, such as Annex 11, the European Union's complementary counterpart to the FDA's CFR 21 Part 11, as well as Japan's ERES guidelines. Furthermore, all countries dealing with exports to the United States must adhere to the FDA's guidelines. Adherence to regulatory requirements has helped drive instrument control software vendors to include packages that ensure that all records are authentic and irrefutable with limited and controlled access. As a result, data control software packages must be validated and undergo periodic audit trail evaluations.

There are a number of software-based systems (for example, Siemens' SiPAT, Optimal Solution's SynTQ, Perceptive Engineering's PharmMV, and Sentronic's SentroSuite). These are designed to meet the established CFR21 Part 11 data integrity criteria. The use of the Open Platform Communications (OPC) communication standard makes PAT systems capable of interfacing with nearly every automation system on the market today including Emerson's delta V, Siemens PCS7 and Rockwell Automation's PlantPAx.

8.7 Process Control Strategy

Controlling a process in a plant environment can range from systems that are simple to implement and operate to complex and expensive. The simplest control system is an operator-mediated control system [27]. In this case, an operator watches various gauges and instruments and makes limited but critical adjustments, such as changing the rate of a reagent addition or increasing or decreasing the temperature of a process that deviates from a set-point (SP). This is a labor-intensive approach. This is also a step above an open-loop system where there is no control beyond a pre-determined SP.

The most common and stable controller is a proportional integral derivative (PID) controller, which uses a control loop feedback mechanism to control process variables and is the most accurate and stable controller [28, 29]. By adding an automated closed-loop control system, the gauges or instrument values, sometimes called process variables (PV), are automatically compared to the pre-determined SP value. The PID controller compares the PV to the SP and calculates the difference, or error (E), between the two signals. The controller calculates the appropriate controller output, based on the E, to enact a change in the process by looping this command back with the goal of focusing the process back towards the set point.

The PID feedback system can be automated so that no operator is directly involved (see Figure 8.3). The downside is that the feedback strategy may start too late and, at best, can only work to minimize any deviation from the ideal production limits as events unfold.

Feedforward controller measures the degree of deviation of a process from the SP and predicts the impact will have on the PV. Algorithm-based computations are made, which leads to changes designed to counteract the predicted deviation, thus aligning the PV with the SP. Feedforward is at its best when the range of process deviations is known. Unfortunately, feedforward systems can be expensive and require considerable input. As a final note, many complex processes use a PID system which couples feedback and feedforward, thus bringing the best of both worlds [30].

Various PID controllers that work directly with instruments or function as an interface when coupled with a data control system are available [31]. Systems such as Simulink can couple directly with spectrometers and incorporate other software packages such as MatLab [32]. Watlow can be interfaced with multiple instruments and is Modbus compatible, while Simex utilizes fuzzy logic as an option for feedforward control. Emerson's Delta V offers wireless controllers to simplify the connectivity of various controllers to a wide range of instruments and software packages [31, 33].

8.8 PAT Installation Logistics

The business need of the plant or project often dictates whether the installation of a process analyzer will be temporary or permanent. Often, a temporary installation is first carried out to solve the problem or to understand the feasibility and caveats. The data acquired in the temporary stage can either be used to justify the subsequent permanent installation or to provide enough process insights and knowledge to make a permanent installation unnecessary.

While temporary installations may be carried out with a relatively short turnaround time, in weeks to months, permanent installation often entails substantially more effort and preparation. This is because there are many more factors that have to be considered for permanent installations such as capital purchase, operator training, instrument calibration maintenance, and data communication strategy. These efforts involve personnel outside the analytical specialist's usual field of influence and thus may take significant time to overcome the financial barriers. A temporary installation, in contrast, usually utilizes existing hardware and can be achieved with a smaller core team of players. It should be emphasized that regardless of the time scale of the installation, safety should always be treated with paramount importance with proper engagement of an expert such as a metallurgist for material compatibility, process engineer for failure-mode analysis, and site safety personnel for awareness and emergency train, especially if the analyzer information is used for controlling or layer of protection purposes [34].

When possible, a temporary installation in an R&D lab or a mini-plant (pilot plant) should be carried out first to acquire feasibility data [35]. Such installation can be as simple as moving a portable instrument right next to the reactor (usually within a fume hood) and inserting a probe into it or installing a flow cell within the process. Such an installation often only requires a simple management of change (MOC) form and can be completed within minutes to weeks, dependent on the complexity of the process and the level of hazard involved. For example, suppose a reaction is carried out at ambient pressure below 200 °C and does not involve strong acid, and appropriate ports are available for the introduction of sampling accessory. In that case, the installation can occur within minutes, provided all the necessary hardware is available and there are no other "red flags" concerning the introduction of the sampling accessory. Such temporary installation occurs on a weekly basis in the authors' laboratory. Once installed, the analytical specialist can train the operator to use the instrument and usually stays involved in the data analysis and result interpretation until the end of the temporary installation phase. Often, sufficient process information can be learned during this phase. If a robust and reproducible process is observed, which can be controlled well by adjusting key input variables (KIV), then such information would conclude that permanent installation is unnecessary. Alternatively, if the key output variables (KOV) are found to be influenced by factors that cannot be easily controlled (e.g., presence of impurities that may deactivate catalysts, trace amount of metal ions that may catalyze a reaction), and if real-time process information extractable from the vibrational spectroscopy provides the only means to monitor and thus control the process, a strong business case can now be built to justify the capital purchase and installation of a permanent analyzer.

8.9 Green Chemistry

Green chemistry is the design of chemical products and processes that reduce or eliminate the use and generation of hazardous substances [36]. Of the twelve principles of green chemistry, five are directly related to the application of spectroscopy in chemical processes, namely, (1) preventing waste, (2) increasing energy efficiency, (3) avoiding chemical derivatives, (4) analyzing in real-time, and (5) minimizing the potential for accidents and exposure.

Spectroscopy-based in-line and off-line methods can replace many of the traditional analytical methods. In most applications, advanced spectral deconvolution software and advanced chemometrics will eliminate sample handling, complex pre-processing, and exposure to reaction mixtures or final products. Off-line methods may require little or no sample preparation, and in the case of non-contact optics, which is typical for Raman spectroscopy, the analyst is never exposed to the sample. Some of the typical green applications where spectrometers are used to replace traditional solvent based methods are listed below:

- Raman (in-line and off-line)
  - ✓ Reaction monitoring and control (*replaces* high-performance liquid chromatography (*HPLC*))
  - ✓ Solvent exchanges (*replaces GC*)
  - ✓ Form control and determination (*replaces* powder X-Ray diffraction (*P-XRD*)); *solid state* nuclear magnetic resonance (*NMR*))
- NIR (in-line and off-line)
  - ✓ Solvent exchanges (*replaces* gas chromatography (*GC*)
  - ✓ Reagent analysis (*replaces titration; HPLC*)
  - ✓ Moisture determination (*replaces Karl Fisher*)
- FT-IR (in-line and off-line)
  - ✓ Reaction monitoring and control (*replaces HPLC*)

The maximum impact of the greenness approach using spectroscopy is via in-line analysis, where control points and CQA are monitored and measured in real time and coupled with feedback and feedforward regulation. This approach is also the backbone of PAT.

8.10 Applications of PAT in Pharmaceuticals

Taking the step from using traditional analytical methods and replacing them with PAT for process control is challenging due to several factors, such as lack of expertise, investment in new equipment, transferability to other sites that do not have PAT capabilities, and the hesitation to file the PAT approaches with the various regulatory agencies. PAT must begin at the development stage to gain fundamental knowledge of the instrumentation and determine whether PAT is appropriate or even needed. The typical approach is based on feasibility studies beginning with in-line and/or off-line spectroscopy-based studies. Gouveia et al. demonstrate some of the challenges in implementing a PAT control strategy for the formation of a key starting material during active pharmaceutical ingredient (API) production [37]. First, the potential of Raman, FT-IR, and NIR spectroscopies was critically evaluated under batch reaction conditions. Second, a monitoring strategy with NIR was proposed for the continuous industrial-scale reaction. The authors used a cross-functional team encompassing several knowledge areas such as chemometrics, spectroscopy, reactor, process design, and information technologies to reach a final control strategy. Ultimately, their work resulted in a rapid and reliable analytical strategy that allowed the following: speeding up the design of the continuous process and a better understanding of the manufacturing requirements to ensure optimal yield and avoid unreacted raw materials and by-products in the continuous reactor effluent.

To meet the demands for greater process control in the manufacturing of pharmaceuticals, companies are implementing PAT and QbD concepts, along with feedback and feedforward controls. Combining these approaches enhances scientific understanding and risk management and leads to a high level of quality and productivity. The combination of control hardware and software integration coupled with an advanced control system have proven successful in tablet production. An application using feeders, a blender, and NIR as a single unit operation enabled enhanced control loop performance for continuous tablet production [38, 39]. Figure 8.4 depicts an advanced hybrid model PID (proportional-integral-derivative) control

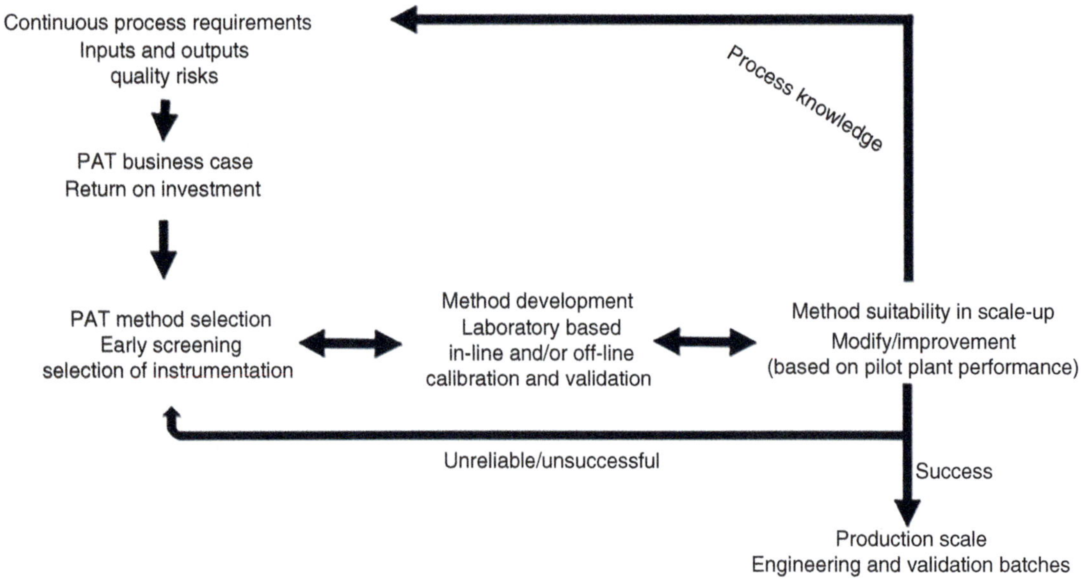

Figure 8.4 Overview of process flow from early-stage development to production. *Source*: John Wasylyk, BMS.

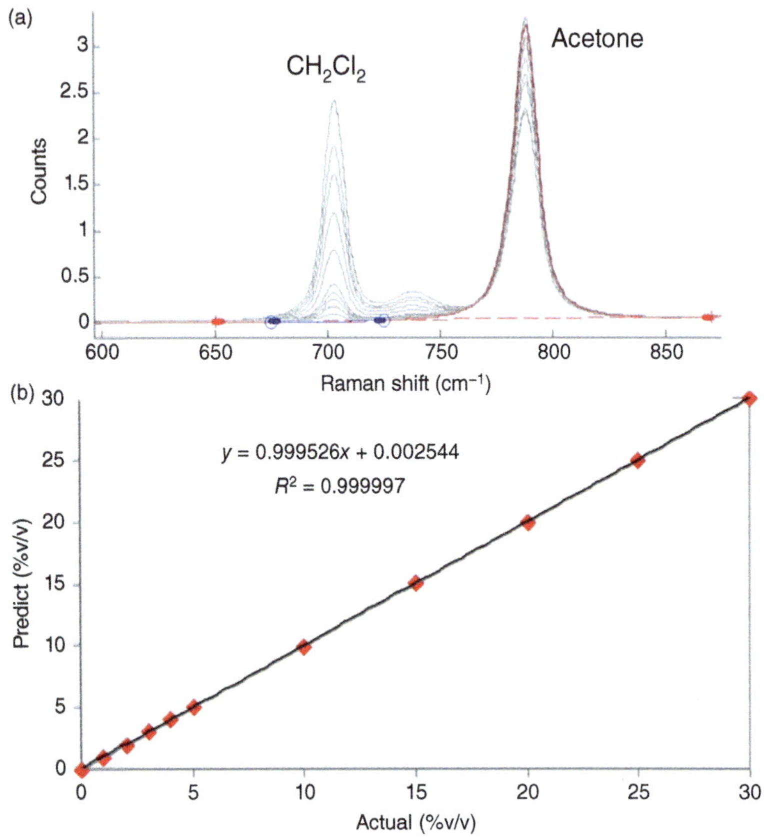

Figure 8.5 Solvent exchange from dichloromethane (CH_2Cl_2) to acetone. (a) Spectral changes; (b) linearity to 30 wt% CH_2Cl_2 in acetone. *Source*: John Wasylyk, BMS.

architecture coupled with real-time NIR monitoring for improved end-product QbD-based efficient manufacturing.

Figure 8.5 shows an example of applying off-line Raman spectroscopy using a 785 nm excitation laser to monitor a distillation process and solvent exchange from dichloromethane (DCM) to acetone. This method was initially developed and utilized as a non-contact off-line method. It was later transferred to an in-line method, which provides maximum safety and productivity enhancement.

An example where in-line spectroscopy and off-line analysis can be combined with modeling tools such as Dynochem® enhances process control and removes the need for excess sampling. This is demonstrated in Figure 8.6, where a solvent exchange from acetonitrile to isopropyl acetate, combined with the remove of excess water is modeled. The initial distillation is followed by a constant volume distillation with the addition of isopropyl acetate. During the unit operation, the residual water is removed with acetonitrile by azeotroping to a point where the residual water remaining is below the acceptable level to proceed to the next synthetic step. The result is only one analysis is required, either in-line or off-line Raman, eliminating the need for Karl Fischer (KF).

The versatility of NIR spectroscopy should not be overlooked. The strong spectral response of water in the NIR region lends itself as an ideal replacement for KF moisture determinations [40]. Figure 8.7 shows how water and three other solvents can be quantified by selecting the appropriate spectral bands for each component, thus eliminating two traditional assays, namely GC and KF.

Off-line spectroscopic methods can eliminate the need for other titration methods, such as acid–base wet chemical analysis. Figure 8.8 presents the linearity of potassium tertiary butoxide (KtBuOx) in tetrahydrofuran (THF) collected on an NIR spectrophotometer. This type of assay is especially important in determining the quality of reagents that are unstable and have a volatile component.

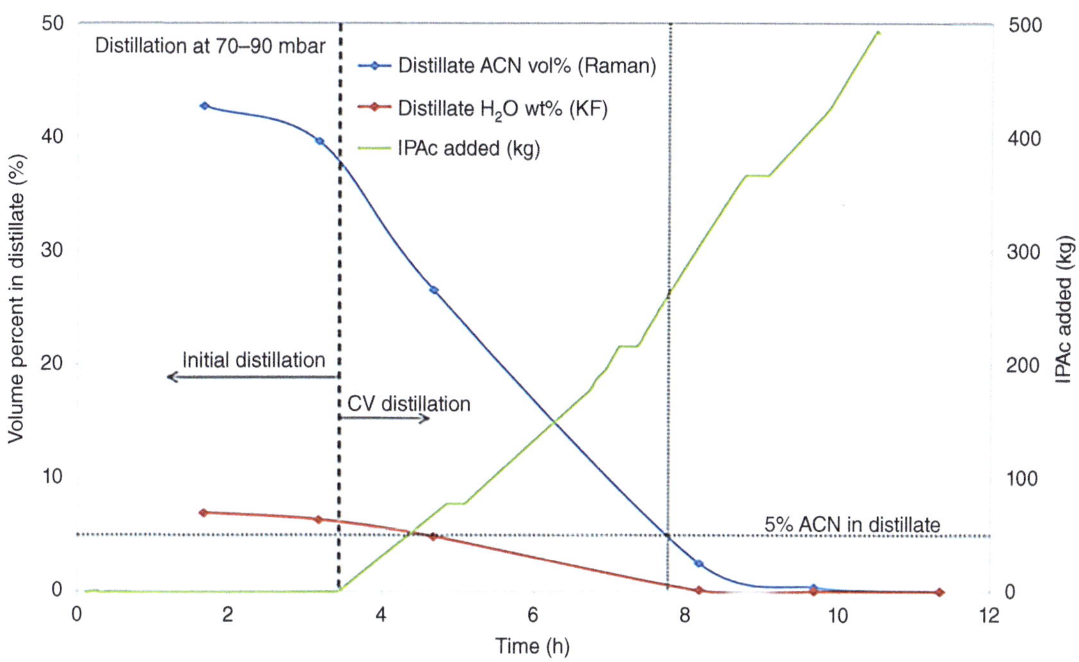

Figure 8.6 Optimization of a distillation unit operation developed using Raman spectroscopy, off-line Karl Fisher, off-line gas chromatography and Dynochem software (Dynochem, Scale-up Suite 2.1, Mettler Toledo, Dublin, Ireland, 2022.) *Source:* John Wasylyk, BMS.

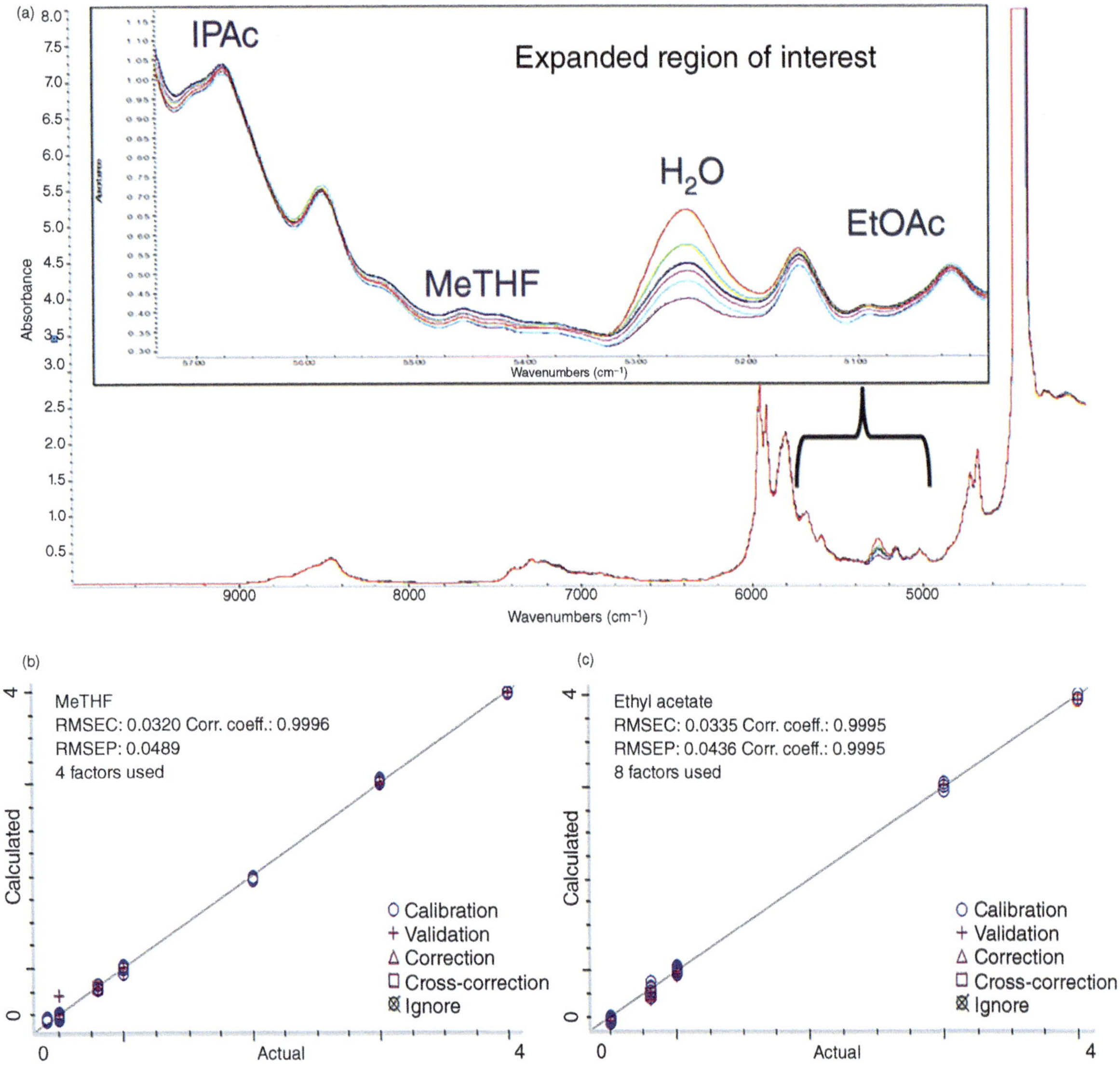

Figure 8.7 (a) FT-NIR spectral regions used for determination of a four-solvent mixture: isopropyl acetate, methyl tetrahydrofuran, water, and ethyl acetate. Calibration curves developed using Thermo TQ Analyst software are shown for (b) MeTHF, and (c) ethyl acetate. *Source*: John Wasylyk, BMS.

8.10.1 Telescope Reactions

The synthesis of chemicals sometimes requires multiple reactions, and tedious work-up between each step is often necessary. Purification between multiple reactions may involve the addition of a quenching reagent, multiple aqueous and organic extractions, filtration, distillation, or crystallization. These operations all require significant input of energy and materials and do not fall in the category of sustainability. One-pot synthesis and telescope processes, coupled with in-line spectroscopy, can circumvent these work-up challenges [41].

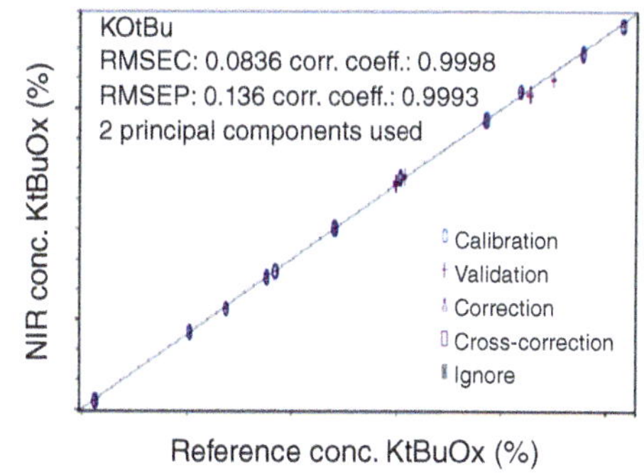

Figure 8.8 Linearity of calibration curve for determination of potassium tertiary butoxide in tetrahydrofuran measured by NIR. *Source*: John Wasylyk, BMS.

An example of a telescope reaction is:

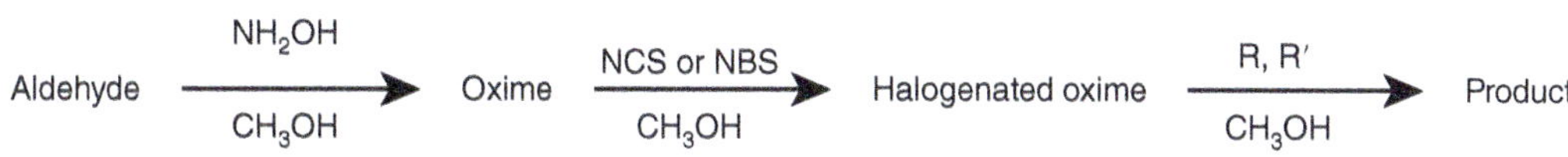

All three steps are amenable to processing in a single reactor and applicable to in-line spectroscopy. An example is the conversion of the oxime to a halogenated oxime, which proceeded in two charges of *N*-chlorosuccinimide (NCS). By utilizing in-line FTIR, the addition and consumption of NCS are observed at 1745 cm^{-1}, while the halogenation of the oxime is observed at 1701 cm^{-1}, as shown in Figure 8.9.

The use of in-line FTIR spectroscopy to support the development and implementation of a two-step telescope reaction in addition to a mid-process distillation step further exemplifies how spectroscopy complements multi-step processes. In this example, a solution of a halogenated benzyl alcohol in *tert*-butyl methyl ether (TBME) with triethylamine and methanesulfonyl chloride (MsCl) yielded a mesylate [42]. The reaction mixture with the new intermediate, mesylate, was quenched, washed, and concentrated at a low volume to remove TBME. Acetone was charged, the mixture was redistilled, and a second portion of acetone was

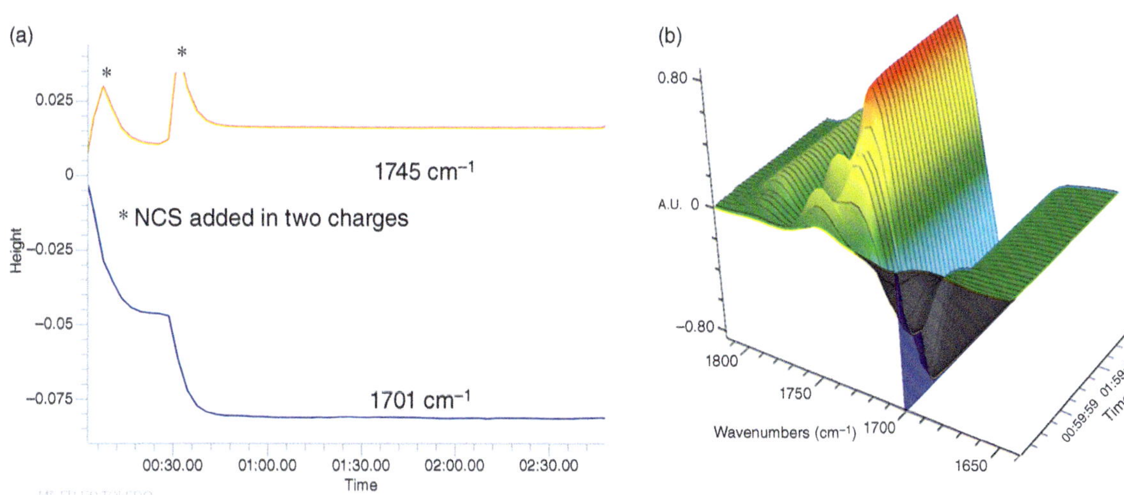

Figure 8.9 (a) Two additions of *N*-chlorosuccinimide (NCS) and the disappearance of the oxime are tracked using in-line FTIR. (b) 3-D plot of the reaction spectra over time. *Source*: John Wayslyk, BMS.

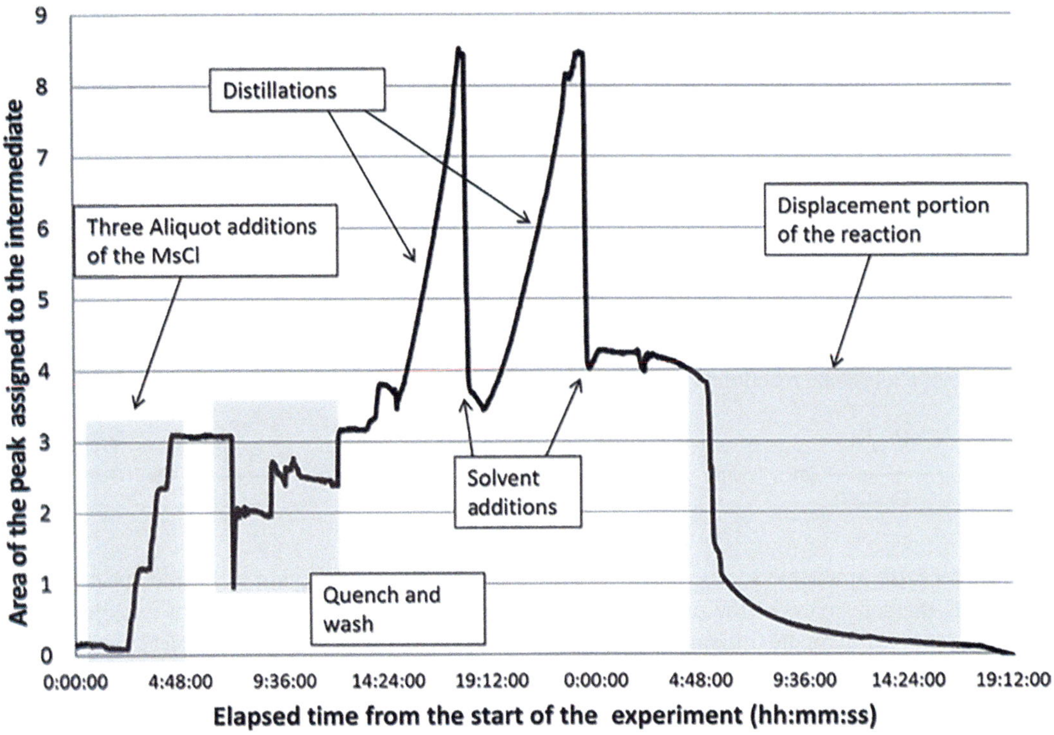

Figure 8.10 Reaction profile for the two-step telescoped reaction and mid-process solvent exchange. *Source*: Clegg et al. [42] / with Permission of SAGE Publications.

added to afford an acetone/mesylate solution. The acetone/mesylate solution was treated with hydroxyl pyridine derivative to afford the product via displacement of the mesylate. Figure 8.10 represents the FTIR data during the entire multi-step process.

When considering the validation of an in-line spectroscopy-based method, the primary method of analysis many times falls on HPLC. A paper published by Hart et al. used PAT for API synthesis complemented with HPLC data [43]. In that study, a heterogeneous etherification reaction involving chloropyrazine and phenol in the synthesis of compound X was monitored by Raman spectroscopy. Raman predictions were used to determine the reaction endpoint. The method and data model were developed at the laboratory scale (250 mL). During method development, solvent charges, different batches of input materials (to model variable fluorescence background), and different baseline correction routines were tested. The end of the reaction was based on %w/w ether. Two pilot-plant-scale (1500 L) reactions were reported. In the first pilot plant batch, Raman predictions were shown to be equivalent to HPLC. Only Raman was used as the analytical control in the second pilot plant batch. As shown in Figure 8.11, Raman predictions of %w/w ether showed that the reaction was completed 600 min before the batch sheet and process description stipulations, indicating more efficient mixing at pilot scales.

8.10.2 Flow Chemistry/Continuous Manufacturing

Recently, there have been enormous advances in the availability of PAT-based instrumentation for flow chemistry challenges. However, analytical products evolved independently of PAT, and so a disconnect remains between the analytical technologies available to a researcher and PAT-enabled tools. One of the primary criteria of qualifying a technology as a PAT tool for flow-analysis lies in its ability to tolerate a wide

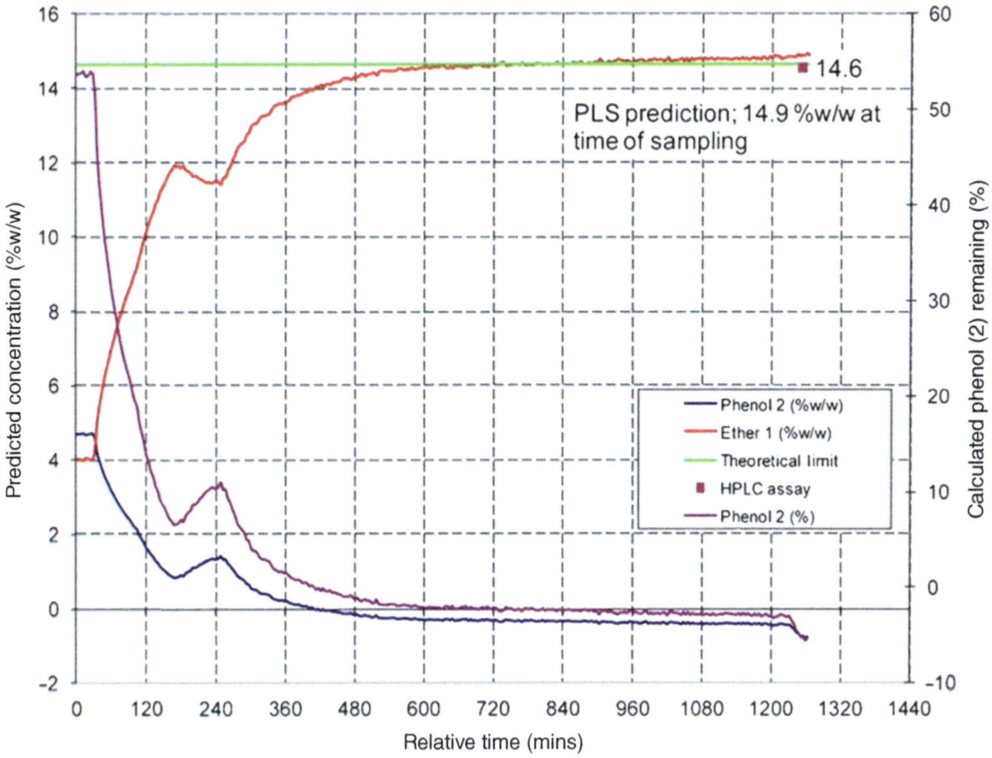

Figure 8.11 Predicted Raman concentrations of phenol and ether. *Source*: Hart et al. [43] / with Permission of American Chemical Society.

range of chemicals and functions in the presence of all phases (i.e., solids, liquids, gases, and mixtures). There are up to two crucial steps for the analysis of a sample from a flow reactor, namely, sampling and analysis. For flow experiments where analysis is performed directly using an inline probe at the flow path of the reactor, a sampling step is not necessary. However, this strategy requires complete compatibility between the detector and the reaction matrices. Various vibrational spectroscopy techniques, coupled with flow cell capabilities, have enhanced the overall field of PAT in flow-based systems. The inclusion of high-definition detectors, which would not only quantitate a target analyte but also confirm the identity of the chemical, is feasible but not popularly applied.

An example of setting up a strategy to ultimately utilize in-line FTIR, a Curtius rearrangement reaction was run in batch mode with an in-line laboratory probe, which provided the chemist with insight into the mechanisms, kinetics of the chemistry being developed and which infrared bands should be monitored. Batch-mode results were then applied to enable the development of the flow chemistry process. An FTIR probe was also incorporated into the flow production process to allow real-time monitoring of the flow chemistry (Figures 8.12 and 8.13). FTIR enabled rapid development and optimization of the Curtius flow chemistry reaction by providing information on the reaction kinetics, flow stream dispersion, impurity profile, and reaction progression and allowed monitoring of the flow system during cleaning verification [44].

After the development of a commercially available flow cell operated with ATR technology, the impact of FTIR in the flow chemistry arena began to flourish [45]. Using standard HPLC connections, this in-line advancement is attached directly to a reaction flow stream and applied to meso- and microscale flow chemistry equipment. A wide range of chemistries can now be closely monitored and controlled using in-line flow cells with the addition of more [46–48].

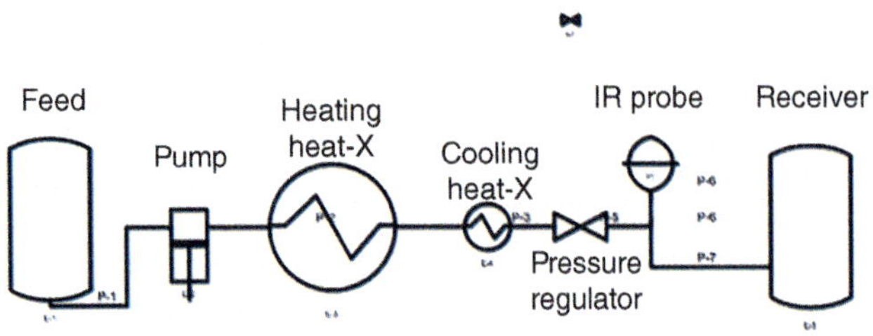

Figure 8.12 Flow system diagram for Curtius reaction [44]. *Source*: Am. Pharm. Review. with Permission of CompareNetworks.

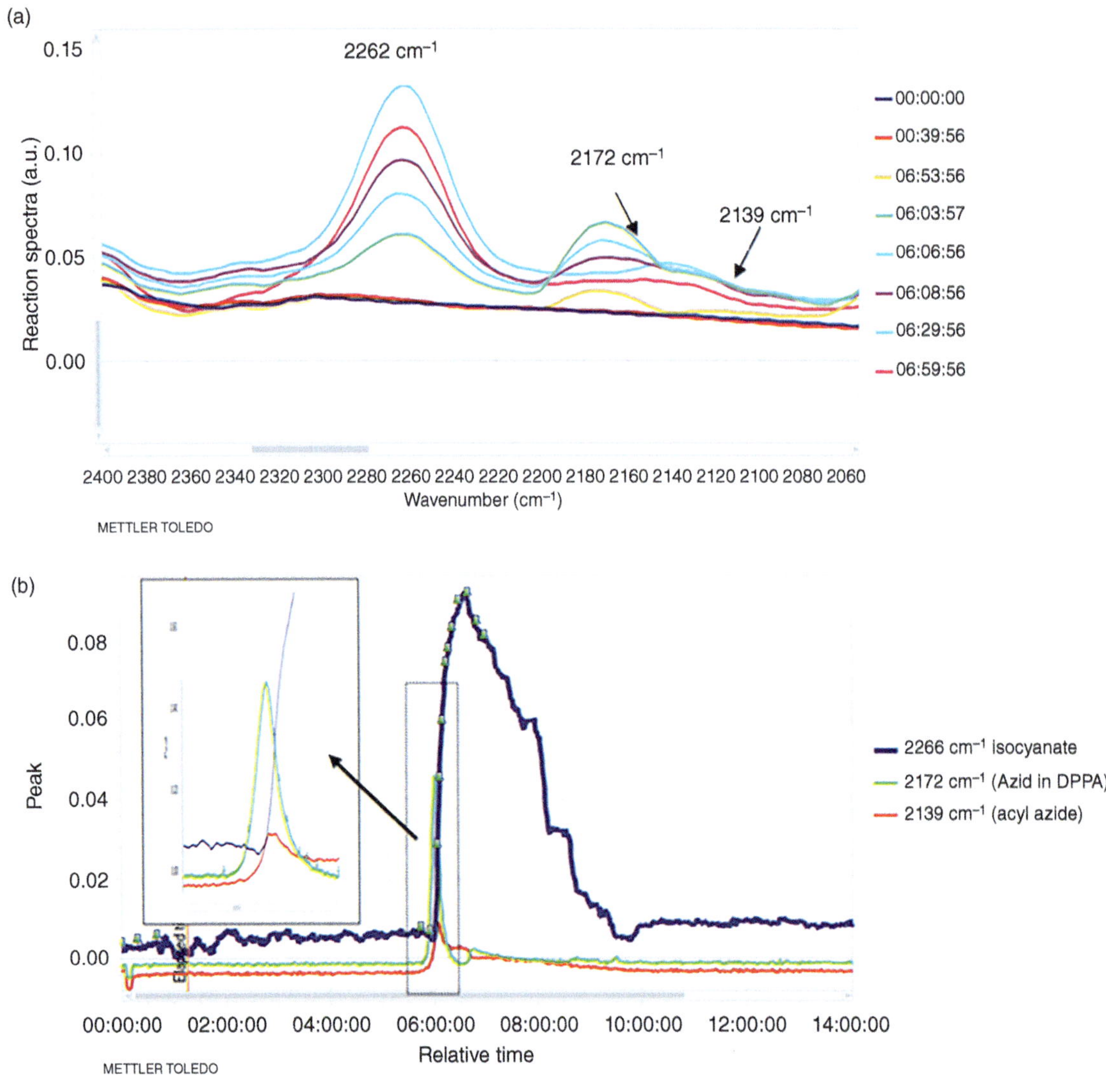

Figure 8.13 (a) FTIR spectral data collected during Curtius rearrangement reaction. (b) Trend analysis using selected bands measured in a flowing system [44]. *Source*: Am. Pharm. Review. with Permission of CompareNetworks.

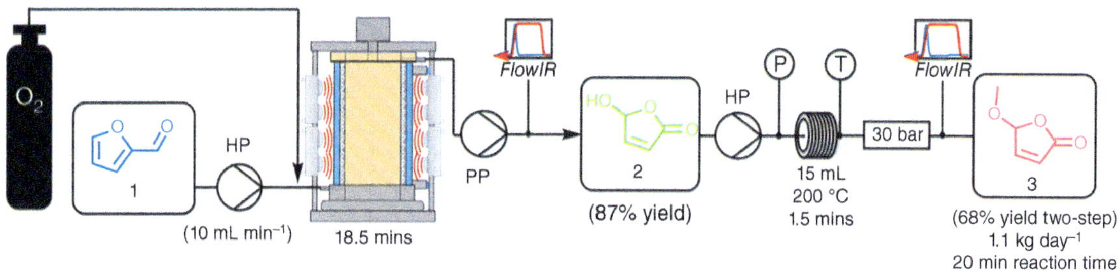

Figure 8.14 A PhotoVortex reactor for the intensification and scale-up of the photo-oxidation of **1** with $^1O_2$ and undertook the first flow synthesis of **3** using superheated reaction solvents to accelerate the reaction and the end result was a telescoped multistep flow process with a potentially significant reduction (sustainable) in solvent waste by eliminating intermediate isolation steps [49]. *Source*: Org. Process Res. Dev with Permission of American Chemical Society.

Martin Edwards et al. recently reported on a scale-up of two consecutive reaction steps for synthesizing alkoxybutenolide monomers [49]. This telescoped route was designed as a kilogram-scale process without isolating a key intermediate between the steps. Control of two photoreactors involved FTIR reaction monitoring and FTIR data analysis via 2D visualization. The first photoreactor was semicontinuous based on a modified rotary evaporator, where FTIR and 2D correlation spectroscopy (2D-COS) revealed the loss of the coproduct. The second step in the telescoped process utilized a continuous Taylor Vortex photoreactor, which enhanced the mass transfer and permitted the use of near-stoichiometric equivalents of O2 and the use of an in-line FTIR and data modeling greatly accelerated process optimization in the Vortex reactor (Figure 8.14).

Utilizing in-line or off-line vibration spectroscopy techniques has advantages over conventional offline characterization techniques (gas chromatography, high-performance liquid chromatography, and mass spectrometry) with respect to sustainability. Coupling spectroscopy with continuous microfluidics, as reviewed by Perro et al., gives access to the microscopic chemical composition of samples, which can be correlated with their macroscopic properties [50]. This approach is widely used in chemistry or biology to get insights into reactive media. Meanwhile, miniaturizing flow chemical reactors offer many advantages, such as small sample size and versatility. Coupling these two approaches creates major opportunities for the analytical field. Small-scale FTIR is one of the leading tools for the analysis of microfluidics, although Raman, fluorescence, and ultraviolet have also come into play.

Not to be excluded, a micro-NIR spectrometer has been applied for in-line monitoring of the synthesis of 5-hydroxymethylfurfural (5-HMF) in a continuous-flow microreactor. Galaverna and colleagues utilized a fixed-bed reactor packed with solid acid catalyst, for the conversion of fructose to 5-hydroxymethylfurfural [51]. Multivariate control charts based on Hotelling's T^2 and Q residuals, coupled with PLS, were also employed for continuous laboratory control over multiple days.

8.10.3 Crystallization

Crystallization is one of the essential unit operations for the separation and purification of solid products in the pharmaceutical industry and is critical to producing high-quality drug substances. The control of the crystallization process determines the polymorph, crystal shape, size, and size distribution of the drug substance [52–55]. The crystallization process can also be governed by kinetics and thermodynamics. Therefore, timely controls are paramount for success. Unit operation control for crystallization has been accomplished by several vibration-based PAT spectroscopies (Raman, NIR, MIR, UV–Vis) and visualization techniques for crystal size, size distribution, and shape, using FBRM, as mentioned earlier. FBRM tracks the number of particles and chord length distribution in real time, providing a quantitative calculation method for crystallization process modeling [56]. When polymorphic control is required, Raman and NIR are most typically used. In some cases, two techniques are utilized and coupled with a feedback/feedforward loop, enabling total process control with the ability to modify the process in real time.

A PAT application for the control of polymorphs and concentration of carvedilol is shown in Figures 8.15 and 8.16. The precise control for crystal forms I and II is accomplished using Raman spectroscopy and

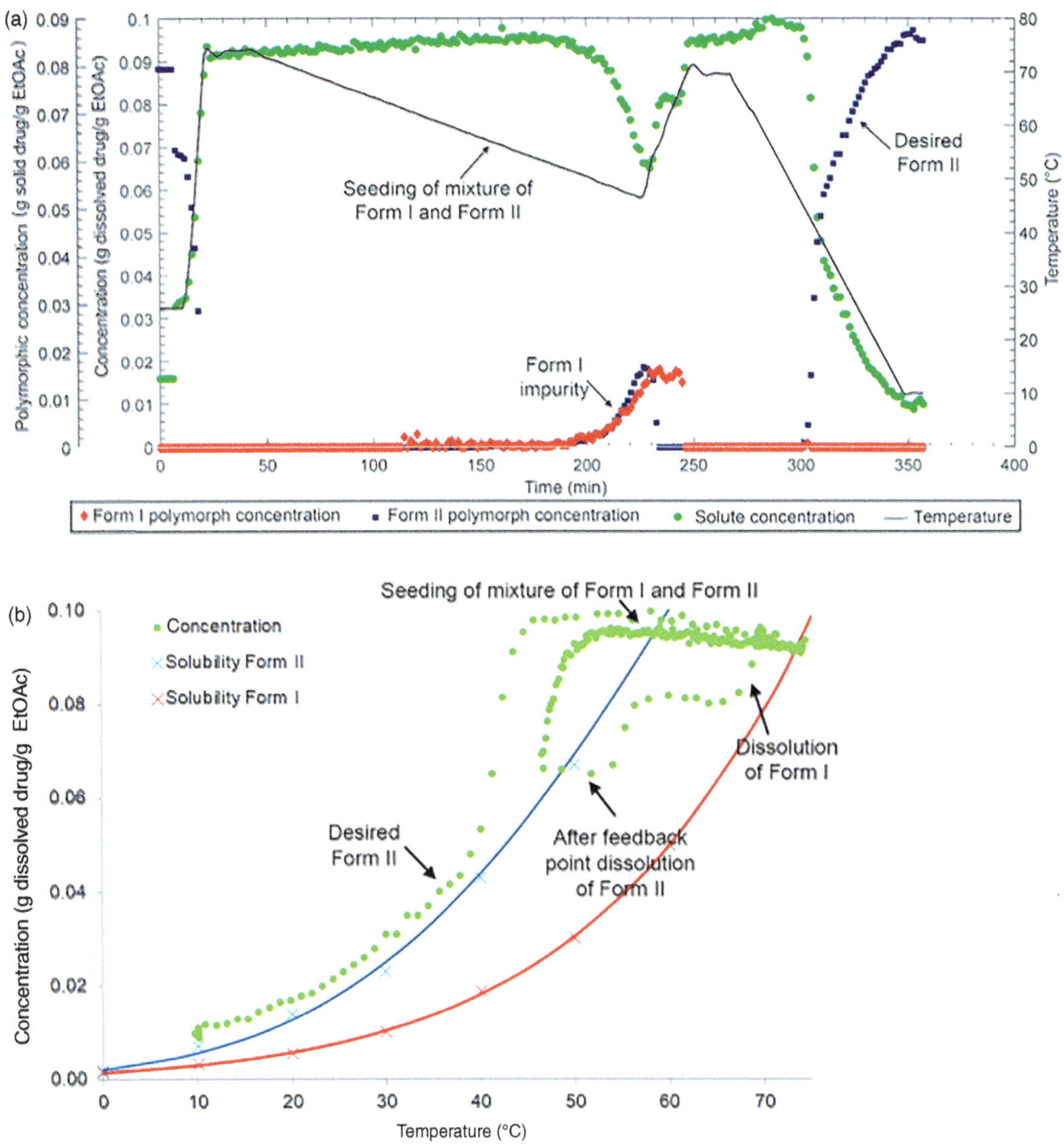

Figure 8.15 (a) The target crystal form is Form I: concentration and temperature profiles; (b) the crystallization phase diagram during the polymorph concentration control process [57]. *Source*: Biotechnology Progress with Permission of American Chemical Society.

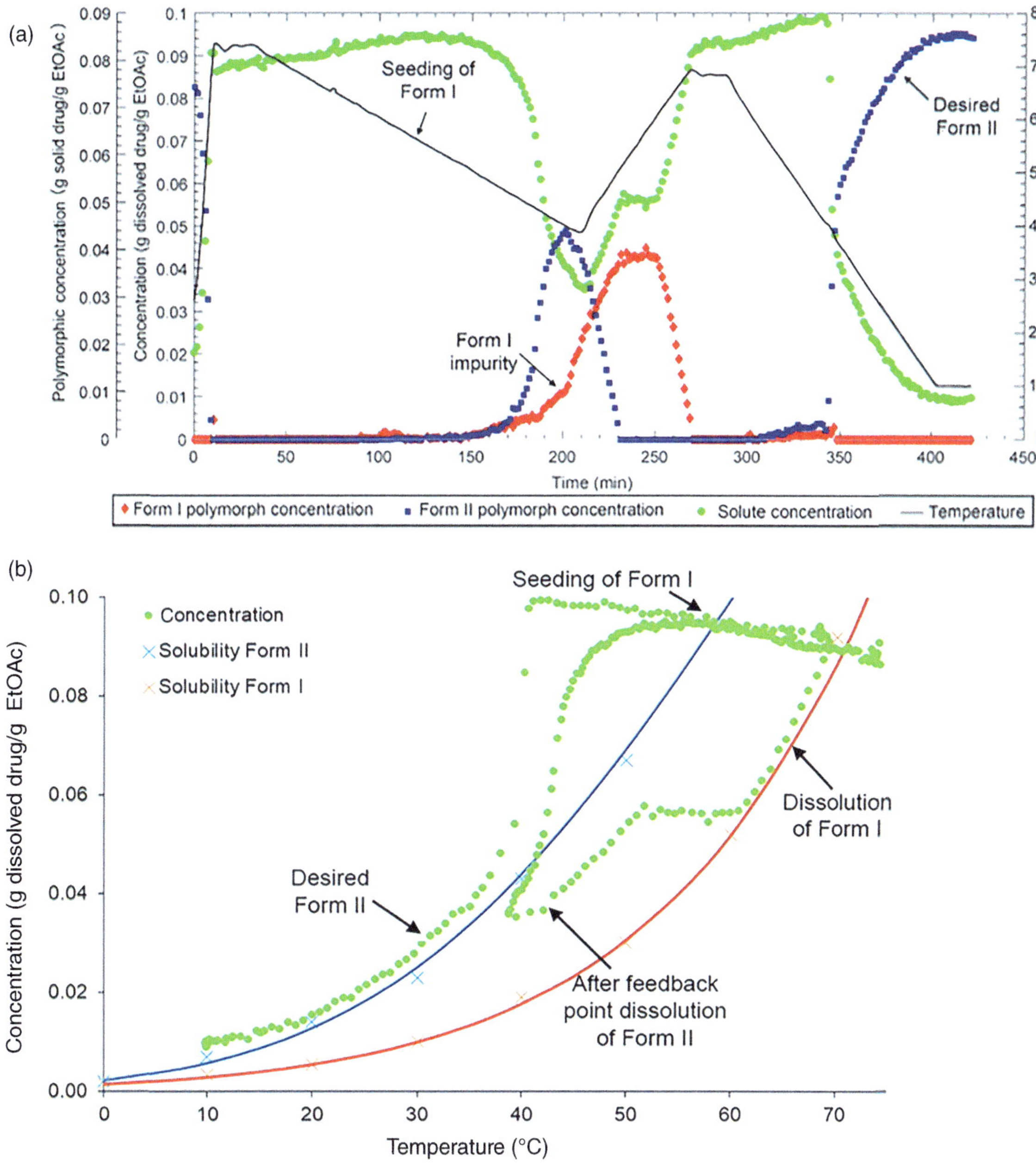

Figure 8.16 (a) The target crystal form is Form II: concentration and temperature profiles; (b) the crystallization phase diagram during the polymorph concentration control process [57]. *Source*: Biotechnology Progress with Permission of American Chemical Society.

temperature control in a feedback design. During the crystallization process, if the desired crystal polymorph is not obtained, the control system will increase the temperature to redissolve the crystals. Once redissolved, the control system will alter the conditions and initiate recrystallization.

This lends itself to a total control unit operation, driving to either form I or form II [57].

An in-line real-time NIR analyzer has been used to measure the moisture content of granules manufactured using a commercial production-scale continuous twin-screw granulator fluid-bed dryer milling process [57]. A central composite response surface design of experiment (DoE) enabled the study of the effect of inlet air temperature and dew point on granule moisture content. The NIR moisture content was compared to KF and loss on drying (LOD) moisture determinations. Using multivariate analysis, the data showed a statistically significant correlation between the conventional methods and NIR. The R^2 values for predicted moisture content by NIR vs. KF and predicted moisture values by NIR vs. LOD were 0.94 ($p < 0.00001$) and 0.85 ($p < 0.0002$), respectively. The application of NIR eliminated the variability which can occur during standard laboratory handling.

8.10.4 Fermentation

A 785 nm laser-based Raman has the advantage of only minimal interference from water when measuring aqueous samples, which makes this method useful for *in situ* monitoring of important industrial bioprocesses. The real-time monitoring of a *Saccharomyces cerevisiae* fermentation process using a Raman spectroscopy was developed to correct the Raman signal for the attenuation caused by light-scattering cell particulate. This enabled the quantification of reaction components and the measurement of yeast cell concentrations. Normalization of the approximated extinction expressions using Raman signal of water as an internal standard permitted correction for the effect of cell scattering. In order to simplify the analysis process, estimations were made with a two-step approach, where initial scattering correction of attenuated signals was followed by linear regression. *In situ* quantification measurements of the fermentation resulted obtaining concentrations for glucose, ethanol, and yeast concentrations [58].

The application of Raman spectroscopy has been expanded to include the simultaneous quantitative determination of glucose, glutamine, lactate, ammonia, glutamate, total cell density (TCD), and viable cell density (VCD) in a CHO fed-batch process [59]. The work was demonstrated *in situ* in 3 L and 15 L bioreactors. In this case, spectral preprocessing and PLS regression were used to correlate spectral data with off-line reference data.

A more recent approach to process analytics for developing an ever-increasing array of biologics involves deploying cutting-edge technologies and cyber-physical systems. Process Analytics 4.0, which entails not only technology aspects such as PAT, assay automation, and high-throughput analytics but also cyber-physical systems that enable data management, visualization, augmented reality, and Internet of things (IoT) infrastructure for real-time analytics in a process development environment. This approach is reviewed by D. Wasalathanthri and is focused on dissecting high-level features of PAT, automation, and data management while not short-changing the business aspects of implementing during process analytical testing in biologics process development [60].

8.11 Non-PAT Applications

Perhaps the most utilized and yet least publicized use of vibrational spectroscopies in both pharmaceutical and chemical industries are QC and unknown identifications. Library spectral search software for foreign material, unknown identification, and counterfeit is readily available, especially for IR spectra. Ease of library maintenance, with ATR-based collected spectra, enables the investigator to identify foreign material during the manufacturing process of clinical drug candidates. Foreign material may be introduced into pharmaceutical products at almost any stage, resulting in non-conformance incidents. Foreign materials could be accidentally brought in from raw materials, manufacturing equipment, manufacturing processes, or deviations introduced by contract manufacturers. All these potential sources could contribute to contaminants in the final product. Investigative analysis aims to identify the foreign materials and find their root causes to prevent further contamination. However, the identification task can be complicated since the origin and nature of foreign matter are various. FTIR microspectroscopy plays a key role by providing information on the functional groups present, leading to the identification of even the smallest of contaminants. Furthermore, unknown materials could be identified by comparing their spectra with library standard spectra since almost every molecule has a unique FTIR spectrum [61, 62].

Despite the increased use of robust quality systems, the product still may contain unidentified defects and discolorations. The foreign matter must be monitored and identified, and its source must be defined to prevent further contamination. However, the identification task can be complicated since the origin and nature of foreign matter are various. This includes the biopharma arena, where FTIR spectroscopy can detect impurities in solution, such as detergents, buffer residues, and protein contaminants in manufacturing, not detectable by reversed-phase HPLC electrospray ionization mass spectrometry(RP-HPLC-ESI-MS) or sodium dodecyl sulfate-polyacrylamide gel electrophoresis (SDS-PAGE). This detection of impurities is possible by comparing the FTIR spectrum of a sample and its subtraction spectrum with an appropriate spectral database of standard reference spectra [63].

8.12 Conclusion

Implementing spectroscopy within the mainstream of all pharmaceutical development and manufacturing stages includes many key considerations. This begins with selecting the appropriate instrument technology, reviewing the infrastructure to support the instrument, its overall path forward and future, and the value proposition. Whether it's for information only during the development cycle or possible implementation in manufacturing, where regulatory considerations must be pre-addressed, every step must be carefully thought through. All steps required for successfully implementing any spectroscopy-based instrument rely on a team effort to manage the wide range of challenges.

References

1. Food and Drug Administration (2004). Draft guidance for industry: PAT – a framework for innovative pharmaceutical manufacturing and quality assurance; availability. Food and Drug Administration.
2. Bakeev, K. (ed.) (2008). *Process analytical technology: spectroscopic tools and implementation strategies for the chemical and pharmaceutical industries*. Chichester, UK: Wiley.
3. Wasylyk, J., Huang, M., Wethman, B., and O'Connor, K. (2019). Spectroscopy facilitates lean analysis. *Pharmaceutical Technology* 43: 35–39.
4. Clegg, I.M., Daly, A.M., Donnelly, C. et al. (2012). Application of mid-infrared spectroscopy to the development and transfer of a manufacturing process for an active pharmaceutical ingredient. *Applied Spectroscopy* 66 (5): 574–579.
5. Ando, M., Fujita, M., Izato, Y.-i., and Miyake, A. (2020). Autocatalytic reaction mechanism of nitric acid and formic acid mixtures based on thermal and in situ Raman spectroscopic analyses. *Journal of Thermal Analysis and Calorimetry* 144: 1–10.
6. Tuchel, D. (2016). Selecting an excitation wavelength for Raman Spectroscopy. *Spectroscopy* 31 (3): 14–23.
7. Larkin, P.J. (2018). *Infrared and Raman Spectroscopy: Principles and Spectral Interpretation*, 2e. Elsevier.
8. Larkin, P., Wasylyk, J., and Raglione, M. (2015). Application of low- and mid-frequency raman spectroscopy to characterize the amorphous-crystalline transformation of indomethacin. *Applied Spectroscopy* 69 (11): 1217–1228.
9. Esmonde-White, K.A., Cuellar, M., Uerpmann, C. et al. (2017). Raman spectroscopy as a process analytical technology for pharmaceutical manufacturing and bioprocessing. *Analytical and Bioanalytical Chemistry* 409: 637–649.

10 Kail, N., Marquardt, W., and Briesen, H. (2009). Process analysis by means of focused beam reflectance measurements. *Industrial & Engineering Chemistry Research* 48: 2936–2946.

11 Fovette, G. (2007). In situ Raman spectroscopy for in-line control of pharmaceutical crystallization and solids elaboration processes: a review. *Chemical Engineering Research and Design* 85 (7): 906–920.

12 IECEx 01A (2018). An Informative Guide comparing various elements of both IECEx and ATEX Ed. 3.

13 European Commission (2014). Directive 2014/34/EU of the European Parliament and of the Council of 26 February 2014 on the harmonisation of the laws of the Member States relating to equipment and protective systems intended for use in potentially explosive atmospheres (recast). OJ L 96, 29.3.2014ATEX 2014/34/EU Guidelines, 26052020.

14 National Fire Protection Association (2021). Recommended practice for the classification of flammable liquids, gases, or vapors and of hazardous (classified) locations for electrical installations in chemical process areas. NFPA 497-2021.

15 National Fire Protection Association (2021). Recommended practice for the classification of combustible dusts and of hazardous (classified) locations for electrical installations in chemical process areas. NFPA 499.

16 National Fire Protection Association (2021). National Electrical Code (NEC). NFPA 70.

17 Alsberg, B.K., Kell, D.B., and Goodacre, R. (1998). Variable selection in discriminant partial least-squares analysis. *Analytical Chemistry* 70: 4126–4133.

18 Forina, M., Casolino, M.C., and De La Pezuela, M.C. (1998). Multivariate calibration: applications to pharmaceutical analysis. *Journal of Pharmaceutical and Biomedical Analysis* 18: 21–33.

19 Read, E., Park, J., Shah, R. et al. (2010). Process analytical technology (PAT) for biopharmaceutical products: Part I. Concepts and applications. *Biotechnology and Bioengineering* 105: 276–284.

20 Pomerantsev, A. and Rodionova, O. (2012). Process analytical technology: a critical view of the chemometricians. *Journal of Chemometrics* 26: 299–310.

21 Jelena Djuris, J., Ibric, S., and Đurić, Z. (2024). *Quality by Design in the Pharmaceutical Development. Computer-Aided Applications in Pharmaceutical Technology*, 2e. Philadelphia: Woodhead.

22 Miller, C.E. (2010). Chemometrics in PAT. In: *Process Analytical Technology: Spectroscopic Tools and Implementation Strategies for the Chemical and Pharmaceutical Industries* (ed. K. Bakeev). Wiley.

23 Setarehdan, S.K., Soraghan, J.J., Littlejohn, D., and Sadler, D.A. (2002). Maintenance of a calibration model for near infrared spectrometry by a combined principal component analysis—partial least squares approach. *Analytica Chimica Acta* 452: 35–45.

24 Mehmood, T. (2016). Hotelling T2 based variable selection in partial least squares regression. *Chemometrics and Intelligent Laboratory Systems* 154 (15): 23–28.

25 Nikzad-Langerodi, R., Lughofer, E., Cernuda, C. et al. (2018). Calibration model maintenance in melamine resin production: Integrating drift detection, smart sample selection and model adaptation. *Analytica Chimica Acta* 1013: 1–12.

26 U.S. Food & Drug Administration (2016). CFR – Code of Federal Regulations Title 21.

27 Asua, J.M. and Leiza, J.R. (2005). Monitoring of emulsion polymerization reactors by Raman spectroscopy: calibration model maintenance. *Applied Spectroscopy* 59 (10): 1280–1285.

28 Li, Q., Chen, M., and Zhang, T. (2025). Closed-loop transfer brightens up artificial intelligence for chemistry. *Green Carbon* 12: 001.

29 Lewin, D. and Scali, C. (1988). Feedforward control in the presence of uncertainty. *Industrial & Engineering Chemistry Research* 27 (12): 2323–2331.

30 Bolton, W. (2009). *Programmable Logic Controllers*. New York: Elsevier.

31 Jeng, J.C. and Ge, G.P. (2018). Data-based approach for feedback-feedforward controller design using closed-loop plant data. *ISA Transactions* 80: 244–256.

32 Quintanilla, P., Navia, D., Moreno, F. et al. (2023). A methodology to implement a closed-loop feedback-feedforward level control in a laboratory-scale flotation bank using peristaltic pumps. *MethodsX* 10: 102081.

33 Al-Saedi, M.I. (2019). Enhancing the feedforward – feedback controller for nonlinear overhead crane using fuzzy logic controller. *IOP Conference Series: Materials Science and Engineering, Baghdad.*

34 Solution Pharmacy (2021). Pilot plant scale-up techniques. https://solutionpharmacy.in/pilot-plant scale-up (accessed 11 June 2021).

35 Gertenbach, D. and Cooper, B.L. (2009). *Scale-Up Issues From Bench to Pilot*. Nashville, TN: AIChE National Meeting.

36 Poliakoff, M. and Licence, P. (2007). Sustainable technology: Green chemistry. *Nature* 450 (7171): 810–812.

37 Gouveia, F.F., Rahbek, J.P., Mortensen, A.R. et al. (2017). Using PAT to accelerate the transition to continuous API manufacturing. *Analytical and Bioanalytical Chemistry* 409: 821–832.

38 Singh, R., Sahay, A., Karry, K. et al. (2014). Implementation of a hybrid MPC-PID, control strategy using PAT tools into a direct compaction continuous pharmaceutical tablet manufacturing pilot-plant. *International Journal of Pharmaceutics* 473: 38–54.

39 Singh, R., Sahay, A., Muzzio, F. et al. (2014). Systematic framework for onsite design and implementation of the control system in continuous tablet manufacturing process. *Computers & Chemical Engineering* 66: 186–200.

40 Wasylyk, J., Mukherjee, S., Wethman, R. et al. (2019). *Spectroscopy: Turning Analytical Support Green.* Ralston: ACSGCE.

41 Wasylyk, J., Mukherjee, S., Wethman, R., and Huang, M. (2020). *Deep Dive into Telescoped Reaction in Pharmaceutical Development*. Virtual: ACSGCE.

42 Clegg, I.M., Daly, A.M., Donnelly, C. et al. (2012). Application of mid-infrared spectroscopy to the development and transfer of a manufacturing process for an active pharmaceutical ingredient. *Applied Spectroscopy* 66 (5): 574–579.

43 Hart, R., Pedge, N., Steven, A., and Kevin Sutcliffe, K. (2015). In situ monitoring of a heterogeneous etherification reaction using quantitative Raman spectroscopy. *Organic Process Research & Development* 19 (1): 196–202.

44 Xiang, Y., Lucas, J., VanAlsten, J. et al. (2012). Using process analytical technology (PAT) tools to support flow chemistry development and production. *American Pharmaceutical Review* 15 (3): 56.

45 Carter, C.F., Lange, H., Ley, S.V. et al. (2010). ReactIR flow cell: a new analytical tool for continuous flow chemical processing. *Organic Process Research and Development* 14 (2): 393–404.

46 Brodmann, T., Koos, P., Metzger, A. et al. (2012). Continuous preparation of arylmagnesium reagents in flow with inline IR monitoring. *Organic Process Research and Development* 16: 1102–1113.

47 Smith, C., Baxendale, I., Lange, H., and Ley, S. (2011). Fully automated, multistep flow synthesis of 5-amino-4-cyano-1,2,3-triazoles. *Organic & Biomolecular Chemistry* 9: 1938–1947.

48 Lange, H., Carter, C., Hopkin, M., and Ley, S. (2011). A breakthrough method for the accurate addition of reagents in multi-step segmented flow processing. *Chemical Science* 2: 765–769.

49 Edwards, M., Pratley, M., Gordon, C. et al. (2024). Process intensification of the continuous synthesis of bio-derived monomers for sustainable coatings using a taylor vortex flow reactor. *Organic Process Research and Development* 28: 1917–1928.

50 Perro, A., Lebourdon, G., Henry, S. et al. (2016). Combining microfluidics and FT-IR spectroscopy: towards spatially resolved information on chemical processes. *Reaction Chemistry & Engineering* 1: 577–594.

51 Galaverna, R., Ribessi, J., Rohwedder, J., and Pastre, J. (2018). Coupling continuous flow microreactors to microNIR spectroscopy: ultracompact device for facile in-line reaction monitoring. *Organic Process Research and Development* 22 (7): 780–788.

52 Chablani, L., Taylor, M., and Mehrotra, A. (2011). Real-time near-infrared granule moisture measurements of a continuous granulation–drying–milling process. *AAPS PharmSciTech* 12: 1050–1055.

53 Kitamura, M., Hara, T., and Takimoto-Kamimura, M. (2006). Solvent effect on polymorphism in crystallization of BPT propyl ester. *Crystal Growth & Design* 6: 1945–1950.

54 Chieng, N., Rades, T., and Aaltonen, J. (2011). An overview of recent studies on the analysis of pharmaceutical polymorphs. *Journal of Pharmaceutical and Biomedical Analysis* 55: 618–644.

55 Bernstein, J. (2011). Polymorphism—a perspective. *Crystal Growth & Design* 11: 632–650.

56 Zhou, G., Moment, A., Cuff, J. et al. (2014). Process development and control with recent new FBRM, PVM, and IR. *Organic Process Research and Development* 19: 227–235.

57 Tacsi, K., Gyürkés, M., Csontos, I. et al. (2019). Polymorphic concentration control for crystallization using Raman and attenuated total reflectance ultraviolet visible spectroscopy. *Crystal Growth & Design* 20: 73–86.

58 Iversen, J., Berg, R., and Ahring, B. (2014). Quantitative monitoring of yeast fermentation using Raman spectroscopy. *Analytical and Bioanalytical Chemistry* 406: 4911–4919.

59 Whelan, J., Craven, S., and Glennon, B. (2013). In situ Raman spectroscopy for simultaneous of multiple process parameters in mammalian cell culture bioreactors. *Biotechnology Progress* 28: 1355–1362.

60 Wasalathanthri, D.P., Shah, R., Ding, J. et al. (2021). Process analytics 4.0: a paradigm shift in rapid analytics for biologics development. *Biotechnology Progress* 37 (4): e3177.

61 Li, G., Torraca, G., Jing, W., and Wen, Z. (2009). Applications of FTIR in identification of foreign materials for biopharmaceutical clinical manufacturing. *Vibrational Spectroscopy* 50 (1): 152–159.

62 Pajander, J., Haugshøj, K., Bjørneboe, K. et al. (2013). Foreign matter identification from solid dosage forms. *Journal of Pharmaceutical and Biomedical Analysis* 80: 116–125.

63 Tiernan, H., Byrne, B., and Kazarian, S. (2020). ATR-FTIR spectroscopy and spectroscopic imaging for the analysis of biopharmaceuticals. *Spectrochimica Acta - Part A: Molecular and Biomolecular Spectroscopy* 241: 118636.

9

Simultaneous Absorbance and Fluorescence Instrument Measuring and Analysis Principles for Quantitative Analysis

Adam Gilmore[1] and Doreen Schober[2,3]

[1] *Horiba Instruments Inc., Piscataway, NJ, USA*
[2] *Concha y Toro, Pencahue, Maule, Chile*
[3] *Organic Agricultural Sciences, Organic Food Quality, University of Kassel, Witzenhausen, Hessen, Germany*

9.1 Introduction

This chapter describes the fundamental principles and analytical advantages of simultaneous analysis of UV–Vis absorbance and fluorescence excitation–emission matrix (EEM) spectroscopic data. The simultaneously acquired UV–Vis and EEM data is significant because it can provide a comprehensive qualitative and quantitative survey of all colored and fluorescing compounds in a sample. This is far more advantageous statistically than collecting individual absorbance and fluorescence spectra as it will be one of the main points of this chapter. It is first acknowledged that in practice, acquiring UV–Vis and EEM data can be accomplished by measuring both sets of data in parallel on separate absorbance and fluorescence spectrophotometers. However, it is also noted there are possible issues and expected complications with matching and processing the data since the instruments may use different optical configurations, data processing steps, and data formats; there would also be a kinetic lag variance between the two measurements, which may be influenced by sample changes such as oxidation. Therefore, we primarily discuss collecting absorbance and EEM data simultaneously with a patented instrument configuration [1, 2]. The data acquisition and processing methods discussed herein can, in principle, apply to parallel acquisition instrument methods.

The chapter begins by schematically outlining the essential optical and instrumental components and configurations required for the simultaneous processing of UV–Vis and EEM data and to distinguish parallel and simultaneous data acquisitions. The synergistic aspects of the simultaneous UV–Vis and EEM analysis are described in detail as these pertain to the fundamental physical and optical behavior of chromophores (colored molecules) and fluorophores in solution. It is well known that most chromophores and fluorophores in solution have broadened spectra and that combining absorbance and fluorescence EEM spectroscopy can significantly increase the specificity for chemical recognition and quantification. The chapter focuses on the key roles that Beer–Lambert linearity, solvent properties, and temperature, among other chemical factors, play in the capability to identify and quantify specific chemicals in solution using the method. The chapter also includes detailed guidelines on instrument and sample handling, including working with automated samplers and cuvettes. The synergistic benefits of simultaneous UV–Vis–EEM measurements, including the ability to correct for concentration-dependent distortions by fluorescence inner filter effects (IFEs) [1, 3], are explained. The collection and processing of the data is described with a focus on organizing the UV–Vis and EEM data into a Multi-Block format and coordinating with the sample class identity (ID) and sample concentration information. Compliance with standard protocols defined by the American Society for Testing

Practical Quantitative Vibrational and Electronic Spectroscopy: A Guide for Developing, Optimizing, and Validating Procedures, First Edition. Edited by Deborah A. Peru.

Companion website: www.wiley.com/go/peruelectronicspectroscopy

Materials (ASTM), International Standards Organization (ISO), and the United States Pharmacopoeia (USP) for empirical model calibration and determining limits of detection and quantification (LOD and LOQ), etc., are also discussed, referenced including cross-referencing to other chapters in this volume. This chapter proceeds to describe, using typical examples, how the UV–Vis–EEM data can be effectively analyzed with univariate and multivariate (both linear and nonlinear) regression models for quantification as well as by quantitative discrimination based on an analyte concentration threshold. Lastly, the examples of both quantitative linear regression and a concentration-threshold based discrimination (Pass/Fail) analysis are described to support the utility of the method for routine laboratory quality control (QC) measurements and applied research problems.

9.2 Simultaneous UV–Vis and Fluorescence Excitation–Emission Matrix Optical Component Configuration

Figure 9.1 schematically outlines the basic elements and organization of the optical components and light path needed for the simultaneous UV–Vis and fluorescence EEM measurement. Noting the elements in ***bold*** represent those embodied in simultaneous instruments compared to those shared with or used in parallel instruments.

A single polychromatic light source or combination of switchable sources with output from the UV–Vis region is generally used. In practice for most UV–Vis instruments the UV component is associated with a deuterium (D2) lamp, while a tungsten (W) lamp provides the longer wavelength components; alternately a pulsed xenon lamp (Xe-Flash) [3] can also emit light over the full UV–Vis spectrum with enriched output in the deep UV. Notably these sources are normally of relatively low power (wattage) and a higher power 150 W xenon arc (Xe-Arc) lamp (ozone producing) can itself provide full spectrum emission from 200 to 1000 nm for both UV–Vis and effective fluorescence excitation. In both UV–Vis and EEM measurements the polychromatic light is typically monochromated with either a single-grating monochromator or for additional

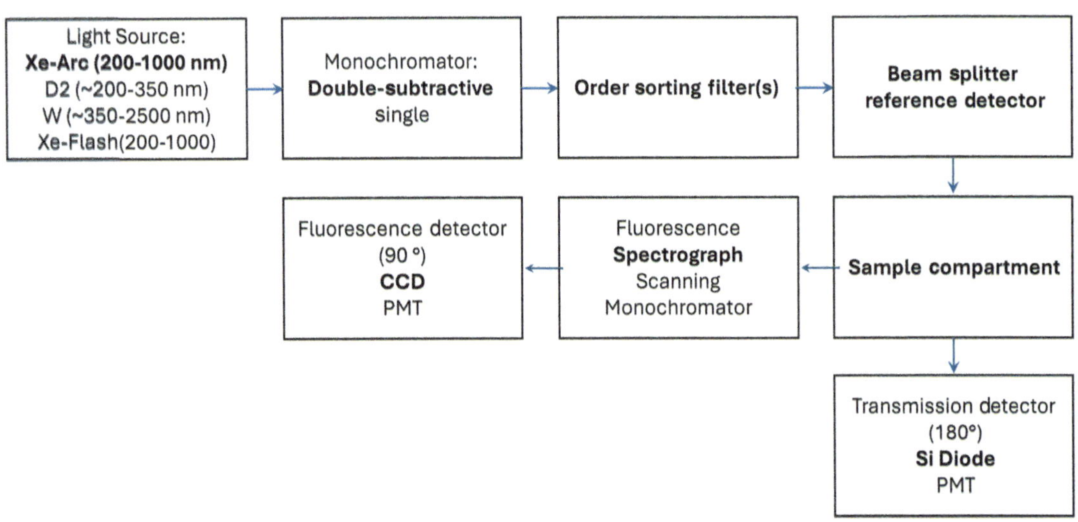

Figure 9.1 Conceptual schematic of light path (arrows) and optical components (boxes) associated with simultaneous absorbance–transmittance and fluorescence EEM spectroscopy instrument. Items in boldface are associated with a patented [2] simultaneous acquisition instrument, while non-bolded items are associated with a parallel (separate) instrument configuration. *Source*: Adam Gilmore.

stray light rejection a double-grating monochromator can be employed in either additive or subtractive mode [2, 3]. Because all grating-based monochromators emit light in 0th, first, second, and higher orders it is essential for accurate absorbance and fluorescence excitation to employ an order-sorting filter mechanism. This is usually accomplished by using an automatic filter selector (typically a wheel) that blocks higher-order transmission, while the monochromator is paused in the correct position to allow the filter(s) to move into position before continuing the scan.

Once the monochromated light has been order sorted the intensity must be measured immediately before it enters the sample compartment and interacts with the sample. This can be accomplished with a reference detector using a pick-off beam splitter that reflects a small, focused portion of the exciting beam onto a reference detector. The spectral characteristics of the beam splitter are radiometrically corrected by a factor measured by placing a calibrated detector in the sample position [4]. The main purpose of the reference correction is to correct for primarily spectral but also kinetic variations in the light source during measurements. For a double-beam UV–Vis measurement the transmitted beam is measured after passing through the sample cell by a transmission detector; the transmitted signal is corrected by dividing its signal by the reference detector signal as a function of wavelength.

The fluorescence emission signal itself can be measured, ideally at 90° to the sample plane as a function of wavelength using either a scanning monochromator and a single channel detector (usually a photomultiplier tube) or with a spectrograph and a multichannel detector such as a charge-coupled device (CCD). In both cases the spectral response of the emission detection (primarily the detector and monochromator or spectrograph-grating combination) needs to be radiometrically corrected for intra- and interlaboratory comparisons. This can be accomplished by comparing the technical emission spectrum to the spectrum of a calibrated light source or by measurement of spectral irradiance standard sample(s) to generate an emission correction factor [1, 4]. To complete the UV–Vis–EEM measuring sequence it is standard practice as with UV–Vis to scan the spectrum from lower-energy (redder) wavelengths to higher-energy (bluer) wavelengths in discrete intervals, dwelling at each interval for a set integration time to either average or sum the signals (depending on the type of detector used). At each absorbance and excitation wavelength the fluorescence emission spectrum is collected by either scanning the emission monochromator or integrating the spectrum from the multichannel detector-spectrograph.

While the above Figure 9.1 description generically explains the concepts and optical elements of UV–Vis–EEM measurements as may be extrapolated to parallel or simultaneous acquisition, Figure 9.2 provides more detail into the specific optical configuration and key functional features of the instrumentation used in the applications discussed in Chapter 10. The Aqualog® instrumentation made by HORIBA is a patented simultaneous absorbance–Transmittance fluorescence EEM instrument. The main advantage of the optical design is the ability to collect absorbance and fluorescence spectra from the same liquid sample placed into the sampling compartment. The primary features of the optical design, exemplified here by the patented instrument, include:

A) The power and wide spectral range of the 150 W vertically mounted ozone-bearing exciting light source to allow excitation of deep UV-absorbing chromophores and fluorophores continuing up to the near-infrared (NIR) range
B) High-stray light rejection for the holographic deep UV (250 nm) blazed gratings and double-subtractive monochromator (fixed 5 nm bandpass) design to remove scattered light interference, especially important in the deep UV spectral region where photon energy is highest
C) Order-sorted excitation optical path to prevent exposure of the sample to higher orders of light while measuring absorbance and fluorescence excitation
D) Geometrically matched all reflective excitation and emission sampling optics to eliminate color-dependent (chromatic) effects associated with lens-based collection optics and ensure optical focus on the sample at all wavelengths to increase sensitivity

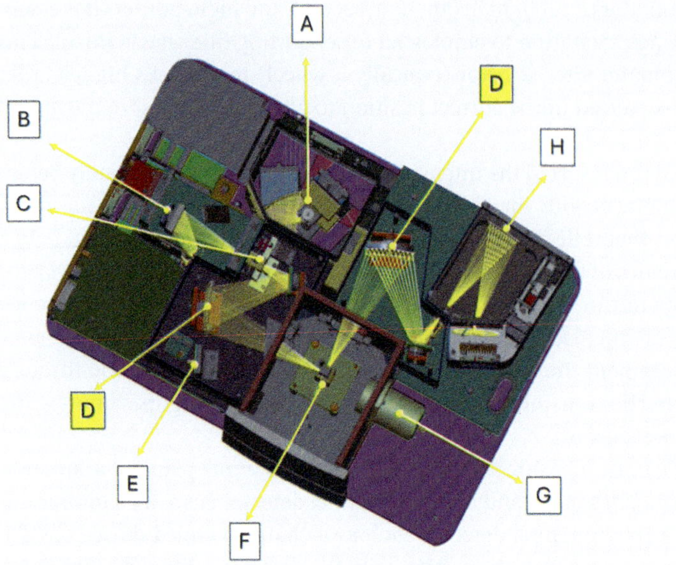

Figure 9.2 Patented optical bench for simultaneous absorbance–transmittance and fluorescence EEM instrumentation. See text for details [2]. *Source*: Adam Gilmore.

E) A spectrally corrected reference diode detector (200–1000 nm) to account for changes in the light source intensity as a function of time and wavelength for signal stability and intra- and interlaboratory comparison
F) A sample compartment with temperature, dry-gas, stirring, and flow-cell compatibility
G) A diode-based transmission detector capable of measuring from 200 to 1000 nm
H) A thermoelectrically cooled, aberration-corrected CCD-spectrograph (250–800 nm) with adjustable binning and gain to maximize the signal to noise and most importantly increase the rate of full emission spectral collection (with integration times ranging from 5 ms to 65 s).

The instrument functions by scanning excitation wavelengths using a designated wavelength interval over the selected range (between 200 and 800 nm). The complex signal processing flow is outlined schematically in Figure 9.3. The dark signals from the reference (R) and absorbance (I) diode and CCD detectors (S) are averaged for the selected integration time before data acquisition. Then for the blank (solvent) at each excitation wavelength interval the sample I, R, and S signals are dark subtracted. The signal processing as described in Steps 2 and 3 includes applying the radiometric spectral correction factors for the wavelength dependence of the excitation R detector (Xcorrect factor) and emission S detector (Mcorrect factor) following the dark signal subtraction as S_c/R_c, where $S_c = (S\text{-dark})^*\text{Mcorrect}$ and $R_c = (R\text{-dark})^*\text{Xcorrect}$.

Step 4 outlines the Blank subtraction process to obtain the final fluorescence signal (Eq. (9.1)). The processing of the I and R detector signals for Transmittance and Absorbance parameters is also defined in conjunction with Steps 1–4.

$$\text{Fluorescence} = (S_c/R_c)\,\text{sample} - (S_c/R_c)\,\text{blank} \tag{9.1}$$

Step 5 lists the post-processing sequence of the EEM data with respect to secondary instrumental and sample-dependent artifacts including emission wavelength interpolation, Rayleigh and Raman scatter masking, Correcting Primary and Secondary IFEs and lastly normalization to Raman scattering unit (RSU)

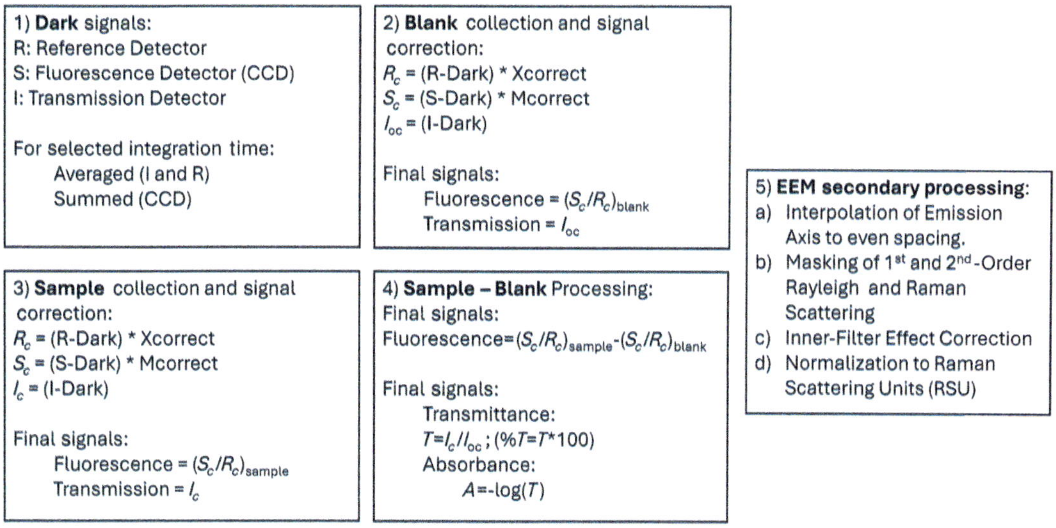

Figure 9.3 The flow of fluorescence, absorbance, and transmittance signal processing and spectral correction steps and formulas. *Source*: Adam Gilmore.

values as defined previously (1). The instrument software uses a batch analysis routine that applies all the aforementioned primary and secondary corrections automatically (on-the-fly) while facilitating control of autosampling and export of absorbance and EEM data files to designated directories/folders for immediate analytical processing, including multivariate chemometric routines that can involve automated wait-for-file functionality and dashboard parametric reporting.

Figure 9.4 illustrates a fully processed UV–Vis–EEM data set for a pure natural compound dissolved in 95% ethanol, as extracted from a common dietary supplement, and acquired with the instrument described in Figure 9.2. Panel A shows the Absorbance (black) and percent (%) Transmittance (blue), while Panel

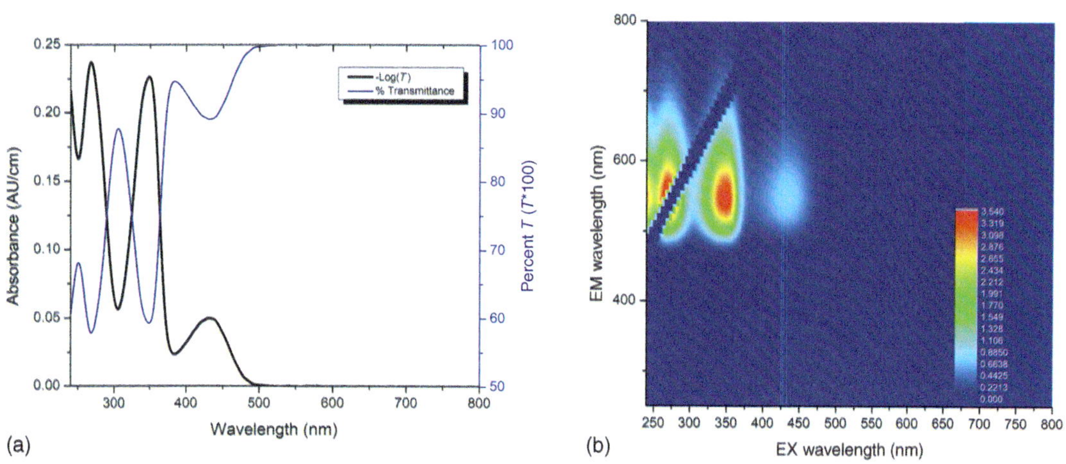

Figure 9.4 Typical sample data for simultaneous absorbance–transmittance (a) and fluorescence excitation emission matrix spectroscopy (b). The sample is the predominant pure chemical component found in a common dietary supplement at 5.5 µg/ml in 95% EtOH. The measurement was made with an Aqualog®-UV800 simultaneous UV–Vis–EEM instrument (HORIBA Instruments Inc.) equipped with the Fast-02 autosampler. The excitation range was 240–700 nm with 5 nm intervals. The emission range was 250–800 nm also with 5 nm spacing. The integration time was 0.1 s, and the CCD gain was medium. *Source*: Adam Gilmore.

B shows the fluorescence EEM. Both the absorbance and fluorescence excitation axes for this compound indicate this spectral range includes 2 major (270 and 350 nm) and one minor band (430 nm) with the emission wavelength orthogonally located at 550 nm. This UV–Vis–EEM sample thus serves as a good example of a chromophoric fluorophore and shows the natural orthogonal correlations between the Absorbance, Transmittance, and the EEM signals.

9.3 Synergistic Relationships Between Absorbance and Fluorescence Excitation–Emission Matrices

The key underlying principles and benefits of the simultaneous UV–Vis–EEM analysis method listed below center on the facts that any chemical in a solution under controlled solvent, pH, and temperature conditions exhibits the following five unique spectroscopic properties:

1) The molar extinction coefficient (ε) that determines the Beer–Lambert linear relationship between concentration and Absorbance (A)

$$A = \varepsilon cl \tag{9.2}$$

where c is the analyte concentration and l is the optical pathlength.[1] It is important to note that the linear relationship must be evaluated over a defined concentration range since exceeding this range may be associated with non-linearity of a component's optical absorbance cross-section (due to aggregation) and/or concentration-dependent changes in the refractive index of the solution.
2) The fluorescence quantum efficiency (ϕf) that defines ratio of the number of photons emitted to the number of photons absorbed. Within the linear Beer–Lambert concentration range both absorbance and IFE corrected fluorescence yields should be linear. It follows that the unique relationship between absorbance and fluorescence is another chemical property that can be useful for component identification and quantification.
3) A unique absorbance spectrum defined by the structural properties and their dependent vibrational behaviors for the molecule in a given solvent.
4) A unique fluorescence excitation spectral shape that is parallel to the absorbance spectrum for a pure compound and also electronically convolved with the emission spectral shape in a purely orthogonal manner.
5) A unique fluorescence emission spectral shape that represents, for aromatic organic molecules, the emission from the lowest lying singlet excited state to the manifold of vibrational ground-state levels.

Based on the above five molecular spectroscopic properties it is easy to understand why the simultaneous UV–Vis–EEM method has the specialized capacity to simultaneously identify and quantify many chromophoric and fluorescent molecules in solutions, even within complex mixtures. It is also important to distinguish chromophoric and fluorescent molecular behavior given that these are distinctly separate molecular properties. Chromophoric activity is associated with the molecules structural ability to absorb light in the UV–Vis range and in many cases chemicals in solution may be chromophoric but weakly or non-fluorescent. Fluorescent molecular behavior is generally associated with the structural properties of

1 The $A=\varepsilon cl$ equation is associated with the primary assumption that A and c are in a linear concentration range according to the Beer–Lambert law while ε is a physical constant for the analyte in the defined solvent condition. Thus, it must be acknowledged that changing the path length l will not alter the chemical behavior or interactions of chemicals within the sample and nonlinear Beer–Lambert conditions associated with such chemical interactions will not be eliminated. The reduced pathlength, however, may allow the user to measure the absorbance of such nonlinear samples within the specification ranges of the instrument.

aromaticity, bond conjugation, or the presence of unbound electrons [3]. Importantly, a wide range of key quality marker compounds in natural product extracts are both chromophoric and fluorescent. Other major components including sugars and carbohydrates, or very small molecules or ions may be both weakly chromophoric and non-fluorescent. However, it is important to consider that non-colored and non-fluorescent molecules may still exert (1) indirect effects on the molecules in solution by their influence on the solution's pH and or refractive index or possibly (2) direct effects by "quenching" of fluorescence.

Here it is a good place to introduce the "Multi-Block" modeling concept to be described in more detail later. Multi-Block modeling allows one to take full advantage of the synergistic relationships among the five molecular spectroscopic properties defined above. The absorbance and EEM spectral variables can be selected, organized, and evaluated simultaneously [5, 6]. In this light, once Multi-Block arrangement is set, analysis of only the absorbance variables, by excluding the EEM information (or vice versa), is also an effective means to evaluate the identity and concentration of chromophoric (or fluorescent) molecules, respectively. Noting it can be the case a highly fluorescent molecule will be detectable well below the absorbance sensitivity level of the instrument; in such cases the absorbance variables and associated noise may advisedly be excluded from the analysis and the IFE correction option can also be disabled.

Another key point relating to the synergistic measurements of dissolved molecules' absorbance and fluorescence behaviors concerns the latter's well-known interactions known as IFEs as mentioned above. IFEs occur when absorbance of the exciting light and or fluorescence within a sample's measuring cell distorts the excitation and emission spectral shapes and intensities compared to when the solution is analyzed under highly diluted (state of infinite dilution) conditions. There are two key interrelated assumptions regarding the use of IFE correction based on absorbance readings, namely:

1) All the molecules in solution should be in a Beer–Lambert linear range over the contiguous measured absorbance and EEM spectral regions.
2) Molecules in solution should be completely dissolved, ideally to a monomeric form, such that there are no excited- or ground-state interactions among components.

These two assumptions are critical because aggregation of molecules can alter their absorbance cross-section and or possibly form ground-state complexes that distort the linear relationship between absorbance and fluorescence of the molecules in solution or suspension (3). Hence, as explained below, dilution by dissolution under a standard solvent condition (pH, temperature, dissolved gas, and solvent) is used to avoid nonlinear behavior, and this condition is assumed to apply to all components in the solution. This is an important consideration since absorbance spectra include information on all dissolved chromophoric components independent of fluorescence activity. It should also be noted that depending on the absorbance and EEM spectral properties of a given compound in the sample, compound-specific regions of absorbance and fluorescence may be selected to eliminate overlap and interference from absorbance of other compounds; the goal would be to isolate the linear absorbance and EEM range for the target compound(s). It is also generally acknowledged that IFE correction has a recognized practical absorbance threshold of >0.05 AU/cm which means that this processing step may in some cases not be recommended as it may propagate the unwanted noise of the weak absorbance signals into the IFE-corrected EEM data.

Another consideration concerning simultaneous measurements of absorbance and fluorescence deals with dynamic or collisional fluorescence quenching phenomena that do not directly affect the ground-state absorbance. Noting dynamic "collisional" quenching primarily affects the dissipation of singlet excited-state energy and is sensitive to temperature and quencher concentration, which influence the collisional rate constants. So-called "static" quenching and combined dynamic-static quenching may also attenuate the fluorescence. Static phenomena generally involve the concentration-dependent formation of a complex between the quencher and fluorophore; static quenching behavior, in most cases, can be differentiated from dynamic quenching as a function of their temperature sensitivity [3]. Absorbance and fluorescence can also be used to differentiate quenching and bleaching phenomena, such as photo-oxidation. Bleaching, which attacks the ground state, proportionally attenuates both absorbance and fluorescence, whereas dynamic quenching primarily attenuates only the fluorescence. Hence, controlling the standard solvent conditions, temperature, and proper dilution are the primary means of mitigating nonlinear distortions through dynamic and/or static fluorescence quenching and IFEs.

9.4 Consensus Standards for Performance and Compliance Testing

Ideally, all absorbance and fluorescence standard measurements should be collected and analyzed to meet the certification requirements in the Standard Chapters USP <857> Ultraviolet–Visible Spectroscopy and USP <853> Fluorescence Spectroscopy.[2] Chapter 7 of this volume provides additional information on USP and ASTM standard practices for UV–Vis and Fluorescence standardization.

The instrument (or vendor) software as described in Figures 9.2 and 9.3 above are equipped with Pass/Fail tests, including an Installation Qualification (IQ) and Operational Qualification (OQ)protocol based on assigned Certified Reference Materials (CRMs) and Standard Reference Materials (SRMs), the latter sourced directly from the National Institute of Standards and Technology (NIST). For UV–Vis, these tests include wavelength accuracy, photometric accuracy, linearity and repeatability, and stray light rejection tests from the UV to Vis region. For fluorescence, these include emission wavelength and spectral shape accuracy and signal to noise [1]. A sealed water CRM from Starna Inc. UK serves as an effective tool for daily tests of wavelength calibration and signal-to-noise. But more importantly, the water CRM can be used as an external standard to normalize EEM signals and quantitatively compensate for changes in CCD binning, gain, and integration time adjustments that may be required when measuring samples with variable signal intensities. Incidentally, for most standard operating protocols (SOPs) the water CRM measurement, known as the RSU test, is the first measurement for a given batch of samples.

9.5 Sample and Signal Optimization

Another key consideration for simultaneous UV–Vis–EEM acquisition is how to optimize the sample preparation and detector signal configuration to maximize the signal-to-noise ratio while monitoring and avoiding detector or parameter saturation. A convenient rule of thumb is to systematically evaluate the sample's solubility, concentration, and dilution factor, usually starting with the highest concentration range, as a function of absorbance, then optimize the fluorescence signal as explained in Figure 9.5 schematically. Step 1 (Figure 9.5) explains that knowing the solvent's UV cutoff, volatility, toxicity, and viscosity properties is important. It is important to establish the Beer–Lambert linear range (primarily the dilution factor requirement), especially in highly concentrated samples that require extraction procedures, such as wine or olive oil. In many cases, the native absorbance of an undiluted extract may exceed hundreds of absorbance units per cm; noting these levels are far beyond the optical specifications of most benchtop spectrophotometers with respect to absorbance detector performance where typically ($AU/cm \leq 3$) is generally the upper limit. It is also important at this stage to select the most effective solvent since a key goal of dilution is to completely solvate (to a monomeric form). Complete dissolution of all molecules helps ensure optical (excited state) and physical interactions among the components including aggregation are maximally reduced or eliminated.

2 Because all international norms are subject to change without notice, the reader is advised to check directly with the international standard bodies (for example, ASTM, USP, and ISO) for periodic updates, as these may influence company policies and/or regulatory compliance. It should also be noted that most international norm document access is subscription-based and subject to strict IP constraints for distribution.

1) Evaluate sample absorbance and Beer–Lambert linearity:

Solvent selection:

- Solubility of analyte
- UV Cutoff
- Optimal Extraction Efficiency
- Volatility Considerations (seal/stopper)
- Refractive Index
- Miscibility
- Viscosity/Density

Dilution factor:

- Linear dilution factor
- Absorbance usually <1 AU/cm

Filtration:

- Typically 0.45 μm Nylon Syringe
- Low UV-absorbing Extractables

2) Optimize fluorescence signal:

- **Integration time (5 ms-65 s)**
- CCD binning (1-8 pixels)
- Gain (High, Medium and Low)
- Target maximum 40,000 cts

- **Measure and use RSU adjustment for batch fluorescence detector conditions (Bin, Gain and integration time)**
- Avoid saturation (Monitor Raw S1 signal)*
- Consider context of sample batches for RSU
- Adjust starting from samples with highest concentration or signal

Figure 9.5 Flow of recommended sample and signal optimization steps and considerations for simultaneous absorbance–transmittance and fluorescence EEM spectroscopy. *Source*: Adam Gilmore.

Dissolution is also important to prevent cuvette and flow cell fouling and carryover which is discussed in more detail below. Care should be exercised to filter or clarify by high-speed centrifugation chemical preparations from solid form or powders since microscopic undissolved particles may be dispersed in the solvent and lead to spurious signals and spikes when they later dissolve and appear as carryover. Following this it is then arguably better to rely on the molar extinction coefficients measured by UV–Vis for evaluating solid or powder-based solution concentrations than simply the weight-to-volume ratio of the preparation.

The sample dilution factor should be evaluated to determine the range where absorbance and dilution factor (or concentration) are in a linear range. For most naturally derived/extracted colored or fluorescence molecules in solution the upper ranges of Beer–Lambert linearity are typically <1 AU/cm; however, for some free organic dyes this may be as high as 2 to 3 AU/cm. It is usually best to use the highest linear range region for final sample preparations to maximize sensitivity and minimize integration time requirements (and exposure to UV radiation). In practice, the best method development also involves determining and using a single dilution factor for all samples in a batch. Keeping in mind signal optimization for sample batches with weaker signals can be compensated for by increasing the integration times (or Gain) and adjusting the RSU factors when compared to batches of samples with higher concentrations.

A key aspect of sample preparation is to reduce particulate light scattering from undissolved particles as well as to standardize the pore-size distribution of colloidal material suspensions. The use of filtration or high-speed centrifugation can help in this regard. ISO 8245 defines the dissolved organic fraction of organic carbon in water as 0.45 μm. Most SOPs for natural product extracts and water commonly employ 0.45 μm syringe filters, usually nylon or glass fiber [7, 8]. Filters must be carefully selected to minimize the effects of UV-absorbing leachates. Thus, to avoid biasing, both the blank and sample should be filtered using the same volume and type of filter.

It follows that once the dilution factors, concentration ranges, and filtration have been established for absorbance the next step (Figure 9.5, Step 2) is to adjust the fluorescence signal detection settings. As mentioned above the CCD sensitivity can be adjusted with respect to the (1) binning or summing of adjacent pixels, (2) by adjusting the gain (electronic count sensitivity threshold) but in most cases primarily by (3) adjusting the integration or shutter-open time. Signal counts will be linear with integration time up to CCD saturation (≥50,000 counts). It is important to keep in mind that because most UV–Vis–EEM model calibrations involve matrix-based chemometrics the matrix dimensions must be conserved for all calibration and validation/application samples. In principle this means that the excitation wavelength range and interval and the CCD binning must remain constant for model calibration and application measurements. Both the Gain and integration times, however, can be adjusted to optimize the fluorescence signal to noise without disturbing the matrix dimensions and corrected using RSU normalization for the final Bin and Gain conditions, as described below.

If one needs to adjust the binning or integration time for signal optimization, then signals would need to be properly compensated by the RSU normalization factor for that condition. Noting that adjusting the Gain can also increase the noise floor since the threshold for count conversion is reduced by increasing Gain. It also means that batches of samples requiring different conditions may best be pre-sorted and separately measured. In many applications where there are large variations among samples in a batch for concentration or signal intensities one can scan key absorbance wavelengths of the samples to identify measuring conditions needed. Alternately a pilot scan using standard Bin and Gain conditions can be run using a short integration time (usually around 0.1 seconds); the resulting data can be evaluated with respect to both absorbance and the uncorrected (raw) S Sample signal which can be used to determine and optimize the signal-to-noise ratio in a purely linear fashion. A good rule of thumb is that for a given set of conditions where the maximum fluorescence count for the most concentrated sample in a batch is near an ideal 40,000 counts (raw S Signal before dark offset subtraction and spectral correction) that one can expect to have useful signal to noise down to 5–10% of this maximum (2000 to 4000 counts), i.e. as a useful batch sample cutoff with a reasonable signal to noise.

Charge-coupled device detectors that have 16-bit resolution will exhibit saturation behavior when a pixel bin selection exceeds $2^{16} = 65{,}536$ counts. Potential CCD saturation warnings will occur when a pixel bin reading exceeds 50,000 counts; the saturation signal is actually evaluated from the raw S detector signal measured before dark subtraction or radiometric correction. This is because the latter may impose a non-linear curvature to the otherwise flat saturated signal region which makes visual interpretation/detection difficult. Also, it has been determined that a warning at 50,000 counts was appropriate since if a maximal allowed binning value of 8 pixels was used then there is a very low probability the binned pixels would respond in a nonlinear manner.

Key factors to keep in mind for UV–Vis–EEM signal optimization can also generally be interpreted by considering their association with the three main noise sources in fluorescence CCD detection being:

1) Dark noise in the CCD detector: this is typically the smallest amplitude of noise being roughly 30 counts per second because the chip is TE cooled, thus reducing Brownian-like motion of electrons within and among the pixels
2) The CCD readout noise, which for weak samples is usually the largest source of noise and is inversely proportional to the pixel binning. Maximizing binning reduces this noise that is associated with the analog-to-digital conversion of the electrons to counts in the CCD chip and
3) The photon noise, which is associated with the Poissonian statistics of the detection where the standard deviation (indicator of the noise) of the signal intensity is equal to the square root of the intensity. Thus, maximizing the counts maximizes the signal-to-noise ratio. In practice, it is always best to try and maximize fluorescence signal counts while being careful to allow for signal variation within a batch of samples to avoid having to re-run a batch due to saturation since experimental conditions are not adjustable once a batch analysis is initiated.

Figure 9.6 shows an example of an optimized UV–Vis–EEM concentration calibration of the compound exhibited in Figure 9.4 obtained using the general guidelines in Figure 9.5. The concentration range for the pure compound was selected by evaluating the signals from extracts of the raw natural product itself. Panel A demonstrates the Beer–Lambert linearity ($R^2 > 0.999$) of the samples from 0 to 0.275 AU/cm over the concentration range of 0 to 5.5 μg/ml measured from the peak absorbance value at 270 nm. Panel B

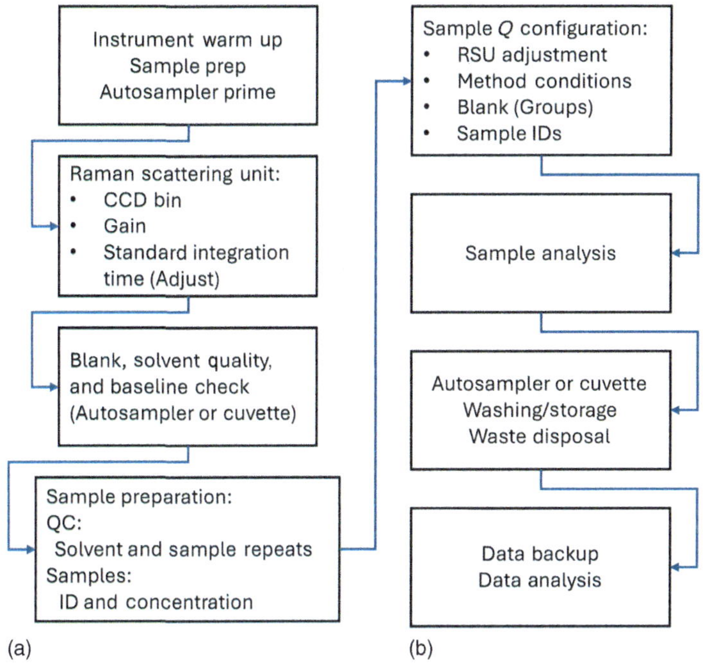

Figure 9.6 Evaluation of Beer–Lambert linearity (a) and Inner Filter Effect correction linearity (b) for the compound described in Figure 9.5 measured at concentrations ranging from 0 to 5.5 μg/ml. Panel A reports the peak absorbance value (270 nm) for each sample as shown for the maximum concentration sample in Figure 9.4. Panel B plots the total integral fluorescence intensity of the EEMs for each sample. The linear regression (red line) and its statistics are shown in the respective legends. *Source*: Adam Gilmore.

shows the corresponding linearity of the IFE-corrected Total Fluorescence (Integral) of the EEMs again with an $R^2 > 0.999$ for the same samples. It should be noted that these data were collected from individual vial preparations for each sample concentration using an autosampler under constant temperature and filtered solvent conditions.

Some materials such as hydrophobic proteins, including monoclonal antibodies [9–11] in a media or buffer, may in addition to requiring large sample volumes to rinse also lead to possibly irreversible fouling of the quartz optical windows and possibly internal components of the system. In such cases of working with proteins, the incorporation of antifouling coated plumbing and optical windows may help to eliminate carryover issues. Another surprising type of material interaction to be aware of involves xanthene dyes, including rhodamines and fluorescein, when these are dissolved in water or aqueous buffers [12].

For a detailed discussion on using the Aqualog® Autosampler, making cuvette measurements, and assessing spectral integrity, please refer to the Chapter 9 Companion Website Manual. Figure 9.7 in the Companion Website contains a flow diagram of a typical standard operating procedure using UV–Vis–EEM with an automated liquid sampling system.

9.6 Considerations of Limits of Detection and Quantification

Because most simultaneous UV–Vis–EEM applications are quantitative, the sensitivity, accuracy, and precision of the methods are generally evaluated with respect to the limits of detection (LOD) and quantification (LOQ). With the described instrument [2] it is important to first distinguish between the limits of a developed method and the absolute limits of the instrument for detection sensitivity. Most developed methods settle on a working range for the analytes of interest with respect to adjusting the instrument conditions, including, for example, the CCD Bin, Gain, and, most importantly, integration time. It is usually the integration time that primarily determines the sensitivity, as explained above, and most methods are developed to save time and thus will not utilize the full available scale of integration time. The influence of other background component signals relative to the target analytes also needs to be considered as these may saturate the detection independent of the target analyte, so a balance in the method is required to keep all component signals in the sub-saturating range for a sample or batch. It should also be noted that for certain compounds it is possible to collect data such that signals from other interfering compounds in distant wavelength regions of the EEM can be saturated without disturbing the linear response of the target compound(s). In these situations, the saturated EEM regions can simply be masked or excluded during analysis. However, it is important to evaluate the linearity of the response to detect and avoid any bleeding/blooming due to adjacent pixel saturation. This situation is analogous to when the Rayleigh scatter may be saturating the CCD, but fluorescence signals closely adjacent to the masked region remain unaffected.

For LOD determination, two main approaches to obtain these values are found in the literature for UV–Vis–EEM applications. For simplicity, they are discussed here as (A) Linear Errors of prediction (including bias correction) and (B) Evaluating the baseline noise standard deviation of a blank sample. Chapter 7 describes Method A in more detail and is based on the Analysis of Variance (ANOVA) of the standard errors for b = intercept and m = slope parameters for the predicted y-variable according to Eq. (9.4).

$$y = mx + b \tag{9.3}$$

Equation (9.4) is rearranged to solve for the predicted LOD concentration, x. This is accomplished by solving for the predicted x from the predicted y using Eq. (9.5).

$$x = (y - b)/m \tag{9.4}$$

When this transformation is made, the predicted slope will equal unity (or a value of 1) and the predicted intercept will be zero independent of the slope and intercept of the predicted y. The slope and intercept errors can then be evaluated and expressed relative to these mean values (m = 1 and b = 0) more directly noting the root mean square error (RMSE) and R^2 values will be the same as Eq. (9.4). Thus, the x to y transformation effectively eliminates both slope and intercept bias for predicted x-values such that they are now both expressed in equal concentration units. Typical cases for this bias correction include when one measures a PARAFAC or Principal Component score as the y-variable to predict a concentration. Alternately, the RMSE of the model (based on the SD of residuals) can be used to determine LOD (see Chapter 7).

It is important to note that the linear-error-based methods are prone to scaling issues and weighting of higher intensity scaled variables can obscure the model's resolution of low concentrations and increase LOD/LOQ. In some standard methods, it is, therefore, customary to make both high and low-concentration range linear calibrations. For low-range calibrations, it is certainly always best to evaluate LOD/LOQ using concentrations as close as possible to the instrumental method's sensitivity limits.

This brings us to Method B, commonly applied to fluorescence spectroscopic applications and defined by ASTM E579-04 for use with quinine sulfate. Noting the method is equally applicable to UV–Vis–EEMs with other chemicals and materials and is arguably less subjective to weighting issues. Here, the baseline noise is evaluated with respect to the standard deviation in the signal region of the analyte for a blank sample, while the signal of a low-concentration sample is also evaluated to gauge the scale of the measurement; the LOD is then equal to $3 \times \text{SD}$ and LOQ is $10 \times \text{SD}$. This method tends to provide truer LOD/LOQ values because the variable weight scaling of higher concentration samples is eliminated. Thus, it is not unusual for the

LOD with this method to be significantly lower than that determined by Method A, although it may still not represent the absolute instrumental detection limit as described above.

With the described instrument [2] and for an integration time of <2 seconds, the detection limits of pure chemicals in solution and chemicals that are well resolved from other components, the LOD and LOQ is primarily a function of its quantum efficiency (ϕf) in the solvent. For chemicals where ϕf is greater than 50% the LOD will generally be in the mid-to-low ng/L (ppt) range. For chemicals in the single digit ϕf range, the limits will approach the low-to-mid mg/L or ppb range. The relationship is fairly linear to ϕf because the correction of the signal by the lamp intensity (i.e., S_c/R_c) closely mimics an ideal light source with uniform intensity across the spectrum.

Evaluating quantitative measurement or model accuracy can also involve a parameter known as the Relative Percent Difference (RPD)

$$\text{RPD} = 100 * \text{Absolute}\ \frac{\text{Predicted} - \text{Measured}}{\left[\frac{\text{Predicted} + \text{Measured}}{2}\right]} \tag{9.5}$$

This is most commonly applied when comparing reference method measurements to empirical model predictions to determine how effectively the combined extraction and analysis method predicts the measured concentration value. In most cases, the RPD and its significance can be conveniently estimated from the R^2 value itself, which represents the overall Relative Error of Prediction or REP = $1-R^2$; this is because R^2 = Sum of Squares of the Modeled Values/Sum of Squares of the Predicted Values.

Another quantitative parameter is analyte recovery, which defines the ability to follow the method's ability to extract and analyze the target molecule efficiency by spiking the sample with a known amount of a target analyte and then following this through the complete extraction method and analysis. An 80–120% recovery value is acceptable in most standard methods. Note that extractions from some raw materials, such as natural plant materials, can be more complex than others due to difficulties with rigid cell walls, vacuoles, chemical degradation, and enzymatic activity, among other factors, including solvent properties.

9.7 Data Evaluation for Linear Trends and Correlations; Correlograms and Univariate Analyses

Because the combined instrument design collects data over a wide wavelength range for both absorbance and fluorescence in many cases the spectral features of specific chemicals may be resolved exclusively, even with samples containing many other compounds. In such cases it is possible and easy to make true univariate regression analyses simply by selecting individual associated variables (or integrated variable regions). The benefits of this are manifold compared to the known issues of inverse multivariate methods such as partial least squares (PLS) which are prone to over-fitting and other pitfalls. The variables (or regions) can be readily identified using a simple process to generate a "correlogram" which can be used to scan the entire absorbance and EEM spectral variable surface to identify variables based on their correlations with analyte concentrations. This is most effectively accomplished with the "Multi-Block" data arrangement where both the absorbance and EEM data are concatenated into a two-way array. Here each variable can be regressed against the analyte concentrations in the calibration data set to yield linear slope, intercept, and R^2 values. The unfolded EEM data variable region can also then be re-folded to view the actual excitation and emission wavelength coordinates associated with the correlations as a means of verifying the analyte's identity. The linear model parameters can then be used to evaluate the EEM variable selection region of unknown samples directly [5]. The correlation region information can also be used to select the most relevant variables for multivariate analysis or classical least squares (CLS) methods as discussed below.

9.8 Multiblock Modeling (Data Fusion) for UV–Vis–EEM Analysis

As described cursorily above fusion of the absorbance and EEM variables helps to leverage the unique capacities of the UV–Vis–EEM method for chemical resolution and quantification. Thus the next step of chemometric data analysis is generally the concatenation of the absorbance and unfolded EEM variables to form a two-way array using the Solo Multi-Block data tool as described schematically in Figure 9.7. Noting, at this stage it is usually best to have already selected the absorbance and EEM variables to be analyzed or excluded using the previously mentioned correlogram analysis. Alternately once can select the EEM variables while the data was still in the three-way form, as described above The three-way analysis may be more visually intuitive because the two-way unfolding pattern (algorithm) will cause overlapping of variable regions that are then difficult to interpret with respect to the actual absorbance and fluorescence excitation emission spectral coordinates. In this context it is for example common when working with natural plant products where the compounds of interest excite and emit in the UV region to exclude the excitation–emission regions of chlorophyll and related red-emitting compounds. A good example is cannabis flower extracts where the chlorophyll content does not correlate with the cannabinoid content or composition ([13]). Importantly, Solo has the capacity to unfold the EEMs and maintain the inclusion/exclusion of selected variables (based on the metadata of the excitation and emission wavelength coordinates). The unfolding process can of course also be reversed to visualize the wavelength coordination of the selected variables.

The Multi-Block model and joined data set should both be saved and properly labeled for use with the predictor toolbox engineered by Eigenvector Inc. as add-ins for Solo and their Partial Least Squares toolbox. This tool is useful for analyzing the data and reporting the results for both regression and discrimination models. One thing to keep in mind is that the Joined spectral data from the Multi-Block model tool will be

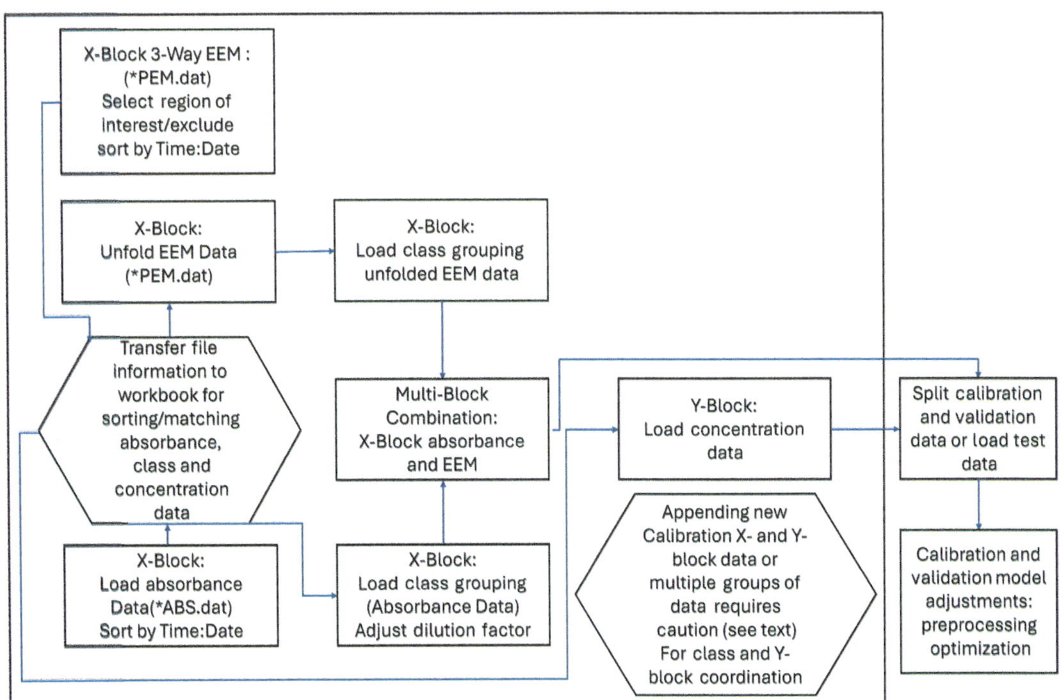

Figure 9.7 Flow of spectral and concentration data organization and processing for building and analyzing a multi-block model for univariate and multivariate analysis of 2-way array data. *Source*: Adam Gilmore.

used as the X-block for model calibration (usually regression or discrimination). It is most important that the concentrations be properly recorded and assigned to the correct samples in the X-block, and this also includes care in assigning the Class ID for each sample's group of interest. Care must be exercised to avoid "scrambling" of the X- and Y-block data when assigning classes or sorting either the X- or Y-block data; always double check the coordination before calibrating or validating/applying a model. Note that in Solo when multiple groups of sample files are sequentially loaded into either the X- or Y-blocks they are assigned unique classes automatically as a function of the sequential loading. Thus one tip is to start with the X-block data (Clear and or leave the Y-block blank) and first delete any such file-loading class assignments for X- then sort the files by the ISO formatted time-date-stamp (Label) field before adding the correctly matching class or concentration information for the Y-block. It is worth noting that the Data Editing tools in Solo can be restrictive for these sorting and coordination operations, so it is generally recommended that sample File, Class, and Concentration assignments and lists be generated and audited carefully in an external workbook such as either Microsoft® Excel and or the instrument's software. The columns of information can then be simply copy-pasted into the appropriate Solo Data Editing entry fields.

9.9 Unsupervised Three-Way Analysis: Parallel Factor Analysis

As mentioned above the least ambiguous quantification methods for UV–Vis–EEM data are based on univariate linear correlations such that problems with inverse method validations, including over- and under-fitting associated with excessive or superfluous pre-processing, are circumvented. That being said, one of the most effective and popular *unsupervised* multivariate analysis tools for EEM data is the tri-linear modeling method known as Parallel Factor Analysis (PARAFAC) [8]. PARAFAC is inherently well conditioned to analyze EEM data because molecular fluorescence is defined by the naturally orthogonal relationship between the excitation and emission spectra and also because fluorescence is a non-negative phenomenon. PARAFAC functions to decompose the three-way EEM matrix data only and cannot be analyzed simultaneously in a Multi-Block arrangement in conjunction with the two-way absorbance data variables. In this regard PARAFAC cannot be used to take full advantage of the synergy between the absorbance and fluorescence signals for chemical identification. PARAFAC generates a trilinear solution for each EEM component as Score* Emission Loading * Excitation Loading. Both the excitation and emission spectral loadings for each component can potentially represent pure chemical component signatures with the score being proportional to the component's intensity or concentration. PARAFAC also cannot be supervised by either class ID or Y-block concentration assignments. In many cases, however, the PARAFAC scores can be used for unequivocal chemical identification and quantification in a secondary regression analysis of the scores [14]. PARAFAC however for complex samples exhibits potentially detrimental limits in the number of components that can be effectively analyzed with respect to statistical significance. This is commonly a factor in complex natural product extracts. PARAFAC also exhibits sensitivity issues associated with the unweighted residual error norm used in most commercially available PARAFAC algorithms. The latter makes it difficult to resolve components with relatively low intensities in the presence of stronger signal components. A common means of dealing with the residual error weighting issue on a sample-by-sample basis is to normalize all EEMs in the data set to the same intensity or area for the PARAFAC decomposition then reconstitute the natural scores by multiplying their normalized values by the total (integral) natural intensity of the EEMs measured independently [8]. This normalization-reversal process however can lead to issues where the noise in weaker samples is magnified such that it hinders resolution and skews the loadings and score results in the stronger samples and components. It should also be noted that low-intensity components may be better resolved by casting a separate PARAFAC model with a limited region of interest variable selection to target the analyte and circumvent the issues with residual error scaling.

As described above it is clear that three-way PARAFAC analysis has some technical advantages with respect to its potential to identify and quantify chemicals and recognize materials based on the component score ratios and clusters. PARAFAC models due to their unique architecture are generally not "cross-validated" in the traditional sense as is common with other multivariate modeling algorithms. The most common and effective PARAFAC validation procedure is known as the "Split-Half" analysis whereby the total calibration set is evaluated for a given number of assumed components then compared to models based on a random splitting procedure [8]. In this manner consistency of the resolved components can be evaluated with a % matching score. This method however is prone to interferences from unique samples with potentially unique components that may not be present in other samples as well as non-random selections of unique samples and groups. In some such cases shuffling the order of the files can help to improve the effectiveness of the split-half algorithm by distributing samples randomly. It is also noteworthy that recent studies have reported that the parameter known as the "core-consistency" also may not provide accurate descriptions of the number of components that can be resolved [15]. In this light it is important to consider that PARAFAC scores and components are best evaluated with respect to correlations with independent spectral and concentration data.

9.10 Empirical Model Calibration and Validation Standardization

The practice and use of empirical multivariate calibrations for purposes of quantitative regression and qualitative discrimination (which may also relate to a concentration threshold) have been extensively documented in international consensus standards published by the ASTM (see E2617 and E2891) and the United States Pharmacopeia (USP) General Chapter <1039>. These references contain comprehensive guidelines on important aspects of experimental design, statistical evaluation relating to precision and accuracy, and model calibration and maintenance schedules.

Some terminology to be familiar with includes (1) the *Calibration data set* and *Cross-Validation data set* used to build the model and (2) the external *Test* or *Validation data set* used to evaluate the model performance. The model calibration data set is most commonly fit to the highest degree, compared to cross-validation and validation, because the model includes noise patterns specific to the calibration data set and its organization pattern. In contrast, the process of validation using an external validation data set is performed in a separate operation. It will have a lower degree of fitting since the noise pattern does not precisely match the calibration data pattern. Below is a list of key concerns when developing and validating multivariate models based on the aforementioned guidance standards:

1) The calibration data and model should include enough examples of all sources and degrees of variation one can expect when applying the model to new, real-world data. Inverse methods such as PLS may lead to erroneous predictions if the cross-validation or test data is outside the concentration range of the calibration model.
2) The validation test set should include all sources and degrees of variation in the calibration data expected in the real-world application.
3) Evaluation statistics should be used to avoid overfitting the calibration model. This can be evaluated by comparing the RMSE of calibration and cross-validation data to check that they are closely spaced in value; a good rule of thumb is that the RMSECV:RMSEC ratio for a given number of latent variables or principal components should be less than 1.4, but the closer to 1, the better.

4) Avoid "Replicate Trapping," which happens when the calibration or validation data sets include multiple replicates. Shared replicates among the calibration and test sets can skew the cross-validation or validation by self-recognition of the same samples tested by the model(s). This leads to an artificially high confidence error and poorer performances with real-world samples. A good way to avoid this issue is to keep replicates (classes) together and exclusive to the calibration, cross-validation, or validation sets. When using Eigenvector Solo for the calibration/validation splitting, the automated protocol can be easily programmed to keep replicates together on a class basis.
5) Validating the calibration model should be done with an independent dataset either (1) collected independently or (2) selected randomly from the sample space of the calibration data, or (3) by a Design of Experiments (DOE) approach to address all variables in the sample space.

9.11 Considerations Using Multi-Block Regression and Discrimination Analysis

In this section we compare and contrast the pre-processing methods and general data treatments associated with Univariate Linear Regression analyses (Correlograms and Gray-CLS) to those used for Multivariate Linear and Nonlinear Regression and Discrimination methods. The Multivariate two-way analyses discussed below can be "supervised" or influenced by both class ID and concentration data with respect to scores and loadings. In some cases, the molecular selectivity and sensitivity of the two-way methods, compared to three-way PARAFAC, can yield lower LODs and tighter, more significant score clusters because the variables can be centered, scaled, and clutter-removal filters can be applied; variable (including region) selection can further enhance the goodness of fits. Table 9.1 summarizes and compares the commonly applied pre-processing steps for X- and Y-block data associated with univariate (Correlogram and Gray-CLS) and multivariate linear (PLS) and nonlinear Extreme Gradient Boost (XGB) regression as well as multi-variate linear (PLS) and non-linear (XGB) discrimination methods. General descriptions of PLS and XGB applications may be found in the following references [5, 6, 16]. As described above Linear Correlogram models may be viewed as direct univariate calibrations not requiring any data scaling, centering or clutter removal filtering. Gray-CLS, as explained further below, generally only requires the application of a residual clutter removal filter for the X-block data. PLS regression can involve centering and scaling of both the X- and Y-blocks in addition to clutter removal filtering of the X-block. Nonlinear XGB on the other hand can invoke a compression of the X-block data using a PLS or principal component analysis (PCA) model with a designated number of latent variable or PCA components in addition to centering and scaling of the X- and Y-block data. Likewise, PLS discrimination is based on the X-block class assignments, and class-dependent preprocessing, including mean centering and scaling, can be applied. Like XGB regression, Discriminant Analysis can invoke an X-block compression step using PLS or PCA, class-dependent centering, scaling, and clutter-removal filtering. Again, caution is advised when applying the inverse PLS and nonlinear methods, such as XGB, to avoid overfitting.

9.12 Multi-Block Quantitative Regression Application: Comparing Partial Least Squares and Gray-Classical Least Squares

As mentioned, there is a relatively new method known as Gray classical least squares(Gray-CLS) that combines the first principles of CLS (white contribution) with the clutter-removal filtering of the CLS residuals (black contribution) to achieve a least squares minimum fit by optimizing only a single variable, namely, the

Table 9.1 Comparison of univariate and multivariate preprocessing steps and functions for regression and discrimination analysis methods commonly applied to the simultaneous analysis of UV–Vis–EEM data.

Univariate regression		Multivariate regression						Discrimination	
Linear correlogram		**Gray-CLS**		**Linear PLS**		**Nonlinear XGB**		**Linear PLSDA**	**Nonlinear XGBDA**
Preprocessing		Preprocessing		Preprocessing		X-Block compression		Preprocessing	X-Block compression
X-Block	Y-Block	X-Block	Y-Block	X-Block	Y-Block	PLS		X-Block	PLS
N/A	N/A	Clutter removal	None	Centering	Centering	PCA		Centering	PCA
				Scaling	Scaling			Scaling	
				Clutter removal		Preprocessing		Clutter removal	Preprocessing
						X-Block	Y-Block		X-Block
						Centering	Centering		Centering
						Scaling	Scaling		Scaling
						Clutter removal			Clutter removal

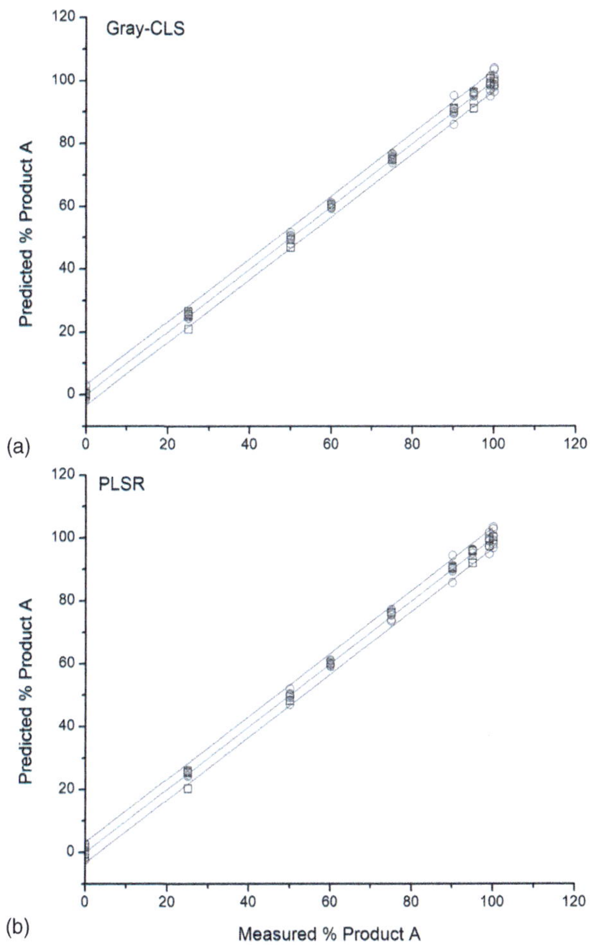

Figure 9.8 Comparison of Gray-CLS (a) and partial least square Regression (b) analysis of the same multi-block data evaluating the % purity of Product A, a dietary supplement mixed with varying concentrations of a common illicit adulterant. The square symbols represent the prediction (test set), and the circles represent the cross-validation of the calibration data. *Source*: Adam Gilmore.

Table 9.2 Quantitative regression analysis statistics for the simultaneous UV–Vis–EEM data presented in Figure 9.8 for PLS and Gray-CLS regression.

	PLS-R		Gray-CLS	
Parameters	**Model (self) prediction**	**Cross-validation prediction**	**Model (self) prediction**	**Cross-validation prediction**
Number of points	16	65	16	65
Degrees of freedom	15	64	15	64
Residual sum of squares	46.86	171.11	51.84	167.18
Pearson's r	0.9997	0.99977	0.99967	0.99977
R^2 value (Coefficient of determination)	0.9994	0.99953	0.99933	0.99954
Adj. R-Square	0.99936	0.99953	0.99929	0.99954
Root mean squared error (±SD)	1.77	1.64	1.86	1.62

weighting factor of the clutter removal filter (https://eigenvector.com/wp-content/uploads/2021/12/CLS_Gray_Models_APACT_2021.pdf). This, therefore, leads to a directly interpretable model. Thus, Gray-CLS does not involve Latent variables, Centering, or Scaling and is less prone to issues of under- and over-fitting compared to inverse chemometric techniques such as PLS.

Figure 9.8 compares the linear regression plots from completely independent ethanolic extracts of Product A, commonly used as a dietary supplement, mixed with another product commonly used as an illicit adulterant and expressed as the measured (*x*-axis) and predicted (*y*-axis) % purity of Product A. The multi-block data in Figure 9.8 included the complete UV–Vis absorbance spectrum and unfolded EEM variables as shown in Figure 9.4 for each sample in order to include all the UV–Vis absorbance end emission spectral features of both the product and adulterant. Figure 9.8a shows the prediction set results (squares) and cross-validation set (circles) results using a Multi-Block model for Gray-CLS regression. Figure 9.8b shows the results for the same data modeled using the PLS regression algorithm. The regression statistics in Table 9.2 show that the Gray-CLS model yields similar statistical goodness to PLS for the same multicomponent data. Therefore, Gray-CLS regression is a practical modeling approach for determining natural product adulteration analysis using UV–Vis–EEM spectral data. Another real-world example of Gray-CLS compared to PLS will be described in Chapter 10 of this book for detecting color additives in wine.

9.13 Multi-Block Method for Classification (Discrimination) Application

As shown above, UV–Vis–EEM regression can be an effective tool for quantitative adulterant detection; however, as shown in earlier publications, simple Pass: Fail tests for adulteration can be constructed based on a concentration threshold and evaluated using multivariate discrimination methods [6, 17]. These earlier studies indicated that one particular discrimination method, based on a Decision Tree method and known as Extreme Gradient Boosting, was particularly effective for this purpose. Discrimination model test results are conventionally displayed as a confusion matrix which displays the accuracy of the model's test class prediction(s) to the known results; this is usually expressed as a function of true positive/negative and false positive/negative ratios and probabilities; see Footnote of Table 9.3 for parameter definitions.

Table 9.3 shows the confusion matrix evaluating the data from Figure 9.8, where a Pass: Fail threshold was set at 90% purity for Product A; this level was selected because current industry standard high-performance liquid chromatography (HPLC) based methods for this adulteration scenario are limited to 90% purity. The confusion matrix was evaluated for the calibration, cross-validation, and prediction data for both a "most probable rule" and a "strict rule" condition. The most probable rule class assignment is made depending on which class the sample's ID is closest to but is independent of the prediction probability. The strict rule has the additional restriction that the prediction probability also has to be above a set threshold, which in this case was 50%. All six test results yielded 100% accuracy for True Positive and True Negative Responses; thus, for this two-class model, the goodness of fit was also described by the Matthews Correlation Coefficient of 1. Thus, it is arguably more convenient and effective for routine QC applications to evaluate these data using such a discrimination model compared to a regression analysis where a secondary evaluation of the concentration value of the predicted data is needed.

Table 9.3 Confusion matrix results for a quantitative discrimination analysis based on a 90% product a purity concentration pass threshold for the data presented in Figure 9.8 and Table 9.2.

Most probable	Class	TPR	FPR	TNR	FNR	N	ERR	P	F1
Calibration	Pass	1	0	1	0	29	0	1	1
	Adulterated	1	0	1	0	36	0	1	1
	Matthew's correlation	1							
Cross-validation	Pass	1	0	1	0	29	1	1	1
	Adulterated	1	0	1	0	36		1	1
	Matthew's correlation	1							
Prediction	Pass	1	0	1	0	7	0	1	1
	Adulterated	1	0	1	0	9	0	1	1
	Matthew's correlation	1							
Strict rule	Class	TPR	FPR	TNR	FNR	N	ERR	P	F1
Calibration	Pass	1	0	1	0	29	0	1	1
	Adulterated	1	0	1	0	36	0	1	1
	Matthew's correlation	1							
CV	Pass	1	0	1	0	29	1	1	1
	Adulterated	1	0	1	0	36		1	1
	Matthew's correlation	1							
Prediction	Pass	1	0	1	0	7	0	1	1
	Adulterated	1	0	1	0	9	0	1	1
	Matthew's correlation	1							

The data was evaluated based on the Most Probable Rule and Strict Rule (at >50% Prediction Probability) using the XGBDA algorithm in Eigenvector Solo.
TPR: proportion of positive cases that were correctly identified (Sensitivity), = TP/(TP + FN).
FPR: proportion of negatives cases that were incorrectly classified as positive, = FP/(FP + TN).
TNR: proportion of negatives cases that were classified correctly (Specificity), = TN/(TN + FP).
FNR: proportion of positive cases that were incorrectly classified as negative, = FN/(FN + TP).
ERR: Misclassification error = proportion of samples which were incorrectly classified, = 1-accuracy, = (FP + FN)/(TP + TN + FP + FN).
P: Precision, = TP/(TP + FP).
F1: F1 Score, = 2*TP/(2*TP + FP + FN).

9.14 Conclusions

This chapter was designed to guide users through the key considerations when using simultaneous UV–Vis–EEM instrumentation for sample analysis. The guidelines for sample preparation, instrument conditions, and methods development should allow users to acquire high-quality spectra for quantitative and discrimination analysis in quality control and applied research applications. Chapter 10 contains a real-world case study using the UV–Vis–EEM technique for grape and wine quality analysis. These chapters are a foundational guide for developing new analytical procedures using simultaneous UV–Vis–EEM instrumentation.

References

1 Gilmore, A.M. (2014). How to collect National Institute of Standards and Technology (NIST) traceable fluorescence excitation and emission spectra. *Methods in Molecular Biology* 1076: 3–27.

2 Gilmore AM, Tong M. (2019). System and method for fluorescence and absorbance analysis. United States Patent US 8,901,513 B2. p. 1–22.

3 Lakowicz, J.R. (2006). *Principles of Fluorescence Spectroscopy*. New York, NY: Springer.

4 DeRose, P., Early, E., and Kramer, G. (2007). Qualification of a fluorescence spectrometer for measuring true fluorescence spectra. *The Review of Scientific Instruments* 78 (3): 033107–033112.

5 Schober, D., Gilmore, A., Chen, L. et al. (2022). Determination of cabernet sauvignon wine quality parameters in Chile by absorbance-transmission and fluorescence excitation emission matrix (A-TEEM) spectroscopy. *Food Chemistry* 392: 133101.

6 Gilmore, A.M., Sui, Q., Blair, B., and Pan, B.S. (2022). Accurate varietal classification and quantification of key quality compounds of grape extracts using the absorbance-transmittance fluorescence excitation emission matrix (A-TEEM) method and machine learning. *OENO One* 56 (4): 107–115.

7 Elcoroaristizabal, S., Bro, R., García, J.A., and Alonso, L. (2015). PARAFAC models of fluorescence data with scattering: A comparative study. *Chemometrics and Intelligent Laboratory Systems* 142: 124–130.

8 Murphy, K.R., Stedmon, C.A., Graeber, D., and Bro, R. (2013). Fluorescence spectroscopy and multi-way techniques. PARAFAC. *Analytical Methods* 5: 6557–6566.

9 Kanthe, A. (2021). No ordinary proteins: Adsorption and molecular orientation of monoclonal antibodies. *Science Advances* 7: 1–14.

10 Oom, A., Poggi, M., Wikstrom, J., and Sukumar, M. (2012). Surface interactions of monoclonal antibodies characterized by quartz crystal microbalance with dissipation: Impact of hydrophobicity and protein self-interactions. *Journal of Pharmaceutical Sciences* 101 (2): 519–529.

11 Hollowell, P., Li, Z., Hu, X. et al. (2020). Recent advances in studying interfacial adsorption of bioengineered monoclonal antibodies. *Molecules* 25 (9): 2047.

12 Mroczek, E.K., Mountain, B.W., editors. (2000). Adsorption of rhodamine WT and Fluorescein onto quartz between 25° and 83 °C. Proceedings of the 22nd New Zealand Geothermal Workshop. New Zealand Geothermal Workshop.

13 Gilmore, A.M., Elhendawy, M.A., Radwan, M.M. et al. (2022). Absorbance-transmittance excitation emission matrix method for quantification of major cannabinoids and corresponding acids: A rapid alternative to chromatography for rapid chemotype discrimination of cannabis sativa varieties. *Cannabis and Cannabinoid Research* 8 (5): 911–922.

14 Gilmore, A.M. and Chen, L. (2020). Improving aromatic water-contaminant detection with machine-learning classification and regression for simultaneous absorbance-transmission excitation emission matrix (A-TEEM) spectroscopy. *Optical Fibers and Sensors for Medical Diagnostics and Treatment Applications* Proc. SPIE 11233, XX, 112330H. https://doi.org/10.1117/12.2556434.

15 Halberg, H.F.F., Bevilacqua, M., and Rinnan, A. (2023). Is core consistency a too conservative diagnostic? *Journal of Chemometrics* 37 (5): e3483.

16 Ranaweera, R.K.R., Capone, D.L., Bastian, S.E.P. et al. (2021). A review of wine authentication using spectroscopic approaches in combination with chemometrics. *Molecules* 26 (14): 4334.

17 Gilmore, A. and Chen, L. (ed.) (2020). *Improving Aromatic Water-Contaminant Detection with Machine-Learning Classification and Regression for Simultaneous Absorbance-Transmission Excitation Emission Matrix (A-TEEM) Spectroscopy*. San Francisco, CA: SPIE BiOS.

10

Analysis of Grape and Wine Quality Features with Simultaneous Absorbance and Fluorescence Technology

Adam Gilmore[1] *and Doreen Schober*[2,3]

[1] *Horiba Scientific, Piscataway, NJ, USA*
[2] *Concha y Toro, Pencahue, Maule, Chile*
[3] *Organic Agricultural Sciences, Organic Food Quality, University of Kassel, Witzenhausen, Hessen, Germany*

10.1 Introduction

Grapes and wines contain several colored and fluorescent components, mostly (poly)phenolic in nature, which determine sensory quality parameters, including color, flavor, and mouthfeel. Fluorescence and ultraviolet spectroscopy have the potential to detect and resolve these components to effectively characterize unique compositional properties of wines from different sources and global regions and as a function of processing, blending, oxidation, and aging, or even spoilage from microbiological and environmental sources. As a rapid and efficient method, ultraviolet-visible and excitation–emission matrix (UV–Vis–EEM) can monitor critical wine color parameters and phenolic concentrations to provide a unique opportunity for improving quality and consistency in wine production facilities.

This chapter focuses on the simultaneous analysis of ultraviolet-visible (UV–Vis) absorbance and fluorescence excitation–emission matrix (EEM) spectroscopy as it is applied to evaluate the chemistry of grapes and wine processing. Theory and practices of modeling UV–Vis–EEM spectral data, including multi-block variable arrangement, are discussed in more detail in Chapter 9 of this book. The technique can facilitate both CIE-Lab and chromaticity analysis important for various sensory and QC-related wine parameters. The UV–Vis–EEM method also incorporates the conventional UV–Vis wavelength analyses based on values at 420, 520, and 620 nm that are commonly used to characterize basic color characteristics of wine, including the wine's redness, brownness, or yellowness, hue, and intensity.

The chapter primarily explains how targeted quantitative analyses of grape and wine phenolic compounds and other basic chemistry parameters, including pH, alcohol, total and volatile acidity, and sulfites, have been measured with the simultaneous UV–Vis and EEM methods. The analytical sections of the chapter show how supervised regression and discrimination using machine learning algorithms, such as Partial Least Squares (PLS) and Extreme Gradient Boosting (XGB), are used for quantitative and qualitative analysis. Non-targeted quantitative analysis examples are also shown to predict sample origin, maturity, sanitary status, and sensory quality. A non-targeted analysis is not related to a specific compound concentration or any supervision but is based on the global spectral properties ("spectral fingerprints"). The chapter gives practical application tips and examples of how these measurements can be optimized and used in daily winery operations. These examples highlight the methodological importance of proper sampling, spectral preprocessing, and chemometric modeling practices for extracting accurate and reproducible chemical information from extracts of finished wine and unfermented grapes. The importance of instrument compliance

Practical Quantitative Vibrational and Electronic Spectroscopy: A Guide for Developing, Optimizing, and Validating Procedures, First Edition. Edited by Deborah A. Peru.

Companion website: www.wiley.com/go/peruelectronicspectroscopy

certification and ongoing performance characterization is also discussed. The chapter concludes with several examples of using UV–Vis–EEM for wine quality analysis. A companion website containing a standard operating procedure (SOP) provides further details on how to set up and use this instrument.

10.2 Recent Applications of Ultraviolet–Visible and Fluorescence Excitation–Emission Matrix Analysis for Quality and Process Control in Grape and Wine Production

Much research has been focused on the development of rapid analysis methods for grape and wine quality parameters suitable for practical use in daily winery operations. It follows that quantification procedures to determine sum parameters for phenolics, tannin-free phenolics, tannins, anthocyanins, and polymeric pigments have become widely distributed for routine analyses. These methods are usually based on spectrophotometry in combination with precipitation and bisulfite bleaching reactions [1, 2]. These methods have replaced more labor-intensive liquid chromatography and mass spectroscopic analyses to a certain extent. However, they are limited to the bulk determination of summed parameter quantifications, and many are still reliant on time-consuming wet chemistry.

Phenolic compounds exhibit intrinsic chemically specific absorbance and fluorescence spectral properties, which makes their direct measurements possible using simultaneous UV–Vis and fluorescence analysis. The UV–Vis–EEM acquisition method for grapes and wines, paired with chemometric data modeling, allows the determination of a wide range of individual phenolic compounds, including non-flavonoids, flavonoids, and polymerized phenolics. The methods described also include common basic chemistry routine parameters due to their secondary spectroscopic optical activities. Non-targeted analysis of the full spectra or selectively determined spectral regions opens up numerous additional possible applications because many quality-related molecules exhibit direct and/ or indirect effects on absorbance and fluorescence properties that the technique can capture. The UV–Vis–EEM method can identify and quantify compounds with high sensitivity, down to ppb ranges, based on their unique absorbance and fluorescence properties.

Numerous production steps require fast and accurate decision-making, which is strongly supported by this technique. Figure 10.1 depicts the steps in the wine-making process and the multiple roles of simultaneous UV–Vis and EEM spectroscopy in evaluating grape growing, harvesting, processing into wine, and storage. This includes product authentication concerning grape variety, growing region, and varietal blending, as well as determination of wine characteristics associated with barrel aging and oak treatments, spoilage, smoke taint, and contamination. Quantitative analysis of the chemistry of grapes and wines provides conclusions on wine sensory characteristics and quality, noting that the processing from grape cultivation to finished wine is complex, and the multiple steps are associated with changing chemical composition. The citation numbers included in Figure 10.1 correspond to the literature references. The next paragraph provides a guide for understanding of the winemaking process and the monitoring the roles of simultaneous UV–Vis and EEM spectroscopy.

To begin with, any wine producer's aim is to achieve the best possible quality for their final product; however, the understanding of sensory quality is highly complex and multi-layered. It is imperative to consider the interpretation of the chemical composition for sensory quality during chemical analysis. Sensory studies suggest positive correlations of certain aroma and flavor descriptors to increased quality scorings, but evaluation of preference and overall quality, in particular, tend to diverge subjectively. The concept of quality can be trained with individual groups of sensory panelists and particular products, but the process is project-specific and time-consuming and, therefore, not ideally suitable for industry needs [27, 28]. According to the cited studies, it is important to consider various intrinsic and extrinsic factors, which include the

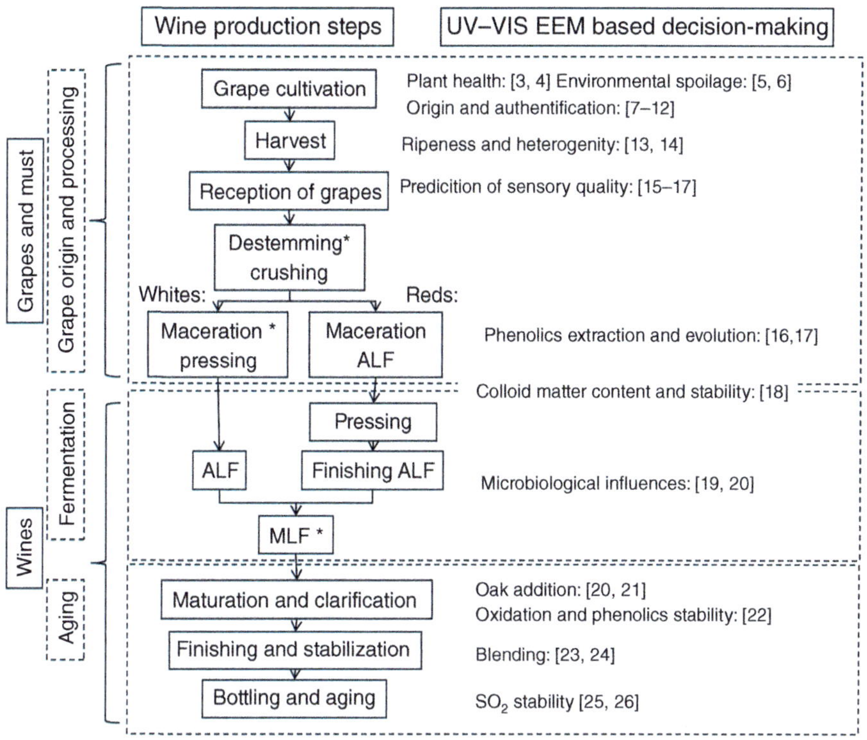

Figure 10.1 Steps in the winemaking process where the simultaneous UV–Vis and EEM technique can be implemented. The citations provide further explanation and details for each step [3–26]. Abbreviations and notations: *optional production step, ALF (alcoholic fermentation), MLF (malolactic fermentation). *Source*: Adam Gilmore / Doreen Schober.

individual's experience and genetics, as well as the testing environment, pricing, and labeling, among other components, respectively [29].

With this knowledge, international wine research and industry agree that chemical measurements provide the most objective evaluations of quality. So-called chemical quality markers are commonly used after being established by comprehensive multivariate data analyses of large chemical and sensorial data sets. Key studies have focused on various internationally recognized major white and red grape varieties like Cabernet Sauvignon, Shiraz, Merlot, and Chardonnay. However, these markers can be understood as indicators for a direct causal, and/or indirect correlative relationship to quality, or so to say as the link between sensory and chemical composition translated into quality prediction [30–32]. They comprise a number of maturity-related basic chemistry parameters and various kinds of phenolic compounds and volatiles, which will be explained in more detail in the following sections as direct and indirect optical effectors. Importantly, these chemical markers are highly determined by environmental and weather conditions and can be modulated over the course of the winemaking process.

Wine production starts in vineyards, and soil and plant health are basic requirements of significant importance to produce high-quality grapes. Among the many factors involved, the amount and composition of dissolved organic matter (DOM) plays a crucial role in root growth, plant–soil interactions, nutrient transfers, and microorganisms. The simultaneous UV–Vis and EEM technique has been used to predict the bioavailability of total soluble copper and free ionic Cu^{2+} as a function of DOM and pH in fungicide-treated vineyards [3] and also to assess further DOM-related effects for other non-grape agricultural crops. *Vitis vinifera L.* grape vines are very susceptible to pathogens with increasing pressure due to the fast global spread of diseases, which makes it an increasingly important topic for wine producers. Grapevine trunk disease (Esca) could be identified by the simultaneous UV–Vis and EEM method with leaf samples; EEM spectra of infected and healthy samples were decomposed by Parallel Factor (PARAFAC) analysis to reveal three unidentified components, which could be used to discriminate the infected samples successfully [4].

In addition to the importance of the soil, viticulture practice, geographical origin, and climate factors, interactions with a distinct vine can influence primary and secondary metabolites of grapes and the later produced wines and, therefore, are critical drivers for wine style and quality. This concept is often referred to as the *Terroir* [33]. Accordingly, and in a more commercial way of thinking, the labeling of wines according to their *Appellation of Origin* is commonly used and is associated with expectations of high-quality products that distinguish certain areas [34]. Rather than an extensive targeted analysis of individual compounds, UV–Vis–EEM spectroscopy paired with innovative chemometric approaches has been shown to have great potential regarding the practical application and sensitivity to ensure the authenticity of wines and to prevent fraud [7, 8]. Correct classification from country to subregion of origin has been demonstrated based on quantitative simultaneous UV–Vis–EEM measurements. Utilizing XGB and DA (discriminant analysis) with the EEM data of Australian and French red wines, Cabernet Sauvignon yielded 100% correct class assignment of the samples for the region of origin and according to grape variety [9, 10]. Shiraz wines from two Australian valleys were furthermore 100% correctly discriminated according to their subregions of origin within the valleys, and the authors strongly suggested using the method for the verification of subregional terroirs by fluorescence fingerprints [11]. A follow-up study including new vintages of Shiraz from Australia obtained an accuracy of 98.8% for the prediction of the vintage year and 93.8% for the same subregions. By PARAFAC analysis of the EEM data, this study additionally identified four underlying components, which could be tentatively associated with (1) flavan-3-ols, (2) anthocyanins, aromatic amino acids, and hydroxybenzoic acids, (3) phenolic acids, phenolic aldehydes, and flavonols, and (4) caffeic acid, coumaric acid, and stilbenes [12].

Having shown the specificity of origin that forms very individual and measurable chemical patterns of grapes and wines, which can be assigned to a distinct place, many other variables within a valley can impact quality, such as vineyard, block, vine, and even berry clusters. So-called vineyard heterogeneity in terms of ripeness can cause production issues and aspire to be monitored as early and hands-on as possible until harvest before the reception of the grapes in the wineries. A composite indicator named "grape heterogeneity index" was suggested, and the complex dataset comprising key indicator chemical compounds for technical-, flavor-, and phenolic maturity was largely supported by simultaneous UV–Vis–EEM, together with CIE lab measurements [13].

A complementary study completely relied on multi-block data[1] to evaluate overall maturity, and XGB–DA-based models of quantile range for red color, 3-isobutyl-2-methoxypyrazine (IBMP), malic acid, and total soluble solids (TSS) achieved 78% correct classification of Cabernet Sauvignon grapes according to overripe, ripe, less ripe, and underripe [14]. Another approach was the prediction of wine sensory attributes based on grape chemical composition. The idea to predict the grape's potential to achieve certain wine quality levels or styles would facilitate several steps throughout the winemaking process, such as the coordination and occupation of fermentation tanks or the precise determination of line, charge, and bottle production numbers, among others. However, Cabernet Sauvignon grape and wine multi-block absorbance and EEM data obtained by experimental fermentations were modeled by means of XGB-R

1 Joined UV–VIS and EEM data, obtained by simultaneous absorbance and fluorescence measurements, see Chapter 9 for a detailed explanation.

(regression) and respective prediction values $R^2 > 0.7$ and $R^2 > 0.5$ were obtained for 5 out of 22 and for 15 out of 22 descriptors. These authors mentioned the complexity of biotransformation from grape to wine during winemaking, which may have limited the predictability of the models. However, analysis of the unsupervised data could identify variables (spectral regions) to improve the models possibly, and PARAFAC of the EEMs pointed out catechins, phenolic aldehydes, and acids, as well as chlorophyll *a* or *b* as underlying compounds [15]. Another study applied Agglomerative Hierarchical Clustering (AHC) for modeling sensory Rate-All-That-Apply (RATA) data of Cabernet Sauvignon wines [35]. AHC identified five characteristic clusters representing different sensory profiles that could be assigned 100% correctly to the EEM data of the same wines by XGB–DA modeling. Subsequent PARAFAC on the obtained EEMs later identified four main underlying fluorophore components, namely catechins/tannins, phenolic aldehydes, anthocyanins, and resveratrol, which were then associated with specific sensory profiles by multifactor analysis (MFA). The PLS-R modeling using EEM data combined with sensory data yielded 68.1% accuracy in predicting the panel score for the perceived astringency of wine [35].

Other approaches directly quantified individual chemical compounds known to be quality markers due to their specifications and properties, as detailed above. Using simultaneous absorbance and fluorescence EEM analysis, an optimized procedure was developed to quantitatively determine 44 individual phenolic compounds and routine chemistry parameters in Chilean red and white wines. A study compared single-block (absorbance or fluorescence) with multi-block data organization modeled with XGB-R and PLS-R algorithms. The XGB-R model using the multi-block data set revealed the highest prediction accuracy and precision for all compounds with a mean $R^2 = 0.94 \pm 0.04$ and relative bias (REP) $= 4.1 \pm 1.8\%$ [17]. A similar approach was developed for several Northern Californian varietal red grapes, and the quantification of individual phenolic compounds utilizing UV–Vis and EEM multi-block data and XGB-R modeling yielded overall predictions of $R^2 = 0.941 \pm 0.025$ and REP $= 5.89 \pm 2.47\%$ [16]. Overall, simultaneous UV–Vis and EEM measurements have been shown to provide numerous possible applications for objective evaluation of grape and wine quality to support decision-making suitable in daily winery operations.

Subsequently, the fermentation process involves several significant steps essential to the final product quality. To produce white wines, grapes are pressed before the fermentation to separate the juice (liquid fraction) from the solids following an optional maceration time. Reds are usually macerated and fermented for several days before pressing and completing alcoholic fermentation (ALF). ALF and subsequent (optional) malolactic fermentation (MLF) to transform malic acid into lactic acid require yeasts and bacteria, and their activity and autolysis may further influence wine composition. Dynamics and integrity of chromophoric colloid matter in fermenting Chardonnay from Burgundy were monitored by simultaneous UV–Vis–EEM non-targeted analysis; the results were congruent with proteomics data and suggested that the more rapid spectrophotometric technique is equally capable of measuring the effects of grape and microbiological proteins on chromophoric colloid matter content and stability. UV–Vis–EEM showed that chromophoric colloidal matter decreased after pressing and during ALF, but increased after MLF [18]. The impact of MLF, changes in acidity, and phenolic compounds could also be discriminated based on the EEM fluorescence fingerprints [19]. Another study utilizing PARAFAC analysis with EEM spectral data verified the impact of different MLF conditions in Chardonnay; *Oenococc us oeni* biofilm (sessile) compared to plankton forms was demonstrated to obtain higher maximum fluorescence intensities (F_{max}) for all four underlying components identified by PARAFAC. This effect was hypothesized to be due to more optimum interactions of the biofilm with the wine matrix, which led to the altered chemical transfers among them. The addition of oak chips was also detectable, highlighting further possible applications for the simultaneous UV–Vis absorbance technique on oak-derived compounds [20]. Another common application in enology is the addition of *Teinturier* grapes or their commercial Color Additive Concentrates (CACs) to achieve a higher color in red wines. These grapes have a different anthocyanin profile than the most common *V. vinifera* L., with di-*O*-glycosides being the primary anthocyanins. To quantify these additions to a wine, a recently developed simultaneous UV–Vis EEM procedure was shown to successfully predict the addition of CACs in the range of 0–0.35% in six different major grape varieties, with a correlation (R^2) of 0.9931 and a root mean square error of prediction (RMSEP) of 0.0158% [23]. Furthermore, the blending of varietal red wines could be deciphered utilizing XGB-R modeling of wine EEM data, which detected up to 1% for Shiraz/Cabernet-Sauvignon and Shiraz/Grenache blends [24]. These examples demonstrate the possibility of monitoring and managing wine stability over the production process. There are many more potential applications, including the determination of an ideal anthocyanin-tannin (A–T) ratio during maceration and ALF of red wines, to achieve wines with the optimum balance of color and mouthfeel extraction.

Once the finished matured (with or without oak barrel contact) wine has been stabilized and bottled, the aging process continues and is subject to further changes in its chemical composition and sensory perception, including color, mouthfeel, and flavor. A study with Italian red wines demonstrated that simultaneous absorbance and fluorescence measurements could monitor quantitative compositional changes in the function of oxidation [22]. Within this study, PARAFAC decomposing the EEMs furthermore identified five underlying components, identified as (1) caffeic acid, (2) flavonol-like, (3) epicatechin, (4) gentisic acid, and (5) anthocyanin-like [22]. Merlot base wines for sparkling wine production from Mexico were also differentiated by non-targeted fluorescence as a function of pH and different aging conditions, including wood barrels and stainless-steel tanks, and PARAFAC of the EEM data identified five key components to distinguish the sample treatments. Simultaneous UV–Vis and EEM spectroscopy in combination with nuclear magnetic resonance (NMR) spectroscopy additionally revealed a novel and potent (poly)-phenolics aromatic region for further non-targeted analysis [21], again pointing out the potential of UV–Vis and EEM spectroscopy for measuring barrel wood-derived phenols.

Another relevant topic associated with barrel aging is contamination with *Brettanomyces* spp.; these yeasts can degrade hydrocinnamic acids to volatile phenols, which at very low concentrations might contribute to wine complexity and maturity, but exceeding the compound's very low odor thresholds leads to negative sensory sensations including spicy, animal-like, medicinal, burnt and woody [36]. Of other origin, but also caused by volatile phenols and their glycosidic precursors, is the so-called smoke taint. The bushfire-derived molecules get stored in the grapes and are hydrolyzed to their free and odorous form during fermentation and aging, causing strong sensations of ashtray-like and burnt in the wines and leading to severe economic losses. Therefore, their early determination in the grapes before grape reception and fermentation in the winery is of utmost importance. As a rapid and high-throughput diagnostics, first research works have shown very promising results for the quantification of volatile phenols and their precursors using the simultaneous UV–Vis and EEM technique [5, 6].

Sulfur dioxide (SO_2) treatments (respecting the legal limits) are crucial for wine microbial and oxidative stability from grape to the aging process of the final wine, and simultaneous UV–Vis–EEM analysis has been presented as a suitable technique to follow up on these treatments; 320 bottled Chardonnay wines from Burgundy were tested that had been treated with 0, 4, and 8 g/hL of SO_2 during grape processing. PARAFAC modeling of EEM data of the wines clearly found three distinct SO_2 dose-aging-dependent aging patterns and identified four underlying components explaining 99.6% of the variability in the data set [25]. In a subsequent study, Fourier transform ion cyclotron resonance mass spectroscopy (FT-ICR-MS) targeted analysis and metabolic fingerprints were modeled with the obtained PARAFAC components of the EEMs. The Spearman rank showed correlations between sulfur-containing compounds and EEM fingerprints, suggesting a potential for simultaneous absorbance UV–Vis and EEM spectroscopy as a rapid method to monitor sulfur treatments and wine stability [26].

10.3 Harnessing the Complementary Nature of Absorbance and Fluorescence Excitation–Emission Matrix Information in Chemometric Models

As discussed in Chapter 9, molecules containing chromophores and fluorophores absorb light in the UV–Vis region, while fluorophores also emit radiation from electronically excited to ground state vibrational levels. An essential requirement for quantification using the simultaneous UV–Vis and EEM technique is that chromophoric molecules in a given solvent and temperature-controlled conditions are characterized by a unique extinction coefficient and absorbance spectrum and that the amount of light absorbed is linearly proportional to its quantity according to the law of Lambert–Beer. Fluorescent compounds follow the same logic and exhibit specific fluorescent quantum efficiencies and individual excitation and emission spectra, albeit fluorescence is more sensitive and complex, often containing broad overlapping of excitation and emission signals, which might affect quantitative and qualitative interpretation. The emission of the same compound compared to excitation is characterized by the typical Stoke's shift to higher wavelengths due to a loss of vibrational energy [37].

Fluorescent compounds are typically aromatics with conjugated double bonds and mobile electrons. The mobility of these electrons is influenced by individual properties like the number of aromatic rings and functional groups, which thereof influence the individual fluorescence patterns and intensity. Several key grape and wine compounds, mainly phenolic compounds, vitamins, amino acids, and chlorophylls, are classified as colored and/or fluorescent, making them applicable to UV–Vis fluorescence analysis methods. The following discussion will mainly focus on phenolic compounds due to their major abundance and sensorial relevance to grape wines [31, 32, 38]. Their detailed biochemical formation can be specifically reviewed elsewhere [39–42], while this chapter will focus on their quality-related optical properties.

Non-flavonoid phenols include phenolic acids (cinnamic-like and benzoic-like), phenolic aldehydes, and stilbene-like compounds, which fluorescence-wise typically extend between the ranges of excitation 260 to 330 nm and emission 320 to 440 nm. Excitation of both phenolic acids families is detected below 285 nm but is separated clearly by their emission wavelength axis. Gentisic acid (benzoic-like) is separated and characterized by EEM wavelengths of 320 to 440 nm. Trans-stilbenes are typically found at excitation wavelengths higher than 290 nm, and emission between 370 and 400 nm, while cis-stilbenes obtain excitation and emission signals of 260 and 400 nm, respectively. Flavonoid phenols are split into two unique regions with flavonols extending between excitation 260 to 268 nm and emission 370 to 422 nm, and flavan-3-ols occurring within excitation 278 to 290 nm and emission 310 to 360 nm. Anthocyanins are typically located at 520 nm and 589 to 590 nm for excitation and emission, respectively. Figure 10.2a provides an overview of a characteristic EEM map of a wine sample, adapted from [43–45]. Absorbance in the UV–Vis region for wine phenols typically includes flavan-3-ols at 280 nm, phenolic acids at 325 nm, flavonols at 340 to 440 nm, and anthocyanins at 520 nm (Figure 10.2b). The compound-specific EEM and absorbance spectral patterns furthermore obtain specific signal intensity-to-compound concentration ratios. They are visualized by univariate correlograms and are obtained by calculating the R^2 of the respective quantitative concentration for the intensity of the spectral signal measured for a specific compound. Figure 10.3c points out the typical patterns for a chemical compound. These patterns are unique to individual compounds and with similar patterns within compound families (Figure 10.3c).

In support of the theoretical synergistic effects of absorbance UV–Vis–EEM data, the empirical results obtained for grape and wine modeling provide further proof of the concept. Figure 10.3 presents

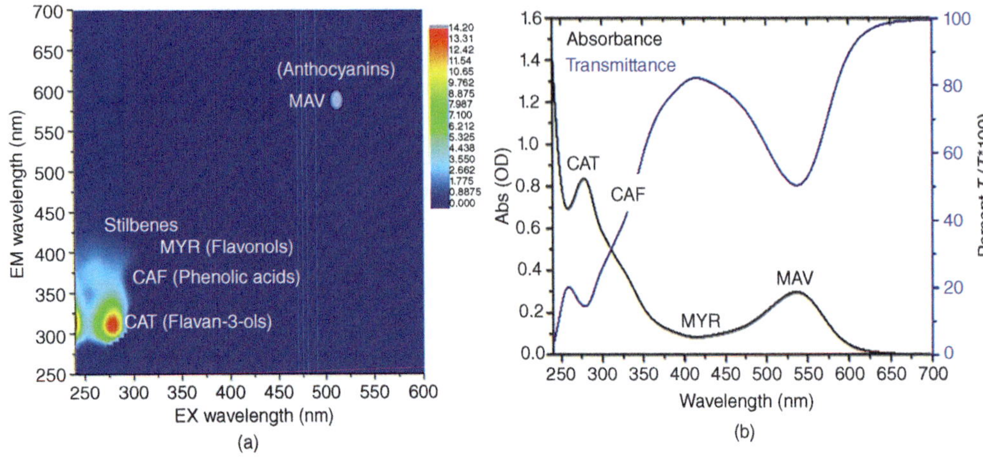

Figure 10.2 (a) Fluorescence EEM diagram for characteristic non-flavonoid and flavonoid phenolic compounds obtained from a typical Cabernet Sauvignon wine. (b) Absorbance typically stretches from 280 to 520 nm from flavan-3-oles to anthocyanins. Abbreviations: CAT (Catechin), CAF (Caffeic Acid), MYR (Myricetin-glucoside), MAV (Malvidin-glucoside); representing their respective groups flavan-3-ols, phenolic acids, flavonols, and anthocyanins. *Source*: Adam Gilmore / Doreen Schober.

characteristic correlation maps for flavan-3-ols, phenolic acids, flavonols and anthocyanins. The R^2 of individual compound concentrations for absorbance and EEM variables (signal intensity of each spectral region) are shown without preprocessing. Figure 10.3 exemplarily depicts anthocyanin malvidin-glucoside, which has a major characteristic absorbance peak between 420 and 620 nm, and three major peaks around the fluorescence excitation/emission of 275/500 nm, 420/450 and 550/600. The maximum peak R^2 of 0.863 and 0.921 for absorbance and EEM, respectively, is reflected by overall high single- and multi-block prediction fits; the multi-block model data organization, however, leads to the highest prediction R^2 of the model for malvidin-glucoside. The same effect is found for all other phenolic compounds summarized in the table in Figure 10.3c, also noting that decreasing maximum peak R^2 leads to slightly lower model fits, which is, however, compensated by the synergistic absorbance and fluorescence MB data organization.

For the model fit statistics, as a general trend, the R^2 of cross-validation (CV) and prediction (P) increased systematically from fitting the absorbance variables to fitting the EEM variables to fitting the multi-block data. In general, single-block predictions were below an R^2 of prediction (R^2P) of 0.90, while multi-block data predictions reached precise fits above R^2P of 0.95 for all compounds. These findings thereafter promote the multi-block data organization, as both absorbance UV–Vis and EEM spectral data were shown to contribute valuable information, and their combination yielded the best regression statistics. Compared to malvidin-glucoside, the model statistics for catechin improve significantly utilizing the EEM spectral information due to the lower characteristic absorbance and higher fluorescence signal intensity for catechin. In contrast, malvidin-glucoside already achieves high-fit models by its strong absorbance, typically at 260280 nm and 490–520 nm alone. The contribution of the fluorescence information also may be more relevant for compounds present at low concentrations. Interestingly, this was true for PLS-R modeling compared to XGB-R, while the latter consistently generated higher R^2 for single- and multi-block modeling [17].

10.4 Major Classes of Optically Active (Colored and Fluorescent) Grape and Wine Chemical Compounds

From a wine quality perspective, the phenolic composition is very complex and diverse. It contains a pool of interacting compounds and is constantly subject to change during the grape growing and winemaking

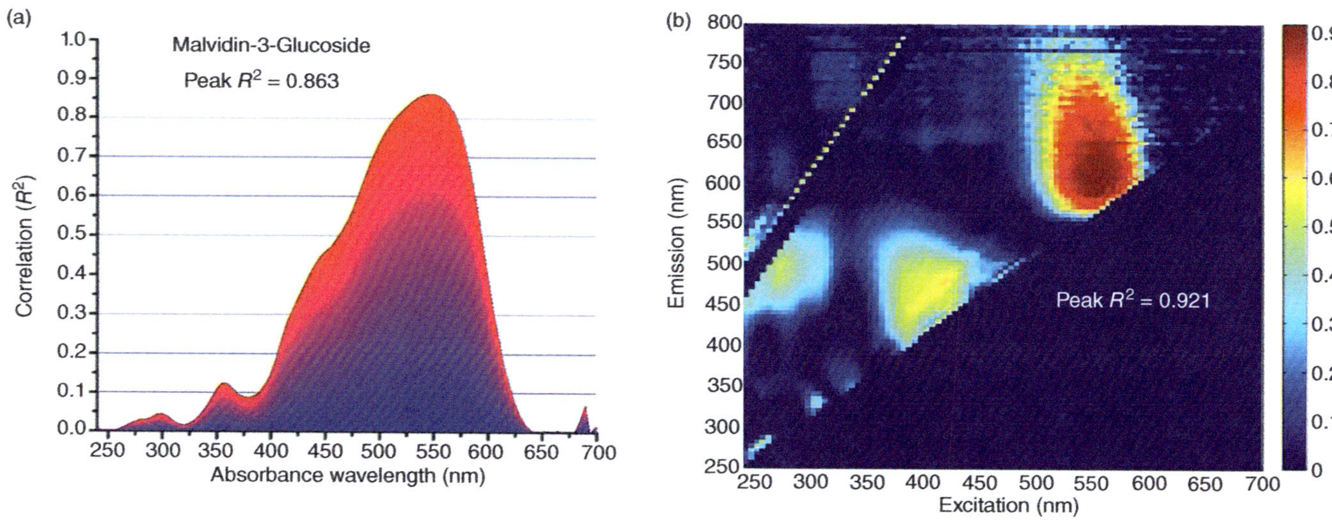

(c)

Compound (family)	Pattern and signal intensity-to-compound concentration				Regres sion type	Data-block organization chemometric model fit R^2		
	Absorbance		Fluorescence EEM					
	>UV abs, nm	Peak R^2	Fl Em, nm	Peak R^2		Abs	EEM	MB
Malvidin glucoside	520	0.863	580–590	0.921	PLS	0.883	0.981	0.992
(Anthocyanins)					XGB	0.985	0.984	0.992
Catechin	278	0.238	312	0.579	PLS	0.319	0.773	0.875
(Flavan-3ols)					XGB	0.870	0.917	0.961
Myricetin glucoside	340–440	0.545	450–550	0.600	PLS	0.417	0.521	0.902
(Flavonols)					XGB	0.761	0.834	0.937
Caffeic acid	325	0.350	425	0.481	PLS	0.642	0.790	0.816
(Phenolic acids)					XGB	0.747	0.873	0.920

Figure 10.3 Univariate intensity-to-compound concentration ratio R^2 correlograms obtained for anthocyanin malvidin-3-glucoside: (a) absorbance UV–Vis, (b) fluorescence EEM, (c) modeling parameters and R^2 for single-block absorbance (ABS) and fluorescence (EEM) or joined multi-block (MB) data. *Source:* Adam Gilmore / Doreen Schober.

process. More importantly, knowing specific individual components and their properties is crucial for tracking processes that affect color stabilization, tannin structure, and flavor.

Phenolic acids are simple non-flavonoid phenols, and both hydroxycinnamic acids (caftaric, coutaric, and ferulic acids) and benzoic acids (gallic acid) have been demonstrated to cause sensations of bitterness and also to contribute to astringency and to act as important copigments binding with anthocyanins in color stabilization. Hydroxycinnamic acids are the main abundant and responsible for browning in white wines following oxidation reactions [46, 47]. Phenolic acids are located mainly in the grape skins and seeds. They are important variables in the winemaking process because they can alter the final chemistry through transfer to the grape must and wine [48]. At wine pH, they are usually stably present as esters, but can be liberated by commercial enzymes and microbiological activity; this is important to control as hydroxycinnamic acids can be subject to formation of volatile phenols by *Brettanomyces* yeasts. Stilbenes, the most commonly known as resveratrol, are of main interest due to their antioxidative properties; albeit their concentrations in grape berries and wine are somewhat limited, their potential to be extracted from other parts of the vine plant and used as an alternative source of potential health additives makes them a highly valuable subject for quantification [49].

Volatile phenols, in contrast, do not directly originate from wine grapes, but their entry and metabolism in grapes and wines are of high impact. The so-called smoke taint is a serious and hard-to-control off-flavor reminiscent of smoke, burnt and ash. Lots of research has been done on the phenomena, and it is now proven that free volatile phenols, including guaiacol, 4-methyl guaiacol, and cresols, among others, are accumulated in the grape berries in the vineyard after smoke exposure formed when lignin in wood is burnt (e.g. bushfires). The compounds have been shown to directly enter the grape berries via their wax cuticle and get stored as glycosides [50]. Similar mechanisms are being discussed for vineyards with geographical proximity to intensive livestock farming, which emits sulfur- and nitrogen-containing components as well as volatile phenols [51, 52] or eucalyptus trees that contaminate grapes with terpene 1,8-cineole [53]. The main problem is that these odorless precursors only get hydrolyzed during winemaking by low pH and enzymatic activities, complicating early diagnostics of contamination by means of sensory tests. Volatile phenols are generally characterized by very low sensory thresholds in one-to-two-digit ppb concentration ranges and are, therefore, highly likely to form prominent aroma-related problems in wine [54]. Later, during wine maturation and storage, spoilage of oak barrels with *Brettanomyces* yeasts, which degrade hydroxycinnamic acids to form 4-ethyl and 4-vinyl derivatives of guaiacol, phenol, and syringol causes concentration-dependent unpleasant sensations of spicy, clover-like, medicinal to horse sweat [36]. However, from the oak barrel itself, phenolic aldehydes including vanillin, furanic compounds, and oak lactones, among other compounds can be extracted and transferred to the wine depending on its origin, grade of toasting, seasoning, barrel size, storage time, and the use of oak chips will lead to a particular aroma profile [55, 56], furthermore the interaction of a particular wine with the oak which may influence the overall phenolic profile including anthocyanin concentrations [57].

Wine and grape flavonoids comprise three major subclasses: anthocyanins, flavan-3-ols, and flavonols. Anthocyanins are specific to red grape varieties and responsible for the typical color; important for *Vitis vinifera* L. are nine different 3-O-monoglucosides, 3-O-acetyl-glucosides, and 3-O-coumaroyl-glucosides of delphinidin, cyanidin, petunidin, peonidin, and malvidin. Usually, malvidin-glucoside is the most abundant anthocyanin, but ratios strongly vary among grape varieties, with Pinot noir possibly one of the major grape varieties with a completely unique anthocyanin profile. The present aglycon and type of sugar residue determine the hue and stability of the red color, albeit long-term stability can only be achieved by complex polymerization and co-pigmentation involving mainly oxygen-mediated reactions [39, 41]. Color, however, is perceived as very important quality criteria, and deep red wines with violet-red nuances are considered to be of higher quality than wines with less intense color or with brick or even brownish tones. *Teinturier*, also known as dyer grapes, are used in winemaking to increase and intensify color. These varieties obtain anthocyanins in skins and pulp, and their anthocyanin profile, including major concentrations of di-O-glycosides, is significantly different from common *Vitis vinifera* L. [58]. Flavan-3-ols in wine are present as four possible catechin derivatives, including two stereoisomers and gallate-esters, and in these monomeric forms, mediate the sensations of bitterness and participate in the stabilization of pigments. Furthermore, they polymerize as oligomers (proanthocyanidins) and polymers (condensed tannins), which are crucial for astringency and mouthfeel perceptions. Tannin expression is strongly determined by their composition, and the quantification of the monomeric building blocks is crucial for decision-making in the production of well-balanced quality wines. Proanthocyanidins from grape seeds, for example, contain more epicatechin-gallate, which at high amounts leads to increased undesirable bitterness. Proanthocyanidins from grape skins, however, are longer-chained than the latter which contributes to a more ripe and velvet astringency [59]. Flavonols finally exhibit a very potent group of quality parameters. Particularly, syringetin and quercetin and their derivatives have been described as indicators of well-balanced vineyards and sunlight exposure. Their concentration increases with a certain intensity of radiation, maximizes, and then begins to decline with increasing intensity. Thus, the named flavonols can be related to optimal conditions and to indicate overall quality of grape and wine [60]. Flavonols can furthermore indicate the presence of other positive-contributing compounds including other phenolic compounds and volatile aromas [61], and have been shown to benefit astringency directly [46]. Overall, rapid quantification using simultaneous UV–vis absorbance and fluorescence EEM spectra of the discussed compounds, which are rightly considered as quality markers, is enormously beneficial for decision-making for grape growing and the winemaking process. Knowledge about individual compounds can support specific conclusions on grape and wine sensory and quality based on the explained correlations above. Additionally, yielded group parameters can provide quick hands-on parameters to more easily monitor winemaking processes, such as total tannin and anthocyanins extraction ratios during ALF. In comparison to the previously discussed grape-derived compounds, SO_2 is a widely used preservative in foods and beverages (added as liquid, gas, or salt). In winemaking, it is commonly applied to the grapes at the reception before further processing, during wine maturation and storage, and finally adjusted before bottling. It can also be produced by fermenting yeasts at different levels [62]. SO_2 is mainly used for microbiological stabilization and protection against oxidation and is a standard internationally regulated parameter [63]. It must be considered that in wine, SO_2 is present in various forms, is pH-dependent, and is in equilibrium with water. The main form at wine pH is usually the hydrogen sulfite anion HSO^{3-} (bisulfite), followed by SO_2 (molecular form) and negligible amounts of $SO_3{}^{2-}$ (sulfite). Free SO_2 is mainly responsible for microbiological control but, in large portions, rapidly and strongly binds to other wine compounds, making it unavailable for other reactions, and the sum of both free and bound SO_2 is commonly referred to as total SO_2. The bisulfite form mainly binds weakly to other carbonyls, phenolics, color compounds, and sugars making it an important antioxidant. However, over time, the concentrations and equilibrium of free and bound SO_2 are subject to changes, and therefore it is important to be constantly measured to be able to adjust free SO_2 to stabilize the wine and to respect the regulation limits. Up to date, three analysis methods are legally accepted for the finished bottled wine; the most common is the iodic titration as described in the Compendium of International Methods of Wine and Must Analysis OIV-MA-AS323-04B [64]. Notwithstanding, as described in this chapter, spectroscopic techniques, including UV–Vis and fluorescence procedures, are valuable and robust options for internal quality control during the winemaking process, including for determining SO_2.

10.5 Grape and Wine Chemical Compounds with Indirect and Solution-Based Effects on Optical Activity

As discussed in Chapter 9 the optical activity of chemicals in solution is often strongly influenced by the solvent properties which constitute the molecular environment. The solvent effects are particularly significant to wine and grape compounds because their absorbance and fluorescence excitation behavior can be strongly influenced by the pH (ionization state) and the solvent's refractive index and polarity. Two of the main purposes of dilution of the wine and grape extracts in an acidified ethanol:water 50:50 pH 2 solution are to eliminate aggregation and co-pigmentation effects to achieve Beer–Lambert linearity for all components as well as to stabilize the ionization states and enhance the fluorescence yield for the majority of the phenolic and anthocyanin compounds [16]. Further, the effects of many of the higher concentration compounds in wine and grape extracts, which are non-colored and non-fluorescent, also need to be considered in the context of their influences, noting that pH, alcohol, sugars, and carbohydrates may affect the refractive index and polarity of the solution. As shown by [17] the pH and alcohol content can be modeled effectively by virtue of their global influences on other colored and fluorescent compounds in solution. For wine and grape extracts, it is also important to consider the longer-term effects of oxidation with respect to solution storage and analysis time, given that stability of the signals is limited to less than three hours at 20 °C before oxidation begins to darken the solution. Here again, de-gassing the solvents used for preparation is important not only to reduce oxidation but also prevent bubble formation in the solution, especially when using an autosampler where a vacuum is used to aspirate the vials.

10.6 Methodology for Sampling, Sample Preparation, Instrument, and Data Process Integration

To begin with, collecting samples in general must ensure statistically representative and randomly selected samples and should follow established protocols. From a practical perspective, sampling of grapes in the vineyard combines with maturity control of the distinct vineyard lots or parcels of a winery. In addition, an advantage can be taken of the logistics already established for the sampling, which are often quite extensive and secure representative sampling. Samples must guarantee a sufficient sample size, and include grapes from different clusters on a vine and from different vines from the entire length of the rows and from different rows covering a plot. A sample size that is practical to handle should be representative but should not have too large a volume because this complicates sample preparation as discussed below.

Sampling of grapes at the moment of reception at the winery, most likely from transport trucks, ideally integrates in the grape processing. This might be possible at the weighing station, while taking advantage of existing sampling routines for basic chemistry parameters (e.g. Brix). However, first and foremost, it is essential to take representative samples and to consider factors such as oxidation and other time-sensitive and transport-related chemical changes. Depending on the harvesting method, transport distances, and the quality of the harvested specimens, grapes may be delivered relatively undamaged, allowing for uncomplicated sampling, which could be carried out directly at grape reception as mentioned above or during unloading onto a destemmer or press. But in other cases, which are not uncommon, especially for larger production facilities, grapes may already be (partly) crushed and with (significant) amounts of juice already running off, which without careful sampling would lead to an artificially increased concentration of phenols, for example, that furthermore accelerates the extraction and oxidation of these components. Sampling of wines from tanks and barrels is generally more accessible; however, especially when sampling from fermenting tanks, the sampling must include different heights of a tank due to commonly occurring heterogeneity and gradients to obtain accurate results, and samples need to be carefully degassed during sample preparation.

Once the representative samples have been obtained, wine and grape samples require different further extraction and dilution protocols before measurement with the simultaneous UV–Vis–EEM technique as depicted in Figure 10.4. An SOP is provided in the online Wiley & Sons companion website and developed for different Chilean *Vitis vinifera* L. varieties. The SOP denotes individual adjustment depending on the sample size and the concentration of each chemical compound. For wine, most target compounds have already been extracted from the berries into a suspension. Still, they are present in concentrations that are

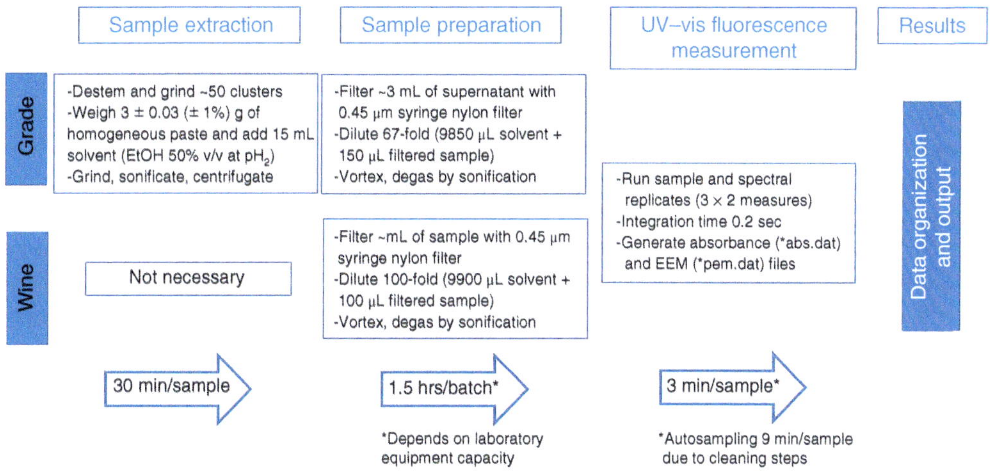

Figure 10.4 Schematic grape and wine sample extraction and preparation, and further adjustment of dilution and UV–Vis fluorescence integration time. *Source*: Adam Gilmore / Doreen Schober.

too high for this evaluation, along with suspended particulate matter that can also obscure optical analysis by light scattering, absorbance, and fluorescence behavior. Thus, wine extraction before UV–Vis EEM analysis is primarily a function of dilution and filtration; it may be advisable first to filter the raw wine before dilution to minimize contributions of leachable UV-absorbing compounds from the filters, and nylon syringe filters (typically 0.45 μm) are generally recommended for exhibiting low UV-extractable materials (see SOP in Companion Website for details). The dilution factor for wine is important for the method development and is key to achieving the Beer–Lambert linear range requirements for accurate UV–Vis EEM analysis. Most intense red wine methods require dilution factors between 100-to-200 fold, whereas most rose and white wines can be diluted less, typically 50-fold. Dilution factors can be adjusted according to the material, and the final dilution factor can be generally evaluated by measuring the absorbance of the solution at 280 nm (flavan-3-ol- and tannin-related saturation) and adjusting so that the most concentrated sample in the batch is <1 AU/cm.

Grapes need to be first mechanically and chemically extracted from the various tissues of the fruit, including the berry skins, pulp, and seeds. Seeds are a special consideration because their extraction level should equal the amount achieved in the winemaking process. Starting with a representative sample of berries (without stems), these are ground to a homogeneous paste. Next, a defined amount is weighed, dissolved with solvent solution, homogenized again, sonicated to break cell walls, centrifuged, and finally filtrated to remove solid materials similar to the wine samples. To yield an optically linear concentration, the final dilution will be a factor of the starting weight of the fruit material and content, concentration, and composition acquired by the extraction method. Good results for red *Vitis vinifera* L. grapes were achieved with a weight of 3 ± 0.03 (± 1%) g of grape material to be extracted and a 67-fold dilution in analysis buffer. This can be determined individually in preliminary tests by extracting different amounts of grapes to achieve an optimum of extracted compounds of interest without saturation. The mechanical crushing can also be varied by the type of mill, ultrasound time, etc. Subsequently, similar to wines, the dilution factor adjustment of the filtered samples should render the most concentrated sample in the batch to be <1 AU/cm at 280 nm. Accurately weighing the grape material from the samples in the extraction replicates is very important for reproducibility. Samples (from fresh or frozen grapes) and their replicates should be further measured immediately after starting the preparation process. Oxidation effects may occur, and the simultaneous absorbance and fluorescence method is very sensitive in this detection; an overall sample stability of approximately three hours has been reported [22]. During the calibration development and validation phases, these fundamental precautions and sampling approaches also apply to the samples used for reference method measurements.

An easy routine day-to-day sample integrity check, also to address the sensitivity of the instrumental technique, can be implemented by monitoring absorbance values at 280, 420, 520, and 620 nm. The relative standard deviation (RSD) of the technical (extraction) as well as measurement (instrument) replicates should be monitored and should include at least triplicates each. As a rule of thumb, for wines, an RSD greater than 5%, and for grapes, an RSD greater than 10% should not be accepted for the measurement to be repeatable.

During each series of measurements, blank subtraction should be applied correctly and controlled by visual inspection of the corresponding EEM spectral contour plots. The processed contour plot should not contain signs of contamination or any other disturbance. Additionally, the processed contour plot for sample measurements should not contain negative values derived from contamination in the blank sample. Periodic blank (and solvent) vials should be strategically interspersed during the measurements to check for sample carryover issues. Further analysis technique and instrument specifications, including optimizing integration time for UV–Vis fluorescence data collection, compound-specific limit of detection (LOD), and limit of quantification (LOQ) methods, as well as evaluation of method recovery, and model fidelity, should be implemented according to the fundamentals of the technique explained in further detail in Chapter 9.

The final step of the UV–Vis–EEM preparation of the wine and grape extracts is primarily associated with filling the vials and/or cuvettes and implementing proper storage conditions, particularly for longer storage time. Low temperatures (−20 to −80 °C) can be critical for storing wine and grape extracts to minimize enzymatic activities and photochemical and atmospheric oxidation for prolonged periods. As discussed above, exposure to room and sunlight should be minimized throughout the extraction and analysis procedures to prevent photochemical processes. Despite the published values, it is also advisable to evaluate the extracts' light, temperature, and oxygen stability during UV–Vis EEM analysis by evaluating select repeats of samples throughout the measurement operation. Another consideration is the storage vessels since the leaching of colored and fluorescent plasticizers and other materials can influence the sample's signals. As with water chemistry, the ideal storage containers would be amber-colored, acid-washed glass or residual-free Nalgene. The remaining sample from before extraction (grape paste, wine) could be stored frozen at −18 °C in small containers.

A well-organized process flow integration of the simultaneous UV–Vis and EEM technique is crucial to its effective application in daily hands-on winery operations, as depicted in Figure 10.5. A realistic scenario first

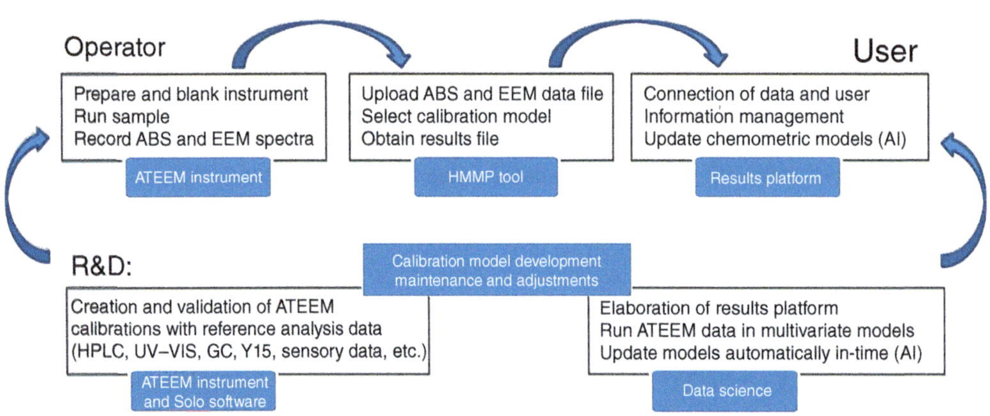

Figure 10.5 Process flow integration of the UV–VIS and EEM technique in practical daily winery operations connects the instrument operator and data user, which is enabled by R&D calibration model elaboration and validation and data science. *Source*: Adam Gilmore / Doreen Schober.

includes the actual instrument operator (could be a technician) who prepares and runs samples, possibly in a winery laboratory. UV–Vis EEM measurements can be effectively made manually with standard 1 × 1 cm quartz cuvettes supplemented with temperature control. To support high sample throughput, an automatic 1 cm flow-cell sampling option is available complementary to the described system and includes an autoinjector capable of running up to 96 two mL vials per load (or 24 ten mL vials) in conjunction with a peristaltic sipper capable of sampling from up to four sources in a continuous and scheduled manner. Respecting the autosampling guidelines for instrumental and sample handling, the implementation has proven to be very effective in terms of repeatability and low-carryover specifications, especially as grape and wine applications exhibit comparatively less problematic matrices to not potentially interact with the system's internal plumbing and flow cell windows surface (see Chapter 9). Subsequent to the measurement, the operator would generate the results from the recorded absorbance (*abs.dat) and EEM (*pem.dat) files. This step is carried out using the device's own software (HMMP tool); a simple interface allows the simultaneous upload of the data files of a full batch, and with this the required calibration model is selected in a drop-down menu application and the final results are generated in the form of easily convertible (comma separated)*.csv-format. A benefit of this application is the possibility to use different calibration models on the same generated absorbance and EEM data to rapidly generate multipurpose results. However, in daily winery operations a large amount of data is generated by measuring a large number of samples. This information needs to be output in an orderly and user-friendly manner. This is supported by using an ISO-formatted time-stamping of the *abs.dat and *pem.dat files to facilitate sample-to-data traceability. Further precise naming of samples according to vineyard lot, delivering transport truck, fermenting tank, replicates, etc. should be standardized for the secure identification of a sample and ideally with a fixed number of digits in the filename to facilitate sample sorting and recognition.

With inputs from a data scientist, identification, traceability, and results output (possibly to a database) enable access to individual relevant information to meet the needs of the user (viticulturist or enologist) to support their decision-making. The HMMP tool generated *.csv-files can be further processed using data scripts (Java, Python, etc.) with the final results being transferred to the final database. Specific data processing enables direct applications on the obtained data, such as extracting particular results from a complete measurement or specific calculations like sum parameters or scores from individual measurements. The results for sample replicates can be stored, and subsequently merged and transferred to the database along with the sample volume and dilution factor and can also include information on measurement status, (i.e. "ongoing" or "complete"). This can be particularly relevant for sample replicates belonging to one target, such as different containers delivered from a wine plot.

In the background, chemometric and further prediction models should be regularly updated and validated by including constantly generated recent data. This may require R&D quality assurance scientists and can be carried out using the Solo software or any other specific chemometric modeling software.

10.7 Model Calibration, Cross-Validation, and Prediction with Multi-Block Absorbance and Fluorescence Data

In simultaneous UV–Vis and EEM data analysis, model calibration is a pivotal step to establish a robust relationship between the predictor variable X-block and the response Y-block. The X-block typically consists of absorbance and fluorescence EEM data. For the fluorescence data, each variable represents the fluorescence intensity of an excitation–emission pair of coordinates. As described in Chapter 9 the multi-block data set is created by the concatenation of the absorbance and unfolded EEM variables to form a two-way array using the Solo software "Multi-Block" data tool. It is recommended before joining absorbance and EEM data to first review and pre-process the spectral data and to evaluate the EEM dataset using PARAFAC to identify possible sample outliers or wavelength regions with anomalies and variables that are not needed (see Chapter 9 for detailed instructions). In contrast, the Y-block comprises concentrations of specific chemical compounds or classes that are obtained from reference analysis methods. The X- and Y-block corresponding data must be carefully aligned, including sample file (names), concentrations, and optionally class assignment.

As the first obligatory basic requirement for effective model calibration, sample materials (i.e. grapes, grape must, and wine) should encompass the entire expected range of response variation for the specific application. That means achieving a representative coverage of the concentration ranges of all compounds of interest and sample matrices, including all desired grape varieties, quality levels, and harvest years. It should also include the process production steps, such as fresh and frozen grapes, macerating and fermenting must and wine, and different wine maturation and storage times. To set up a new calibration, it is recommended to separate calibrations according to grapes and wines, albeit due to the high dilution factors the effects of sample matrices should be vanishing. Additionally, it might also be easier due to the possibly different dilution factors required for the mentioned samples. However, these can be adjusted in later sample evaluation from the generated absorbance *abs.dat and fluorescence *pem.dat files because the HMMP tool of the Solo software enables the adjustment of dilution factors.

The selection of the training set for the development of the calibration is a critical step, as well as the validation set, which will subsequently be used to evaluate the goodness of the calibration more realistically. Various methodologies can be employed to construct an appropriate training set, including structured, random, and manual approaches [65]. In a structured approach, Design of Experiments (DOE) plays a crucial role in preparing a sample list. For example, a full-factorial design can be used to assemble samples containing all possible combinations of values. This approach ensures comprehensive coverage of the calibration space, enhancing the model's ability to capture complex relationships. Conversely, the random approach involves randomly selecting samples throughout the calibration space to achieve a uniform distribution of samples or a normal distribution around specific points in the concentration space. This approach is particularly suitable for samples with complex matrices, like wine. Manual design is often used to complement training sets initially constructed using structured or random methods. For example, if a randomly designed training set lacks sufficient representation of samples at the periphery of the calibration space, additional samples can be selected using the factorial design approach. This augmentation enhances the training set's ability to capture variability across the entire calibration space, resulting in a more robust calibration model. In some cases, circumstances dictate the use of whatever samples are available, leading to the "take it or leave it" (TILI) method. Here, it is generally preferable to "leave it" rather than construct a calibration that may prove inaccurate or unreliable.

In addition, the analytical performance of a calibration is always subject to some uncertainty and variability arising from sample preparation, instrument day-to-day variability, and other sources. Thus, it is advisable to replicate samples with repetitive preparations or to perform repeated measurements on different days, at different times, or by different analysts. Furthermore, increasing the quality and accuracy of the respective spectroscopy technique and the accuracy of the reference analysis determinations are key factors affecting calibration model development. Important parameters include the method recovery (starting from sample extraction to analysis efficiency), instrumentation, application-related measurement deviations, and method LOD/LOQ. Simultaneous absorbance and fluorescence EEM performance has been shown to mainly depend on the absorbance and fluorescence properties of the individual compounds as discussed above. However, HPLC, UV–Vis enzymatic, and other routine chemical analyses have been demonstrated to also significantly contribute to final calibration model goodness. Among wine phenols analyzed by HPLC, anthocyanins have been shown with the generally lowest RSD and highest recovery rates, followed by phenolic acids and flavonols, and vice versa, flavan-3-ols with the highest RSD and lowest recovery rates. HPLC compared to UV–Vis enzymatic assays in general was characterized by higher precision and accuracy; however, these effects were notable in the finally developed calibrations and are important to be considered [17].

Ideally spectral and reference analysis performance should be, from the beginning of method development, tested to validate all used methods and to possibly define necessary adjustments with the overall aim to achieve prediction models with the best possible accuracy and precision. Finally, extractions from raw materials such as grape material can lead to increased measurement deviations due to difficulties with tough cell walls, vacuoles, chemical degradation, oxidation, and enzymatic activity, among other factors.

As mentioned above, same as for calibration, method validation is crucial for successful method development. It should start with cross-validation as the first critical tool for estimating the model's performance and fine-tuning algorithm parameters, such as the number of latent variables in PLS. This process involves iteratively splitting the whole data set (X-block and Y-block) into a calibration set for model development and a test (or internal validation) set. Common cross-validation methods include Venetian Blinds, where test sets are determined by selecting every *s*th object in the data set; Contiguous Blocks, which selects contiguous blocks of n/s objects in the data set for test sets; Random Subsets, where s different test sets are determined through random selection of n/s objects in the data set; and Leave-One-Out, where every single sample in the data set is used as a test set (https://wiki.eigenvector.com/index.php?title=Using_Cross-Validation). An easy and reliable approach is a random split keeping the sample replicates together. Depending on the overall size of the sample set, common split ratios are calibration:test 80:20. However, at least 20 individual samples are needed for a final validation following ASTM standards [66].

Cross-validations will generally fit a higher degree than validations with an independent sample set and entail the danger of model overfitting because not all foreign sample conditions may be covered, including, for example, unique concentration and matrix properties. It is important, therefore, to validate models by assessing the prediction performance using an independent test set, and depending on the method validation parameters, the choice of independent test set samples for a regression model may include samples of different harvest years, geographical origins, grape varieties, quality levels, among others. Furthermore, the size of the validation set should be as extensive as possible; a general rule is that the more samples available for covering all potential properties, especially when working on a new calibration, the better. There is, however, the possibility the calibration set exceeds memory limits and thus becomes unmanageable. So overall, it is more important that all the sources of variance are accounted for than simply having as many samples as possible. Furthermore, samples should be continuously included to evaluate accuracy, precision, specificity, linearity, range, and robustness. In general, it is recommended that the fundamental guidelines for model validations follow the official ASTM standards [66], which are described in Chapters 7 and 9 of this book.

10.8 Statistical Evaluation of Absorbance Ultraviolet–Visible and Fluorescence Excitation–Emission Matrix Applications

Most UV–Vis–EEM publications on wine and grape extract analysis center on multivariate regression methods to predict concentrations of phenolic compounds and basic chemistry parameters central to wine quality evaluation. The regressions generally use reference parameters from HPLC for phenolics and other analytical methods for basic chemistry parameters, as described above. The most common multivariate regression methods include PLS-R and XGB-R but also, more recently, Gray-CLS, as described earlier. The model goodness-of-fit criteria for these regression methods are generally quite similar to model calibration and validation evaluations described in Chapters 7 and 9 as well as being referenced in the USP 857 (UV–Vis) and 853 (Fluorescence) and ASTM E2617 methods cited in the previous chapters. For regression, the RMSE values and R^2 for the calibration, cross-validation, and prediction (or test set) are inarguably the central basis for model validation. Other evaluation criteria can include the Q and T residual statistics that can be useful for identifying outliers especially as these may be associated with model pre-processing and hypertuning adjustments as well as general methodological issues such as sampling errors. As described earlier the

ANOVA of the linear models can provide information on detection and quantification limits as well as bias with respect to the predicted slope and intercept parameters. Further spike and recovery statistics can be evaluated to gauge the accuracy of the method from extraction to final concentration reports.

Discrimination studies for wine and grape analysis with the UV–Vis–EEM method generally center around supervised multivariate methods such as PLS–DA and XGB–DA. SIMCA or Soft Independent Modeling by Component Analogy can also be used and is based on comparing separate principal component analyses for each sample class. Typical applications of discrimination modeling for wine and grapes include determining grape varieties, regions and sub-regions for growth, and concentration-based thresholds during blending and contamination detection, as introduced above. For all these methods, model evaluation and validation generally center on using a "confusion matrix" which summarizes the correct identification results from the calibration, cross-validation and prediction (test set). It is important to note that when implementing a confusion matrix, all samples must be identified properly and correctly classified since the test can only report on defined class groups. Thus, a confusion matrix is actually only a method validation tool and cannot be used for unknown and undefined samples. The confusion matrix (see Chapter 9 for parameter definitions) compares the ratios of the True and False positive and negative sample class assignments. The confusion matrix may include several computed parameters based on these ratios, including the error, probability, and $F1$ statistics (see Chapter 9) as well as Matthew's Correlation coefficient (similar to a chi-square statistic) for two-class or Pass–Fail models. As implemented in Eigenvector Solo, confusion matrices are commonly evaluated at two levels: the "Most Probable" rule and the "Strict" rule. In the Most Probable rule the sample ID is always assigned to the closest matching class independent of the Prediction Probability, and thus, there is no opportunity for unassigned classes. For the Strict rule a Prediction Probability threshold is selected, typically 50%, such that this threshold represents an additional criterion limiting class assignment. Unassigned values are an optional result of the Strict Rule evaluation. The confusion matrix report will also include data on misassigned samples and samples with multiple class assignments. In conclusion, before a discrimination model is used in a real-world application, it is important to carefully plan, execute, and evaluate the validation set and create a confusion matrix with fully populated and representative calibration and validation data sets.

In addition to supervised regression and discrimination statistics, which are directly implemented in the instrumental software Solo, data obtained by UV–Vis EEM analyses can be subject to other types of modeling. Spectral absorbance and unfolded EEM data may be exported as *.csv and computed as untargeted information in whatever statistical program desired such as R Studio. Furthermore, readily generated results such as from regression, discrimination, PARAFAC, or other implemented Solo methods (see Chapter 9) may also be implemented into further modeling such as the reported grape heterogeneity index [13].

10.9 Application Case Studies for Grapes and Wine

10.9.1 Real-World Regression Methods for Grape and Wine Quality Determination

With the previously discussed fundamental considerations of simultaneous absorbance and fluorescence spectroscopy, in combination with the direct and indirect optical properties exhibited by grape and wine phenols and basic chemistry, quantification procedures have been implemented using the UV–VIS EEM instrumental technique. An advantage gained over other industry methods (UV–Vis, enzymatic assays, etc.), the UV–Vis EEM technique can measure (poly)phenols, a diverse set of quality-related chemical markers, down to one-digit ppm range with an averaged respective R^2 and relative bias (REP) of 0.94 ± 0.04 and $4.1 \pm 1.8\%$ in wines [17] and 0.941 ± 0.025 and $5.89 \pm 2.47\%$ for grapes [16]. The measurement results can be used directly, or to obtain summed parameters (compound groups, attribute groups, etc.), ratios like (anthocyanin-to-tannin), scores (mouthfeel potential, quality potential etc.), or even be subject to additional multivariate modeling to support decision-making in viticulture and enology daily operations.

The approach was first developed for wine samples [17] because wines require less sample extraction and preparation, which are a potential source of additional method development technical issues. Nevertheless, spectral and reference analysis conditions were carefully analyzed and optimized with data processing and selected regression algorithms to achieve the most precise application. Respective absorbance and fluorescence single-block predictions were generally below an R^2 of prediction (R^2P) of 0.90. In contrast, multi-block data predictions reached increased precise fits above R^2P of 0.95 for all compounds (as shown in Figure 10.3 and related discussion). The XGB regression method yielded correspondingly better fits than partial least squares regression (PLS-R) methods. Additionally, for the 44 phenolic and basic chemistry compounds, apart from their respective individual spectral properties, the goodness of their reference analysis methods, including HPLC, UV–Vis precipitation assays, and basic standard methods, played a significant role in the precision and accuracy of the later prediction models. As a result, the highest fits were consistently obtained for the anthocyanin family and color with R^2P values >0.96 for both XGB-R and PLS-R. Flavan-3-ols yielded lower fits than other phenolic compounds overall; however, the benefits of the multi-block data for the regression algorithm increased significantly. As a further general trend for all compounds, concentrations >10 mg/L were found with precise fits R^2P between 0.929 to 0.992 (XGB-R) and 0.875 to 0.992 (PLS-R), and XGB-R compared to PLS-R was more robust toward concentration and compound-specific patterns (see Figure 10.6). Consequently, consistently lower prediction root mean square errors (RMSE) for XGB-R compared to PLS-R were obtained, indicating systematically higher accuracy and precision of the prediction of the former. Overall, the relative bias ranged between 0.7 and 9.2% for XGB-R and 0.9 and 12% for PLS-R and was compound-specific, lower for anthocyanins while increasing for flavonols, hydroxycinnamic acids, and flavan-3-ols. The correlations obtained for the basic chemistry parameters (TSS, acids, pH, etc.) were more likely caused by indirect effects on the colored and fluorescent compounds in solution, possibly quenching, ionization, and changes to refractive index rather than being fluorophores or being active in the UV–Vis regions. However, they exhibit higher concentrations up to grams per liter and overall robust predictions for these parameters were also yielded. The advantage of them being included in this application (although

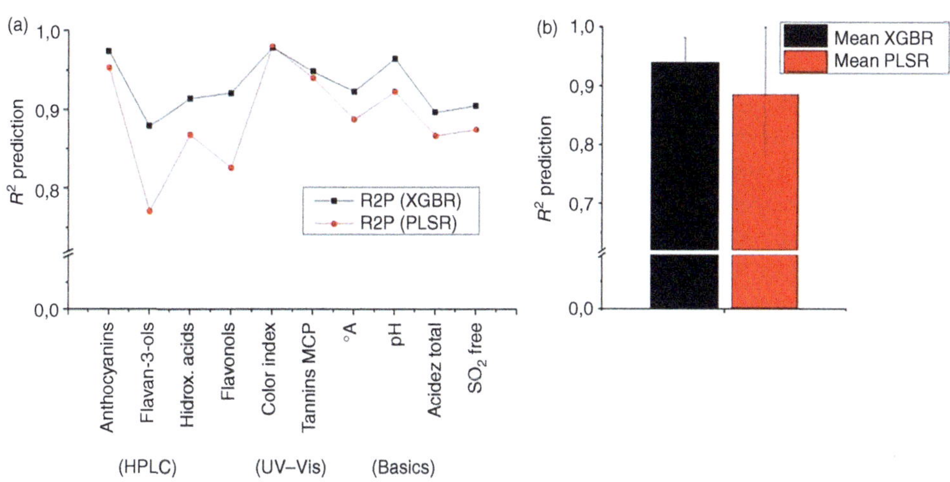

Figure 10.6 XGB-R compared to PLS-R multi-block models obtained systematically higher R^2 of prediction fits for the 44 compound models, including the HPLC parameters anthocyanins, flavan-3-ols, hydroxycinnamic acids, and flavonols, and the precipitation UV–Vis and basic chemistry (a), leading to respective overall mean increased R^2P values of XGB-R vs. PLS-R (b). *Source*: Adam Gilmore / Doreen Schober.

other rapid methods exist like FTIR, etc.) together with the phenolic composition in the same UV–VIS EEM measurement run is that these complementary results can be obtained in one measurement cycle. However, the application was first developed based on 126 experimental and commercial Cabernet Sauvignon, Carmenere, Malbec, Merlot, Pinot noir, Syrah, Chardonnay, and Sauvignon blanc wines, including different quality levels and harvest (storage) years, and then successfully blind validated with a smaller independent set of wines of another year. Still, it is recommended to routinely test and update the existing models over time as a standard practice to follow [66] to achieve a robust, practical tool for the wine industry application, which allows distinct conclusions on wine properties and quality.

In another application, the grape phenolic and basic chemistry of 226 grape samples sourced from different vineyards, maturation stages, and several different monovarietals, were calibrated as previously discussed, using XGB-R for modeling absorbance and fluorescence multi-block data. Grape berries were extracted using the modified Iland method [67]. In brief, grapes were destemmed and then homogenized with a Vitamix T&G® 2 commercial blender to obtain grape homogenate. One gram of each grape homogenate was acutely weighed in a 15 mL test tube, to which 10 mL 50% ethanolic solution (pH 2.0 adjusted with 37% HCl) was added. The obtained XGB regression models yielded an average bias slightly increased compared to wine of $5.89 \pm 2.47\%$, albeit at values in the acceptable range at significantly <10%. As mentioned, both the repeatability of the grape extraction method and HPLC reference method likely constituted the major sources of variation (5 to 8% RSD) compared to the repeatability of the UV–VIS EEM measurement (<2% RSD). However, like for wine, the grape simultaneous absorbance fluorescence EEM method is considered a useful tool in the wine industry for rapid decision-making. The quantitative analysis of grapes using XGB–DA and Agglomerative hierarchical cluster analysis (HCA) and key grape varieties such as Cabernet Sauvignon, Grenache, Malbec, Merlot, Petite Syrah, Pinot noir, Syrah, Teroldego and Zinfandel were 100% correctly classified and resolved, respectively [16]. Details for both the grape and wine methods and results can be found in the cited publications.

10.9.2 Supervised Discrimination for Varietal, Regional, Vintage, and Blending Classification

Several studies have reported successful discrimination models for wine and grape extracts due to the richness of information available from the multi-block (data fusion) methods for the UV–Vis–EEM method. Topics of interest include demonstrating the advantage of the UV–Vis–EEM method to Inductively Coupled Plasma Mass Spectroscopy (ICP-MS) for identifying growth regions in Australia and France. In the context of regional authentication, the UV–Vis–EEM method was also shown to have the capacity to resolve sub-regional growing areas in Australia [9, 10]. The most effective algorithm for these discrimination studies was XGB–DA, augmented with supporting data from principal component analysis (PCA) and PARAFAC cluster analysis methods. Interestingly, the regional and varietal discrimination models showed consistent patterns with the phenolic compound composition for both the Australian and Northern California studies, as reported for grape varieties by [16].

Consistent with the regional and varietal discrimination studies, it was also shown that the UV–Vis–EEM method could be used to discriminate not only wine varieties and growth regions but also three different stages of fermentation from selected growth regions [19]. The XGB–DA models allowed effective discrimination of the post-primary fermentation, post-MLF, and pre-blending stages. That study also showed that the blending percentages could be accurately assayed using XGB regression.

Another related study showed that UV–Vis–EEM discrimination by XGB–DA yielded highly accurate predictions of assigned classes to suggest that the wine's phenolic composition was the main driver for sensory differences associated with the clusters determined from the RATA analysis [35]; a significant correlation was also drawn between the UV–Vis–EEM prediction and measured perceived astringency. Another related study showed that XGB–DA and XGB regression with comparative data-fusion (multi-block) schemes could significantly predict sensory attributes associated with grape-to-wine processing, including aroma, flavor, taste, mouthfeel, and color.

10.9.3 Unsupervised Cluster Modeling: A Different Approach for Grape and Wine Quality Prediction

In contrast to discriminant analysis, cluster analysis can be applied without known sample identities. As introduced above in this chapter, wine quality, in terms of sensory characteristics, is a subjective metric that may vary based on the panelist but can be standardized to a certain point using chemical analysis by instrumental methods. Unsupervised clustering is a suitable approach for chemical data to create quality categories without including subjective entry and output variables. The method may be applied to untargeted or targeted analysis data. Among several cluster methods, spectral clustering has been shown to be the most powerful tool for grape and wine quality prediction. When working with samples from different vintages, the data sets may get normalized by range or by z-score [68].

As a first trial, spectral clustering was applied to Cabernet Sauvignon grape and wine respective datasets, and the obtained clusters were analyzed by Analysis of Variance (ANOVA) and Tukey's post hoc test (unpublished data, a short version was published as conference participation [69]). The main drivers for the cluster formation for grapes and wines were non-volatile phenolic compounds, including anthocyanins, flavonols, flavan-3-ols, hydroxycinnamic acids, and tannins. Interestingly, grape samples were observed to be more clearly separated into clusters than wine samples, presumably due to fermentation processes that change the chemical composition by polymerization, or liberation of precursors, among others [42].

These results suggest that chemical data may be sufficient, if not even better, to yield quality categories for grapes and wines without the need for extensive sensory evaluation. It may also be much more precise than estimating quality, style, or even grape pricing by using viticulture parameters like yield or trellis system, which often lead to inaccuracy and less precise classifications in daily winery operations, thus complicating the production processes.

Similar to the discussed approach, utilizing UV–Vis–EEM data in combination with HCA and RATA analysis, sensory style categories were established, and in combination with XGB–DA ripeness categories, including overripe, ripe, less ripe, and underripe, could be determined [14]. However, while the exact relationship between chemistry and sensory quality parameters is complex and cannot be satisfactorily linked; the advantage and innovation of clustering methods is that they provide fixed group information from chemical data, which can be subsequently interpreted independently as quality(related) categories.

10.10 Conclusion

This chapter, in context with the theme of this book on quantitative spectroscopy applications, provides practical guidance on the development and hands-on daily use of simultaneous absorbance UV–Vis and fluorescence EEM spectroscopy for research and field application in the grape wine industry. Fluorescent compounds important to grape and wine quality are examined regarding their sensory-chemical and optical properties. Using phenolic compounds and basic chemistry parameters, the reader is guided through specific spectroscopic and chemometric information, from representative sampling and sample selection to working with combined multi-block absorbance and fluorescence data sets to elaborate calibrations and validations using different chemometric modeling approaches. The theory to achieve a well-organized winemaking production process by integrating UV–VIS and EEM instrument measurements and data management is presented along with case studies for several successfully implemented applications and suggestions for further potential approaches.

References

1 Thorngate, J. (2006). Methods for analyzing phenolics in research. *American Journal of Enology and Viticulture* 57 (3): 269–279.

2 Luis Aleixandre-Tudó, J., Buica, A., Nieuwoudt, H.H. et al. (2017). Spectrophotometric analysis of phenolic compounds in grapes and wines: A review. *Journal of Agricultural and Food Chemistry* 65 (20): 4009–4026.

3 Ouédraogo, F., Cornu, J.-Y., Janot, N. et al. (2022). Do DOM optical parameters improve the prediction of copper availability in vineyard soils? *Environmental Science and Pollution Research* 29 (20): 29268–29284.

4 Moret, F., Lemaître-Guillier, C., Grosjean, C. et al. (2019). Clone-dependent expression of Esca disease revealed by leaf metabolite analysis. *Frontiers in Plant Science* 9: 1960.

5 Brandt, B., Perez, I., Cave, J. et al. (2023). *Rapid Smoke Impact Evaluation using Simultaneous Absorbance and Fluorescence Spectroscopy*. American Society of Viticulture and Enology.

6 Gilmore, A., Jeffery, D., Ranaweera, R. et al. (2019). *Analyses of Smoke-Tainted Wine Using Simultaneous Absorbance-Transmittance and Fluorescence Excitation-Emission Mapping*. American Society of Viticulture and Enology.

7 Ranaweera, K.R.G., Adam, M., and Jeffery, D.W. (2024). Fluorescence spectroscopy for red wine authentication. In: *Wine Analysis and Testing Techniques*, vol. 1 (ed. M.Á. Pozo-Bayón and C.M. González), 23–38. New York, NY: Springer.

8 Ranaweera, R.K.R., Capone, D.L., Bastian, S.E.P. et al. (2021). A review of wine authentication using spectroscopic approaches in combination with chemometrics. *Molecules* 26 (14): 4334.

9 Ranaweera, R.K.R., Gilmore, A.M., Capone, D.L. et al. (2021). Spectrofluorometric analysis combined with machine learning for geographical and varietal authentication, and prediction of phenolic compound concentrations in red wine. *Food Chemistry* 361: 130149.

10 Ranaweera, R.K.R., Gilmore, A.M., Capone, D.L.B. et al. (2021). Authentication of the geographical origin of Australian Cabernet Sauvignon wines using spectrofluorometric and multi-element analyses with multivariate statistical modelling. *Food Chemistry* 335: 127592.

11 Ranaweera, R.K.R., Bastian, S.E.P., Gilmore, A.M. et al. (2023). Absorbance-transmission and fluorescence excitation-emission matrix (A-TEEM) with multi-block data analysis and machine learning for accurate intraregional classification of Barossa Shiraz wine. *Food Control* 144: 109335.

12 Wang, H. and Jeffery, D.W. (2024). Machine learning model stability for sub-regional classification of Barossa Valley Shiraz wine using A-TEEM spectroscopy. *Food* 13 (9): 1376.

13 Armstrong, C.E.J., Previtali, P., Boss, P.K. et al. (2023). Grape heterogeneity index: Assessment of overall grape heterogeneity using an aggregation of multiple indicators. *Plants* 12 (7): 1442.

14 Armstrong, C.E.J., Gilmore, A.M., Boss, P.K. et al. (2023). Machine learning for classifying and predicting grape maturity indices using absorbance and fluorescence spectra. *Food Chemistry* 403: 134321.

15 Armstrong, C.E.J., Niimi, J., Boss, P.K. et al. (2023). Use of machine learning with fused spectral data for prediction of product sensory characteristics: The case of grape to wine. *Food* 12 (4): 757.

16 Gilmore, A.M., Sui, Q., Blair, B., and Pan, B.S. (2022). Accurate varietal classification and quantification of key quality compounds of grape extracts using the absorbance-transmittance fluorescence excitation emission matrix (A-TEEM) method and machine learning. *OENO One* 56: 107–115.

17 Schober, D., Gilmore, A., Chen, L. et al. (2022). Determination of Cabernet Sauvignon wine quality parameters in Chile by absorbance-transmission and fluorescence excitation emission matrix (A-TEEM) spectroscopy. *Food Chemistry* 392: 133101.

18 Coelho, C., Parot, J., Gonsior, M. et al. (2017). Asymmetrical flow field-flow fractionation of white wine chromophoric colloidal matter. *Analytical and Bioanalytical Chemistry* 409 (10): 2757–2766.

19 Pacheco, M., Winckler, P., Marin, A. et al. (2022). Multispectral fluorescence sensitivity to acidic and polyphenolic changes in Chardonnay wines – The case study of malolactic fermentation. *Food Chemistry* 370: 131370.

20 Coelho, C., Gougeon, R.D., Perepelkine, L. et al. (2019). Chemical transfers occurring through *Oenococcus oeni* biofilm in different enological conditions. *Frontiers in Nutrition* 6: 95.

21 Herbert-Pucheta, J.E., Padilla-Maya, G., Milmo-Brittinham, D. et al. (2019). Multivariate spectroscopy for targeting phenolic choreography in wine with A-TEEMTM and NMR crosscheck non-targeted metabolomics. *BIO Web of Conferences* 15: 02006.

22 Quatela, A., Gilmore, A.M., Gall, K.E.S. et al. (2018). A-TEEMTM, a new molecular fingerprinting technique: Simultaneous absorbance-transmission and fluorescence excitation-emission matrix method. *Methods and Applications in Fluorescence* 6 (2): 027002.

23 Gilmore, A. and Chen, L. (2024). Is my wine red enough? *Chemistry in Australia – Un-wined.*

24 Ranaweera, R.K.R., Gilmore, A.M., Bastian, S.E.P. et al. (2022). Spectrofluorometric analysis to trace the molecular fingerprint of wine during the winemaking process and recognise the blending percentage of different varietal wines. *OENO One* 56 (1): 189–196.

25 Coelho, C., Aron, A., Chloé, R.-G. et al. (2015). Fluorescence fingerprinting of bottled white wines can reveal memories related to sulfur dioxide treatments of the must. *Analytical Chemistry* 87 (16): 8132–8137.

26 Roullier-Gall, C., Hemmler, D., Gonsior, M. et al. (2017). Sulfites and the wine metabolome. *Food Chemistry* 237: 106–113.

27 Hopfer, H., Nelson, J., Ebeler, S.E., and Heymann, H. (2015). Correlating wine quality indicators to chemical and sensory measurements. *Molecules* 20 (5): 8453–8483.

28 Hopfer, H. and Heymann, H. (2014). Judging wine quality: Do we need experts, consumers or trained panelists? *Food Quality and Preference* 32 (Part C): 221–233.

29 Charters, S. and Pettigrew, S. (2007). The dimensions of wine quality. *Food Quality and Preference* 18 (7): 997–1007.

30 Cleary, M., Chong, H., Ebisuda, N. et al. (2015). Objective chemical measures of grape quality. *ACS Symposium Series* 120: 365–378.

31 Niimi, J., Tomic, O., Næs, T. et al. (2020). Objective measures of grape quality: From Cabernet Sauvignon grape composition to wine sensory characteristics. *LWT* 123: 109105.

32 Niimi, J., Tomic, O., Næs, T. et al. (2018). Application of sequential and orthogonalised-partial least squares (SO-PLS) regression to predict sensory properties of Cabernet Sauvignon wines from grape chemical composition. *Food Chemistry* 256: 195–202.

33 Van Leeuwen, C. and Seguin, G. (2006). The concept of terroir in viticulture. *Journal of Wine Research* 17 (1): 1–10.

34 Atkin, T. and Johnson, R. (2010). Appellation as an indicator of quality. *International Journal of Wine Business Research* 22 (1): 42–61.

35 Souza Gonzaga, L., Bastian, S.E.P., Capone, D.L. et al. (2021). Modelling Cabernet-Sauvignon wine sensory traits from spectrofluorometric data. *OENO One* 55 (4): 19–33.

36 Heresztyn, T. (1986). Metabolism of volatile phenolic compounds from hydroxycinnamic acids by Brettanomyces yeast. *Archives of Microbiology* 146 (1): 96–98.

37 Lakowicz, J.R. (2006). *Principles of Fluorescence Spectroscopy*, 954. Springer.

38 Merkyte, V., Longo, E., Windisch, G., and Boselli, E. (2020). Phenolic compounds as markers of wine quality and authenticity. *Foods* 9 (12): 1785.

39 Casassa, L.F. (2017). Flavonoid phenolics in red winemaking. In: *Phenolic Compounds – Natural Sources, Importance and Applications* (ed. M. Soto-Hernández, M. Palma-Tenango, and R. Garcia-Mateos M del). InTech.

40 Cheynier, V. (2012). Phenolic compounds: From plants to foods. *Phytochemistry Reviews* 11 (2): 153–177.

41 Cheynier, V., Dueñas-Paton, M., Salas, E. et al. (2005). Structure and properties of wine pigments and tannins. *American Journal of Enology and Viticulture* 57 (3): 298–305.

42 Waterhouse, A.L. (2002). Wine phenolics. *Annals of the New York Academy of Sciences* 957 (1): 21–36.

43 Airado-Rodríguez, D., Durán-Merás, I., Galeano-Díaz, T., and Wold, J.P. (2011). Front-face fluorescence spectroscopy: A new tool for control in the wine industry. *Journal of Food Composition and Analysis* 24 (2): 257–264.

44 Airado-Rodŕiguez, D., Galeano-Díaz, T., Durán-Merás, I., and Wold, J.P. (2009). Usefulness of fluorescence excitation-emission matrices in combination with parafac, as fingerprints of red wines. *Journal of Agricultural and Food Chemistry* 57 (5): 1711–1720.

45 Santos, I., Bosman, G., Aleixandre-Tudo, J.L., and du Toit, W. (2022). Direct quantification of red wine phenolics using fluorescence spectroscopy with chemometrics. *Talanta* 236: 122857.

46 Hufnagel, J.C. and Hofmann, T. (2008). Orosensory-directed identification of astringent mouthfeel and bitter-tasting compounds in red wine. *Journal of Agricultural and Food Chemistry* 56 (3): 1376–1386.

47 Hufnagel, J.C. and Hofmann, T. (2008). Quantitative reconstruction of the nonvolatile sensometabolome of a red wine. *Journal of Agricultural and Food Chemistry* 56: 9190–9199.

48 Kennedy, J. (2008). Grape and wine phenolics: Observations and recent findings. *Ciencia e Investigación Agraria* 35 (2): 107–120.

49 Pawlus, A.D., Waffo-Téguo, P., Shaver, J., and Mérillon, J.-M. (2012). Stilbenoid chemistry from wine and the genus Vitis, a rewiev. *Journal International des Sciences de la Vigne et du Vin* 46 (2): 57–111.

50 Krstic, M.P., Johnson, D.L., and Herderich, M.J. (2015). Review of smoke taint in wine: Smoke-derived volatile phenols and their glycosidic metabolites in grapes and vines as biomarkers for smoke exposure and their role in the sensory perception of smoke taint. *Australian Journal of Grape and Wine Research* 21 (S1): 537–553.

51 Lo, Y.C., Koziel, J.A., Cai, L. et al. (2008). Simultaneous chemical and sensory characterization of volatile organic compounds and semi-volatile organic compounds emitted from swine manure using solid phase microextraction and multidimensional gas chromatography-mass spectrometry-olfactometry. *Journal of Environmental Quality* 37 (2): 521–534.

52 Matsui, K. (2016). A portion of plant airborne communication is endorsed by uptake and metabolism of volatile organic compounds. *Current Opinion in Plant Biology* 32: 24–30.

53 Capone, D.L., Jeffery, D.W., and Sefton, M.A. (2012). Vineyard and fermentation studies to elucidate the origin of 1,8-cineole in Australian red wine. *Journal of Agricultural and Food Chemistry* 60 (9): 2281–2287.

54 Milheiro, J., Filipe-Ribeiro, L., Vilela, A. et al. (2019). 4-Ethylphenol, 4-ethylguaiacol and 4-ethylcatechol in red wines: Microbial formation, prevention, remediation and overview of analytical approaches. *Critical Reviews in Food Science and Nutrition* 59 (9): 1367–1391.

55 Rodrıguez-Delgado, M.A., Malovana, S., Perez, J.P. et al. (2001). Separation of phenolic compounds by high-performance liquid chromatography with absorbance and fluorimetric detection. *Journal of Chromatography A* 912 (2): 249–257.

56 Lv, Y., Wang, J.N., Jiang, Y. et al. (2023). Identification of oak-barrel and stainless steel tanks with oak chips aged wines in Ningxia based on three-dimensional fluorescence spectroscopy combined with chemometrics. *Molecules* 28 (9): 3688.

57 Del Fresno, J.M., Morata, A., Loira, I. et al. (2020). Evolution of the phenolic fraction and aromatic profile of red wines aged in oak barrels. *ACS Omega* 5 (13): 7235–7243.

58 Kőrösi, L., Molnár, S., Teszlák, P. et al. (2022). Comparative study on grape berry anthocyanins of various teinturier varieties. *Food* 11 (22): 3668.

59 Peleg, H., Gacon, K., Schlich, P., and Noble, A.C. (1999). Bitterness and astringency of flavan-3-ol monomers, dimers and trimers. *Journal of the Science of Food and Agriculture* 79 (8): 1123–1128.

60 Martínez-Lüscher, J., Brillante, L., and Kurtural, S.K. (2019). Flavonol profile is a reliable indicator to assess canopy architecture and the exposure of red wine grapes to solar radiation. *Frontiers in Plant Science* 10: 1–15.

61 Song, J., Smart, R., Wang, H. et al. (2015). Effect of grape bunch sunlight exposure and UV radiation on phenolics and volatile composition of *Vitis vinifera* L. cv. Pinot noir wine. *Food Chemistry* 173: 424–431.

62 Fleet, G.H. (1993). *Wine Microbiology and Biotechnology*. CRC Press.

63 Ribéreau-Gayon, P., Glories, Y., Maujean, A., and Dubourdieu, D. (2006a). *Handbook of Enology. Handbook of Enology*, vol. 1. West Sussex, England: John Wiley and Sons, Ltd.

64 Sect. (2009). "Sulfur dioxide" (Resolution Oeno 377/2009). Method OIV-MA-AS323-04B.

65 Kramer, R. (1998). *Chemometric Techniques for Quantitative Analysis*, 1e. Boca Raton: CRC Press.

66 Sect. (2017). ASTM methods E2617. Standard Practice for Validation of Empirically Derived Multivariate Calibrations.

67 Iland, P., Bruer, N., Edwards, G. et al. (2013). *Chemical Analysis of Grapes and Wine: Techniques and Concepts*, 2e (ed. N. Bruer, G. Edwards, S. Caloghiris, and E. Wilkes). Campbelltown, SA: Patrick Iland Wine Promotions Pty Ltd.

68 Hastie, T., Tibshirani, R., and Friedman, J.H. (2009). *The Elements of Statistical Learning: Data Mining, Inference and Prediction*. Springer.

69 Schober, D., Legues, M., Guidez, H., et al. (2020). How geographical origin and vineyard management influence cv. Cabernet Sauvignon in Chile-Machine learning based quality prediction. XIII International Terroir Congress; Adelaide, Australia (online).

11

Surface-Enhanced Raman Spectroscopy Theory

Stuart Farquharson and Chetan Shende

Real-Time Analyzers, Inc, Middletown, CT, USA

11.1 Introduction

Surface-enhanced Raman spectroscopy (SERS) has become a powerful analytical tool due to its superior specificity and extreme sensitivity. The rich molecular information provided by Raman scattering and the increased sensitivity provided by surface enhancement from silver or gold nanoparticles has been successfully used for biological [1, 2] chemical [3], environmental [4, 5], homeland security [6, 7] military [8, 9], medical [10–12], and pharmaceutical applications [13–15]. Since its prediction [16, 17], discovery [18], recognition [19, 20], and quantification [21, 22], the mechanism responsible for the large increase in Raman scattering has been the subject of considerable research [23].

11.2 Discussion

11.2.1 Basic Theory of Surface-Enhanced Raman Spectroscopy

Two mechanisms have been identified that contribute to this large increase in scattering efficiency, an electromagnetic mechanism involving surface plasmons, and a chemical mechanism involving charge transfer between the metal particle and the molecule being enhanced. Irradiating a nanoscale particle composed of materials with free (or almost free) electrons, such as copper, gold, and silver, with an electromagnetic wave having a field strength, E_o, and wavelength, λ_o, much greater than the particle radius, r, can cause the conducting electrons of the particle to oscillate collectively producing dipolar surface plasmons. If a laser is used and is in resonance with the dipolar plasmon, the particle will radiate light with a surface plasmon field strength, E_{sp}, characteristic of dipolar radiation (Figure 11.1). This is a coherent process, and E_{sp} will be greatest perpendicular to the irradiation axis (at $\theta = 0$ or $180°$) with a gain, g (due to the dielectric), approximated by:

$$E_{sp} = g \cdot E_o(1 + 3\ \cos^2\theta) \tag{11.1}$$

A molecule in the plasmon field of the metal particle will be excited by both the E_{sp} and E_o and generate the surface-enhanced Raman scattered radiation, E_{SERS}, which is proportional to the molecule's polarizability, α_R, according to Eq. (11.2) (here the Raman stokes shift is assumed small, see Reference [25] for a complete treatment):

$$E_{SERS} = k \cdot \alpha_R \cdot g^2 \cdot E_o \tag{11.2}$$

Practical Quantitative Vibrational and Electronic Spectroscopy: A Guide for Developing, Optimizing, and Validating Procedures, First Edition. Edited by Deborah A. Peru.

Companion website: www.wiley.com/go/peruelectronicspectroscopy

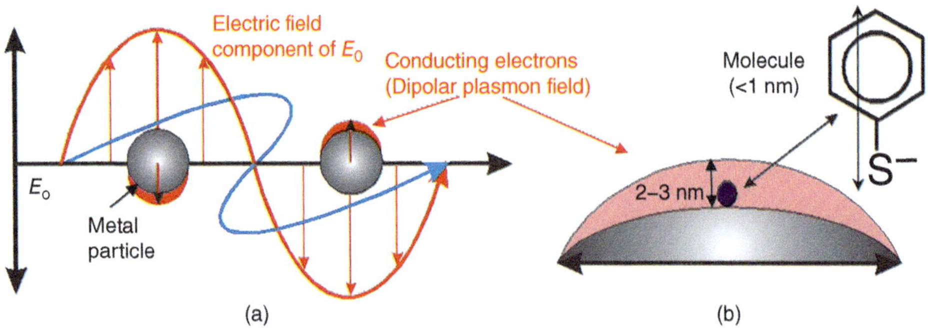

Figure 11.1 (a) Illustration of an electromagnetic wave with magnitude E_o (electric and magnetic field components are shown in (red) and perpendicular (blue) to the plane of the page, respectively), propagating left to right, causing the conducting electrons to oscillate (up and down) and produce a dipolar plasmon field (modified from Reference [24]). (b) Illustration of molecule within a plasmon field of an ~50 nm diameter silver particle showing relative scale. *Source*: RTA authors.

Furthermore, the SERS intensity, is proportional to the field squared, or

$$I_{\text{SERS}} = k' \cdot \alpha_{\text{R}}^2 \cdot g^4 \cdot I_o \tag{11.3}$$

In essence, this fourth power dependence of g indicates that the electromagnetic mechanism is the dominant contribution to the SERS scattering intensity.

The ability for a metal to produce and sustain a dipole, d, is dependent on its polarizability and the electromagnetic field strength, i.e.:

$$d = \alpha_{\text{metal}} \cdot E \tag{11.4}$$

The polarizability of small particles, α_{metal}, with radius, r, in turn can be described in terms of its optical constants:

$$\alpha_{\text{metal}} = r^3 \cdot (\varepsilon_{\text{p}} - 1)/(\varepsilon_{\text{p}} + 2) \tag{11.5}$$

and for very small particles, g can be approximated by:

$$g = (\varepsilon_{\text{p}} - \varepsilon_{\text{m}})/(\varepsilon_{\text{p}} + 2\varepsilon_{\text{m}}) \tag{11.6}$$

where;

$$\varepsilon = m^2 = (n + \text{Im}k)^2 \tag{11.7}$$

ε_{p} and ε_{m} are the dielectric constants for the particle and the matrix, respectively, m and n are the complex and real refractive indices, and k is the absorptive index (Im signifies the imaginary component of the complex expression). Note that as ε_{p} becomes equal to $-2\varepsilon_{\text{m}}$, g will become very large, signifying a plasmon resonance.

In general, the plasmon spectrum (the field as a function of wavelength) can be used to determine the laser excitation wavelength that will generate the most intense SERS. The plasmon or extinction spectrum can be calculated using Mie scattering theory. Mie theory, which is derived from Maxwell's equations, gives a general description of light scattering by particles and allows prediction of the extinction spectrum for a collection of j particles given their sizes, shapes, chemical compositions, and associated optical constants [26].

The extinction spectrum (measured by a visible–near infrared spectrophotometer) has become the standard method of characterizing a SERS-active medium, where the extinction can be written in terms of the

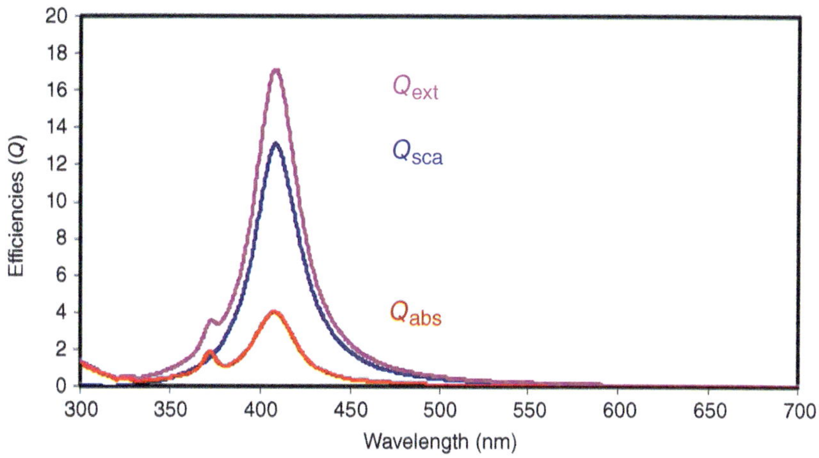

Figure 11.2 Mie calculated plasmon spectra for plasmon efficiencies, Q (as labeled). For 25 nm radius silver particle in water. *Source*: RTA authors.

extinction cross-section, $C_{ext,j}$ and n_j, the number of particles of type j per cm^2 of incident beam summed over the collection of j-type particles:

$$\text{Extinction} = \Sigma_j(n_j \cdot C_{ext,j,\lambda})/2.303 \tag{11.8}$$

where C_{ext} is a the sum of light absorption and scattering ($C_{ext} = C_{abs} + C_{sca}$). These terms are simply the extinction, absorption, and scattering efficiencies (Q_{ext}, Q_{abs}, Q_{sca}) per cm^2 of particle surface area. In predicting extinction spectra from theory, it is more convenient to calculate the efficiencies. The absorption and scattering efficiencies can be determined throughout the visible–near infrared spectrum using the wavelength dependent optical constants according to:

$$Q_{abs} = 4X \cdot \text{Img} \tag{11.9}$$

$$Q_{sca} = (8/3) \cdot X^4 \cdot g^2 \tag{11.10}$$

where:

$$X = 2\pi r(\varepsilon_m)^{1/2}/\lambda \tag{11.11}$$

for the case of a sphere. Variations in particle geometry require more complex definitions for X and ε which has resulted in numerous, substrate-specific, equations [23].

For a basic theoretical treatment, the particles are assumed to be a single size, spherical, and composed of silver or gold. As an example, Q_{sca} is calculated for a 25 nm radius silver sphere in water excited at 410 nm. From tables, n and k can be found for water (1.33 and 2×10^{-9}, respectively) [27, 28], which allows approximating ε_m as n^2 yielding a value of 1.77. And from literature, ε_p for silver can be found (ca. −4.2) [29]. According to Eqs. (11.6) and (11.10), $g = 9$ and $Q_{sca} = 14.6$ (Figure 11.2). Note that this suggests that I_{SERS} will be $(9)^4$ as strong as I_0 or nearly four orders of magnitude greater.

In the early 1980s, it was found that the maximum enhancement for SERS experiments was observed to the red of that predicted by theory [21]. Common explanations included non-spherical metal shapes [25], aggregation in the case of colloids [30], and the formation of a charge-transfer complex between the analyte and the metal [31]. In all cases, there is supporting experimental data in the form of the plasmon spectra, in which a secondary peak appeared to the red of the primary plasmon peak. The charge-transfer explanation forms the basis of the chemical mechanism.

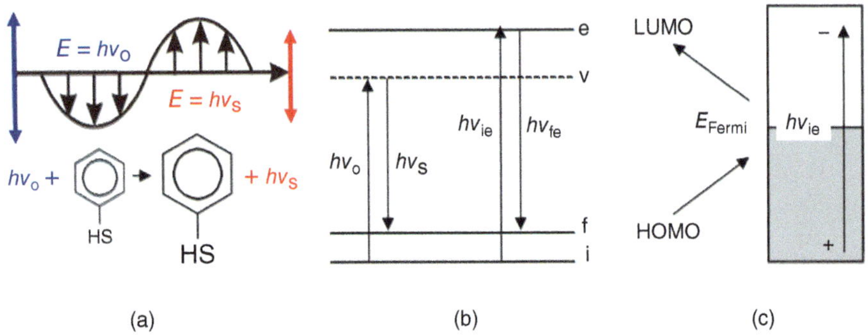

Figure 11.3 (a) Illustration of laser radiation inducing a dipole in benzenethiol that re-radiates at an energy minus that of a molecular vibration ($h\nu_s$), (b) an energy level diagram illustrating the same process for the pre-resonance and resonance cases ($h\nu_o = h\nu_{ie}$), and (c) an energy diagram indicating the relative energies for the Fermi level in a metal in relation to molecular orbitals (charge transfer may also occur between the HOMO and the metal holes). *Source*: RTA authors.

The basic equation relating the intensity of the Raman scattered radiation, I_{Raman}, to the polarizability of a molecule, α, can be defined by:

$$\alpha_{\rho\sigma} = \sum_e \left[\frac{\langle f|\mu_\sigma|e\rangle \langle e|\mu_\rho|i\rangle}{(h\nu_{ie} - h\nu_o - \text{Im}\Gamma)} + \frac{\langle i|\mu_\rho|e\rangle \langle e|\mu_\sigma|f\rangle}{(h\nu_{fe} + h\nu_o - \text{Im}\Gamma)} \right] \tag{11.12}$$

which describes the transition from some initial vibrational state, i, to a final state f, summed over all possible excited electronic states (real or virtual, e or v in Figure 11.3), where μ represents the dipole moment for the transition (using the bra-ket formalism to describe each wave function), either along the laser or Raman field axes, ρ and σ (Γ is the damping rate of the electronic states).

It is easy to see that if the laser excitation energy is chosen to match an electronic transition energy ($h\nu_o = h\nu_{ie}$), that the denominator of the first term becomes very small and consequently the polarizability becomes very large. This is the resonance Raman condition. Typically, molecules only have electronic transitions that occur in the ultraviolet part of the spectrum (below 350 nm). But there are two exceptions germane to SERS: (1) dye molecules that have extended π orbital systems supporting π to π^* transitions well into the red, and (2) metal–ligand complexes that support charge-transfer transitions.

Since charge-transfer complexes are known for silver and silver clusters, it is reasonable to assume that such electronic transitions can occur when a molecule forms a bond with a silver particle. In this case, some of the metal plasmon electrons can escape from the surface, propagate through space (ballistic electrons), and couple to the molecular orbits of chemisorbed or even physisorbed molecules. In essence a transient charge transfer occurs, analogous to the metal–ligand complex. The magnitude of the transient charge transfer is especially large if the relative energies of the lowest-unoccupied molecular orbital (LUMO) and highest-occupied molecular orbital (HOMO) are close in energy to the metal Fermi level.

Pyridine on silver represents a thoroughly studied system. It has been shown that a charge-transfer band appears at 600 nm for this system, and wavelength-dependent enhancement profiles, obtained by exciting at various laser wavelengths, show the greatest enhancements at this wavelength [32]. By changing particle size it is possible to detune the electromagnetic (plasmon) contribution, and the chemical contribution has been estimated at 10^3 (albeit, the particle sizes were far from optimum and the electromagnetic contribution was estimated at 10^2).

The most ideal condition occurs if the particle plasmon absorption (appropriate metal particle size or features), charge-transfer absorption, and molecular absorption can all be accessed by the same laser wavelength. This is essentially the case when dye molecules, such as rhodamine 6G (abs_{max} = 530 nm) or crystal violet (abs_{max} = 560 nm) are examined. It is therefore not surprising that enhancement factors (EFs) of 10^{12-14} and single-molecule detection are reported for these systems.

11.2.2 Substrates and Structures

Numerous methods have been developed to produce surfaces or solutions containing metals with optimum roughness or diameter to promote SERS [33, 34]. The initial work employed electrodes in electrolytic cells [18, 19, 21] was quickly followed by silver and gold colloidal solutions [35–38], and vapor-deposited metal-coated substrates [39–45]. While electrodes are rarely used today, colloids and substrates are widely used.

Electrodes are "roughened" in electrolytic solutions by changing the applied potential to dissolve (oxidize) and redeposit (reduce) the active metal. The greatest advantage of this method is that the surface can be regenerated and when used with a flow system, continuous measurements are possible [38, 46]. Unfortunately, the oxidation–reduction process does not produce the same SER-active structures each time and this approach is highly irreproducible. Furthermore, the method requires introducing the sample into an electrolytic solution and sample cell, severely limiting applications.

Colloids are prepared by reducing a metal salt solution to produce metal nanoparticles, which at high concentrations can form aggregates. The greatest advantage of colloids is that they are easy to prepare and simple to use. However, the particle and aggregate size are strongly influenced by the initial chemical concentrations, temperature, pH, and rate of mixing, and as a result are somewhat difficult to reproduce [37]. Nevertheless, several companies have refined early preparations to the point of consistent colloids suitable for SERS. Yet, analysis of a given chemical requires repeat measurements to ensure spectral accuracy.

Substrates continue to be one of the most active area of SERS research with the goal of producing patterned structures that provide repeatable EFs. There have been several hundred publications since the 1990s describing new substrates and/or preparations [34]. Initial structures were prepared by evaporating or sputtering silver or gold onto glass slides [40, 41, 47] and chromatographic plates [48]. These substrates proved non-uniform in that EFs on the order of 10^6 were only obtained at a few locations across the surface, known as "hot spots" [47, 49]. In an effort to improve reproducibility, photolithography was used to produce precise, repeatable structures, which were then coated with silver or gold (see product description below). Due to the equipment expense of photolithography, a number of alternative deposition procedures were pursued using various templates, such as graphene [50], black phosphorus [51], and structured crystals of boron and titanium nitride [52]. In many cases, EFs of 10^8 to 10^{12} were reported for measurements of dyes, such as crystal violet, methylene blue, Rhodamine B and 6G. These large enhancements include the resonance Raman mechanism provided by employing a laser with a wavelength that overlaps the dye absorption, as described in the previous section. Consequently, such enhancements and extreme sensitivity don't often apply to real-world applications, such as the detection of drugs.

One of the more successful and relatively cost-effective substrates is vapor deposition of silver on polystyrene spheres, known as silver film over nanospheres (AgFON) [53, 54]. This substrate was used to monitor several drugs, including glucose [55]. The greatest advantage of substrates is that measurements do not have to be performed in solution, and they work well with Raman-coupled-microscope systems [56, 57]. The greatest disadvantage of substrates is that the sample should be dried to concentrate it on the surface of the SERS-active metal to achieve maximum sensitivity. This can take time, depending on the solvent, and reproducibility has proven difficult [58].

As described above, there are numerous research publications about new SERS substrates, and except for colloids, there are few commercial products available to perform SERS measurements. The availability of silver and gold nanoparticles has grown with the lateral flow assay test strip market driven by COVID.

Both silver and gold colloids are available at various sizes and concentrations from a number of suppliers. A few suppliers advertising SERS colloids include Alpha Chemistry (Holbrook, NY), CD Bioproducts (Shirly, NY), Fortis Life Sciences (Montgomery, TX), Nanopartz (Loveland, CO), Sigma-Aldrich (Allentown, PA), StellarNet (Tampa, FL), and Ted Pella (Redding, CA).

Suppliers of silver and gold substrates include 4n4Nano (Korneuburg, Australia), D3-Technologies (Glasgow, United Kingdom), NikaLite (Bicester, United Kingdom), PiCO Foundry (Daejeon, South Korea), SERSative (Warsaw, Poland), and Silmeco (Copenhagen, Denmark). 4n4Nano sells 4n4SERS substrates coated with silver or gold nanoparticles. The substrate serves as a heatsink, so that high-powered lasers, up to 500 mW, can be used to achieve part-per-million sensitivity without damaging the substrate. D3-Technologies sells Klarite™, a gold-coated, lithographically produced inverted pyramid structure, with wells 1.8 μm on a side. Since this size is too large to generate a plasmon field, it is more likely that the modest enhancement (e.g. 10^{3-4}) is due to nanoscale roughness of the gold surface (Personal communication with R. Van Duyne.). NikaLite coats glass substrates with gold nanoparticles, and achieves modest enhancements. PiCO Foundry offers small cross-hatched grids sputter coated with silver or gold nanoparticles. SERSative produces photolithographic substrates coated with silver designed to fit in the wells of a 96-well plate. Silmeco sells substrates coated with gold or silver nanopillars, typically on the order of 100 nm in diameter.

In an effort to overcome many of the limitations for the various SERS substrates cited above, Simple SERS Sample Vials were introduced by Real-Time Analyzers Inc. (RTA) in 2005 [59]. The 2 mL glass vials are internally spin-coated with a porous glass containing silver or gold. The porous sol–gel silica network offers a unique environment for stabilizing SERS-active metals. The porosity of the sol–gel is on the scale of hundreds of microns, sufficient to pass both chemicals and small biologicals, such as pathogens, while the metal is a nanoscale network structure that does not change as a function of pH, solvent, or analyte concentration. RTA personnel also developed and patented 1-mm glass capillaries internally coated with the metal-doped sol–gels [60, 61]. These are attached to a syringe so that a sample can be drawn into the capillary and immediately measured. This design ensures interaction between the analyte and metal particles and typically improves detection limits by 20 to 100 times. Overall, we have measured over 500 chemicals, biochemicals, and pathogens with enhancements ranging from 10^4 to 10^{10}.

The overall structure of our SERS substrates is composed of interconnected nodules approximately 50 to 200 nm in diameter (Figure 11.4a). The interconnecting junctions provide further amplification, similar to the gap between two spheres that, theoretically, increase enhancement by a factor of 10^4 compared to single spheres (Figure 11.4b) [62]. Furthermore, the interconnecting structure ranges from 20 to 500 nm, providing a broad extinction spectrum extending into the near-infrared (Figure 11.4c). This structure appears to provide an ideal match between the Fermi level and the molecular orbitals of a wide range of analytes to generate strong charge-transfer states.

11.2.3 Enhancement Factors and Detection Limits

The surface EF defines the amount that the bulk-phase Raman signal intensity of a molecular vibration is increased by the combined electromagnetic and chemical SER mechanisms according to Eq. (11.13):

$$\mathrm{EF} = I_{\mathrm{SER}}/I_{\mathrm{R}} \cdot N_{\mathrm{R}}/N_{\mathrm{SER}} \tag{11.13}$$

where I is the vibrational mode intensity, N is the number of molecules in the laser scattering volume, and SER and R have their usual meaning (if the laser power is different for the two measurements it must be taken into account). The intensity can be calculated by integrating the area under the peak or more often, simply by measuring the peak height above the baseline.

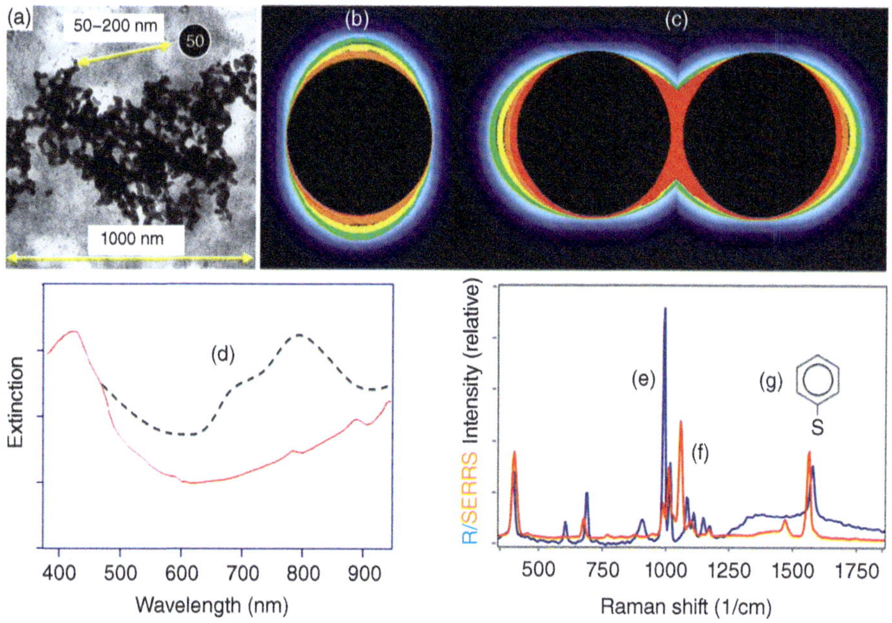

Figure 11.4 (a) 20,000X transmission electron micrograph of silver-doped sol–gel substrate, ~1–2 μm thick, containing several hundred connected nodules. Illustration of increasing electric field strength (green to red) for (b) a single colloidal sphere and (c) paired colloidal spheres. (d) Extinction spectrum of RTA's aggregated nanoparticles without and with (dashed line) benzenethiol. (e) Normal Raman (blue) and (f) SERS (red) spectra of (g) of benzenethiol at 10 M and 1 nM, respectively, with an ~ EF ~10^{10}. Both benzenethiol samples were measured using 80 mW at 785 nm with a ~200 μm diameter illumination spot. *Source*: RTA authors.

From a theoretical standpoint, N_R can be determined if the exact scattering volume for the spectrometer and sample concentration, C_R, are known. In the case of metal colloids and the metal-doped sol–gels, the percent of this volume occupied by the metal must be known along with C_{SER} to determine N_{SER}. Typically, the metal concentration is 0.1 to 1% for colloids or sol–gels. In the case of substrates, where a drop of the sample is dried on the surface, N_{SER} is the concentration of the sample times the drop volume and the ratio of the dried drop to laser beam areas. Since these calculations may not be straightforward (e.g. accounting for surface roughness and irradiated area), a more practical approach to determining EFs is to simply use the actual sample concentrations measured, i.e. C_R and C_{SER}. However, it should be noted that EFs for colloids and sol–gels will be *under*estimated by at least two orders of magnitude, while dried samples on substrates will be *over*estimated by as much as 4 orders of magnitude.

The enhancing capabilities of the silver-coated sol–gel structure are exemplified by measuring benzenethiol with and without the sol–gel (Figure 11.4 e and f). In both cases, the same Raman spectrometer, power (80 mW of 785 nm), acquisition time (10 seconds), and sample holder (1 mm glass capillary) were used. For normal Raman measurements, 10 M benzenethiol was added to the capillary. For the sol–gel-SERS measurement, 1 nM benzenethiol was added to an internally sol–gel filled capillary. The normal Raman and SERS for the two benzenethiol peaks at 410 and 1565 cm^{-1} are approximately equivalent in intensity, indicating an EF of ~10^{10}, is equal to the difference of the sample concentrations. While the illuminated volumes were identical, the amount of sample was approximately twice as much for the Raman measurement since the sol–gel in the capillary occupied half of the volume, based on the measured mass of a capillary with and without the sol–gel plug. It should also be noted that there are some minor differences in the peak positions and intensities for the SERS, caused by vibration mode interaction with the plasmon field.

11.3 Conclusion

It is clear from the above discussions that the detection limit is more important than the EF for real-world applications. It is standard practice for Raman spectroscopists to estimate the detection limit based on a concentration that would give a signal-to-noise ratio of three. While this is reasonable since the signal intensity is proportional to the analyte concentration, this is not necessarily the case for SERS, where the signal depends on surface coverage and "hot spot" density. Since the latter is tough to quantify, a practical approach is to report the lowest measured concentration of the analyte of interest.

While developing new SERS structures is valuable, it is more important to use this technology to solve critical problems. To that end, RTA has developed and applied SERS-active materials in analytical detection methods with the goal of reducing deaths due to drug overdoses and the consumption of pathogen-laden food. The next two chapters describe the detection of fentanyl in patients and *Listeria monocytogenes* in food and on surfaces.

References

1 Zheng, X.-S., Jahn, I.J., Weber, K. et al. (2018). Label-free SERS in biological and biomedical applications: Recent progress, current challenges and opportunities. *Spectrochimica Acta Part A: Molecular and Biomolecular Spectroscopy* 197: 56–77. https://doi.org/10.1016/j.saa.2018.01.063.

2 Cotton, T.M., Kim, J.-H., and Chumanov, G.D. (1991). Application of surface-enhanced Raman spectroscopy to biological systems. *Journal of Raman Spectroscopy* 22 (12): 729–742. https://doi.org/10.1002/jrs.1250221203 22.

3 Fan, M. and Gustavo, & Brolo, A. G. (2020). A review on recent advances in the applications of surface-enhanced Raman scattering in analytical chemistry. *Analytica Chimica Acta* 1097: 1–29. https://doi.org/10.1016/j.aca.2019.11.049.

4 Terry, L.R., Sanders, S., Potoff, R.H. et al. (2022). Applications of surface-enhanced Raman spectroscopy in environmental detection. *Analytical Science Advances* 3 (3–4): 113–145. https://doi.org/10.1002/ansa.202200003.

5 Shende, C., Inscore, F., Sengupta, A. et al. (2010). Rapid extraction and detection of trace Chlorpyrifos-methyl in orange juice by surface-enhanced Raman spectroscopy. *Sensing and Instrumentation for Food Quality and Safety* 4 (3–4): 101–107. https://doi.org/10.1007/s11694-010-9100-6.

6 Golightly, R.S., Doering, W.E., and Natan, M.J. (2009). Surface-enhanced Raman spectroscopy and homeland security: A perfect match? *ACS Nano* 3 (10): 2859–2869. https://doi.org/10.1021/nn9013593.

7 Farquharson, S., Shende, C., Smith, W. et al. (2014). Selective detection of 1000 *B. anthracis* spores within 15 minutes using a peptide functionalized SERS assay. *The Analyst* 139 (24): 6366–6370. https://doi.org/10.1039/c4an01163e.

8 Hakonen, A., Andersson, P.O., Stenbæk Schmidt, M. et al. (2015). Explosive and chemical threat detection by surface-enhanced Raman scattering: A review. *Analytica Chimica Acta* 893: 1–13. https://doi.org/10.1016/j.aca.2015.04.010.

9 Farquharson, S., Gift, A., Maksymiuk, P., and Inscore, F. (2005). Surface-enhanced Raman spectra of VX and its hydrolysis products. *Applied Spectroscopy* 59 (5): 654–660. https://doi.org/10.1366/0003702053946100.

10 Lane, L.A., Qian, X., and Nie, S. (2015). SERS nanoparticles in medicine: From label-free detection to spectroscopic tagging. *Chemical Reviews* 115 (19): 10489–10529. https://doi.org/10.1021/acs.chemrev.5b00265.

11 Farquharson, S., Gift, A., Shende, C. et al. (2008). Surface-enhanced Raman spectral measurements of 5-fluorouracil in saliva. *Molecules* 13 (10): 2608–2627. https://doi.org/10.3390/molecules13102608.

12 Farquharson, S., Dana, K., Shende, C. et al. (2017). Rapid identification of buprenorphine in patient saliva. *Journal of Analytical & Bioanalytical Techniques* 8 (3): 368. https://doi.org/10.4172/2155-9872.1000368.

13 Sacre, P.-Y., De Bleye, C., Netchacovitch, L. et al. (2018). Critical review of surface-enhanced Raman spectroscopy applications in the pharmaceutical field. *Journal of Pharmaceutical and Biomedical Analysis* 147: 458–472. https://doi.org/10.1016/j.jpba.2017.06.056.

14 Shende, C. and Farquharson, S. (2022). Detection of tacrolimus in saliva using a lateral flow assay and surface-enhanced resonance raman scattering. *Analytical and Bioanalytical Chemistry* 13 (447).

15 Farquharson, S., Brouillette, C., Smith, W., and Shende, C. (2019). A surface-enhanced raman spectral library of important drugs associated with point-of-care and field applications. *Frontiers in Chemistry* 7: 706. https://doi.org/10.3389/fchem.2019.00706.

16 Philpott, M.R. (1975). Effect of surface plasmons on transitions in molecules. *The Journal of Chemical Physics* 62 (5): 1812–1817. https://doi.org/10.1063/1.430708.

17 Chew, H., McNulty, P.J., and Kerker, M. (1976). Model for Raman and fluorescent scattering by molecules embedded in small particles. *Physical Review* 13 (1): 396–404. https://doi.org/10.1103/physreva.13.396.

18 Fleischmann, M., Hendra, P.J., and McQuillan, A.J. (1974). Raman spectra of pyridine adsorbed at a silver electrode. *Chemical Physics Letters* 26 (2): 163–166. https://doi.org/10.1016/0009-2614(74)85388-1.

19 Jeanmaire, D.L. and Van Duyne, R.P. (1977). Surface Raman spectroelectrochemistry. *Journal of Electroanalytical Chemistry and Interfacial Electrochemistry* 84 (1): 1–20. https://doi.org/10.1016/s0022-0728(77)80224-6.

20 Albrecht, M.G. and Creighton, J.A. (1977). ChemInform abstract: Anomalously intense Raman spectra of pyridine at a silver electrode. *Chemischer Informationsdienst* 8 (44): 1103–1105. https://doi.org/10.1002/chin.197744044.

21 Farquharson, S., Weaver, M.J., Lay, P.A. et al. (1983). Observation of a reversible adsorbed redox couple using surface-enhanced Raman scattering: Pentaammine(pyridine)osmium(III)/(II) at silver electrodes. *Journal of the American Chemical Society* 105 (10): 3350–3351. https://doi.org/10.1021/ja00348a073.

22 Weaver, M.J., Farquharson, S., and Tadayyoni, M.A. (1985). Surface enhancement factors for Raman scattering at silver electrodes. Role of adsorbate–surface interactions and electronic structure. *The Journal of Chemical Physics* 82 (11): 4867–4874. https://doi.org/10.1063/1.448657.

23 Kerker, M. and Thompson, B. (1990). *Selected Papers on Surface-Enhanced Raman Scattering*, vol. MS10 (ed. M. Kerker), 225. SPIE Press.

24 Haynes, C.L., McFarland, A.D., and Van Duyne, R.P. (2005). Surface-enhanced Raman spectroscopy. *Analytical Chemistry* 77 (17): 338 A–346 A. https://doi.org/10.1021/ac053456d.

25 Wang, D.S. and Kerker, M. (1981). Enhanced Raman scattering by molecules adsorbed at the surface of colloidal spheroids. *Physical Review B* 24 (4): 1777–1790. https://doi.org/10.1103/physrevb.24.1777.

26 Bohren, C.F. and Huffman, D.R. (2009). *Absorption and Scattering of Light by Small Particles*. Wiley-Vch.

27 Hale, G.M. and Querry, M.R. (1973). Optical constants of water in the 200-nm to 200-μm wavelength region. *Applied Optics* 12 (3): 555. https://doi.org/10.1364/ao.12.000555.

28 Schiebener, P., Straub, J., Levelt Sengers, J.M.H., and Gallagher, J.S. (1990). Refractive index of water and steam as function of wavelength, temperature and density. *Journal of Physical and Chemical Reference Data* 19 (3): 677–717. https://doi.org/10.1063/1.555859.

29 Johnson, P.B. and Christy, R.W. (1972). Optical constants of the noble metals. *Physical Review B* 6 (12): 4370–4379. https://doi.org/10.1103/physrevb.6.4370.

30 Takenaka, T., Eda, K., Mabuchi, M. et al. (1984). Surface enhanced Raman scattering by isomeric mono-bromopyridines adsorbed on gold and silver sol particles. *Bulletin of the Institute for Chemical Research, Kyoto University* 62 (4): 219–232.

31 Ueba, H. (1983). Theory of charge transfer excitation in surface enhanced Raman scattering. *Surface Science* 131 (2–3): 347–366. https://doi.org/10.1016/0039-6028(83)90283-2.

32 Yamada, H., Nagata, H., Toba, K., and Nakao, Y. (1987). Charge-transfer band and SERS mechanism for the pyridine-Ag system. *Surface Science* 182 (1–2): 269–286. https://doi.org/10.1016/0039-6028(87)90101-4.

33 Norrod, K.L., Sudnik, L.M., Rousell, D., and Rowlen, K.L. (1997). Quantitative comparison of five SERS substrates: Sensitivity and limit of detection. *Applied Spectroscopy* 51 (7): 994–1001. https://doi.org/10.1366/0003702971941377.

34 Guselnikova, O., Lim, H., Kim, H., et al. (2022). New trends in nanoarchitectured SERS substrates: Nanospaces, 2D materials, and organic heterostructures. *Small* 18(25): 2107182. https://doi.org/10.1002/smll.202107182.

35 Lee, P.C. and Meisel, D. (1982). Adsorption and surface-enhanced Raman of dyes on silver and gold sols. *The Journal of Physical Chemistry* 86 (17): 3391–3395. https://doi.org/10.1021/j100214a025.

36 Sutherland, W.S., Laserna, J.J., Angebranndt, M.J., and Winefordner, J.D. (1990). Surface-enhanced Raman analysis of sulfa drugs on colloidal silver dispersion. *Analytical Chemistry* 62 (7): 689–693. https://doi.org/10.1021/ac00206a008.

37 Laserna, J.J., Berthod, A., and Winefordner, J.D. (1988). Evaluation and optimization of experimental conditions for surface-enhanced Raman detection of analytes in flow injection analysis. *Microchemical Journal* 38 (1): 125–136. https://doi.org/10.1016/0026-265x(88)90011-2.

38 Angel, S.M., Katz, L.F., Archibald, D.D. et al. (1988). Near-infrared surface-enhanced Raman spectroscopy. Part I: Copper and gold electrodes. *Applied Spectroscopy* 42 (8): 1327–1331. https://doi.org/10.1366/0003702884429698.

39 Weitz, D.A., Garoff, S., and Gramila, T.J. (1982). Excitation spectra of surface-enhanced Raman scattering on silver-island films. *Optics Letters* 7 (4): 168. https://doi.org/10.1364/ol.7.000168.

40 Seki, H. (1982). Surface enhanced Raman scattering of pyridine on different silver surfaces. *The Journal of Chemical Physics* 76 (9): 4412–4418. https://doi.org/10.1063/1.443556.

41 Vo-Dinh, T., Hiromoto, M.Y.K., Begun, G.M., and Moody, R.L. (1984). Surface-enhanced Raman spectrometry for trace organic analysis. *Analytical Chemistry* 56 (9): 1667–1670. https://doi.org/10.1021/ac00273a029.

42 Vo-Dinh, T., Stokes, D.L., Li, Y.-S., and Miller, G.H. (1991). Fiber optic sensor probe for in-situ surface-enhanced Raman monitoring. *Proceedings of SPIE, the International Society for Optical Engineering* 1368: 203–209. https://doi.org/10.1117/12.47693.

43 Wachter, E.A., Storey, J.M.E., Sharp, S.L. et al. (1995). Hybrid substrates for real-time SERS-based chemical sensors. *Applied Spectroscopy* 49 (2): 193–199. https://doi.org/10.1366/0003702953963779.

44 Sutherland, W.S. and Winefordner, J.D. (1991). Preparation of substrates for surface-enhanced Raman microprobe spectroscopy. *Journal of Raman Spectroscopy* 22 (10): 541–549. https://doi.org/10.1002/jrs.1250221002.

45 Li, Y.-S. and Wang, Y. (1992). Chemically prepared silver/alumina substrate for surface-enhanced Raman scattering. *Applied Spectroscopy* 46 (1): 142–146. https://doi.org/10.1366/0003702924444506.

46 Farquharson, S., Smith, W.W., Elliott, S., and Sperry, J.F. (1999). Rapid biological agent identification by surface-enhanced Raman spectroscopy. *Proceedings of SPIE, the International Society for Optical Engineering/Proceedings of SPIE* 3855: 110–110. https://doi.org/10.1117/12.371269.

47 Maya, L., Vallet, C.E., and Lee, Y.H. (1997). Sputtered gold films for surface-enhanced Raman scattering. *Journal of Vacuum Science & Technology a Vacuum Surfaces and Films* 15 (2): 238–242. https://doi.org/10.1116/1.580518.

48 Szabo, N.J. and Winefordner, J.D. (1997). Evaluation of two commercially available TLC materials as SER substrates. *Applied Spectroscopy* 51 (7): 965–975. https://doi.org/10.1366/0003702971941629.

49 Nie, S. (1997). Probing single molecules and single nanoparticles by surface-enhanced Raman scattering. *Science* 275 (5303): 1102–1106. https://doi.org/10.1126/science.275.5303.1102.

50 Zhang, E., Xing, Z., Wan, D. et al. (2021). Surface-enhanced Raman spectroscopy chips based on two-dimensional materials beyond graphene. *Journal of Semiconductors* 42 (5): 051001. https://doi.org/10.1088/1674-4926/42/5/051001.

51 Kundu, A., Rani, R., and Hazra, K.S. (2019). Controlled nanofabrication of metal-free SERS substrate on few layered black phosphorus by low power focused laser irradiation. *Nanoscale* 11 (35): 16245–16252. https://doi.org/10.1039/c9nr02615k.

52 Chen, M., Liu, D., Du, X. et al. (2020). 2D materials: Excellent substrates for surface-enhanced Raman scattering (SERS) in chemical sensing and biosensing. *TrAC Trends in Analytical Chemistry* 130: 115983. https://doi.org/10.1016/j.trac.2020.115983.

53 Jensen, T.R., Duval, M.L., Kelly, K.L. et al. (1999). Nanosphere lithography: Effect of the external dielectric medium on the surface plasmon resonance spectrum of a periodic array of silver nanoparticles. *The Journal of Physical Chemistry B* 103 (45): 9846–9853. https://doi.org/10.1021/jp9926802.

54 Dick, L.A., McFarland, A.D., Haynes, C.L., and Van Duyne, R.P. (2001). Metal film over nanosphere (MFON) electrodes for surface-enhanced Raman spectroscopy (SERS): Improvements in surface nanostructure stability and suppression of irreversible loss. *The Journal of Physical Chemistry B* 106 (4): 853–860. https://doi.org/10.1021/jp013638l.

55 Shafer-Peltier, K.E., Haynes, C.L., Glucksberg, M.R., and Van Duyne, R.P. (2002). Toward a glucose biosensor based on surface-enhanced Raman scattering. *Journal of the American Chemical Society* 125 (2): 588–593. https://doi.org/10.1021/ja028255v.

56 Palonpon, A.F., Ando, J., Yamakoshi, H. et al. (2013). Raman and SERS microscopy for molecular imaging of live cells. *Nature Protocols* 8 (4): 677–692. https://doi.org/10.1038/nprot.2013.030.

57 Kneipp, J., Seifert, S., and Gärber, F. (2024). SERS microscopy as a tool for comprehensive biochemical characterization in complex samples. *Chemical Society Reviews* 53 (15): 7641–7656. https://doi.org/10.1039/d4cs00460d.

58 Grys, D., Chikkaraddy, R., Kamp, M. et al. (2020). Eliminating irreproducibility in SERS substrates. *Journal of Raman Spectroscopy* 52 (2): 412–419. https://doi.org/10.1002/jrs.6008.

59 Farquharson, S., Chad, N., and Lee, Y. (2003). Material for surface-enhanced Raman spectroscopy, and SER sensors and method for preparing same. U.S. Patent No. 6623977, U.S. Patent and Trademark Office.

60 Farquharson, S., and Maksymiuk, P. (2005). Simultaneous chemical separation and surfaceenhanced Raman spectral detection using metal-doped sol-gels. U.S. Patent No. 6943031, U.S. Patent and Trademark Office.

61 Farquharson, S., and Maksymiuk, P. (2008). Apparatus for effecting chemical separation and plural point SERS detection using metal-doped sol-gels. U.S. Patent No. 7462492, U.S. Patent and Trademark Office.

62 Hao, E. and Schatz, G.C. (2004). Electromagnetic fields around silver nanoparticles and dimers. *Journal of Chemical Physics* 120 (1): 357–366. https://doi.org/10.1063/1.1629280.

12

Detecting Fentanyl in Patient Samples Using Surface-Enhanced Raman Spectroscopy

Chetan Shende[1], *Albert Arias*[2], *and Stuart Farquharson*[1]

[1] *Real-Time Analyzers, Inc., Middletown, CT, USA*
[2] *Virginia Commonwealth University School of Medicine, Richmond, VA, USA*

12.1 Introduction

Opioids are unequaled for their ability to treat pain. They also stimulate the reward (pleasure) pathway in the brain, which, unfortunately, has contributed to the current opioid crisis involving the overuse of prescription opioids and the use of powerful synthetic opioids, such as fentanyl. In the United States, the use of fentanyl began an exponential growth in ~2012, used both by itself and mixed with other drugs [1]. Figure 12.1 shows the number of deaths involving fentanyl overdose with and without stimulants compared to prescription opioids and heroin. Fentanyl, a Schedule II prescription drug, is often used as an anesthetic during surgery. Unfortunately, it is currently mixed with recreational drugs like cocaine or heroin for illicit use [2]. Due to its potency, which is ~100 times as strong as morphine [3], only a small amount is required for a "high," making it ideal for smuggling across borders. The amount of fentanyl crossing the borders increased from 11,200 lbs. in 2021 to 27,000 lbs. in 2023 [4]. Its use by itself and with other drugs resulted in 123,563 hospital emergency department visits in 2021 [5], and an estimated 71,000 overdose deaths [2].

Most of the opioid users entering the emergency departments survive because the hospitals have laboratory equipment, such as gas chromatography coupled with a mass spectrometer (GC/MS), to determine the drug type and concentration in biofluids, such as blood plasma or urine. The lab analyses allow the medical staff to administer the appropriate medication, such as naloxone for fentanyl overdose [6]. In contrast [7, 8], emergency responders, such as police and ambulance personnel, do not have such lab facilities. And, although portable drug test kits are easy to use, they are not reproducible or quantitative. We believe that a point-of-care device, based on surface-enhanced Raman spectroscopy (SERS), could be developed so that emergency responders could rapidly determine if a person is suffering from opioid overdose, abuse, or use and then administer treatment that potentially saves their life.

A decade ago, portable Raman spectrometers were investigated for their ability to identify powders suspected of being explosives or drugs [9–13]. Today, they are a relatively common tool several government agencies use for border security and drug seizures [14, 15]. However, the sensitivity of Raman spectroscopy is insufficient for detecting drugs in biofluids. To overcome this limitation, samples mixed with gold colloids can amplify Raman signals by as much as 6 orders of magnitude [16, 17]. This method, known as SERS, can identify and quantify drugs in body fluids at ng/mL concentrations. We have been investigating the ability of SERS to measure drugs in saliva for more than 15 years [18–22], including recent measurements of opioids and opioid treatment drugs in the saliva of US military veterans [23–25]. However, these previous

Practical Quantitative Vibrational and Electronic Spectroscopy: A Guide for Developing, Optimizing, and Validating Procedures, First Edition. Edited by Deborah A. Peru.

Companion website: www.wiley.com/go/peruelectronicspectroscopy

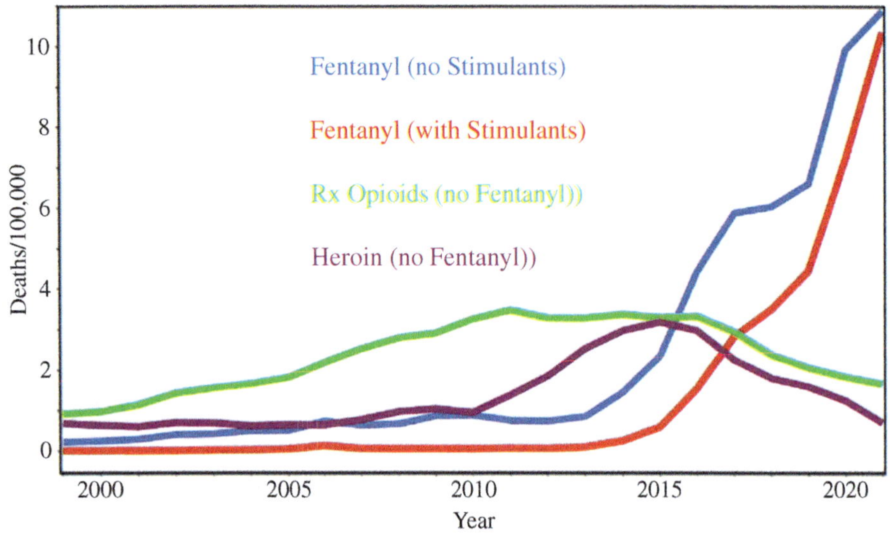

Figure 12.1 Plot of fentanyl with and without stimulants compared to prescription opioids and heroin. *Source*: [1] / John Wiley & Sons.

measurements required the use of liquid- or solid-phase extraction methods, limiting the ability to perform the analysis at the point-of-care.

Here, we describe the development of a simple-to-use SERS flow separation strip capable of identifying opioids, specifically fentanyl, in patient saliva samples. The SERS strips are based on lateral flow assays, in which the antibodies have been substituted by a SERS-active pad for detection. Twenty-eight patient saliva samples were collected at Virginia Commonwealth University (VCU), measured using a portable Raman spectrometer, and compared to GC/MS urinalysis. The SERS strips with a portable Raman spectrometer form the basis of a point-of-care drug identification device that can quickly assess the drug form required for doctors to treat the patient quickly.

12.2 Experimental

Analytical-grade phosphate-buffered saline (PBS), sodium citrate, chloroauric acid, cetyltrimethylammonium bromide, and HPLC-grade water were all purchased from Sigma Aldrich (St Louis, MO). De-identified pooled saliva, certified drug-free (liquid), was purchased from Lee BioSolutions (Maryland Heights, MO). Forensic-grade standards of amphetamine, cocaine, codeine, diazepam, fentanyl, hydrocodone, methadone, methamphetamine, morphine, nicotine, and tetrahydrocannabinol (THC) at 1 mg/mL in methanol or acetonitrile were purchased from Cerilliant Corp (Round Rock, TX). 10 μL of each forensic drug standard was initially diluted first with PBS then with saliva to produce two sets of 10 to 0.5 μL/mL samples for measurements. Eppendorf plastic centrifuge tubes, used to hold samples, were purchased from VWR International (Radnor, PA). Glass membrane filter sheets, backing material for the SERS strips, nitrocellulose membrane, conjugate pad, and wicking pad materials were purchased from Sigma Millipore (Burlington, MA) to produce the SERS-active pads. The gold nanoparticles and the SERS-active pads were synthesized in-house according to previously published procedures [26]. Following standard safety precautions, all Raman measurements were performed at Real-Time Analyzers (RTA) inside a Biosafety Level 2 cabinet. All used samples were autoclaved and packed in biohazard bags before disposal.

The VCU personnel recruited 44 patients to provide saliva samples (Samples numbered 1–44) under an Independent Review Board (IRB) and Human Research Protection Office (HRPO) approval. The following patients decided against using their samples for studies (Samples 27–36 and 38–41). Samples 13 and 15 did not have sufficient saliva for measurement. In total, 28 saliva samples from VCU were shipped to RTA. VCU also supplied GC/MS analysis of urine for all 28 samples.

Surface-enhanced Raman spectroscopy measurements were performed using a modified 5 lb. Raman spectrometer (Wasatch Photonics, model FPR-785-WS, Orlando, FL, USA). The spectrometer employed a 785 nm diode laser and a room-temperature 512 channel silicon array detector. The SERS strips were inserted into a simple enclosure attached to the spectrometer that aligned the focal point of the laser on the SERS pad. Spectra were recorded from ~200 to 2300 cm^{-1} with ~20 cm^{-1} resolution. Peak positions are reported to the nearest 5 cm^{-1}. Each spectrum employed ~40 mW of laser excitation power focused to ~200 μm at the sample. Acquisition times varied from 1 to 10 sec. A fieldable tablet computer was used to control the laser power and acquisition time and perform spectral analysis. RTA's Chem-ID software was used to identify drugs. A complete surface-enhanced Raman spectral analysis of the drugs described here has been published [23].

12.3 Results and Discussion

The design of the SERS flow separation strips is based on the components of lateral flow assay (LFA) strips, in which the standard antibody test line was replaced by a SERS-active pad consisting of a glass-fiber pad impregnated with gold nanoparticles (Figure 12.2a). Specifically, the SERS strips consist of (1) a backing material on which (2) the nitrocellulose membrane is attached with the following components: (3) a sample pad consisting of a membrane filter to separate drugs from biological fluid, (4) a conjugate pad or the SERS pad that contains the immobilized gold nanoparticles, and (5) wicking pad that absorbs the sample and promotes capillary flow through the membrane (Figure 12.2b). To obtain a uniform distribution of gold nanoparticles and pad-to-pad consistency, larger strips of conjugate pad were soaked with gold nanoparticles and allowed to dry at room temperature. The strips were then inserted under the sample collection filter pad section of the LFA card stock. The thin layer of glue under the edge of the LFA filter strip was used to "fix" the SERS strip. The card stock with SERS strips was then cut into ¼″ × 3″ strips and used as is.

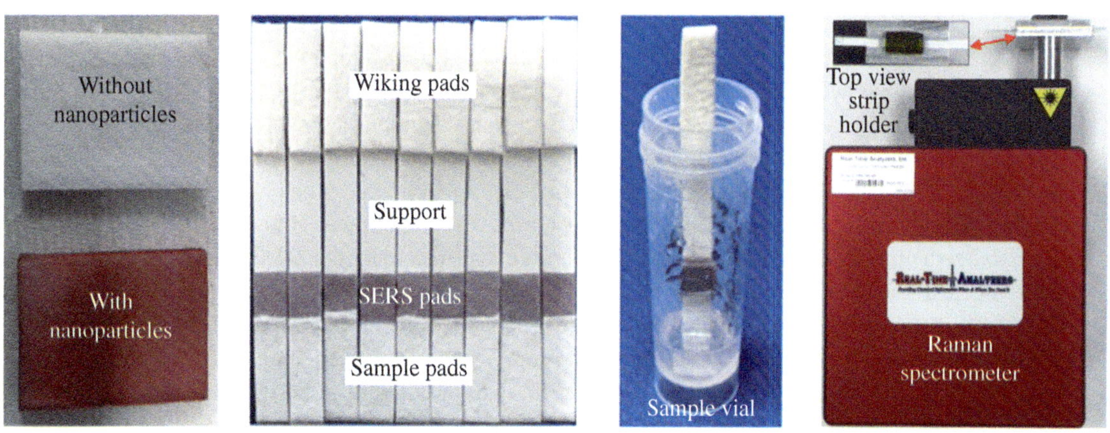

(a) (b) (c) (d)

Figure 12.2 Photographs of (a) 5 × 7 mm collection pads before and after impregnation with gold nanoparticles. (b) SERS flow separation strip used to measure saliva. (c) ~1 mL saliva sample in a 5 mL sample tube and SERS test strip. (d) Wasatch Raman spectrometer with a SERS strip holder attached above the laser. *Source*: RTA authors.

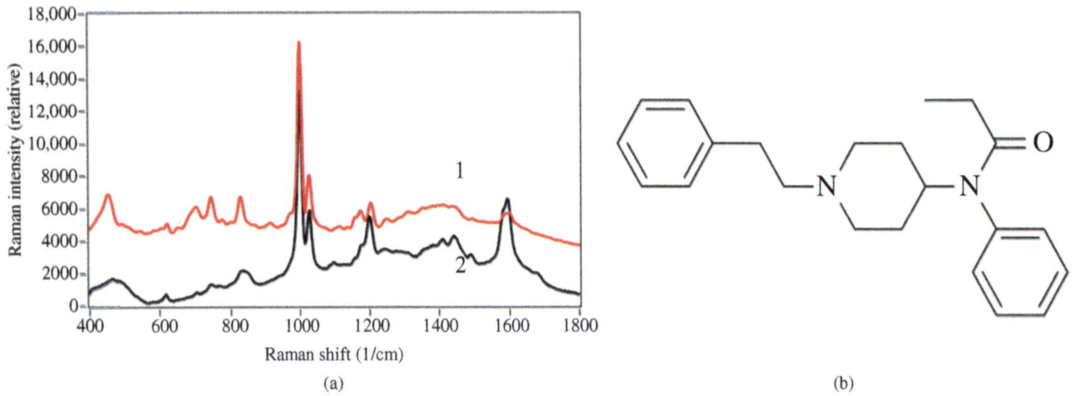

Figure 12.3 (a) SERS of 10 μg/mL fentanyl in PBS using 785 nm excitation on (1) SERS gold colloidal suspension (red) and (2) SERS pads (black). (b) Chemical structure of fentanyl. Spectral conditions for A: 40 mW at 785 nm, 3-sec acquisition. *Source*: RTA authors.

Basic optimization steps for preparing the SERS strips included testing combinations of different lengths of the membrane filter, conjugate pad, or the SERS pad, and wicking pad type and thickness. The SERS pad optimization involved synthesizing gold nanoparticles with different sizes, concentrations, and soaking times of the conjugate pads into the nanoparticle suspension.

The capabilities of the SERS strips were initially tested using 10 μg/mL samples of a forensic fentanyl standard diluted in PBS. For each measurement, a 100 μL sample was added to a 5 mL plastic tube, and a SERS strip was inserted into the tube until the sample reached the wicking pad (Figure 12.2c). The SERS strip was then air-dried for 5 minutes, placed in the Raman spectrometer, and measured (Figure 12.2d). For these first measurements, the sample flowed to the wicking pad in ~1 minutes. Spectra with very good signal-to-noise ratios (S/N) were obtained with only a 3-seconds acquisition, matching previously reported SERS for these drugs using a gold colloidal suspension (Figure 12.3a) [25]. The fentanyl spectrum is dominated by the symmetric and asymmetric phenyl ring breathing modes at 1000 and 1030 cm^{-1}, respectively (Figure 12.3b). It also contains several modest intensity peaks. The 830 cm^{-1} peak is due to out-of-plane phenyl CH bending, the 1180 cm^{-1} peak is due to phenyl ring CC stretching, the 1200 cm^{-1} peak is due to phenyl-C symmetric stretching, the 1330 cm^{-1} peak is due to piperidine CH wagging, and a doublet at 1585 and 1600 cm^{-1} is due to CC stretching of the two phenyl rings [27, 28].

The uniformity of Raman intensity using the SERS pads was compared to that of the gold colloidal suspension. Nine repeat measurements of 10 ppm fentanyl in PBS added to a uniform gold colloidal suspension at different positions yielded consistent spectra with an average 1000 cm^{-1} peak height of 11,582 and a standard deviation of 257 (2%RSD) in Figure 12.4a. In contrast, nine repeat measurements of a single SERS pad that has variable structure at different positions yielded an average peak height of 3455 and a standard deviation of 1014 (29%RSD), as illustrated in Figure 12.4b.

The effectiveness of the vertical flow SERS strip to separate fentanyl from saliva was dramatically shown by adding a patient saliva sample to a SERS strip and comparing it to the same sample added to a SERS pad (Figure 12.5a). Next, the expected sensitivity of the SERS strips was determined by preparing and measuring fentanyl spiked in de-identified, pooled saliva at 10, 5, and 1 μg/mL (Figure 12.5b). The SERS strips were placed in sample vials containing 1 mL of each prepared sample. Vertical flow of the samples on the SERS strips took 1–2 minutes. The signal intensity decreased by a factor of ~2 compared to samples prepared in PBS, while the noise level increased (data not shown). This can be explained by the fact that the membrane filter is not expected to trap 100% of the glycoproteins, and some fouling of the SERS substrate is expected,

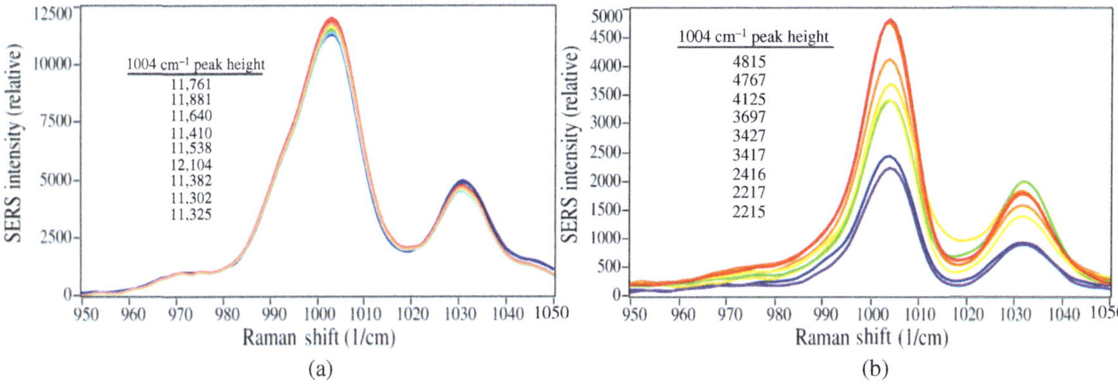

Figure 12.4 Nine measurements of 10 µg/mL fentanyl in PBS (a) in a gold colloidal suspension at nine positions and (b) on a SERS pad at nine positions. Note that the intensity in the colloidal suspension is ~3 times greater. Spectral conditions as in Figure 12.3. *Source*: RTA authors.

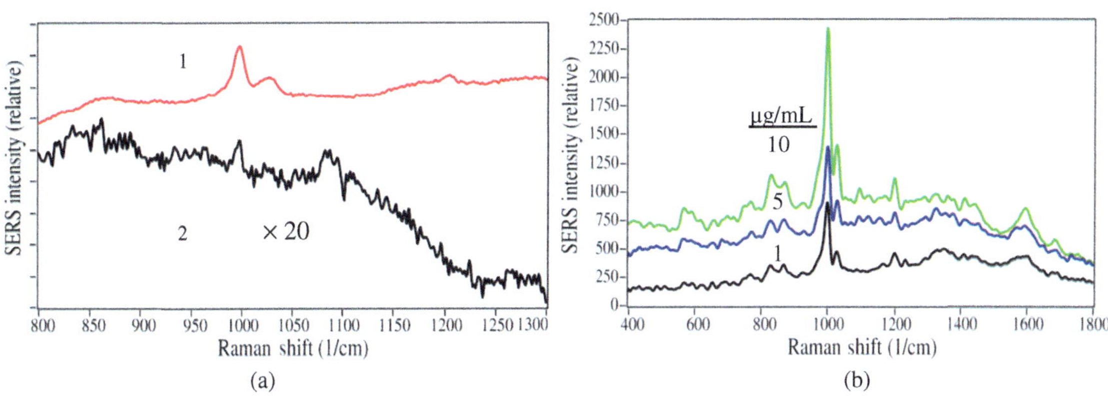

Figure 12.5 (a) SERS of a VCU patient saliva sample measured using (1) a SERS strip and (2) just a SERS pad. (b) SERS of fentanyl spiked at 10, 5, and 1 µg/mL in purchased de-identified pooled saliva on SERS strips. Spectral conditions as in Figure 12.3. *Source*: RTA authors.

especially due to smaller molecular weight proteins. The results nevertheless demonstrate that the SERS vertical flow test strips can be used to measure drugs in saliva at 1 µg/mL, and possibly lower.

Virginia Commonwealth University patients already undergoing treatment, generally with buprenorphine, were instructed to drool ~1 mL of saliva into 5 mL plastic tubes and cap them. VCU personnel added labels to the sample vials containing the date and a patient-coded identifier and mailed them to RTA for SERS analysis (Figure 12.6). At the same time as saliva collection, a urine sample was collected from the patients during the scheduled visit and used to perform GC/MS analysis.

The following procedural steps were established to optimize the performance of the SERS strips: (1) dip the strip into a ~ 0.5 cm depth saliva sample for 1 minute, (2) remove the sample and allow the strip to dry for 10 minutes, (3) mount the strip above the laser, and (4) measure the SERS.

Our previous studies have shown that many patients have several drugs in their system at any one time, such as amphetamine, cocaine, methadone for treatment, and, more often than not, nicotine from smoking. The Raman spectra for amphetamine, cocaine, methadone, and fentanyl are all dominated by the symmetric and asymmetric phenyl ring breathing modes at 1000 and 1030 cm^{-1}, respectively (Figure 12.7). Complicating the spectral analysis is nicotine, in that its concentration is often considerably higher in saliva than in

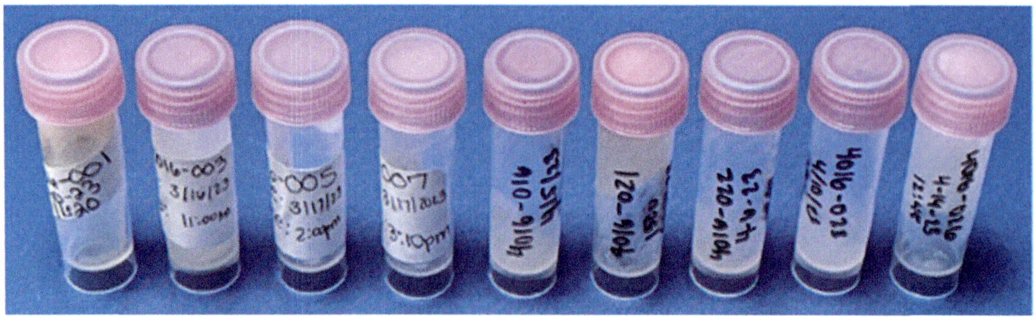

Figure 12.6 Photograph of the first nine samples measured. *Source*: RTA authors.

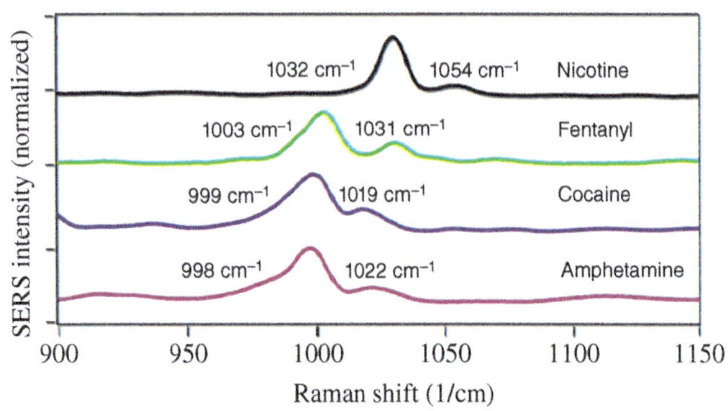

Figure 12.7 SERS of the pyridine ring of nicotine and the phenyl ring symmetric and trigonal stretching modes of several drugs included in this analysis. *Source*: RTA authors.

these drugs. This represents a significant challenge for automated software. Consequently, critical to this analysis is a good spectral library. For this study, ~60 drugs at 10 ppm on SERS pads were measured similar in concentration to the expected sample concentrations. However, to obtain high S/N spectra, the library drugs were measured in PBS, not saliva.

Post-analysis of the SERS spectra found that 19 of 28 sample spectra had a broad overlapping doublet peak at ~700 cm^{-1}, which could be a SERS pad artifact. Spectra with reasonable S/N were obtained from 16 of 28 (57%) samples. The other 12 samples either had very low drug content or, if present, did not separate adequately from the saliva mucins. Refer to the Chapter 12 Companion Website for the complete set of spectra for the 28 patient samples measured by SERS.

In an effort to discern the overlapping ring modes in the SERS spectra, two of RTA's spectral analysis software programs were used in tandem to identify the drugs in the saliva samples. RTA's Chem-ID software was used to identify the primary drugs in each sample, while RTA's S-Quant software was used to determine the percentage of the top 2 or 3 drugs in the sample if there was more than one. Since 21 of 28 samples contained a broad and relatively intense peak at ~700 cm^{-1}, the spectral range was confined from 800 to 1700 cm^{-1}. The two-software approach using a gold colloidal suspension SERS spectral library is demonstrated below for samples 1–5 in order of increasing complexity.

Sample 2 produced a high-quality spectrum with an S/N of 235 for the ~1000 cm^{-1} peak, suggesting a relatively high concentration of the identified drug, in this case, fentanyl. The Chem-ID software was used to rank the library spectra in terms of a "hit quality index," with 0 for a perfect match to 1 for no match [29]. The first derivative spectrum was used to eliminate differences in the baseline between the sample

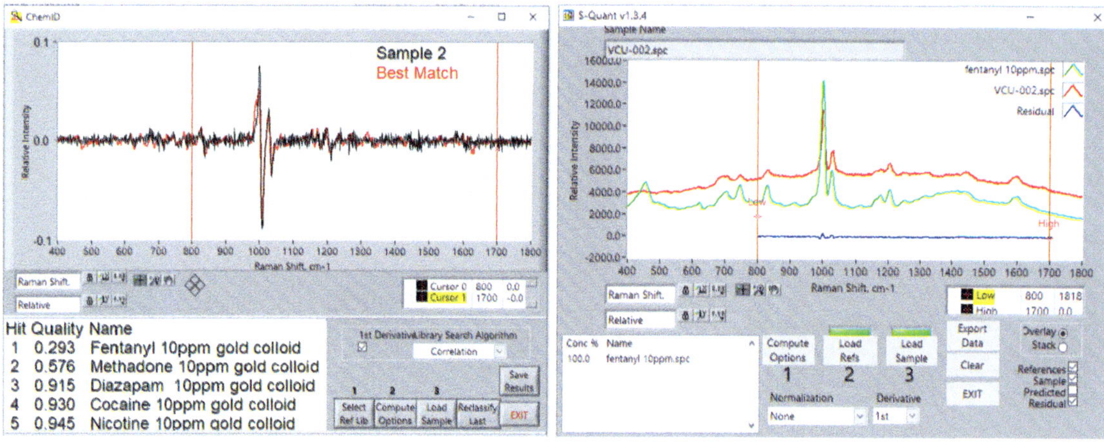

(a) (b)

Figure 12.8 Sample 2 spectral analysis using (a) Chem-ID that shows the first derivative of the sample (black) and fentanyl (red) as the best library match with a hit quality index of 0.293, and using (b) S-Quant that shows the sample spectrum (red), the fentanyl spectrum (green) as the best match, and the residual spectrum (blue). *Source*: RTA authors.

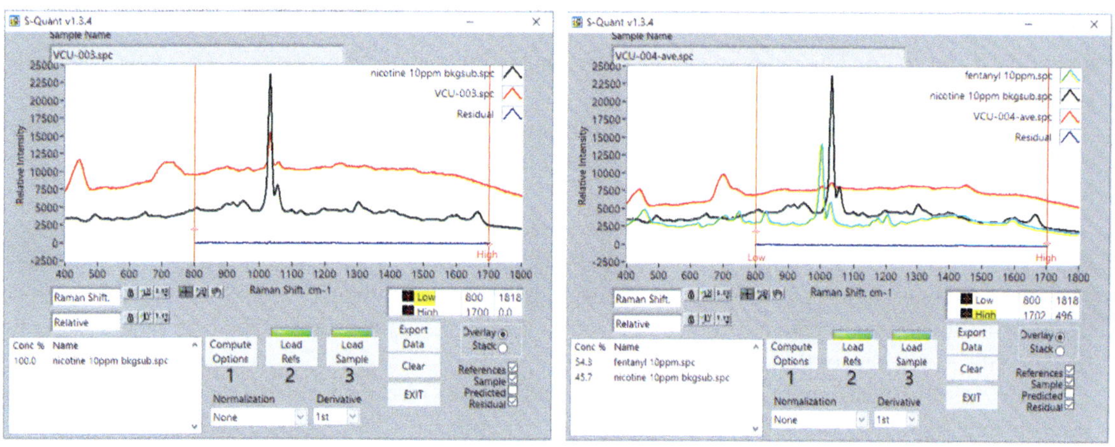

(a) (b)

Figure 12.9 Spectral analysis using Chem-ID and S-Quant for (a) Sample 3 showing nicotine, and (b) Sample 4 showing fentanyl and nicotine as the best matches. Note the flat residual spectra. *Source*: RTA authors.

and the library spectra. Only the top two library drugs, fentanyl and methadone, had scores less than 0.9 (Figure 12.8a). The S-Quant software determined the best reference spectral fits for the measured spectrum and the relative % concentration. The best match was fentanyl alone, as any inclusion of methadone or other drugs increased the residual spectrum, defined as the sample spectrum minus the best-fit spectrum (Figure 12.8b).

Sample 3 also produced a high-quality spectrum with a ~1000 cm^{-1} peak S/N of 114. The same combined software analysis identified nicotine as the only drug present. Note the residual spectrum (Figure 12.9a). Sample 4 produced a low-quality spectrum with 1000 cm^{-1} peak S/N of 38. Nevertheless, the combined software analysis identified and quantified fentanyl at 53.3% and nicotine at 45.7%. Note the residual spectrum (Figure 12.9b).

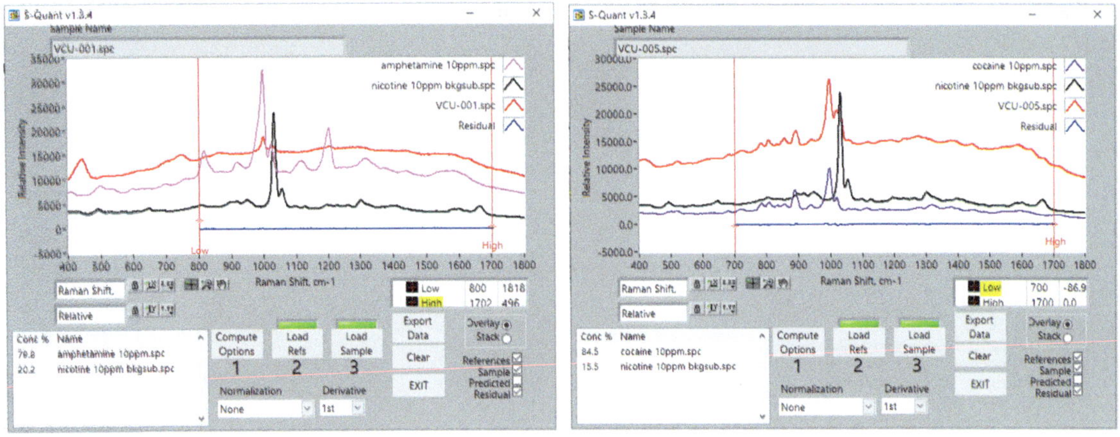

(a) (b)

Figure 12.10 Spectral analysis using Chem-ID and S-Quant for (a) Sample 1 showing amphetamine and nicotine, and (b) Sample 5 showing cocaine and nicotine as the best matches. Note the flat (blue) residual spectra. *Source*: RTA authors.

Sample 1 produced a reasonable quality spectrum with 1000 cm^{-1} peak S/N of 65 (Figure 12.10a). The combined software analysis identified amphetamine and nicotine as the likely drugs, while S-Quant yielded a very good spectral match composed of 79.8% amphetamine and 20.2% nicotine, with a near 0 residual spectrum.

Sample 5 produced a reasonable quality spectrum with 1000 cm^{-1} peak S/N of 180 (Figure 12.10b). The combined software analysis identified cocaine and nicotine as the likely drugs, while S-Quant yielded a very good spectral match composed of 84% cocaine and 15.5% nicotine, with a near 0 residual spectrum.

The SERS spectral analysis for the 28 samples collected by VCU is compared to the GC/MS urinalysis in Table 12.1. While nicotine was not part of the GC/MS urinalysis, the patients indicated if they were smokers or not on their consent forms. According to the forms 26 of 28 were smokers. GC/MS identified nicotine in 22 (85%) of the samples, while SERS detected nicotine in 24 (92%) of the samples, and both methods identified nicotine in 19 (79%) of the same patient samples.

The GC/MS identified fentanyl in 13 samples (46%), while SERS identified fentanyl in 12 of these 13 patient samples and eight additional patient samples (71%). In summary, GC/MS only detected fentanyl in 60% (12/19) of the same subject samples found positive by SERS test strips.

The significantly higher number of detected fentanyl samples in patient saliva by SERS compared to patient urine samples by GC/MS is most likely due to the fact that most users either inhale or take fentanyl orally, and trace amounts likely remain in saliva for several hours. Whereas it takes several hours for the drug to be metabolized and excreted in urine, which will then be at a much lower concentration when analyzed by GC/MS. It is also possible that the SERS spectral analysis misidentified other drugs, such as amphetamine and cocaine, and other opioids, as fentanyl, since they also have intense SERS peaks at 1000 cm^{-1}. While plausible, the Chem-ID software library contains the same drugs used in the GC/MS library (i.e., amphetamine, methadone, 3,4-methylenedioxyamphetamine (MDA), methylenedioxy-methamphetamine (MDMA), morphine, oxycodone, and oxymorphone). Including these drugs in the Chem-ID software analysis generated negative peaks in the residual spectra, suggesting that these drugs were not present in the samples. Furthermore, it is unlikely that opioids not found in the library could have been identified as fentanyl.

Table 12.1 Nicotine and fentanyl were identified in patient samples by SERS (blue boxes) and by GC/MS (Yes in boxes).

Cross-referenced SERS/S-Quant data and GC/MS data					
Sample	**Nicotine**	**Fentanyl**	**Sample**	**Nicotine**	**Fentanyl**
1	Yes		19	Yes	
2			20	Yes	
3	Yes		21	Yes	
4	Yes	Yes	22	Yes	Yes
5			23	Yes	
6			24	Yes	
7	Yes	Yes	25	Yes	Yes
8	Yes	Yes	26		Yes
9	Yes		37		Yes
10	Yes	Yes	42	Yes	Yes
11			43	Yes	Yes
12	Yes		44	Yes	Yes
14	Yes		**Total GC/MS:**	**22**	**13**
16	Yes	Yes	**Total SERS:**	**24**	**20**
17	Yes	Yes	**Overlap:**	**19 of 24**	**12 of 20**
18	Yes		**% Overlap:**	**79.2%**	**60.0%**

Bold represent summary of positive responses and overlap between techniques.

12.4 Conclusions

A novel SERS vertical flow separation strip was developed that successfully separated drugs from saliva, specifically fentanyl, which was detected in 20 of 28 (71%) patient samples, using a portable 785 nm laser excitation Raman spectrometer equipped with a 60 drug library, and identification software. The patient samples were collected by clinicians at Virginia Commonwealth University, who provided GC/MS urinalysis that indicated fentanyl in only 13 of 28 samples (46%). The SERS flow separation strip employed a SERS-active pad that allowed a detection limit of fentanyl at 0.1 ng/mL. The value of a field-usable Raman-based drug analyzer is that, unlike GC/MS, it can provide point-of-care analysis. As such, this technique could be used by police, emergency responders, and medical practitioners to rapidly identify the drug(s) involved in an overdose and thereby provide appropriate treatment that could potentially save lives. Specifically, if a person tests positive for fentanyl, the administration of naloxone by emergency responders reverses the drug's effects within minutes and can often save lives [6]. Furthermore, such a point-of-care analyzer would also be valuable to Homeland Security and the North Atlantic Treaty Organization (NATO) military countries, as fentanyl and its analogs, such as carfentanil, are considered potential agents of terrorism and war [30, 31]. Furthermore, as new analogs of fentanyl and other drugs are developed, the SERS lateral-flow test strip requires little more than measuring the new drug for identification and a concentration series in the biofluid for quantitation. This is in contrast to standard antibody-based tests, which require years to develop a specific antibody to distinguish a new drug from those of the same type, such as carfentanil and remifentanil, found in the present study.

Acknowledgments

This research was partially funded by the Department of Defense, Chemical and Biological Defense, Contract No. W911QY19C0008, and the National Institute of Health, National Institute of Drug Abuse, Contract No. 2R44DA032178-02. The authors thank Tiffany Pignatello and John Hurley at the Virginia Commonwealth University School of Medicine for their contributions to this work.

References

1 Friedman, J. and Shover, C.L. (2023). Charting the fourth wave: Geographic, temporal, race/ethnicity and demographic trends in polysubstance fentanyl overdose deaths in the United States, 2010–2021. *Addiction* 118 (12): 2477–2485. https://doi.org/10.1111/add.16318.

2 Frank, R.G. and Pollack, H.A. (2017). Addressing the fentanyl threat to public health. *New England Journal of Medicine* 376 (7): 605–607. https://doi.org/10.1056/nejmp1615145.

3 Gorodeizky, C.W. and Martin, W.R. (1965). A comparison of fentanyl, droperidol, and morphine. *Clinical Pharmacology & Therapeutics* 6 (6): 731–739. https://doi.org/10.1002/cpt196566731.

4 U.S. Customs and Border Protection (2021, December 15). Drug Seizure Statistics. U.S. Customs and Border Protection. https://www.cbp.gov/newsroom/stats/drug-seizure-statistics (accessed 14 December 2024).

5 Substance Abuse and Mental Health Services Administration (2022). Drug abuse warning network: Findings from drug-related emergency department visits, 2021 (HHS Publication No. PEP22-07-03-002). Rockville, MD: Center for Behavioral Health Statistics and Quality, Substance Abuse and Mental Health Services Administration. https://www.samhsa.gov/data/.

6 Walter, K. (2023). Fentanyl overdose. *Journal of the American Medical Association* 329 (2): 184. https://doi.org/10.1001/jama.2022.22462.

7 Klapheke, M. and Pasarica, M. (2017). Opioid risk mitigation strategies and overdose resuscitation. *MedEdPORTAL Publications* 13: 10621. https://doi.org/10.15766/mep_2374-8265.10621.

8 White, N.D. (2018). Increasing naloxone access and use to prevent opioid overdose death and disability. *American Journal of Lifestyle Medicine* 13 (1): 33–35. https://doi.org/10.1177/1559827618803874.

9 Santillán, J.D., Brown, C.D., and Jalenak, W. (2007). Advances in Raman spectroscopy for explosive identification in aviation security. *Proceedings of SPIE, the International Society for Optical Engineering/Proceedings of SPIE.* 6540: 124–131. https://doi.org/10.1117/12.723067.

10 Donahue, M., Huang, H., Brouillette, C., et al. (2011). Detecting explosives by portable Raman analyzers: A comparison of 785, 976, 1064, and 1550 nm (Retina-Safe) laser excitation. Spectroscopy, April supplement.

11 de Veij, M., Deneckere, A., Vandenabeele, P. et al. (2008). Detection of counterfeit Viagra® with Raman spectroscopy. *Journal of Pharmaceutical and Biomedical Analysis* 46 (2): 303–309. https://doi.org/10.1016/j.jpba.2007.10.021.

12 Hargreaves, M.D., Page, K., Munshi, T. et al. (2008). Analysis of seized drugs using portable Raman spectroscopy in an airport environment-a proof of principle study. *Journal of Raman Spectroscopy* 39 (7): 873–880. https://doi.org/10.1002/jrs.1926.

13 West, M.J. and Went, M.J. (2009). The spectroscopic detection of drugs of abuse in fingerprints after development with powders and recovery with adhesive lifters. *Spectrochimica Acta Part A: Molecular and Biomolecular Spectroscopy* 71 (5): 1984–1988. https://doi.org/10.1016/j.saa.2008.07.024.

14 Izake, E.L. (2010). Forensic and homeland security applications of modern portable Raman spectroscopy. *Forensic Science International* 202 (1–3): 1–8. https://doi.org/10.1016/j.forsciint.2010.03.020.

15 Weyermann, C., Mimoune, Y., Anglada, F. et al. (2011). Applications of a transportable Raman spectrometer for the in situ detection of controlled substances at border controls. *Forensic Science International* 209 (1–3): 21–28. https://doi.org/10.1016/j.forsciint.2010.11.027.

16 Weaver, M.J., Farquharson, S., and Tadayyoni, M.A. (1985). Surface enhancement factors for Raman scattering at silver electrodes. Role of adsorbate–surface interactions and electronic structure. *The Journal of Chemical Physics* 82 (11): 4867–4874. https://doi.org/10.1063/1.448657.

17 Jeanmaire, D.L. and Van Duyne, R.P. (1977). Surface Raman spectroelectrochemistry. *Journal of Electroanalytical Chemistry and Interfacial Electrochemistry* 84 (1): 1–20. https://doi.org/10.1016/s0022-0728(77)80224-6.

18 Farquharson, S., Shende, C., Inscore, F.E. et al. (2005). Analysis of 5-fluorouracil in saliva using surface-enhanced Raman spectroscopy. *Journal of Raman Spectroscopy* 36 (3): 208–212. https://doi.org/10.1002/jrs.1277.

19 Farquharson, S., Gift, A., Inscore, F., and Shende, C. (2008). SERS method and apparatus for rapid extraction and analysis of drugs in saliva. U.S. Patent No. 7393691, U.S. Patent and Trademark Office.

20 Inscore, F., Shende, C., Sengupta, A. et al. (2011). Detection of drugs of abuse in saliva by surface-enhanced Raman spectroscopy (SERS). *Applied Spectroscopy* 65 (9): 1004–1008. https://doi.org/10.1366/11-06310.

21 Farquharson, S., Shende, C., Sengupta, A. et al. (2011). Rapid detection and identification of overdose drugs in saliva by surface-enhanced Raman scattering using fused gold colloids. *Pharmaceutics* 3 (3): 425–439. https://doi.org/10.3390/pharmaceutics3030425.

22 Dana, K., Shende, C., Huang, H., and Farquharson, S. (2015). Rapid analysis of cocaine in saliva by surface-enhanced Raman spectroscopy. *Journal of Analytical & Bioanalytical Techniques* 6 (6): 1. https://doi.org/10.4172/2155-9872.1000289.

23 Farquharson, S., Dana, K., Shende, C. et al. (2017). Rapid identification of buprenorphine in patient saliva. *Journal of Analytical & Bioanalytical Techniques* 08 (03): 368. https://doi.org/10.4172/2155-9872.1000368.

24 Farquharson, S., Brouillette, C., Smith, W., and Shende, C. (2019). A surface-enhanced Raman spectral library of important drugs associated with point-of-care and field applications. *Frontiers in Chemistry* 7: 706. https://doi.org/10.3389/fchem.2019.00706.

25 Farquharson, S., Shende, C., Newcomb, J. et al. (2023). Analysis of drugs in saliva of US Military veterans treated for substance use disorders using supported liquid extraction and surface-enhanced Raman spectral analysis. *Molecules* 28 (5): 2010. https://doi.org/10.3390/molecules28052010.

26 Shende, C., Brouillette, C., and Farquharson, S. (2019). Detection of codeine and fentanyl in saliva, blood plasma and whole blood in 5-minutes using a SERS flow-separation strip. *The Analyst* 144 (18): 5449–5454. https://doi.org/10.1039/c9an01087d.

27 Dollish, F.R., Fateley, W.G., and Bentley, F.F. (1974). *Characteristic Raman Frequencies of Organic Compounds*. New York NY: John Wiley & Sons.

28 Leonard, J., Haddad, A., Green, O. et al. (2017). SERS, Raman, and DFT analyses of fentanyl and carfentanil: Toward detection of trace samples. *Journal of Raman Spectroscopy* 48 (10): 1323–1329. https://doi.org/10.1002/jrs.5220.

29 Farquharson, S., Gladding, Z., Brouillette, C. et al. (2022). Efficacy of medications in space determined by Raman spectroscopy: Aspirin degradation. *Journal of Analytical & Bioanalytical Techniques* 13 (9): 472.

30 Wax, P.M., Becker, C.E., and Curry, S.C. (2003). Unexpected "gas" casualties in Moscow: A medical toxicology perspective. *Annals of Emergency Medicine* 41 (5): 700–705. https://doi.org/10.1067/mem.2003.148.

31 Riches, J.R., Read, R.W., Black, R.M. et al. (2012). Analysis of clothing and urine from Moscow theatre siege casualties reveals carfentanil and remifentanil use. *Journal of Analytical Toxicology* 36 (9): 647–656. https://doi.org/10.1093/jat/bks078.

13

Detection of Foodborne Pathogens on Surfaces and in Food by SERRS

Chetan Shende[1], Wayne Smith[2], and Stuart Farquharson[1]

[1]*Real-Time Analyzers, Inc. Middletown, CT, USA*
[2]*WSmithConsulting, Inc. South Glastonbury, CT, USA*

13.1 Introduction

Each year in the United States, approximately 48 million people (1 in 6) get sick, 128,000 are hospitalized, and 3000 die from foodborne pathogens, according to the Centers for Disease Control and Prevention [1]. To reduce these numbers, the Food and Drug Administration (FDA) Food Safety Modernization Act (FSMA) shifted the focus from responding to foodborne illness outbreaks to preventing them [2]. Consequently, all food manufacturers are required to test production lots before shipment, and equipment between production runs. A lot is cleared for delivery when it meets the United States Department of Agriculture (USDA) statistical probability of detection (POD), indicating no pathogens have been detected for a set of samples. This involves collecting food and equipment swab samples and adding them to enrichment broths to grow the potential pathogens to the point of detection. Traditional detection includes cell counts in agar Petri dishes, absorption, fluorescence spectroscopy, and turbidity measurements of growth broths. While both cell count and absorption are still common, polymerase chain reactions (PCR) have become just as common. Although PCR amplification takes only 2–4 hours, it still requires 10–30 hours of pathogen-dependent growth in a broth before the DNA sequence is isolated from the sample and amplified [3]. The greatest limitations of PCR are that only one sample can be replicated at a time, and primers and equipment are expensive. All of these test methods are limited by the growth period, which is pathogen-dependent. For example, to grow and test *Escherichia coli* O157:H7 [1] takes ~24 hours, and as much as 36 hours for *Listeria monocytogenes* [4].

In an effort to reduce the time that food shipments are delayed and equipment remains idle, we investigated the extreme sensitivity of surface-enhanced resonance Raman spectroscopy (SERRS) to detect dye-labeled, pathogen-specific probes. The probes consist of silver spheres coated with a reporter molecule, such as a dye, which are functionalized with an antibody specific to the pathogen of interest. Probes were developed and tested for four common pathogens: *Campylobacter jejuni*, *E. coli* O157:H7, *L. monocytogenes*, and *Salmonella typhimurium* added to food samples. In-house tests achieved the detection of these pathogens in chicken, ground beef, cantaloupe, and cheese, respectively, at 4 CFU/g within seven hours using these probes. Verification of the method was performed by an independent USDA-certified laboratory, which confirmed the detection of *L. monocytogenes*, one of the slowest-growing pathogens, at 170 CFU on 4″ × 4″ stainless-steel plates (1.7 CFU/cm^2) in less than nine hours. We believe the described measurement

Practical Quantitative Vibrational and Electronic Spectroscopy: A Guide for Developing, Optimizing, and Validating Procedures, First Edition. Edited by Deborah A. Peru.

Companion website: www.wiley.com/go/peruelectronicspectroscopy

method and analyzer could reduce equipment downtime, food recalls, and, most importantly, illnesses and deaths.[1]

13.2 Experimental

13.2.1 Reagents and Equipment

Analytical-grade phosphate-buffered saline (PBS), sodium silicate, high-performance liquid chromatography (HPLC) grade water, Eppendorf centrifuge tubes, benzenethiol, and the dyes were all purchased from Sigma–Aldrich (Milwaukee, WI). Wash buffer and positive and negative control samples 50–90–90 and 50–90–05 were purchased from Sera Care Life Sciences (Milford, MA). 3 M Quick Swabs with letheen broth were purchased from SOS Cleanroom (Dallas, TX). The following kit components: swabs, pipettes, and glass vials were purchased from World Bioproducts (Libertyville, IL), Globe Scientific (Mahwah, NJ), and Glassvials, Inc. (Baltimore, MD), respectively.

C. jejuni, *E. coli* O157:H7, *L. monocytogenes*, and *S. typhimurium*, and monoclonal and polyclonal antibodies for these respective pathogens were obtained from KPL, Inc. (Gaithersburg, MD) as lyophilized, heat-killed cells, and then stored at 4 °C until used. The lyophilized cells were suspended in 1 mL of HPLC-grade water to produce stock solutions of 10^9 CFU. These stock solutions were diluted in water or PBS to produce additional concentrations at 10^8, 10^7, 10^6, 10^5, and 10^4 CFU for SERRS measurements. A blender and homogenizer bags were purchased from Interscience (France).

The SERRS measurements were performed using a 5 lb. Raman spectrometer described in Section 12.2 of Chapter 12. An in-house 2 mL vial holder was used to align the laser focal point at the center of the vial bottom. Furthermore, spectral analysis was confined to the 900 to 1700 cm^{-1} region for sample analysis. Using in-house software, a Lenovo ThinkPad computer was used to control the Raman spectrometer, laser power, acquisition time, and data analysis. Each spectrum employed ~40 mW of 785 nm laser excitation power focused to ~200 μm at the sample. Acquisition times varied from 1 to 10 seconds (see figure captions). All measurements at RTA and those at the certification laboratory were performed in a Biosafety Level 2 cabinet following standard safety precautions. All used samples were autoclaved and packed in biohazard bags before disposal.

13.2.2 Probe Design

Silver nanoparticles, 70 to 75 nm in diameter, were prepared following a modified Lee–Meisel method [5]. The silver nanoparticles were mixed with 3-mercaptopropyl trimethoxysilane, followed by the addition of the reporter molecule. The reporter molecule-coated silver was then encapsulated with a silica layer by adding sodium silicate. This formed aggregates containing an estimated 5 to 10 silver nanoparticles. Next, the reporter molecule-coated silver aggregates, encased in silica, were functionalized by mixing with the pathogen-specific polyclonal antibodies. After incubation, the antibody-conjugated, reporter-labeled nanoparticles (SERRS Probes) were purified by centrifugation and re-suspended in borate buffer. The inside bottom of 2 mL glass vials, used to hold samples, were functionalized with complementary antibodies specific to one of the four pathogens. Figure 13.1a illustrates the probe design containing the silver nanoparticles coated with a reporter molecule, encased in silica, and functionalized with antibodies. Details of the system accessories used in this work are shown in Figure 13.1b.

---

1 O157:H7 is as a particular strain of *E. coli* that causes a severe intestinal infection in humans.

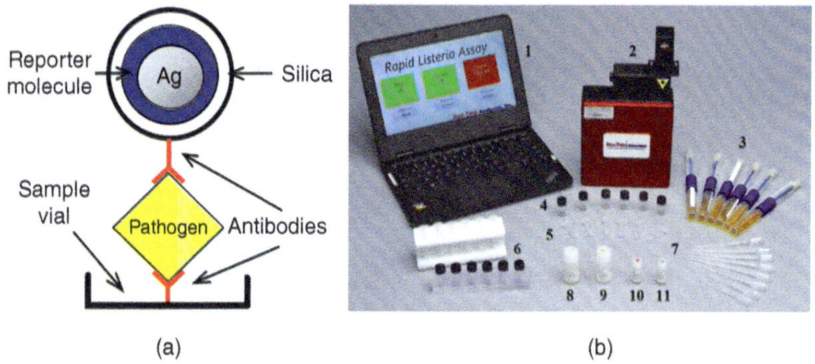

Figure 13.1 (a) Illustration of probe design: silver nanoparticles coated with a reporter molecule, encased in silica, and functionalized with antibodies specific to one of the following pathogens: *C. jejuni*, *E. coli* O157:H7, *L. monocytogenes*, or *S. typhimurium*. Glass sample vial functionalized with a complementary antibody. (b) Photograph of the IFPA with kit components for six samples: (1) Laptop computer with analysis software, (2) Raman spectrometer, (3) swabs with growth media, (4) 2 mL sample vials, (5) 2 mL centrifuge tubes, (6) tube and vial holders, (7) pipettes, (8) probes, (9) wash buffer, (10) negative control sample, and (11) positive control sample. *Source*: RTA authors.

13.3 Results and Discussion

13.3.1 In-shift Food Pathogen Analyzer

The development of the in-shift food pathogen analyzer (IFPA) in Figure 13.1b was motivated by the need to reduce infections, illnesses, deaths, and financial losses due to foodborne pathogen outbreaks. The financial consequences of outbreaks include the cost of identifying the source of food outbreaks, the loss of revenue due to recalled food products, recertification of production equipment, and brand-equity loss. The IFPA is based on our previous development of a surface-enhanced Raman spectrosocpy (SERS) analyzer used to detect anthrax spores on surfaces, such as those found in the US Senate Hart Senate Office Building, after dissemination of letters in 2001 [6, 7]. In addition to SERS signal enhancement, the required sensitivity of 1000 spores/cm^2 was achieved by employing a peptide to selectively bind and isolate the spores for measurement [8]. Similarly, specific antibodies were used to bind and isolate the foodborne pathogens. However, this previous approach was insufficient to achieve the measurement goal of 1 CFU/cm^2 on surfaces and 1 CFU/25 g in food. Consequently, reporter molecules were included in the foodborne pathogen probes described here. Dye reporter molecules were chosen since they have extended π to π^* transitions well into the red part of the visible spectrum, allowing for resonance Raman spectroscopy (RRS) using standard laser wavelengths, such as 532 and 638 nm. RRS typically adds a 10^2 to 10^4 additional enhancement to the SERS signal. These dye transitions are further shifted by 50 nm or more when the dye molecules are within the plasmon field of a SERS active metal [9]. This allows measurements to be made with a 785 nm laser, the most often used wavelength in commercial Raman spectrometers.

While dyes were originally used as the probe reporter molecules for all four pathogens, the use of benzenethiol as the reporter molecule is described in keeping with Chapter 11 on SERS theory. Benzenethiol was also chosen for its intense SERRS obtained on silver surfaces [10]. Specifically, it forms a metal–ligand complex with the probe silver nanoparticles producing a broad wavelength charge transfer transition centered at ~750 nm (see Figure 11.4d, Chapter 11). Excitation with a 785 nm laser produces a very intense spectrum, as shown in Figure 13.2.

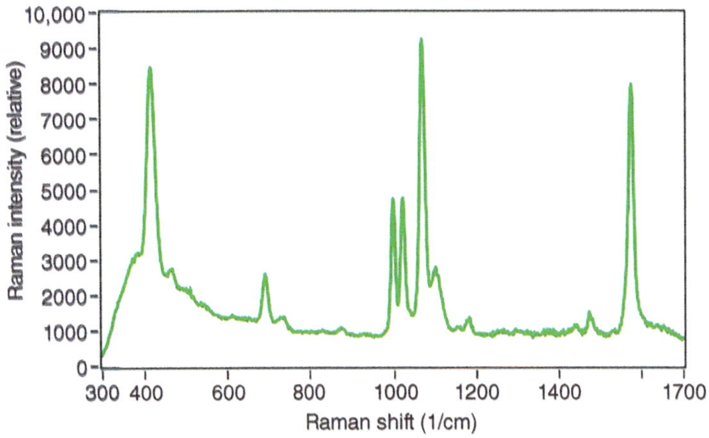

Figure 13.2 SERRS of benzenethiol-coated silver nanoparticles encapsulated with silica. Spectral conditions: 30 seconds acquisition using 40 mW at 785 nm. *Source*: RTA authors.

Table 13.1 Pathogen in food sample conditions (all 25 g inoculated with 100 CFU).

Pathogen	*E. coli*	*S. typhimurium*	*L. monocytogenes*	*C. jejuni*
Food	Ground beef	Cheese	Cantaloupe	Chicken
Broth (mL)	100	100	225	225
Total/incubation time (hrs)	6.5/5	6.5/5	6/4.5	6/4.5

13.3.2 Surface-Enhanced Resonance Raman Spectroscopy Detection of Pathogens in Food

Four pathogens in foods that caused numerous outbreaks were chosen to demonstrate the initial capabilities of the IFPA: *E. coli* O157:H7, *S. typhimurium*, *L. monocytogenes*, and *C. jejuni* in ground beef, cheese, cantaloupe, and chicken, respectively (Table 13.1). In addition to benzenethiol, three dyes were used as reporter molecules for the other three pathogens.

Although 1 CFU/25 g is the goal for quantification, in the Association of Official Agricultural Chemists (AOAC) validation study, we tested the food samples at 100 CFU/25 g to demonstrate the initial feasibility of our method. Additionally, spiking 100 CFU in a 25 g food sample yields a statistically significant sample. If spiked at 1 CFU/25 g, the heterogeneous distribution of the bacterial cells, as described by the Poisson distribution equation, becomes significant.

First, a sampling method was developed using dead cells of the four pathogens in PBS to establish expected sensitivity. The following procedure was used.

1) Mix 10^4 *dead* cells in 1 mL of PBS with 20 μL of SERRS Probes in a plastic centrifuge tube and allow to bind for 30 minutes.
2) Add the 1 mL solution to a vial previously functionalized on the inside-bottom with the corresponding antibody and allow to bind for an additional 30 minutes.
3) Draw off and discard the top ~1 mL solution using a pipette, and add 1 mL of wash buffer and gently swirl for 1 minutes.
4) Repeat step 3 with wash buffer two more times.

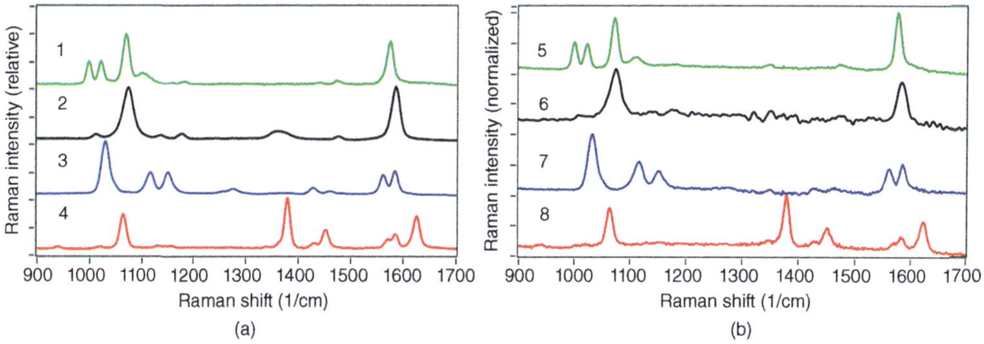

Figure 13.3 (a) SERRS of the probes used for (1) *E. coli* O157:H7 (green), (2) *S. typhimurium* (black), (3) *L. monocytogenes* (blue), and (4) *C. jejuni* (red). (b) SERRS of 100 CFU/25 g in food for (5) *E. coli* O157:H7 in ground beef (green), (6) *S. typhimurium* in cheese (black), (7) *L. monocytogenes* in cantaloupe (blue), and (8) *C. jejuni* in chicken (red). *Source*: RTA authors.

5) Remove the ~1 mL of wash buffer and place the vial in the IFPA vial holder and measure the Raman spectrum. The result is reported as "Presumptive Positive or Negative Pathogen Detected based on the SERRS intensity of the reporter molecule."

The analysis includes measuring (1) a "Blank" sample vial of wash buffer that was automatically subtracted from the Control and Sample spectra to remove the weak vial glass and buffer spectra; (2) a "Control" sample containing unbound probes that were not removed during the wash steps; and (3) a "Sample." The total sample preparation and measurement time was less than 90 minutes for each pathogen. High signal-to-noise spectra, unique to each reporter molecule, were obtained for each of the four pathogen probes measuring 10^4 *dead* cells/mL in PBS (Figure 13.3a).

The IFPA's capabilities were further tested by measuring the four pathogens in food, one per food type. Each sample was prepared by inoculating 25 g of food with 100 CFU live cells, in accordance with Table 13.1.

Food sample preparation and analysis involved the following steps:

1) A 25 g sample of a pathogen-inoculated food sample was placed in a homogenizer bag containing 100 or 225 mL of the appropriate enrichment broth (see Table 13.1).
2) The broth was homogenized for 30 seconds in a lab blender.
3) The broth was incubated in an oven for five hours at 37 °C.
4) 1 mL of the incubated broth was placed in a 2 mL centrifuge tube and centrifuged for one minute.
5) The supernatant was removed by pipette, and 0.3 mL of the SERRS Probes were added to the sample and incubated for an additional 15 minutes.
6) The incubated sample was added to a 2 mL glass sample vial functionalized with the corresponding antibody and incubated for an additional 15 minutes.
7) 1 mL of wash buffer was added to the vial, swirled and removed, and repeated two times.
8) After first measuring the SERRS of a Blank sample of wash buffer in a vial, and second, a Control sample of SERRS Probes without the pathogen in a vial, the SERRS of the Sample was measured, and the sample was reported as Positive or Negative.

The total time of analysis was 6.5 hours for *E. coli* in ground beef, and *S. typhimurium* in cheese, and 6.4 hours for *L. monocytogenes* in cantaloupe and *C. jejuni* in chicken (Table 13.1). Again, very good signal-to-noise spectra, unique to each reporter molecule, were obtained for the four pathogen probes at 100 CFU in the 25 g food samples (Figure 13.3b).

13.3.3 Surface-Enhanced Resonance Raman Spectroscopy Comparison with Independent Lab Results for Listeria on Surfaces

An independent USDA-certified laboratory performed an independent study using the AOAC guidelines for food pathogen detection. They performed a series of tests to determine the time required to detect *L. monocytogenes* on stainless-steel plates at 1 CFU/cm^2. This is the target concentration required by the AOAC for surface measurements. This pathogen was chosen as it has the slowest growth time and, thus, the longest time for detection using standard methods, and consequently, a successful short time to detection using SERRS will have the greatest impact.

Sixteen different samples were prepared as follows: *L. monocytogenes* at ~100 CFU on 10 plates, ~300 CFU on three plates, and 0 CFU on three plates. A matching set of 16 reference plates was also prepared for measurements according to the USDA Food Safety Inspection Service (FSIS) Microbiology Laboratory Guidebook. Specifically, samples prepared in PBS buffer were pipetted onto a 4-in. by 4-in. center area of a 6-in. by 6-in. stainless-steel plate and spread using a sterile inoculation loop. Note that 16 in^2 is equal to 103 cm^2. The same method was used for the blank reference samples. The samples on all 32 plates were allowed to dry for 16–24 hours at 25 °C in a Biosafety Level 2 hood. The test samples were random and blind-coded, so the analyst did not know the pathogen levels.

The sample preparation and SERRS analysis involved the following steps:

1) The valve on top of a 3 M Quick Swab was opened and the bulb squeezed to transfer all of the letheen enrichment broth, ~1 mL, into the tube with the swab.
2) The wet swab was removed from the tube, and used to swab the 4 in. by 4 in. center area using horizontal then vertical sweeps of the surface.
3) The swab was placed back into the tube.
4) The tube was vortexed vigorously for 10 seconds to release pathogens from swab tip into the letheen broth.
5) After one hour, the tube was vortexed again for 10 seconds.
6) 1 mL of the letheen broth was added to 9 mL of the *L. monocytogenes* selective enrichment broth and incubated for 7 and 8 hours at 37 °C.
7) 1 mL of the incubated sample was added to a SERRS-antibody vial and let sit for 15 minutes.
8) After first measuring the SERRS of (A) a "Blank" sample of wash buffer in a SERRS vial, and (B) a "Control" sample prepared according to steps 1–8 in a SERRS vial, then (C) a pathogen sample was measured and reported as presumptive positive or presumptive negative. Each SERRS measurement took less than one minute.

Figure 13.4a and b shows the SERRS results for samples incubated for 7 and 8 hours, respectively.

The results were rated according to the POD. The POD is defined as the ratio of the number of presumptive positive pathogen test samples divided by the number of confirmed positive pathogen detections for the same test samples by traditional petri dish growth. Presumptive positive is defined as any value from 0.25 to 0.75. Ratios below 0.25 indicate a failed test, while ratios above 0.75 indicate a confirmed positive pathogen detection. For the ten ~100 CFU plates measured by SERRS after eight hours of preparation, swabs from three plates produced intense spectra of the reporter molecule, Dye #1, while swabs from seven plates did not. The intensity of the latter seven swabs, at ~1/20th the intensity of the positive samples, was equivalent to the Control samples, and therefore, four of these SERRS tests are false negatives. Nevertheless, the POD is 0.3 (3/10), the sample would be labeled as a presumptive positive, and the equipment needs further disinfecting before being used.

In contrast, for the same ~100 CFU plates, the SERRS measurements after eight hours of preparation found that six of the plates were positive. Four were negative, matching the plate-to-plate results of the

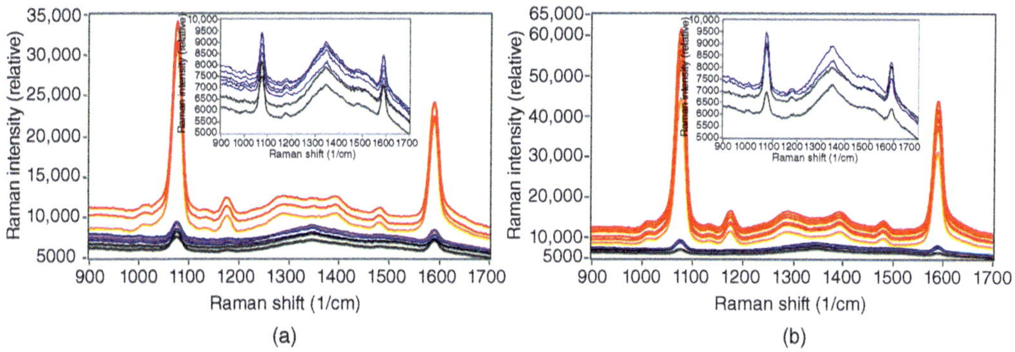

Figure 13.4 SERRS of 170 *L. monocytogenes* CFU functionalized with Dye #1, swabbed from 4″ × 4″ stainless-steel plates, incubated in broth for (a) 7 hours and (b) 8 hours according to the above procedure. Spectral conditions: 30-sec spectra per pathogen and intensities normalized to 1. *Source*: RTA authors.

Table 13.2 Comparison of the SERRS IFPA to standard growth methods for *L. monocytogenes* on 16 stainless-steel plates.

IFPA vs. confirmation tests	IFPA:<7 hours presumptive positive	IFPA:<8 hours confirmed positive	56 hours FB confirmation positive	56 hours MOX confirmation positive	72 hours USDA/FSIS positive
170 CFU (10 replicates)	3	6	6	6	4
310 CFU (3 replicates)	3	3	3	3	3
Uninoculated (3 replicates)	0	0	0	0	0

standard 56-hour Fraser Broth (FB) and Modified Oxford agar (MOX) methods and the 72-hour traditional petri dish growth method (USDA/FSIS). The POD is 1.0 (6/6), and the sample would be labeled confirmed positive, again requiring equipment disinfecting.

It is also worth noting that the swabs from the three plates inoculated with ~300 CFU all produced intense dye spectra at both 7 and 8 hours. Similarly, the un-inoculated three plates only produced weak dye spectra equivalent to the Control samples. Furthermore, the three plates inoculated with ~300 CFU and the three un-inoculated plates perfectly matched the FB, MOX, and USDA/FSIS test results (Table 13.2).

It is worth noting that all of the 16 plates that did *not* include the addition of *L. monocytogenes* produced very weak reporter molecule spectra equivalent to the Control samples, indicating the SERRS method could be used. These negative results are critical in demonstrating that the eight-hour method could replace both the traditional methods and PCR.

Furthermore, the reference methods are used to determine the exact amount of pathogen that was added to the plates, in this case, 170 and 310 CFU/16 in^2 or 1.7 and 3.1/cm^2. These concentrations would be listed in the performance documents.

13.4 Conclusions

Surface-enhanced resonance Raman probes were developed that could be used to identify pathogens in food and on food-processing equipment. The probes incorporated reporter molecules on silver to promote SERRS and antibodies to bind to specific foodborne pathogens. In-house measurements of *C. jejuni* in chicken, *E. coli* in ground beef, *L. monocytogenes* in cantaloupe, and *S. typhimurium* in cheese, detected these pathogens within seven hours, albeit at relatively high concentrations of 4 CFU/g of food.

Independent measurements by a USDA-certified lab detected 170 CFU of *L. monocytogenes* on 4″ × 4″ stainless-steel plates using SERRS and standard growth methods in nine hours. This represents a substantial improvement over three standard detection and growth methods. Specifically, FB, MOX, and petri dish growth took 56, 56, and 72 hours, respectively. The positive and negative results were identical plate-per-plate for the 16 samples measured using SERRS and standard methods.

The results suggest that the IFPA could shorten the time required to put manufacturing equipment back into service and the time to release food from refrigerated storage to market. Most importantly, the IFPA's sensitivity could detect an outbreak sooner than current methods, which could reduce illnesses and deaths.

Acknowledgments

This research was partially funded by the National Science Foundation, Contract no. IIP-1330886. The authors also thank Carl Bruoillette of RTA for contributing to this work.

References

1 USAFacts Team (2023). How many people get sick from foodborne illness in the US? https://usafacts.org/articles/how-many-people-get-sick-from-foodborne-illnesses/ (accessed 14 November 2024).

2 US Food & Drug Administration (2024). Food Safety Modernization Act (FSMA). www.fda.gov/food/guidance-regulation-food-and-dietary-supplements/food-safety-modernization-act-fsma

3 Hatiboruah, D., Devi, D.Y., Namsa, N.D., and Nath, P. (2020). Turbidimetric analysis of growth kinetics of bacteria in the laboratory environment using smartphone. *Journal of Biophotonics* 13 (4): e201960159. https://doi.org/10.1002/jbio.201960159.

4 Costa-Ribeiro, A., Lamas, A., Prado, M., and Garrido-Maestu, A. (2023). Evaluation of the novel mTA10 selective broth, MSB, for the co-enrichment and detection of Salmonella spp., *Escherichia coli* O157 and *Listeria monocytogenes* in ready-to-eat salad samples. *Food* 13 (1): 63–63. https://doi.org/10.3390/foods13010063.

5 Lee, P.C. and Meisel, D. (1982). Adsorption and surface-enhanced Raman of dyes on silver and gold sols. *The Journal of Physical Chemistry* 86 (17): 3391–3395. https://doi.org/10.1021/j100214a025.

6 Farquharson, S., Shende, C., Gift, A., and Inscore, F. (2012). *Detection of Bacillus Spores by Surface-Enhanced Raman Spectroscopy*. InTech EBooks. https://doi.org/10.5772/31859.

7 GOA-636-03 (2003). Capitol Hill Anthrax Incident, US General Accountability Oversite Report, June.

8 Farquharson, S., Shende, C., Smith, W. et al. (2014). Selective detection of 1000 *B. anthracis* spores within 15 minutes using a peptide functionalized SERS assay. *The Analyst* 139 (24): 6366–6370. https://doi.org/10.1039/c4an01163e.

9 Kitching, H., Kenyon, A.J., and Parkin, I.P. (2014). The interaction of gold and silver nanoparticles with a range of anionic and cationic dyes. *Physical Chemistry Chemical Physics* 16 (13): 6050–6059. https://doi.org/10.1039/c3cp55366c.

10 Biggs, K.B., Camden, J.P., Anker, J.N., and Van, R.P. (2009). Surface-enhanced Raman spectroscopy of benzenethiol adsorbed from the gas phase onto silver film over nanosphere surfaces: Determination of the sticking probability and detection limit time. *Journal of Physical Chemistry A* 113 (16): 4581–4586. https://doi.org/10.1021/jp8112649.

14

Development and Validation of Handheld Vibrational Spectroscopy Methods for the Quantitation of Active Ingredients in Pharmaceutical Solid Dosage Forms

Pierre-Yves Sacré[1], Charlotte De Bleye[2], Pierre Beckers[2], Philippe Hubert[2], and Eric M. Ziemons[2]

[1]*Research Support Unit in Chemometrics, Department of Pharmacy, University of Liege (ULiege), Liege, Belgium*
[2]*ViBra-Sante Hub, Laboratory of Pharmaceutical Analytical Chemistry, Department of Pharmacy, University of Liege (ULiege), Liege, Belgium*

14.1 Introduction

Since the launch of the Process Analytical Technology (PAT) initiative by the US Food and Drug Administration (US FDA) in 2004 [1], vibrational spectroscopy techniques (i.e. Raman and near-infrared (NIR) spectroscopy) have gained popularity in the pharmaceutical industry. This is partly explained by their non-destructive character, the large variety of interfaces, and the fast acquisition speed. In addition, both NIR and Raman spectroscopy are sensitive to the physical and chemical properties of the sampled matter and can inform about a wide range of quality attributes of pharmaceuticals (e.g. nature and amount of active ingredient, humidity, and polymorphism). The first developments mainly involved benchtop instruments to monitor different manufacturing steps or to perform atline; online or inline controls, the recent technological progresses and miniaturization encouraged the development of handheld instrumentation [2]. As in many industrial sectors, the pharmaceutical sector experiences an actual paradigm shift from the "Sample to Laboratory" to "Laboratory to Sample" [3].

Tablets are the most widely used pharmaceutical dosage forms. They are solid dosage forms in which an active ingredient (API) is blended with excipients (inactive compounds) and compressed into the final dosage. They may exhibit a wide variety of sizes, shapes, and colors [4]. Before and during the marketing phase, to ensure the safety and efficacy of pharmaceutical product, several critical quality attributes (CQAs) must be controlled. These CQAs are summarized in the quality target product profile (QTPP) and cover the identity, strength, and purity of the drug product [5]. For most active ingredients, both NIR and Raman spectroscopy are ideal tools to control their identity and quantity either as intermediate or finished drug products. Raman spectroscopy has the advantage of being easier to interpret thanks to its sharp spectral features. This may partly explain its popularity in analyzing raw, semi-finished, and finished pharmaceutical samples.

Because of the diversity of tablet formulations, several difficult situations may be encountered among which are low-dosed active ingredients (typically less than 1% (w/w)), co-formulations (multiple APIs in the same tablet), homothetic formulations (same API/excipients ratio but different tablet mass), multilayer tablets, sugar-coated tablets, and fluorescence of the coating (especially for Raman spectroscopy). Technological solutions exist in handheld spectroscopy for some of these challenges [6]: surface-enhanced Raman spectroscopy (SERS) for the analysis of low-dosed APIs [7, 8], spatially offset Raman spectroscopy (SORS) analysis of samples getting rid of the container/coating signal [9, 10], fluorescence of coating is not an issue for NIR spectroscopy and may be mitigated in Raman spectroscopy using a 1064 nm laser [11] or dedicated technologies such as the sequentially shifted excitation (SSE) spectroscopy [12, 13]. The other

Practical Quantitative Vibrational and Electronic Spectroscopy: A Guide for Developing, Optimizing, and Validating Procedures, First Edition. Edited by Deborah A. Peru.

Companion website: www.wiley.com/go/peruelectronicspectroscopy

challenges have not yet been tackled with commercial handheld devices and must be analyzed by benchtop spectrophotometers or using specific lab-made techniques. An additional comment regarding the presence of coating is the appropriate choice of optics to collect the Raman spectra. Indeed, depending on the architecture of the spectrophotometer and the collection optics, the focal point of the laser beam may be at the surface of the sample or a few millimeters inside the sample for measurements in a backscattering configuration. If the focal point is at the surface of the sample, the presence of the coating will have a preponderant influence on the collected spectrum (with possibly no information from the tablet core) [6]. The difference in the collection optics will also influence the sampled volume and consequently the measurement representativity.

The present chapter will describe the development and validation of a handheld Raman spectroscopy procedure for the quantitation of an active ingredient in immediate-release tablets. Doing so allows us to focus on the development and validation steps of the quantitative model, avoiding the technological challenges that may be encountered. The presented approach follows the Analytical Quality by Design (AQbD) approach described in ICH Q14/Q2(R2) [14, 15] and the USP <1220> [16] guidelines and may be applied in most situations with slight adaptations. Although not covered by the present chapter, the development, validation, and routine use of the spectroscopy method requires a fully qualified and controlled spectrophotometer to ensure good functioning of the device, which directly impacts the quality of the spectral data. Generally, the manufacturer performs the installation, operational, and performance qualification (IQ/OQ/PQ) operations, and kits exist to conduct the PQ tests during routine use. Recent reviews cover the generalities of regulatory guidelines and standards [17] and the data analysis in Raman spectroscopy [18].

14.2 Procedure Design and Development

14.2.1 Analytical Target Profile and Technology Selection

The first step in the development of a procedure is the definition and formalization of its objective and intended purpose. In our case, the purpose of the procedure has been described as the "in-field quantitation of ibuprofen at 20% (w/w) in immediate release tablets of fixed composition." Now that the purpose is identified, the required procedure performances may be defined in the Analytical Target Profile (ATP). The ATP is defined as *"A prospective summary of the performance characteristics describing the intended purpose and the anticipated performance criteria of an analytical measurement."* [14] It includes the measurement requirements for one or more quality attributes and drives the choice of analytical technology.

Therefore, the ATP, by formalizing the expected performance of the analytical procedure, will guide the whole process of technology selection, analytical procedure development, validation, and ongoing procedure performance verification (OPPV). Table 14.1 shows the minimum requirements for the considered procedure.

Regarding the objective and availability of equipment, a handheld Raman spectrometer was selected as the instrument of choice: Thermo Fisher™ TruScan™ instrument with the Trutools™ add-on toolbox. This handheld device provides in-field analysis of samples and has pharmaceutical quality assurance certifications. The Trutools™ add-on enables us to fix the acquisition parameters, build chemometric models (e.g. partial least squares, PLS), import, and run the developed model on the equipment.

14.2.2 Risk Assessment

The risk assessment step identifies and scores the causes that could impact the analytical results during routine use. It also identifies the parameters that should be controlled during routine use and those that cannot be controlled and should be included in the procedure development and validation phases.

Table 14.1 Analytical target profile (ATP).

Parameter	Desired procedure performances
Specificity	Main API features present in the first latent variables and/or in the regression vector and/or in the variable importance in projection (VIPs)
Range	70–130% of theoretical concentration (20% (w/w))
Linearity	Coefficient of determination above 0.95, slope 95% confidence interval contains the value of 1, intercept 95% confidence interval contains the value of 0
Combined accuracy and precision	Maximum ± 10% of relative total error expressed as the 95% beta-expectation tolerance interval
Accuracy	No limit required (controlled via the combined criterion)
Precision	No limit required (controlled via the combined criterion)
Analysis time	Maximum 1 minute total analysis time
Robustness	The validation dataset should be representative of potential natural variations of the measured materials (particle size, compression force during tableting, etc.) and experimental conditions (temperature, humidity, etc.)

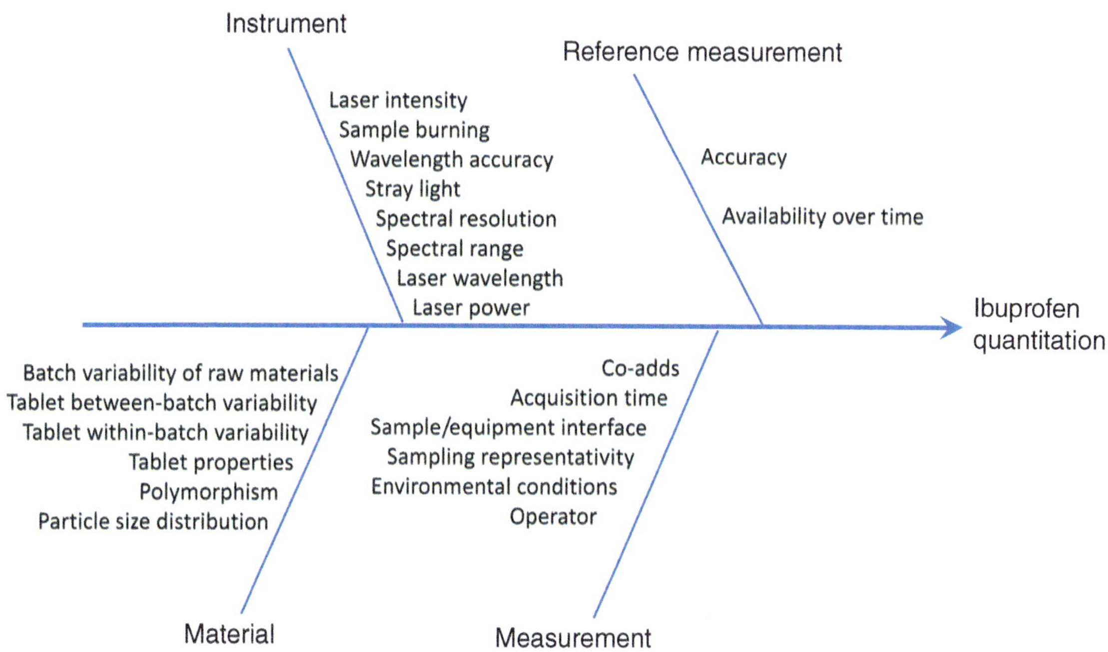

Figure 14.1 Ishikawa or fishbone diagram for the ibuprofen quantitation procedure. *Source:* PY Sacré.

Figure 14.1 shows an Ishikawa or fishbone diagram with some of the identified risks. The main identified categories are the material variability, instrument parameters, measurement parameters, and the reference analytical procedure. The instrument parameters are either fixed (e.g. spectral resolution or spectral range) or checked through instrument qualification. They are, therefore, considered as having no impact on the procedure development/validation. Material variability should be included in the method development and validation, which will be further discussed in the following paragraphs. Some of the measurement parameters will be investigated and fixed during procedure development (acquisition time, number of co-adds, sample/equipment interface), and the other parameters will be included in the multivariate model development and validation (environmental conditions and operator).

Since the development of the multivariate regression model relies on modeling the relationship between the measured spectral data and quantitative values obtained using a reference analytical procedure (typically liquid chromatography), the latter must be validated and must have an uncertainty sufficiently low considering the intended performance of the procedure being developed. In addition, the reference procedure should ideally be available during the routine use of the spectroscopic procedure to enable the analysis of periodical quality control samples and/or out-of-trend (OOT)/out-of-specifications (OOS) samples.

In our case, the reference procedure exhibited less than ±4% of relative total error (expressed as the 95% beta expectation tolerance interval) over the 70–130% of theoretical ibuprofen concentration (20% (w/w)) which is compatible with our final objective of ±10% of relative total error for the spectroscopic procedure.

14.2.3 Acquisition Parameters Settings

A procedure can be designed and developed once the ATP and the risk analysis are performed. The measurement parameters for spectral acquisition using a handheld Raman device for qualitative analysis of raw materials are often optimized automatically to obtain a sufficient signal-to-noise ratio (SNR). To improve the instrument's performance for quantitative measurements and ensure reliable results over time, acquiring Raman spectra with identical parameters for each analysis is necessary. This is why the Trutools™ add-on was used in the present case. This option allows the analyst to fix optimal combinations of laser power, acquisition time, and number of co-adds (a.k.a., co-added scans). Since the procedure is intended to analyze white tablets and no component is particularly sensitive to light degradation, the laser power was set to its maximum to decrease the analysis time while keeping a satisfactory SNR.

Regarding the number of co-adds and acquisition time, a design of experiments (DoE) has been performed to investigate the effect of each factor and their potential interaction. A full factorial design with the variables of acquisition times (2 and 5 seconds) and co-adds (1, 2, and 5) was used to measure the Raman spectra of tablets at the target ibuprofen concentration. Each combination of factors was measured three times, resulting in 18 experiments. The data were eventually analyzed by ANOVA Simultaneous Component Analysis (ASCA [19]) in the PLSToolbox software [20]. ASCA allows the user to analyze multivariate data acquired by designed experiments and visualize and quantify the different sources of variability included in the DoE. The ASCA results obtained after normalization of the data by standard normal variate (SNV) normalization showed that "time" was the most important factor (78.4%) followed by the "co-adds" (17%). However, there was no significant interaction between "time" and "co-adds" (the effect explained only 1.2% of the total variability and exhibited a p-value of 0.06 computed on 10,000 permutations). This information indicates that the values for "acquisition time" and "co-adds" may be optimized independently. As shown in Figure 14.2, a longer acquisition time increases the signal intensity above 700 cm^{-1} but results in a noisier signal below 700 cm^{-1}. The increase in co-adds decreased the noise linearly. Therefore, the best acquisition parameters were found using the combination of 2 seconds acquisition time and 5 co-adds.

14.2.4 Quantitative Regression Modeling

Now that the acquisition parameters are set, the calibration phase of the model may begin. As stated in ICH Q14 and USP <1220>, the robustness of the analytical procedure should be built-in since the development phase. This is especially true for multivariate regression models since, unlike liquid chromatography procedures, the regression model is calibrated once and used in future routine testing of unknown samples

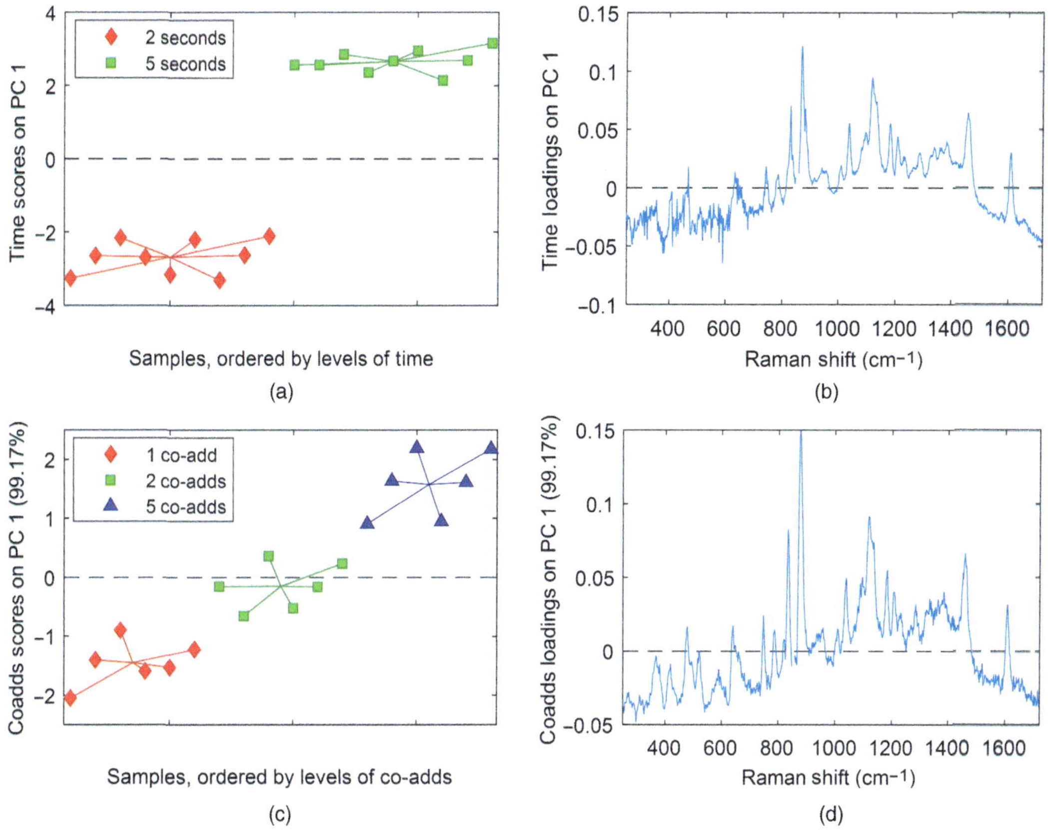

Figure 14.2 Results of the ASCA analysis. (a) Scores of the first principal component of the "time" effect (78.4% of the total variability); (b) loading of the first principal component of the "time" effect; (c) scores of the first principal component of the "co-adds" effect (17% of the total variability), (d) loadings of the first principal component of the "co-adds" effect. *Source*: PY Sacré.

without calibration samples. It is, therefore, of utmost importance to prepare the calibration dataset carefully to include as many potential variation sources as possible to improve the robustness of the calibration model. The sources of variation were detailed during the risk assessment analysis. In our case, such variability may consist in the use of different batches of raw materials, and different particle sizes (considering the acceptable particle size variation that might be encountered in routine production). The polymorphism of ibuprofen was not included explicitly but was considered as covered by the different batches of raw materials. The tablet properties (e.g. different compression forces) were also considered as covered by the manufacturing of different independent batches. Still, they could also be explicitly included by producing different tablets with varying forces of compression. Some samples were measured outside the laboratory to account for the environmental conditions (e.g. humidity and light).

The calibration dataset covered a range of concentrations between 70 and 130% of the target concentration (14–26% (w/w) of ibuprofen). To ensure the specificity of the multivariate model, the two major excipients (microcrystalline cellulose with a target concentration of 29% (w/w) and mannitol with a target concentration of 42.5% (w/w)) were varied relative to the API concentration to avoid collinearity. A D-optimal mixture design with three repeated central points was used to select the formulations, resulting in 15 formulations (see Figure 14.3). The minor excipients' concentration was kept constant. The DoE was computed with the JMP software [21].

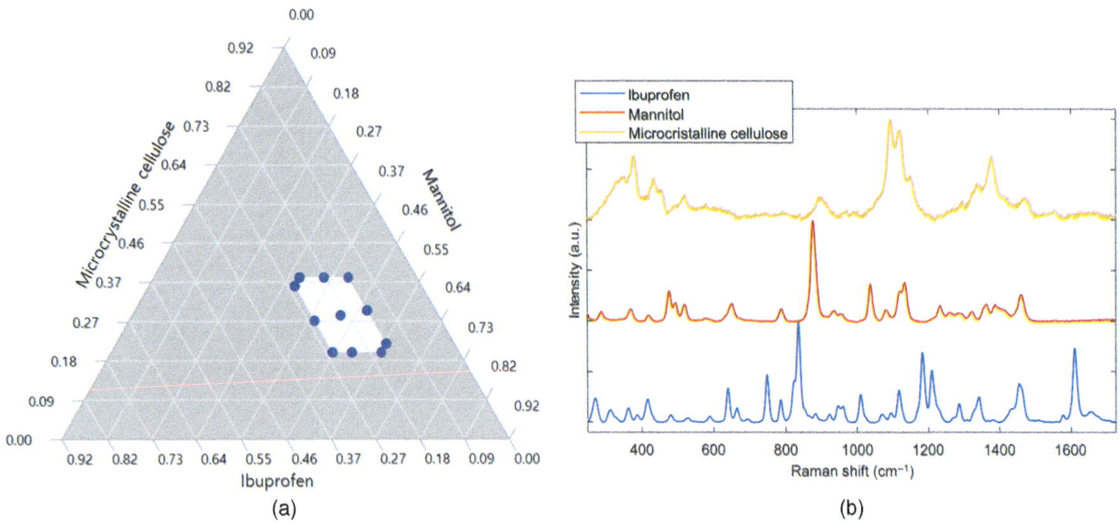

Figure 14.3 (a) Ternary plot showing the different combinations of ibuprofen (active ingredient) and the two major excipients (values expressed as % (w/w)). The dots show the different combinations realized for the calibration dataset; (b) Reference Raman spectra of the three major components. *Source*: PY Sacré.

Two different batches of 15 formulations were produced, three tablets were analyzed per formulation, and three analyses at different positions per tablet were analyzed, leading to a total of 270 calibration spectra. The Truscan™ first analyzed each of the 90 calibration tablets with the nose cone accessory, and then the tablets were analyzed by high-performance liquid chromatography (HPLC). The choice of the nose cone for the tablet measurement is based on practical aspects among which is the ease of changing the sample position while ensuring a constant focal distance.

Before starting the regression analysis of the data, it is a good practice to perform an exploration of the data to ensure that no outlier is present. Indeed, outlying spectra exhibit undesired unusual variability that may prevent the development of an accurate regression model. However, one must ensure the sample is a true outlier before excluding spectra. If there is no apparent reason for the outlying character of the spectrum, the spectrum may exhibit an under-sampled variability that will very likely be encountered in routine. In this case, new calibration samples may be included to improve the robustness of the model.

For unsupervised exploration of the data, we excluded the non-informative noisy part of the spectrum. Since we did not detect any signal from the tablets' component above 1725 cm^{-1}, we pursued the 250–1725 cm^{-1} spectral range for ongoing processing (Figure 14.4a). Looking at the raw spectra, one can see that an important shift in the baseline is present (Figure 14.4b). This is explained by both scattering effects and the auto-fluorescent character of microcrystalline cellulose. Therefore, an SNV preprocessing was applied before visualization of the data by principal component analysis (PCA). PCA is a useful dimension reduction tool that builds new latent variables (called principal components) that maximize the explained variance of the data in an orthogonal and degressive way. Figure 14.4c shows three panels where the PC1-PC2, PC1-PC3, and PC2-PC3 plots for the main ingredients are color-coded based on concentration levels.

As one can see, the main source of variability (64.9% of the SNV normalized data) in the data is linked to the concentration of microcrystalline cellulose while ibuprofen concentration is mainly captured by the second principal component (26.7% of the captured variance). Mannitol-related variation is captured by a combination of PC1 and PC2. The inspection of the residuals of the PCA showed no outlying spectrum is observed in the calibration dataset. This was confirmed by a deeper investigation of the data using tools such as a distance plot computed from a robust PCA analysis [22, 23].

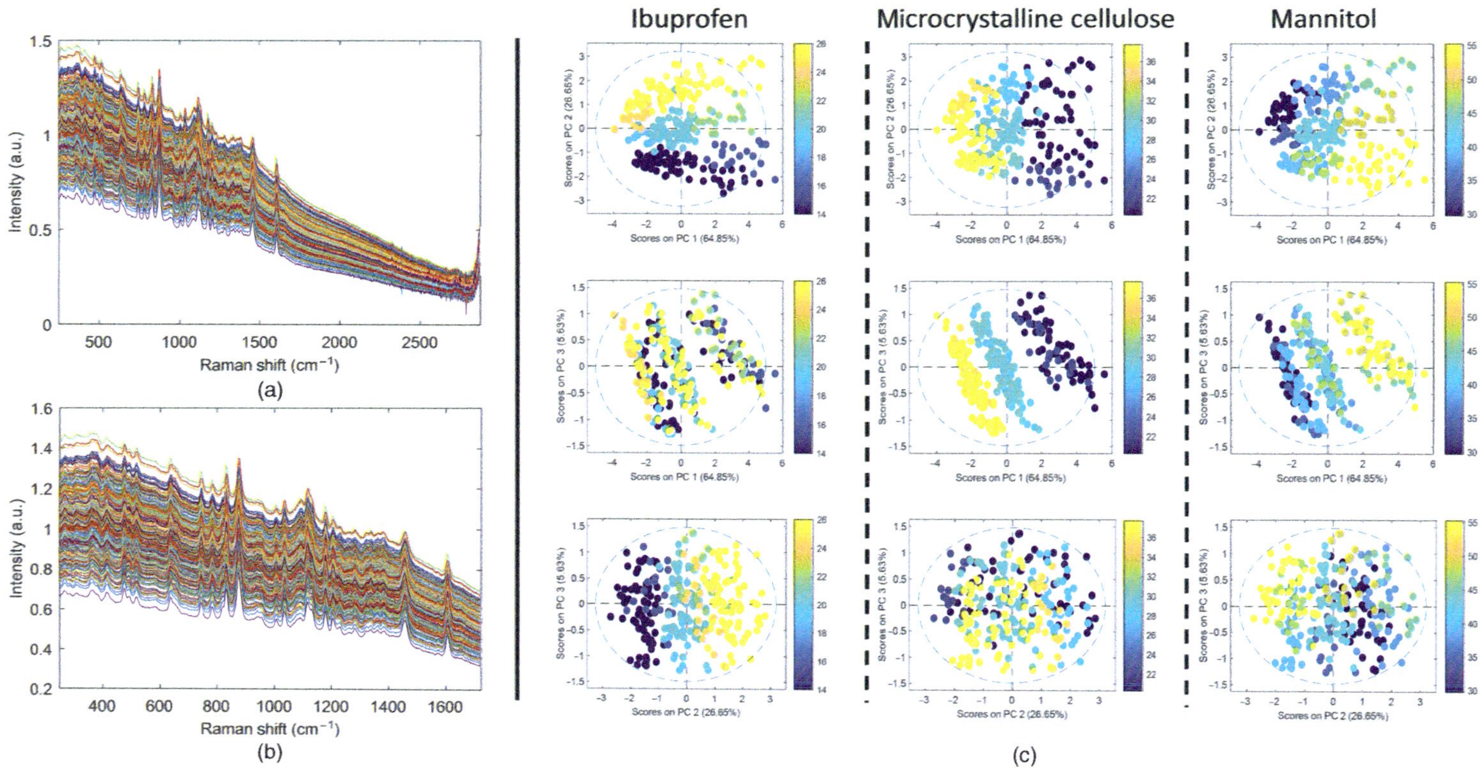

Figure 14.4 (a) Raw calibration spectra with the original 250–2875 cm^{-1} spectral range; (b) raw spectra with the reduced 250–1725 cm^{-1} spectral range; C: PC1-PC2, PC1-PC3, and PC2-PC3 score plot of the PCA analysis of the SNV normalized spectra. (c) The plots are colored according to the main ingredients' concentration (ibuprofen, microcrystalline cellulose, and mannitol from left to right). *Source*: PY Sacré.

The multivariate regression model that has been considered in this work is the PLS regression [24, 25]. This supervised model builds new latent variables as linear combinations of the original variables (Raman intensities) that maximize the *covariance* between the Raman signal and the measured concentration. PLS has the advantage of being easily interpretable while showing good accuracy in a wide variety of applications. In our case, we will use the interpretability of the PLS to demonstrate the selectivity of the model. The PLS algorithm also returns metrics such as Hotelling's T^2 (score distance) and the Q residuals (orthogonal distance) that may be used in the analytical control strategy (see Section 14.4) [26].

The building of a PLS regression model requires the optimization/selection of several parameters, such as the number of latent variables (controlling the trade-off between bias and variance), the choice of the data pre-processing, and the variable (or spectral range) selection. For our example, the spectral range was preliminarily set as 250–1725 cm^{-1}. Considering that no disruptive variables or spectral artifacts are evidenced and the informative variables are widespread over the whole spectral range, this spectral range is maintained for the final model.

The preprocessing of the spectral data is a crucial step that is used to remove (or decrease) the influence of non-informative variability. Indeed, since most multivariate statistical tools analyze the variability of the data, unwanted additional variability may hinder the development of an accurate regression model or at least lower its performance. Therefore, several approaches have been developed to remove this unwanted variability and ease access to the relevant ones. In our data, both additive and multiplicative effects were evidenced by plotting the spectra as a function of the mean spectrum. Taking this information into account, we investigated the raw data, normalization (standard normal variate, SNV and multiplicative scatter correction, MSC) [27], Savitzky–Golay smoothing (moving window of 3 to 15) [28], Savitzky–Golay first derivative (polynomial order 2 and moving window of size 5 to 25), baseline correction (asymmetric least squares with p: 0.001 and lambda: 10^5) [29], combination of smoothing + baseline correction and combination of smoothing + normalization (SNV or MSC). The optimization of both the preprocessing and the number of latent variables should be performed concomitantly since there is a strong interaction between both (the optimal number of latent variables may be different depending on the preprocessing). Therefore, a common approach is to select a set of preprocessing, build regression models of varying complexity for each preprocessing, and then select the model having the best performance. This has been performed using the "model optimizer" of the PLS Toolbox software with the test of the different preprocessing over model complexities of 2 to 10 latent variables and a Venetian blinds cross-validation strategy. The choice of Venetian blinds was dictated by the fact that it is generally a safe choice when there are enough samples and no order (e.g. time series) present in the data. However, replicates must be kept together to avoid placing replicates in both the calibration and validation datasets. This would lead to overly optimistic results. This tool allowed us to easily build 1062 PLS models and compute their performances measured as root mean squared error of calibration (RMSEC) and cross-validation (RMSECV), among others. Once the models are computed, the RMSECV/RMSEC ratio is plotted as a function of RMSECV [30]. This plot allows us to observe the models' performances together with the possible overfitting. Indeed, a low RMSECV indicates the models' ability to predict new samples, while high RMSECV/C ratio indicates possible overfitting. The "best model" is located at the lower left corner.

As shown in Figure 14.5a, the use of derivative or baseline correction as preprocessing (alone or combined to SNV, MSC, or smoothing) returned worst results compared to the raw data. The zoom on the lower left corner (Figure 14.5b) highlights the fact that a combination of smoothing with normalization (either MSC or SNV) exhibited the best prediction results. The larger the smoothing window, the better the results. However, there is no clear difference between SNV and MSC. The final preprocessing was selected as a combination of Savitzky–Golay smoothing with a window size of 15 and SNV. The preference of SNV over MSC was dictated by the fact that SNV is calibration independent (each sample is normalized considering its own data) which is not the case for MSC whose parameters are computed from the calibration dataset. This

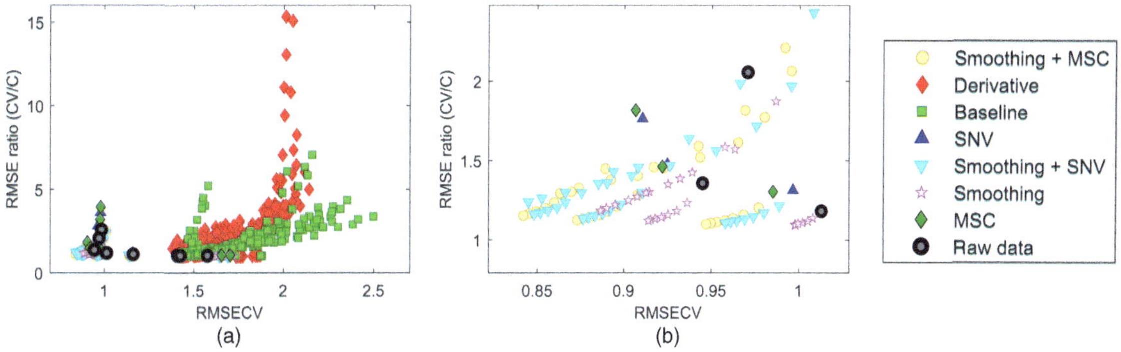

Figure 14.5 (a) RMSECV/C vs. RMSECV plot of the 1062 models built for the selection of the preprocessing; (b) zoom on the lower left corner of figure A. *Source*: PY Sacré.

is an interesting property considering the possible change of calibration dataset over time during model maintenance (e.g. calibration update or augmentation). The best performances were obtained for models of complexity between 6 to 9 latent variables.

The number of latent variables has been optimized in a second phase. For this purpose, the calibration dataset was split in a training and a test set. This enables the testing of the chosen setting with "external" data without depending too much on cross-validation. Indeed, depending on the cross-validation strategy (e.g. leave-one out, leave k-out, and Venetian Blinds), different optimal latent variables could be chosen, and the observed RMSECV could be too optimistic or too pessimistic. Therefore, complementary to the cross-validation, a training/test set splitting could be useful to select to correct the complexity of the regression model. However, as we will see in the next paragraph, it is important to stress that neither cross-validation nor the training/test set splitting constitutes a formal external validation [31].

Several approaches may be used to split a dataset in training/test sets (e.g. Kennard–Stone [32], Duplex [33], Onion [34], D-optimal [35], Sample set Partitioning based on joint x–y distances (SPXY) [36], and Combined analytical signal (CAS) [37]) when the dataset is sufficiently large. The main objective is to select a "representative" subset of samples to constitute the "test set" (or "internal test set" as described in ICH Q14) that will mimic an external dataset during model development. A major risk would be to remove informative samples from the training set and not to place replicates of the same sample in both training and test set, which would lead to overoptimistic and biased results. Once the split is done, it is good practice to look at how well the test set covers the training set or the range of the modeled property (i.e. in our case, how uniformly the concentration range is covered). Other approaches may be used for smaller datasets to keep the calibration dataset complete and avoid removing informative samples from the training set [38]. In our example, the Onion approach was used keeping 70% (n = 192) of samples in the training set and 30% (n = 78) in the test set while keeping replicates (the three spots per tablet) together. This choice was justified since the Kennard–Stone split didn't cover uniformly the concentration range compared to the Onion split. This would have led to biased RMSE values and, therefore, to a wrong model complexity selection.

Once the split is performed, several PLS models are built with an increasing number of latent variables, and their performances are compared. In this case, we will have access to both the RMSE of calibration, cross-validation, and prediction (from the test set). The "best" complexity will be selected as the one that minimizes the prediction error (RMSEP), and that is present in the lowest left corner of the RMSECV/C ratio vs. RMSECV plot. The obtained results are shown in Figure 14.6. As one can see, the model with 6 latent variables provided the best results while not being too complex. The RMSECV/C ratio shows that after six latent variables, the model starts to overfit the training set. An inspection of the six *T–U* plots [31]

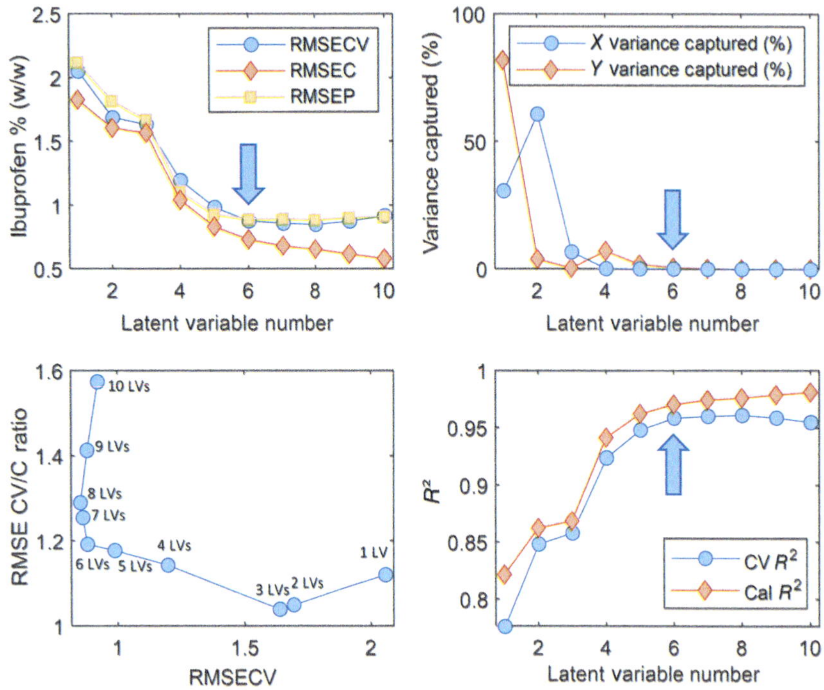

Figure 14.6 Figures of merit of the developed PLS model after training/test set splitting. The arrow shows the selected complexity (6 LVs). The selected preprocessing consisted of Savitzky–Golay smoothing (window: 15), SNV, and mean-centering. *Source*: PY Sacré.

didn't show any worrying pattern and confirmed that six latent variables was a suitable dimension (data not shown). The final model that will eventually be validated has the following characteristics:

- Spectral range (250–1725 cm^{-1})
- Preprocessing: Savitzky–Golay smoothing (window: 15) + SNV + mean center
- Six latent variables.

14.3 Procedure Validation

Considering the analytical procedure lifecycle, the first stage ("procedure design" in USP <1220>) is now completed. Now the analyst can enter in stage 2 related to the "performance qualification" in USP or to the "validation" in ICH Q14/Q2 (R2). This next step is important before using the procedure in routine to ensure the ATP is met or, in other words, that the procedure is fit for its intended purpose. It is also during this step that the analytical control strategy for the ongoing procedure performance monitoring will be confirmed and finalized.

Based on the ATP described earlier, a validation protocol containing the intended purpose of the analytical procedure, the preparation of validation standards, the experiments to perform, and how each ATP item will be tested is edited.

Although there are several possible practical ways to perform the validation study, we have followed the "combined validation of accuracy and precision" approach [15, 39]. The rationale for this choice will be further developed. In practice, the experimental work is designed so that specificity/selectivity, accuracy, and precision over the reportable range are considered simultaneously to provide sound, overall knowledge of the performance of the analytical procedure and to reflect the routine use of the procedure.

The validation work will be conducted to mimic routine analyses. The analyses are grouped in different series. In a series, the samples are analyzed in a short amount of time by a single operator. This informs us on the "repeatability" or "intra-assay precision" of the analytical procedure. Between the series, several variations corresponding to the random events that could impact precision during routine use are applied. In our example, only one equipment was available and included, but different operators analyzed samples coming from different independent production batches. The between-day effect was mimicked by turning the equipment off, letting it cool down for two hours, and switching it on again. A total of three series were analyzed, but this number may be increased depending on the variability of the procedure. For each series, a newly produced independent batch of 15 formulations of tablets was analyzed over 3 different positions per tablet (reaching to a total of 405 spectra). The formulations were the same as for the calibration set (coming from the mixture design). Once analyzed, the tablets were identified and analyzed with the reference procedure. The 15 formulations were grouped in 5 concentration levels, and, because of the mixture plan, the levels are not perfectly balanced. The first and last concentration levels have 12 independent repetitions each, the second and fourth levels have 6 independent repetitions each, and the target level has 9 independent repetitions. Each independent repetition is a different tablet. It is important to remember that validation samples are completely independent from calibration samples and that no models' parameter tuning has been performed with the validation samples. The acquired validation spectra are eventually projected onto the previously built PLS model, and the predicted concentrations are processed to decide on the validity of the analytical procedure.

The first validation criterion is the specificity/selectivity criterion. It is needed to prove that the regression model predicts the compound of interest and that if samples have a different composition from the targeted one, the model will reject them. Therefore, the specificity/selectivity criterion may be fulfilled by comparison of the active ingredient's spectrum and loadings/regression vector of the model. In this case, we decided to compare the ibuprofen preprocessed spectrum in the same way as the PLS model and the VIPs. The latter is defined as variable importance in the projection (VIPs), often used as a variable selection tool since it helps to understand which **X** space predictor variables (Raman intensities) best explain the **y** (concentration) variance [40, 41]. Indeed, in our case, the regression vector was difficult to interpret because of the variability included in the model and as the contribution of ibuprofen was dispersed over several latent variables (as one can see in Figure 14.6). VIPs scores showed an unambiguous connection to the ibuprofen spectrum (see Figure 14.7). Tablets having different composition (other excipients, other API, placebos, ...) have also been projected onto the model, and their outlier behavior has been confirmed since both their Hotelling's T^2 and Q residuals were far from the 99.9% critical limits (data not shown).

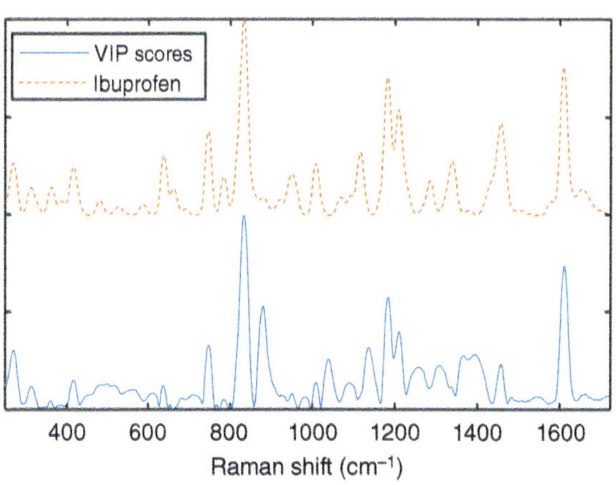

Figure 14.7 Overlay of the preprocessed ibuprofen Raman spectrum and the variable importance in projection (VIP) scores of the PLS regression model to fulfill the specificity/selectivity criterion. *Source:* PY Sacré.

Regarding the quantitative criteria (accuracy, precision, linearity, and reportable range), it has been demonstrated that conventional chemometrics figures of merit (e.g. RMSEP or ratio of performance to deviation, RPD [42]) are not sufficient to guarantee the validity of the multivariate regression methods [43]. Therefore, we followed the "combined accuracy and precision" strategy and decided on the validity of the method based on the predictive quality of the results that will be obtained in future routine use. To achieve this goal, β-expectation tolerance intervals ($\beta = 95\%$) were computed over the investigated concentration range (70–130% of the target concentration) and summarized in a so-called total error profile. If the β-expectation tolerance intervals (describing both the systematic error or accuracy and the random error or precision) is fully comprised in the acceptance limits set at ±10% in our case, we can conclude that the procedure is valid since it demonstrates that 95% of the future results will have a total analytical error of less than ±10% compared to the reference method. This performance was considered sufficient for the intended purpose of the procedure (see ATP in Table 14.1). The building of the total error profiles is described in [39, 44–47]. This validation strategy also allows the analyst to visually assess the validity of the procedure and the valid concentration range (or reportable range). The total error profiles were computed with the e·noval software [48] and the linearity plot was computed with the JMP software [21].

Two approaches were followed during validation: (1) the use of a single spot/spectrum per tablet (models 1S1P-A, -B, and -C for each of the three spots/coordinate locations on the tablet analyzed, respectively); and (2) the average of the predicted concentrations from the three spots/spectra of the same tablet (model 3S3P). The latter should ensure a better representativity of the measured tablet. For the models 1S1P-A, -B, -C, the spectra were collected at the same coordinate locations used for model building and validation.

The models 1S1P-A, -B, -C, and 3S3P had RMSEP (% (w/w)) values of 0.83, 0.78, 0.82, and 0.63, respectively, and RPD values of 5.5, 5.9, 5.5, and 7.2, respectively. Both RMSEP and RPD seem to indicate "good" to "very good" models [49]. However, when looking at the total error profiles of the four models (Figure 14.8), one can see that none of the 1S1P models are valid over the whole concentration range. Only the 3S3P model shows a total analytical error of less than ±10% over the 13.96–25.63% (w/w) reportable range. Therefore, the reported routine analysis results will be the average of the predictions of three spots per tablet. These three spots are acquired in a total of 30 seconds, allowing us to analyze a tablet in less than one minute considering the handling and processing of the data. The validation performances are summarized in Table 14.2.

The linearity of the method is demonstrated by the predicted vs. actual plot shown in Figure 14.9. The coefficient of determination, R^2 is 0.981, the intercept [±95% confidence interval] is 0.463 [−0.01; 0.936], and the slope [±95% confidence interval] is 0.983 [0.960; 1.007]. These values agree with the previously defined ATP confirming the validity of the method. The residual plot shows no pattern in the residuals over the concentration range.

Table 14.2 Validation results for the 3S3P model.

Actual concentration (% (w/w))	Average predicted concentration (% (w/w))	Relative bias (%)	Repeatability (% RSD)	Intermediate precision (% RSD)	Relative beta-expectation tolerance limits (%)	Risk of falling outside acceptance limits (%)
13.96	14.11	1.08	3.60	3.97	[−7.52, 9.67]	2.83
17.47	17.78	1.77	3.00	3.06	[−4.93, 8.47]	1.05
19.70	19.89	0.97	3.62	3.62	[−6.62, 8.55]	1.39
22.74	22.83	0.41	3.05	3.29	[−6.99, 7.82]	1.22
25.63	25.64	0.04	2.37	2.37	[−4.84, 4.92]	0.02

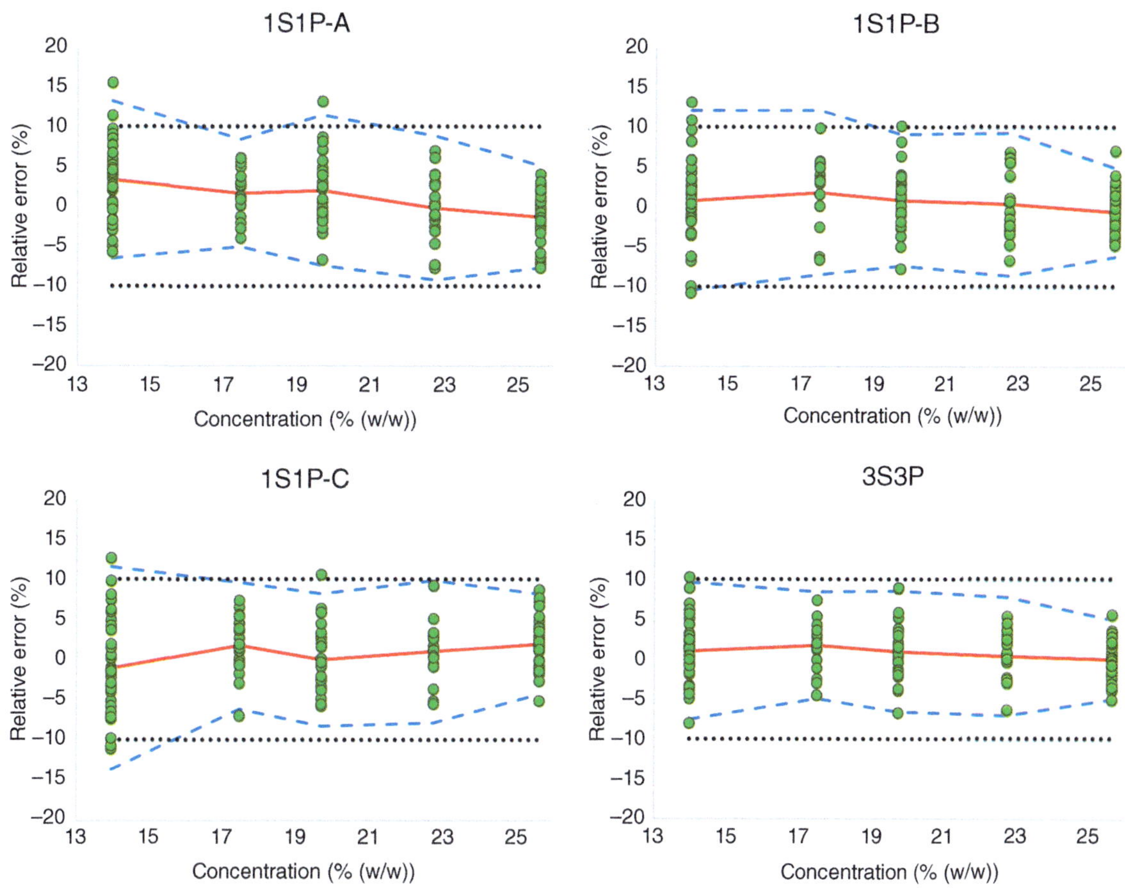

Figure 14.8 Total error profiles of the 1 spot models and the averaged predictions model (3S3P). The acceptance limits are set at ±10% relative error (black dotted lines). The dots represent the sample relative errors, the plain line represents the bias (accuracy), and the dashed lines represent the β-expectation tolerance intervals ($\beta = 95\%$). *Source*: PY Sacré.

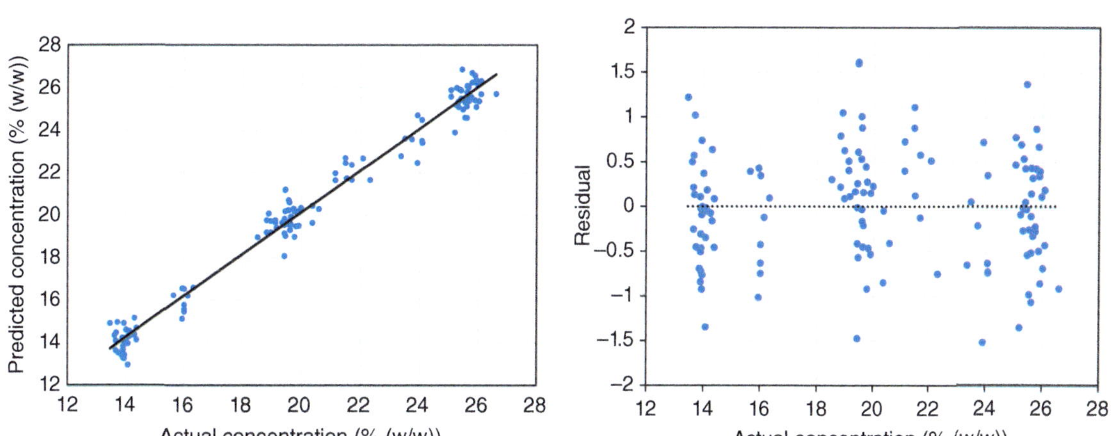

Figure 14.9 Linearity plot of the 3S3P model showing the predicted vs. actual results together with the residuals. *Source*: PY Sacré.

14.4 Ongoing Procedure Performance Verification

Once the analytical procedure is validated, the analyst may use it in routine. However, the performance of the procedure should be monitored during routine use to ensure that the reportable values generated are fit-for-purpose. This is especially true for multivariate regression methods since the calibration is performed once before the routine use. The OPPV phase of the procedure lifecycle should inform the analyst of potential performance issues or adverse trends and indicate that maintenance of the regression model is needed [50]. The OPPV covers several strategies, including routine monitoring using system suitability tests (SSTs) to ensure the generated results are trustworthy.

In our case, the SSTs include qualification tests of the equipment and use of 99.9% (~ 4 standard deviations) critical limits for Hotelling's T^2 and the Q residuals. If a routine sample has its Hotelling's T^2 and/or Q residuals values outside the 99.9% critical limit, it could be considered as a potential outlier whose variability is not well fitted by the PLS model. Its predicted value should be discarded, and its status should be confirmed by the reference method. The individual T^2 and Q values may be replaced by a total distance such as the one described in [23, 51], and both T^2, Q or total distance values should be monitored by control charts (see ISO 7870) to detect trends or significant differences associated with procedure changes.

If changes are detected and the model goodness becomes not "fit-for-purpose," maintenance or complete model rebuilding should be considered. This step is out of the scope of the chapter, but the interested reader may read references [52–54]. Depending on the changes that occurred (e.g. new equipment and reparation of the equipment), another possibility is the use of calibration transfer strategies. These will be developed in the next paragraph.

14.5 Calibration Transfer

The performance of the validated regression model depends on the quality of the spectral data generated by the instrument. Spectroscopic data are composed of signal and noise. In Raman spectroscopy, a major source of noise comes from the instrument [55]. Indeed, variability in wavenumber or intensity axis, laser power, and laser spot size influence the quality of the data collected. This is, of course, very important if the model is intended to be deployed on different instruments from different manufacturers but has also been demonstrated to be critical for similar instruments (same model from the same manufacturer) [8, 56, 57]. Therefore, if the model should be deployable on several instruments or if a new instrument should replace the older one, a transfer of the calibration model should be envisaged and validated [18]. Indeed, another possibility is building separate regression models for each instrument. However, this would require huge modeling and validation resources, which becomes prohibitive when considering the maintenance of the different models. Complicating this scenario is the consideration that models must also be maintained during use when instruments are replaced due to aging and repairs.

Several strategies may be considered for deploying a regression model on multiple instruments: direct transfer, global modeling, or univariate or multivariate calibration transfer. The direct transfer relates to the direct use (without modification) of the developed model on new instruments. To be effective, the preprocessing and the modeled spectral range should be carefully selected during model development. However, examples of successful direct transfer may be found in the literature [56]. The global calibration is a strategy where several instruments are included in the calibration phase of the model. This approach may be interesting when the differences between instruments are not too large. If differences between instruments are too large, too much variability may be included in the model which may lead to excessively complex or inaccurate regression models [58]. Of course, a wise selection of spectral range and preprocessing may reduce this inter-instrument variability. Therefore, global modeling may be seen as an extension of the direct transfer strategy. A successful example of qualitative global modeling with Truscan RM instruments may be found in [8].

The calibration transfer may be performed in several ways, either correcting the predicted concentrations (univariate transfer, e.g. slope/bias correction) or transforming the spectra so that the spectra collected on the new "slave" instrument are as similar as possible to the "master" instrument. A lot of algorithms are available for this spectral transformation task among which Direct Standardization (DS), Piecewise Direct Standardization (PDS), Spectral Space Transformation (SST), Generalized Least Squares Weighing (GLSW), Orthogonal Signal Correction (OSC), and Alignment of Matrices are available in the PLS Toolbox software and exportable to instruments. Although the algorithmic details will not be covered, the general scheme is the same.

First, it is mandatory to select representative samples carefully. Indeed, these samples will condition the success of the calibration transfer. If the transfer is directly envisaged during the model building, these samples may come from the calibration dataset. It is important to mention that the validation dataset cannot be used during the building of the transfer models. The number of transfer samples should not be large but should cover the variability of each instrument as representatively as possible. In this task, the use of the previously mentioned "combined analytical signal" may help in selecting the most representative samples [37]. Once the samples are selected, their spectra are measured on both instruments, and the transfer parameters are computed [59].

It is important to assess the success of the transfer before using the transferred models in routine. If the transfer is performed during the validation of the regression model, the validation samples may be analyzed by different instruments applying the transfer model, and their accuracy profile may be built. If the transfer is performed during the routine use of the analytical procedure, the analyst must follow the transfer of analytical procedures guidelines [60]. In this context, the use of total analytical error and tolerance intervals are recommended to perform a risk-based transfer as described in [61, 62].

14.6 Conclusion

In this chapter, we presented the development and validation of a quantitative procedure for the determination of ibuprofen in tablets. The whole process followed the requirements and guidelines of USP <1220> and ICH Q14/Q2(R2) also referenced to AQbD. Although the chapter focused on a simple formulation, the described approach and the provided references should interest analysts developing vibrational spectroscopy-based quantitative models. The general idea behind this framework is to accumulate knowledge and understanding of both the analytical procedure and the analyzed materials. This knowledge should enable the generation of trustworthy, fit-for-purpose, reportable results and facilitate the maintenance operations of the quantitative multivariate model. Even if the workload seems to be higher under the AQbD framework, we believe the time spent during the development and validation of the procedure will help save time and money during routine use and maintenance operations. We hope this chapter has provided the reader with essential information to follow for the development, validation, and implementation of new procedures.

References

1 U.S. Food and Drug Administration (2004). Guidance for industry, PAT-A framework for innovative pharmaceutical development, manufacturing and quality assurance. FDA-2003-D-0032, September.

2 Crocombe, R.A. (2018). Portable spectroscopy. *Applied Spectroscopy* 72 (12): 1701–1751.

3 Popping, B. and Diaz-Amigo, C. (2020). A paradigm shift: From "Sample to Laboratory" to "Laboratory to Sample". *Journal of AOAC International* 104 (1): 1–6.

4 General Chapter (2023). <1151> pharmaceutical dosage forms. In: *USP-NF*. United States Pharmacopeia: Rockville, MD.

5 ICH Q8 (R2) (2009). International Council for Harmonisation of Technical Requirements for Pharmaceuticals for Human Use (ICH). Pharmaceutical Development.

6 Deidda, R., Sacre, P.Y., Clavaud, M. et al. (2019). Vibrational spectroscopy in analysis of pharmaceuticals: Critical review of innovative portable and handheld NIR and Raman spectrophotometers. *TrAC, Trends in Analytical Chemistry* 114: 251–259.

7 Kimani, M.M., Lanzarotta, A., and Batson, J.S. (2021). Rapid determination of eight benzodiazepines in suspected counterfeit pharmaceuticals using surface-enhanced Raman scattering with handheld Raman spectrometers. *Journal of Forensic Sciences* 66 (6): 2167–2179.

8 Lanzarotta, A., Kimani, M.M., Thatcher, M.D. et al. (2020). Evaluation of suspected counterfeit pharmaceutical tablets declared to contain controlled substances using handheld Raman spectrometers. *Journal of Forensic Sciences* 65 (4): 1274–1279.

9 Mansouri, M.A., Sacré, P.Y., Coïc, L. et al. (2020). Quantitation of active pharmaceutical ingredient through the packaging using Raman handheld spectrophotometers: A comparison study. *Talanta* 207: 120306.

10 Eliasson, C., Macleod, N.A., Jayes, L.C. et al. (2008). Non-invasive quantitative assessment of the content of pharmaceutical capsules using transmission Raman spectroscopy. *Journal of Pharmaceutical and Biomedical Analysis* 47 (2): 221–229.

11 Kang, Y., Zhou, Y., Wu, Q. et al. (2020). Low-content quantitation in Entecavir tablets using 1064 nm Raman spectroscopy. *Journal of Spectroscopy* 28: 1–11.

12 Cooper, J.B., Abdelkader, M., and Wise, K.L. (2013). Sequentially shifted excitation Raman spectroscopy: Novel algorithm and instrumentation for fluorescence-free Raman spectroscopy in spectral space. *Applied Spectroscopy* 67 (8): 973–984.

13 Conti, C., Botteon, A., Bertasa, M. et al. (2016). Portable sequentially shifted excitation Raman spectroscopy as an innovative tool for in situ chemical interrogation of painted surfaces. *The Analyst* 141 (15): 4599–4607.

14 ICH Q14 (2023). International Council for Harmonisation of Technical Requirements for Pharmaceuticals for Human Use (ICH). Analytical Procedure Development.

15 ICH Q2 (R2) (2023). International Council for Harmonisation of Technical Requirements for Pharmaceuticals for Human Use (ICH). Validation of Analytical Procedures.

16 General Chapter (2023). <1220> Analytical Procedure Lifecycle. In: *USP-NF*. United States Pharmacopeia: Rockville, MD.

17 Ntziouni, A., Thomson, J., Xiarchos, I. et al. (2022). Review of existing standards, guides, and practices for Raman spectroscopy. *Applied Spectroscopy* 76 (7): 747–772.

18 Barton, B., Thomson, J., Lozano Diz, E., and Portela, R. (2022). Chemometrics for Raman spectroscopy harmonization. *Applied Spectroscopy* 76 (9): 1021–1041.

19 Bertinetto, C., Engel, J., and Jansen, J. (2020 Nov). ANOVA simultaneous component analysis: A tutorial review. *Analytica Chimica Acta: X* 6: 100061.

20 PLS_Toolbox 9.2 (2023) [Internet]. Manson, WA USA 98831: Eigenvector Research, Inc.; Available from: http://www.eigenvector.com (accessed 20 January 2024).

21 JMP *JMP Pro 17.2.0 [Internet]*. Cary, NC: SAS Institute Inc.; 2023. Available from: www.jmp.com (accessed 25 Jan 2024).

22 Rodionova, O., Kucheryavskiy, S., and Pomerantsev, A. (2021). Efficient tools for principal component analysis of complex data—A tutorial. *Chemometrics and Intelligent Laboratory Systems* 213: 104304.

23 Rodionova, O.Y. and Pomerantsev, A.L. (2019). Detection of outliers in projection-based modeling. *Analytical Chemistry* 92 (3): 2656–2664.

24 Wold, S., Sjöström, M., and Eriksson, L. (2001). PLS-regression: A basic tool of chemometrics. *Chemometrics and Intelligent Laboratory Systems* 58 (2): 109–130.

25 Ali, H., Muthudoss, P., Ramalingam, M. et al. (2023). Machine learning–enabled NIR spectroscopy. Part 2: Workflow for selecting a subset of samples from publicly accessible data. *AAPS PharmSciTech* 24 (1): 34.

26 Bu, D., Wan, B., and McGeorge, G. (2013). A discussion on the use of prediction uncertainty estimation of NIR data in partial least squares for quantitative pharmaceutical tablet assay methods. *Chemometrics and Intelligent Laboratory Systems* 120: 84–91.

27 Fearn, T., Riccioli, C., Garrido-Varo, A., and Guerrero-Ginel, J.E. (2009). On the geometry of SNV and MSC. *Chemometrics and Intelligent Laboratory Systems* 96 (1): 22–26.

28 Abraham, S. and Golay, M.J.E. (1964). Smoothing and differentiation of data by simplified least squares procedures. *Analytical Chemistry* 36 (8): 1627–1639.

29 Eilers, P.H.C. (2003). A perfect smoother. *Analytical Chemistry* 75 (14): 3631–3636.

30 Wise, B.M. (2022). Evaluating models: Hating on R-squared [Internet]. Eigenvector [cited 2024 Jan 12]. Available from: https://eigenvector.com/%EF%BF%BCevaluating-models-hating-on-r-squared/ (accessed 10 Feb. 2024).

31 Esbensen, K.H. and Geladi, P. (2010). Principles of proper validation: Use and abuse of re-sampling for validation. *Journal of Chemometrics* 24 (3–4): 168–187.

32 Kennard, R.W. and Stone, L.A. (1969). Computer aided design of experiments. *Technometrics* 11 (1): 137–148.

33 Snee, R.D. (1977). Validation of regression models: Methods and examples. *Technometrics* 19 (4): 415–428.

34 Gallagher, N.B., O'Sullivan, D. Selection of representative learning and test sets using the onion method. EigenVector Research Inc [Internet]. Available from: https://eigenvector.com/wp-content/uploads/2020/01/Onion_SampleSelection.pdf (accessed 20 Feb. 2024).

35 Ferré, J. and Rius, F.X. (1997). Constructing D-optimal designs from a list of candidate samples. *TrAC, Trends in Analytical Chemistry* 16 (2): 70–73.

36 Galvao, R., Araujo, M., Jose, G. et al. (2005). A method for calibration and validation subset partitioning. *Talanta* 67 (4): 736–740.

37 Pomerantsev, A.L. and Rodionova OYe. (2023). Subset selection using combined analytical signal. *Microchemical Journal* 190: 108654.

38 Kucheryavskiy, S., Rodionova, O., and Pomerantsev, A. (2023). Procrustes cross-validation of multivariate regression models. *Analytica Chimica Acta* 1255: 341096.

39 General Chapter (2023). <1210> Statistical tools for procedure validation. In: *USP-NF*. United States Pharmacopeia: Rockville, MD.

40 Chong, I.G. and Jun, C.H. (2005). Performance of some variable selection methods when multicollinearity is present. *Chemometrics and Intelligent Laboratory Systems* 78 (1–2): 103–112.

41 Farrés, M., Platikanov, S., Tsakovski, S., and Tauler, R. (2015). Comparison of the variable importance in projection (VIP) and of the selectivity ratio (SR) methods for variable selection and interpretation. *Journal of Chemometrics* 29 (10): 528–536.

42 Bellon-Maurel, V., Fernandez-Ahumada, E., Palagos, B. et al. (2010). Critical review of chemometric indicators commonly used for assessing the quality of the prediction of soil attributes by NIR spectroscopy. *TrAC, Trends in Analytical Chemistry* 29 (9): 1073–1081.

43 De Bleye, C., Chavez, P.F., Mantanus, J. et al. (2012). Critical review of near-infrared spectroscopic methods validations in pharmaceutical applications. *Journal of Pharmaceutical and Biomedical Analysis* 69: 125–132.

44 Hubert, P., Nguyen-Huu, J.J., Boulanger, B. et al. (2004). Harmonization of strategies for the validation of quantitative analytical procedures. *Journal of Pharmaceutical and Biomedical Analysis* 36 (3): 579–586.

45 Hubert, P., Nguyen-Huu, J.J., Boulanger, B. et al. (2007). Harmonization of strategies for the validation of quantitative analytical procedures. *Journal of Pharmaceutical and Biomedical Analysis* 45 (1): 70–81.

46 Hubert, P., Nguyen-Huu, J.J., Boulanger, B. et al. (2007 Sep). Harmonization of strategies for the validation of quantitative analytical procedures. *Journal of Pharmaceutical and Biomedical Analysis* 45 (1): 82–96.

47 Hubert, P., Nguyen-Huu, J.J., Boulanger, B. et al. (2008). Harmonization of strategies for the validation of quantitative analytical procedures: A SFSTP proposal. *Journal of Pharmaceutical and Biomedical Analysis* 48 (3): 760–771.

48 e-noval (2023). *e-noval 4.1 [Internet]*. Mont-Saint-Guibert: Cencora Pharmalex Available from: https://www.arlenda.com/enoval4.1/ (accessed 27 Feb 2024).

49 Esbensen, K.H., Geladi, P., and Larsen, A. (2014). The *RPD* myth.... *NIR News* 25 (5): 24–28.

50 Borman, P.J., Guiraldelli, A.M., Weitzel, J. et al. (2024). Ongoing analytical procedure performance verification using a risk-based approach to determine performance monitoring requirements. *Analytical Chemistry* 96 (3): 966–979.

51 Avohou, T.H., Sacré, P.Y., Hamla, S. et al. (2022). Optimizing the soft independent modeling of class analogy (SIMCA) using statistical prediction regions. *Analytica Chimica Acta* 1229 (August): 340339.

52 Wise, B.M. and Roginski, R.T. (2015). A calibration model maintenance roadmap. *IFAC-PapersOnLine* 48 (8): 260–265.

53 Miyano, T., Nakagawa, H., Watanabe, T. et al. (2015). Operationalizing maintenance of calibration models based on near-infrared spectroscopy by knowledge integration. *Journal of Pharmaceutical Innovation* 10 (4): 287–301.

54 Wang, H.P., Chen, P., Dai, J.W. et al. (2022). Recent advances of chemometric calibration methods in modern spectroscopy: Algorithms, strategy, and related issues. *TrAC, Trends in Analytical Chemistry* 153: 116648.

55 Pelletier, M.J. (2003). Quantitative analysis using Raman spectrometry. *Applied Spectroscopy* 57 (1): 20A–42A.

56 Steinbach, D., Anderson, C.A., McGeorge, G. et al. (2017). Calibration transfer of a quantitative transmission Raman PLS model: Direct transfer vs. global modeling. *Journal of Pharmaceutical Innovation* 12 (4): 347–356.

57 Brouckaert, D., Uyttersprot, J.S., Broeckx, W., and De Beer, T. (2017 Sep). Calibration transfer of a Raman spectroscopic quantification method for the assessment of liquid detergent compositions between two at-line instruments installed at two liquid detergent production plants. *Analytica Chimica Acta* 984: 1–18.

58 Fearn, T. (2001). Standardisation and calibration transfer for near infrared instruments: A review. *Journal of Near Infrared Spectroscopy* 9 (4): 229–244.

59 Workman, J.J. (2017). A review of calibration transfer practices and instrument differences in spectroscopy. *Applied Spectroscopy* 72 (3): 340–365.

60 General Chapter (2023). <1224> transfer of analytical procedures. In: *USP-NF*. United States Pharmacopeia: Rockville, MD.

61 Dewé, W., Govaerts, B., Boulanger, B. et al. (2007). Using total error as decision criterion in analytical method transfer. *Chemometrics and Intelligent Laboratory Systems* 85 (2): 262–268.

62 Rozet, E., Dewé, W., Ziemons, E. et al. (2009). Methodologies for the transfer of analytical methods: A review. *Journal of Chromatography B* 877 (23): 2214–2223.

15

Statistics of Microplastics Analysis: How to Achieve Healthy Statistical Representation

Eunah Lee[1], *Sanga Kim*[2], *and Julie Y. Chen*[3]

[1] *HORIBA Reno Technology Center, Reno, NV, USA*
[2] *HORIBA Korea Ltd., Manan-Gu, Anyang-Si, Gyeonggi-Do, Republic of Korea*
[3] *HORIBA Instruments Inc., Irvine, CA, USA*

15.1 Introduction

It has been reported that approximately 60% of plastic products are discarded into landfills and the environment [1]. The total production over 65 years (1950 to 2015) amounts to 8300 million metric tons of virgin plastics. With nearly 60% of this waste contributing to microplastics in the environment, the quantities amount to billions of tons polluting the world in this period. It is not surprising that recent reports are finding microplastics everywhere including in human bodies [2]. It is critical to assess, monitor, and regulate microplastics to get any chance of reversing widespread pollution and preventing harmful impacts on ecosystems and human health.

Microplastics and other environmental pollutants are a pressing global concern that necessitates a unified consensus on how to monitor the scale of the problem. This involves agreement on collecting, processing, and analyzing samples, as well as reporting meaningful and comparable data. It is crucial to develop consensus standards and adhere to standard operation procedures (SOPs) compiled and published by trusted organizations such as ASTM International, the International Organization for Standardization (ISO), and local regulatory agencies like the California State Water Resources Control Board (CA Waterboard) [3–5]. The CA Waterboard defines microplastics in drinking water as "solid polymeric materials to which chemical additives or other substances may have been added, which are particles with at least two dimensions greater than 1 nm and less than 5,000 micrometers (μm). Polymers that are derived in nature that have not been chemically modified (other than by hydrolysis) are excluded" [6].

In 2020, ASTM International updated ASTM D8332-20, "Collection of Water Samples with High, Medium, or Low Suspended Solids for Identification and Quantification of Microplastic Particles and Fibers" [7]. This standard practice contains procedures for collection and measurement of a high volume (1500 L/1.5 km) of water sample (Section 7.3). The volume of the collected sample matrices (e.g. kiloliters of drinking water) is many orders larger (c.f. $1\ \text{kL} = 10^9\ \text{mm}^3 = 10^{18}\ \mu\text{m}^3 = 10^{27}\ \text{nm}^3$) than the target microplastic pollutant size range (e.g. mm^3, μm^3, and nm^3). Developing and deploying dedicated methods to separate, fractionate, screen, and select subsets of samples in succession is necessary.

Many of the SOPs and agency procedures for measuring microplastics involve using spectroscopy techniques for chemical identification and quantitative statistical counting procedures. For example, the CA Waterboard has published SOPs for analyzing microplastics in drinking water using Raman [8] and

Practical Quantitative Vibrational and Electronic Spectroscopy: A Guide for Developing, Optimizing, and Validating Procedures, First Edition. Edited by Deborah A. Peru.

Companion website: www.wiley.com/go/peruelectronicspectroscopy

mid-infrared [9] spectroscopy in collaboration with the Southern California Coastal Water Research Project (SCCWRP) [10] based on academic research [11].

Suppose the water contains an excess of inorganic and organic materials. In that case, it may be necessary to prepare samples for analysis to remove non-microplastic backgrounds that can interfere with spectroscopic (and other) technologies before analysis. For example, ASTM International published an SOP to prepare water samples for analysis with Raman spectroscopy, IR spectroscopy, and Pyrolysis–GC/MS [12]. The procedure includes separating microplastics from the sample matrices through techniques such as centrifugation, Section 7.2 "Low to Very Low Suspended Solids").

Quantitative analysis of microplastics is necessary to assess the degree of pollution. However, microplastics are collections of discrete targets such as particles, fibers, films, fragments, and other forms. Typical methods of quantitative analysis using a calibration curve or a calibration model are not compatible with discrete targets [13, 14]. Statistical analysis based on counts, volume, or mass is the most effective way to quantitatively analyze discrete targets like microplastics [15].

Analyzing the "whole population" in statistical analysis is ideal, but it is nearly always impractical and often impossible. Proper sampling, therefore, is the most critical first step in statistical counting for quantitative analysis. The *Merriam-Webster Dictionary* defines sampling as

> "The act, process, or technique of selecting a representative part of a population for the purpose of determining parameters or characteristics of the whole population." (https://www.merriamwebster.com/dictionary/sampling)

Sampling (or sub-sampling) can be achieved in many ways. For example, CA Waterboard instructs to size-fraction using filters and sieves of different pore or mesh sizes in its SOP for Raman (Section 11.1 "Filtering") [8].

Sampling can be combined or augmented with similar processes, such as screening and fractioning, and can be repeated through successive sub-samplings. In this chapter, we will explore sub-sampling models proposed and published in an ISO guidance (ISO/DIS 16094-2) that specifies a methodology for the qualitative and quantitative analysis of microparticles of plastic or elastomeric materials in water using a microscopy technique coupled with vibrational spectroscopy [16].

15.2 Microplastics Analysis Workflow

Figure 15.1 shows a typical microplastic analysis workflow. The process begins with sample collection from the source, followed by sample preparation to remove non-plastic materials and extract micro-particulates. This is followed by sample treatment to selectively tag or stain microplastics and obtain size fractions with sieves and filters [17]. Subsequent quantitative assessment of the samples for microplastics is obtained by counts, sizes, shapes, etc., and to identify polymers.

The goal of microplastics analysis is to assess and characterize microplastics in their entirety quantitatively. Ideally, the measurement should include every possible characteristic: optical (e.g. reflective vs. diffusive, transparent vs. opaque), textural (e.g. rough vs. smooth), chemical (e.g. polymer identification), and statistical (e.g. counts, sizes, shapes, and their distributions). Optical microscopy provides imaging data for assessing the counts, the sizes, and the morphology of microplastics. Raman spectroscopy is a versatile technique used to identify the chemical makeup of matter, such as the presence of minerals, biologicals, organics, and polymers. Raman microscopy combines two techniques, provides a wide range of measurement characteristics within one platform, and can comprise analysis down to submicron scales. In this chapter, we will follow the workflow that uses a Raman microscope to analyze microplastics down to a sub-micron scale. We will introduce nanoparticle tracking analysis (NTA) to determine nanoplastic particles in tandem analysis with Raman microscopy [18].

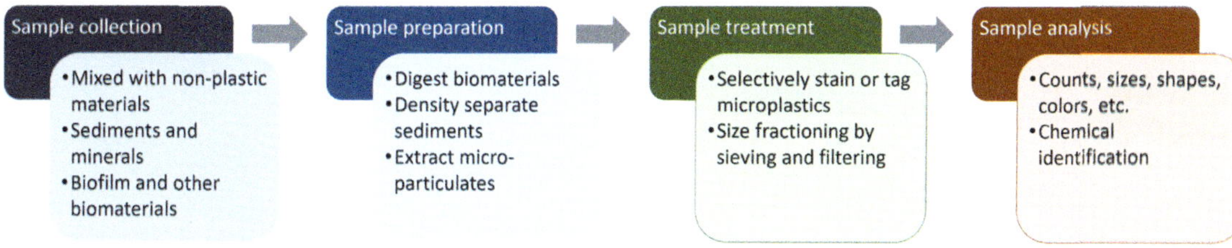

Figure 15.1 Schematics of the typical microplastics analysis workflow. *Source*: Eunah Lee author.

15.3 Sample Treatment (Selective Staining)

The American Chemistry Council (ACC) reported that microplastics represent only 4% of particles collected from the environment on average, and in some cases, much less [19]. The other 96% comprises natural materials like minerals, dirt, and sand, insect parts, pollen, and more. Whether we accept the percentage or not, it is clear that not all of the particles found in the environment are microplastics, as naturally occurring particles also exist. Conversely, by screening particles to include only microplastics for analysis, the throughput of analysis would be significantly improved. Screening samples is a method of subsampling (selecting a subset of a sample), but there is no impact on statistical representativeness because the whole population of microplastics is thoroughly analyzed.

Maes et al. proposed selectively tagging microplastics with Nile Red [20]. The proposed method involves observing particles stained with Nile Red solution with an optical or Raman microscope through a yellow-orange mask (529 nm) and under blue/green light (450–510 nm). This highlights Nile Red-stained microplastics for subsequent analysis, such as acquiring counts, sizes, and shapes from the micrographs. For nanoplastics ranging from 1 nm to 1 μm where diffraction-limited optical microscopy is impractical, NTA technique can be used to obtain similar information.

The NTA uses lasers to determine particle movement and infers particle size using the Stokes–Einstein relationship. NTA illuminates the sample (e.g. particles suspended in water) with a red, green, and/or blue laser and detects elastically scattered light (i.e. Rayleigh scattering) from particles moving by Brownian motion. NTA tracks particle displacements and analyzes them to determine size distribution and number concentration as results. Since the illuminated sample volume and the number of particles imaged are well known, particle number concentration is readily determined. Thus, two critical pieces of information are determined from a single measurement: particle size distribution and particle concentration, even for polydisperse samples.

The NTA can be expanded to selectively detect fluorescing particles by applying optical filters, as demonstrated in Figure 15.2. Since Nile Red excites in green and emits in red, NTA can selectively analyze Nile Red stained particles by using a green laser and applying a long-pass filter that only passes red light. Nile Red fluorescence emission profiles vary depending on the solvent [21, 22]. Thus, it is necessary to select the wavelength of the long-pass filter (and emission mask) accordingly.

We prepared a target sample by mixing the same volumes of unstained microspheres (Polysciences, Warrington, Pam USA) and Nile Red stained microspheres (Spherotech, Lake Forest, IL, USA). The sample was then diluted by 1000× with reverse osmosis (RO) water. The control sample consisted of unstained microspheres diluted the same way. We recorded the fluorescing particle size distributions of the target and control samples using ViewSizer 300 (HORIBA, Irvine, CA, USA) with a long-pass filter (NewPort, Irvine, CA, USA) that passes light longer than 650 nm only.

The particle size distribution plot in Figure 15.3 shows the mean diameter of fluorescing particles is 960.88 nm in the target sample with a concentration of 7.00×10^9 particles/ml. The results can also serve to screen samples for further analysis. The number of samples for subsequent chemical analyses with,

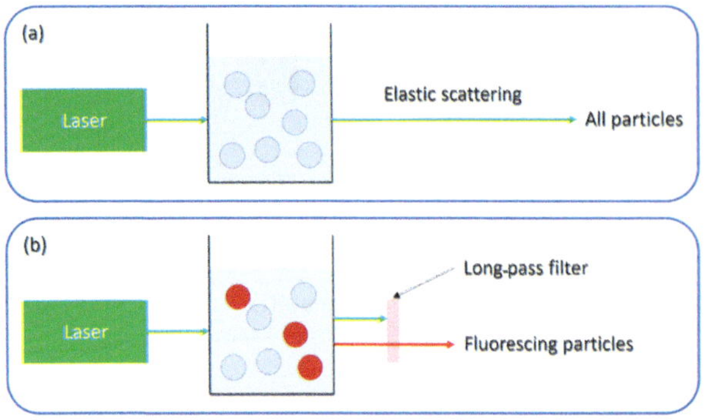

Figure 15.2 Schematics comparing (a) standard NTA to analyze all particles in the sample and (b) selective NTA to analyze fluorescing particles. Fluorescing particles are microplastics that have been tagged selectively with Nile Red. *Source*: Eunah Lee author.

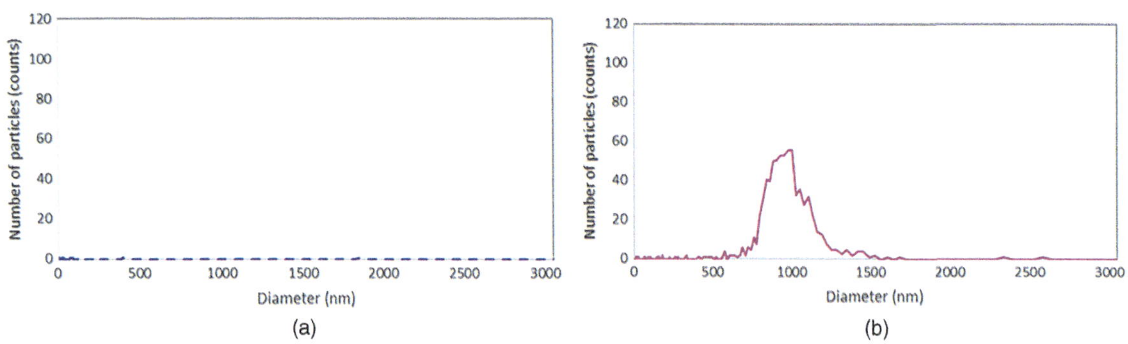

Figure 15.3 Size distributions fluorescing particles in (a) the control sample containing unstained microspheres only and (a) the target sample containing unstained and Nile Red stained microspheres in equal volumes. Excited with a green laser (@ 520 nm) and measured with the long-pass filter (@ 650 nm). *Source*: Eunah Lee author.

for example, Raman microscopy, can be reduced by proceeding with samples containing high fluorescing particle concentrations and thus containing large amounts of nanoplastics. This will improve the speed and effectiveness of the overall analysis. Raman microscopy can acquire spectra from nanoparticles such as carbon nanotubes for chemical analysis even if the target size is smaller than the diffraction limit [23].

Several other studies describe alternative methods to improve the selectivity and detection limits using different types of dyes, fluorophores, and near-infrared spectroscopy. The principle remains the same: leveraging spectroscopic properties to screen and isolate microplastics for analysis to minimize the time and effort spent on non-microplastic materials.

15.4 Subsampling

ISO/DIS 16094-2 proposes four (4) models for selecting the surface area of a filter as a statistical representation in Section 9.6.2 when analyzing particles retained on the filter with a microscope (e.g. optical or Raman microscope). These are the Total Model, the Random Model, the Cake Model, and the Spiral Model.

To assess the statistical representativeness of these models, we created simulated data to represent the entire sample. We then selected a subset according to the sampling instructions per model and compared the results (e.g. counts and average sizes) to calculate deviations. The smaller the deviations, the better the statistical representations. Deviations are calculated with the equations below. For examples and exercises in this chapter, we will accept deviations of 5% or less.

$$\text{Deviation in average size} = \frac{|\text{average size of all points} - \text{average size of selected points}|}{\text{average size of all points}} \tag{15.1}$$

$$\text{Deviation in counts} = \frac{|\text{expected counts per sampling ratio} - \text{counts of selected points}|}{\text{expected counts per sampling ratio}} \tag{15.2}$$

15.4.1 Total Model

The total model does not subsample. The total model instructs to analyze all particles retained on the filter. Statistical representativeness is not relevant. Its use is designated to small quantity samples with 1000 particles or less. A small number of particles ensures that analysis is completed within a reasonable time.

15.4.2 Random Model

The random model creates a subsample by counts and establishes a subsample using a predetermined number of randomly selected particles. The model is designated for large quantity samples with 1000 particles or more. This model requires a constraint to avoid repeat selections, as truly random selection does not exclude repeat selections.

To assess statistical representativeness, we created a simulation data set of 10,000 "particles." Using the RANDARRAY function of Excel, we assigned arbitrary sizes to each particle, ranging from 1 μm to 5000 μm in diameter to match the defined size range of microplastics. We then selected 1000 particles randomly using the INDEX, RANDARRAY, and SEQUENCE functions of Excel, avoiding repeat selections [24]. This represents a 10% subsampling by counts. Using the equation above, we calculated the deviation in average sizes (all particles vs. selected particles). We then repeated the exercise nine more times for a total of 10 reproductions. Count deviations are irrelevant to this model as the number of counts in subsampling is designated but not determined.

Table 15.1 compares the average size of selected particles using the Random Model method with the average size of all particles. The close agreement demonstrates the effectiveness of subsampling using the Random model and that it is statistically representative of the whole. Correspondingly, the deviations shown in Figure 15.4 range from 0.12% to 2.60%, which, as noted above, is within acceptable limits.

Statistical representativeness holds well even when the subsampling ratio decreases, as shown in Figure 15.5. We created two more sets of simulation data the same way with 50,000 particles and 100,000 particles. Selecting 1000 particles in both simulations, they represent 2% and 1% subsampling by counts.

Table 15.1 Random model simulation results.

Reproductions	1	2	3	4	5	6	7	8	9	10
Average sizes of all particles (μm)	2489.9	2501.8	2519.7	2509.9	2500.9	2497.3	2515.8	2507.0	2509.0	2492.1
Average sizes of selected particles (μm)	2457.5	2498.0	2501.9	2554.2	2565.9	2525.2	2512.7	2461.4	2522.8	2499.5

The simulation was reproduced 10 times. In each simulation, the average size of all particles represents the expected result, and that of selected particles is the determined result.

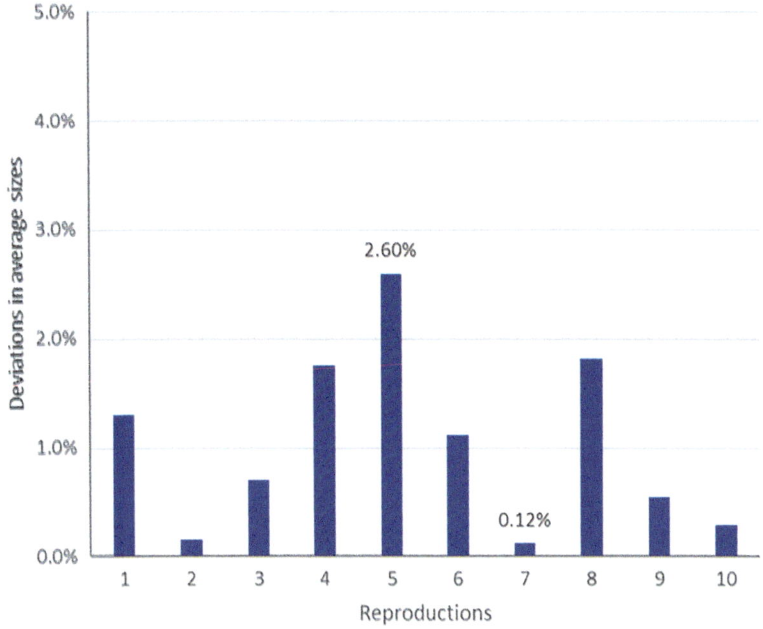

Figure 15.4 Random model simulation deviation results. The simulation was reproduced 10 times. In each simulation, the deviation in average size was calculated using Eq. (15.1). *Source*: Eunah Lee, Author.

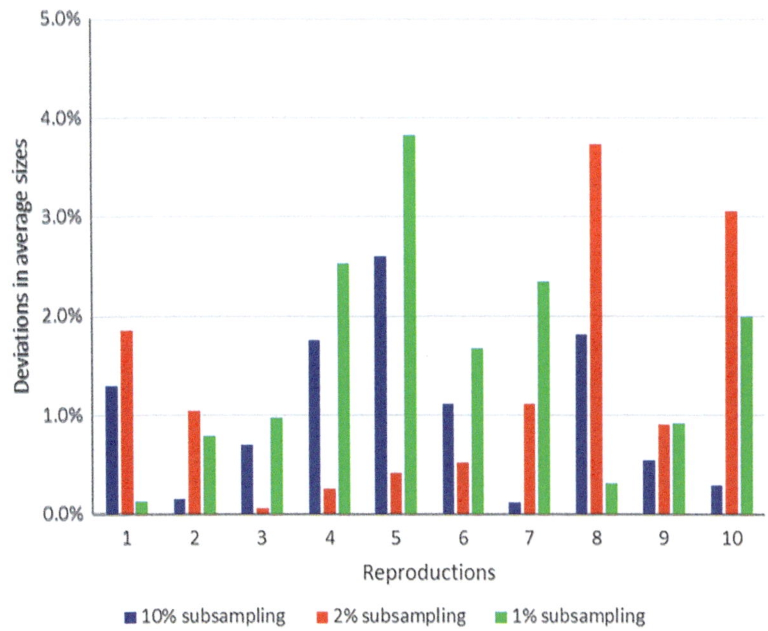

Figure 15.5 Random model simulation results with 10%, 2%, and 1% subsampling ratios. For each case, the simulation was reproduced 10 times. In each simulation of each case, the deviation in average size was calculated using Eq. (15.1). *Source*: Eunah Lee, author.

15.4.3 Cake Model

The cake model subsamples by area and assumes that the filters are circle-shaped, as depicted in Figure 15.6. It instructs the analysis of particles in a subset of the filter area in the shape of a slice of cake. It is designated for large-quantity samples with 1000 particles or more. It requires analyzing 20% or more for particles down to 5 μm and 1% to 4% or more for particles down to 1 μm of the entire filter area.

To determine the statistical representativeness for particles down to 5 μm, we created a simulation of 5000 particles distributed on a circle-shaped filter of 50 mm in diameter (25 mm radius) as shown in Figure 15.7.

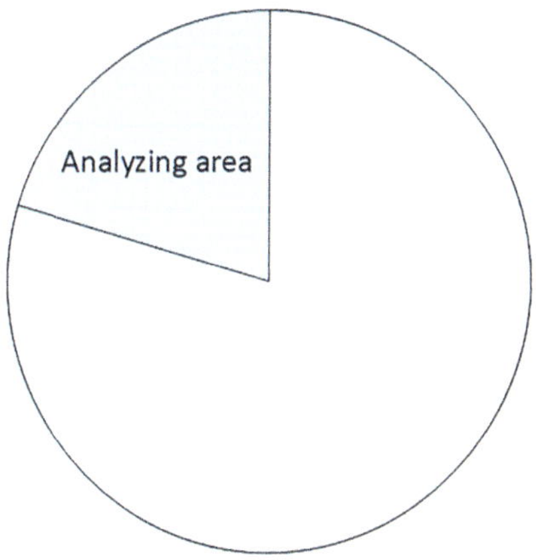

Figure 15.6 Schematic diagram of cake model. *Source*: Eunah Lee, Author.

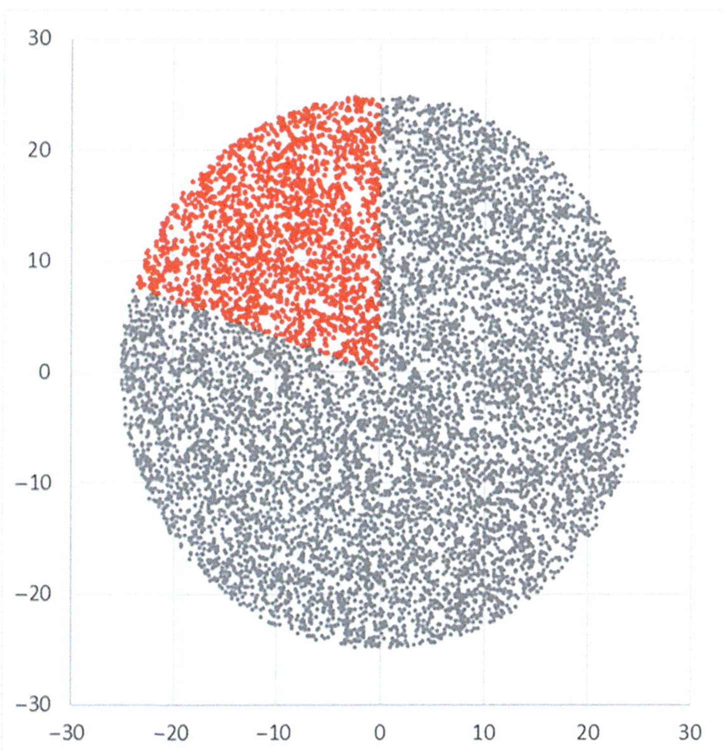

Figure 15.7 Cake model simulation schematics. Red dots in the "slice of cake" represent particles selected for analysis. *Source*: Eunah Lee, Author.

For particles down to 5 μm, we assigned arbitrary sizes to particles ranging from 5 to 5000 μm. We then selected particles within a "slice" of the filter, 20% by area and 72° in angle. Gray dots represent all particles on the "filter" in this simulation data. Red dots represent selected particles in subsampling.

We counted the number of particles within the area. If the statistical representativeness is ideal, the expected value is 2000. We calculated the average size of selected particles and compared it to that of all particles. We calculated deviations in counts and average sizes. Table 15.2 and Figure 15.8 show results from the simulation process reproduced 10 times.

In this simulation, the statistical representativeness is good. Deviations in average particle sizes range from 0.10% to 1.86%. Those in counts range from 0.10% to 2.00%.

Table 15.2 Cake model simulation results.

Reproductions	1	2	3	4	5	6	7	8	9	10
Average size of all particles (μm)	2498.4	2497.7	2507.9	2487.9	2516.2	2499.7	2490.1	2510.7	2504.1	2495.0
Average size of selected particles (μm)	2485.1	2483.6	2475.6	2456.2	2513.7	2453.1	2478.4	2490.0	2486.7	2497.9
Expected counts of selected particles	2000	2000	2000	2000	2000	2000	2000	2000	2000	2000
Actual counts of selected particles	1970	1976	1979	1962	1971	1978	1998	1960	2016	2022

In each simulation, the average size of all particles represents the expected result, and that of selected particles is the measured result. In each simulation, the expected counts of selected particles is constant by design and compared to the actual counts of selected particles.

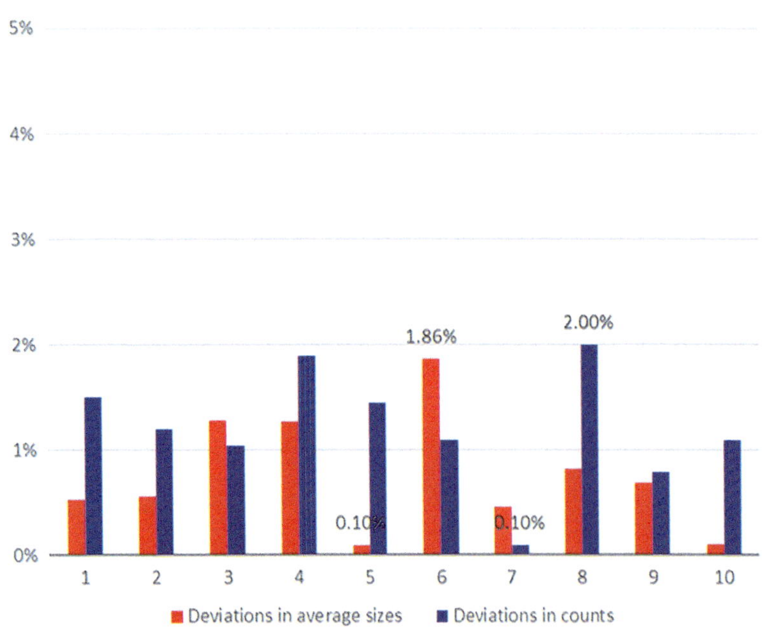

Figure 15.8 Cake model simulation results. The simulation was reproduced 10 times. In each simulation, the deviation in average size was calculated using Eq. (15.1), and that in counts using Eq. (15.2). *Source*: Eunah Lee, author.

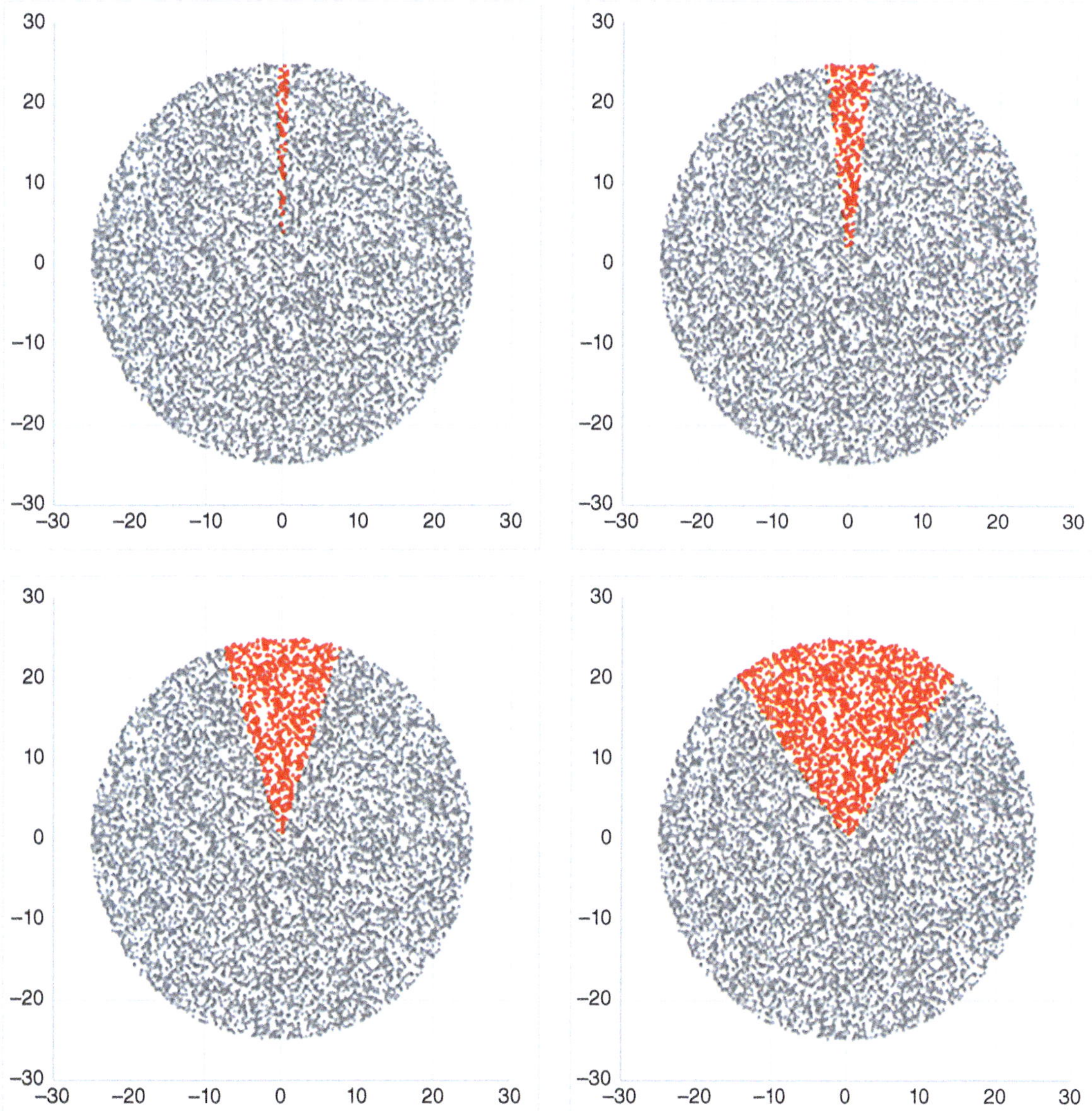

Figure 15.9 Cake model simulation schematics. Red dots in the "slice of cake" represent particles selected for analysis. The size of slice increases from left to right to subsample 1% (= 3.6°), 4% (= 14.4°), 10% (= 36°), and 20% (= 72°) of all particles, respectively. *Source*: Eunah Lee, author.

Figures 15.9 and 15.10 show results for particles down to 1 μm. In this range, we reassigned particle sizes between 1 and 5 μm instead of 5 to 5000 μm. Then selected particles within slices of 1% (= 3.6°), 4% (= 14.4°), 10% (= 36°), and 20% (= 72°).

The results show that 1% subsampling is not satisfactory. Deviations from expected values often exceed the threshold (defined for this chapter) of 5% for average sizes and counts. The results demonstrate that 4% subsampling, at the very least, is recommended to achieve acceptable statistical representativeness.

Based on these simulation results, statistical representativeness improves as the subsampling ratio increases. So why not go for a high subsampling ratio at all times? Because of the time penalty. For example,

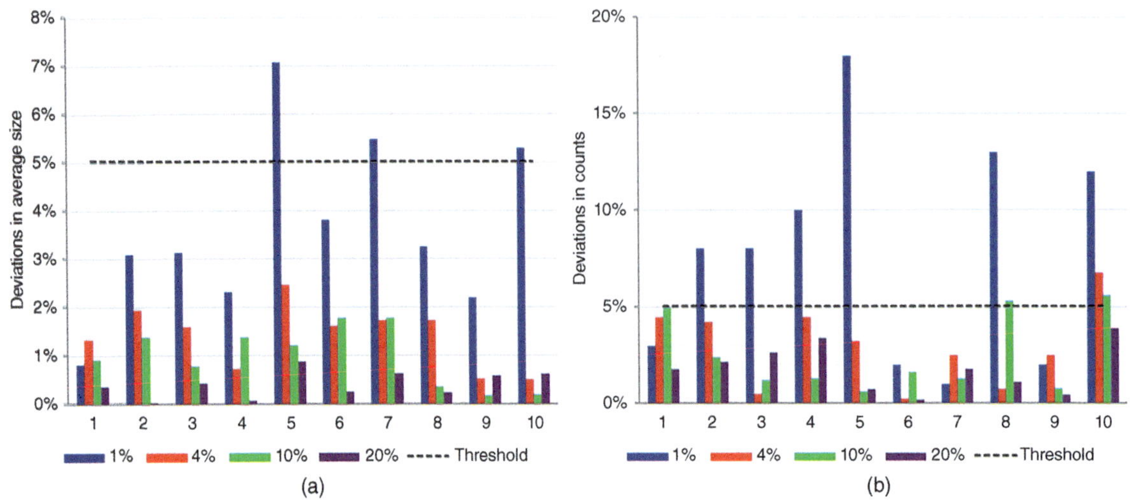

Figure 15.10 Cake model simulation results with 1%, 4%, 10%, and 20% subsampling ratios. (a) Deviations in average sizes. (b) Deviations in counts. For each case, the simulation was reproduced 10 times. In each case simulation, the deviation in average size was calculated using Eq. (15.1), and that in counts using Eq. (15.2). *Source*: Eunah Lee, author.

analyzing 20% of the filter area takes five times longer than analyzing 4%. Deviations do decrease, indicating the statistical representativeness does improve. However, is it worth the time invested when results from 4% subsampling show "acceptable" statistical representativeness? It is ultimately the analysts' discretion based on the analytical objective (e.g. high throughput at high speed vs. detailed characterization at high accuracy), but it is worth considering the return on the investment carefully.

15.4.4 Spiral Model

The spiral model subsamples by area. The spiral model instructs the analysis of particles in several small areas distributed in a spiral pattern, as depicted in Figure 15.11.

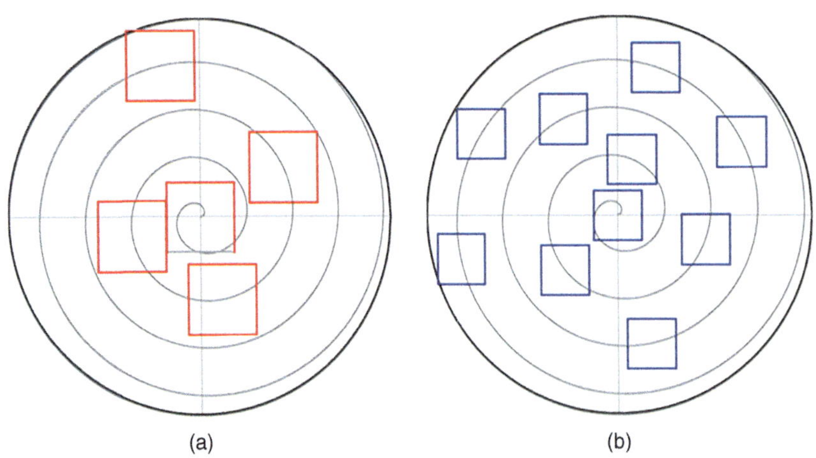

Figure 15.11 Schematic diagrams of spiral model. It is possible to achieve the same subsampling ratio in different schematics. For example, one can achieve 20% subsampling with (a) five areas of 4% (annotated as red squares) or (b) 10 areas of 2% (annotated as blue squares). *Source*: Eunah Lee, author.

Its use is designated for large-quantity samples with 1000 particles or more. For microplastics down to 5 μm, the model requires analyzing five or more "sub-areas" that collectively cover 20% or more of the entire filter area. For microplastics down to 1 μm, the model specifies the collective area of analysis to be 1% or more of the whole filter area. There are no constraints to the size or shape of sub-areas. It is the analysts' discretion to analyze, for example, five "large" sub-areas, each of which is 4% of the filter area, or 10 "small" sub-areas, each of which is 2% of the filter area, both scenarios meeting the requirement of analyzing 20% of the filter area.

To compare the statistical representativeness of the two scenarios, we created a simulation data set made of 10,000 "particles," sizes ranging from 5 to 5000 μm, evenly distributed on a 25 mm diameter circular filter. The filter area is 1963.5 mm^2, 20% of which is 392.7 mm^2. The simulation is performed for two scenarios – five large sub-areas (78.54 mm^2) vs. 10 small sub-areas (39.27 mm^2). Each large sub-area, annotated as red squares in Figure 15.12a, is 78.54 mm^2. Each small sub-area, annotated as blue squares in Figure 15.12b, is 39.27 mm^2. We repeated the simulation 10 times. The results for the average size are presented in Table 15.3, and the results for counts in Table 15.4.

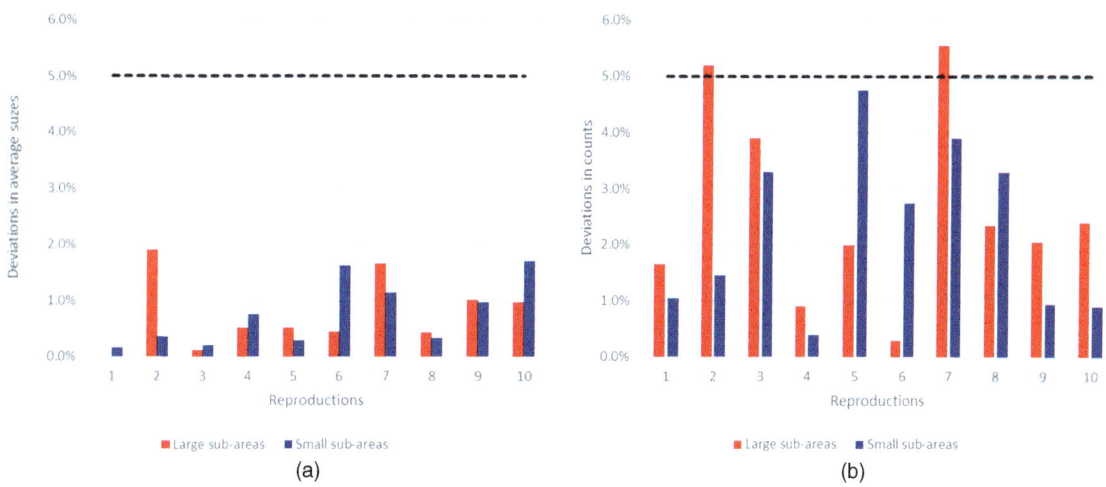

Figure 15.12 Spiral model simulation results with five large sub-areas and 10 small sub-areas. (a) Deviations in average sizes. (b) Deviations in counts. For each case, the simulation was reproduced 10 times. In each scenario simulation, the deviation in average size was calculated using Eq. (15.1), and that in counts using Eq. (15.2). *Source*: Eunah Lee, author.

Table 15.3 Spiral model simulation results of average sizes.

Reproductions	1	2	3	4	5	6	7	8	9	10
Average sizes of all particles (μm)	2498.4	2497.7	2507.9	2487.9	2516.2	2499.7	2490.1	2510.7	2504.1	2495.0
Average sizes of particles selected by large sub-areas (μm)	2498.0	2449.9	2511.0	2500.7	2529.1	2511.0	2531.4	2521.5	2529.7	2470.8
Average sizes of particles selected by small sub-areas squares (μm)	2502.8	2488.7	2502.8	2468.9	2523.5	2459.0	2518.6	2519.2	2479.6	2537.7

In each simulation, the average size of all particles represents the expected result, and that of selected particles the measured result.

Table 15.4 Spiral model simulation results of counts.

Reproductions	1	2	3	4	5	6	7	8	9	10
Expected counts of selected particles*	2000	2000	2000	2000	2000	2000	2000	2000	2000	2000
Counts of particles selected by large sub-areas	1967	1896	2078	1982	1960	2006	1889	2047	1959	2048
Counts of particles selected by small sub-areas	1979	1971	1934	2008	2095	2055	1922	1934	1981	1982

In each simulation, the expected count of selected particles is constant by design and compared to the actual count of selected particles.

Both scenarios presented acceptable and comparable statistical representativeness in average sizes. However, the scenario where five large sub-areas were selected fails in its statistical representativeness in counts per the threshold deviation limit set for this simulation. The simulation results indicate the scenario of 10 small sub-areas is a better choice than five large sub-areas to achieve acceptable statistical representativeness.

Based on these simulation scenarios, the random subsampling model provides the best statistical representativeness. It maintained acceptable statistical representativeness down to 1% subsampling.

15.5 Visual Examination

Visual examination is usually the first step in microplastic analysis. It can become systematic and automated by capturing and analyzing a "good" image (or micrograph) with an optical microscope equipped with a digital camera. Commercial software is available to analyze micrographs to report count, size, color, and morphology [25].

15.5.1 Objective Lens Selection

The size range of microplastics is often three orders of magnitude wide (1 to 5000 μm). It is impossible to examine microplastics of all sizes with the same methods. ASTM and CA Waterboard instruct to fraction microplastics by size with filters and sieves as a part of sample preparation. They specify size fractions be <1 μm (nanoplastics), 1 to 20 μm, 20 to 212 μm, 212 to 500 μm, and 500 μm < with 5000 μm as the upper cut-off for microplastics by definition. Microplastics in the size fraction of 1 to 20 μm are subvisible and require a high magnification objective lens to examine. On the other hand, microplastics in the size fraction of 212 to 500 μm can be examined with a low magnification objective lens. Selecting the "correct" objective lens to record a good micrograph and perform accurate, effective, and efficient analysis is critical. This section will review the considerations for selecting the correct objective lens based on optical and practical criteria. Please note that scenarios in this section assume filters are directly presented to the microscope with retained microplastics. ASTM and CA Water board procedures allow microplastics to be transferred from filters to substrates at the discretion of the analyst.

Commercially available filters include polycarbonate membranes (e.g. circular-shaped, diameter = 25 mm or 47 mm), Si wafers (e.g. square-shaped, 10 mm × 10 mm), and SiN membranes (e.g. square-shaped, 3.5 mm × 3.5 mm, embedded in circular disk). The field of view (FOV) size varies from one microscope to the next, depending on the optical design and configuration. Still, it is typically on the order of millimeters

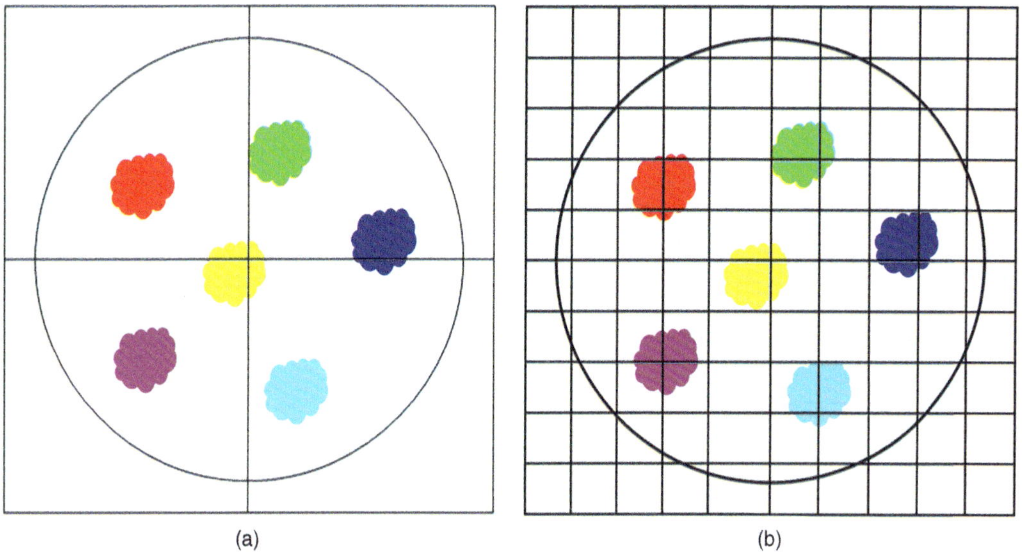

Figure 15.13 Schematics of acquiring an image of the entire filter by recording and combining multiple FOV images. (a) When using a low magnification objective lens (e.g. 10×), it takes a small number of images (e.g. 4) and thus a short time to acquire the image of the entire filter. (b) When using a high magnification objective lens (e.g. 50×), it takes a large number of images (e.g. 100) and thus a long time to acquire the image of the entire filter. *Source*: Eunah Lee, author.

(e.g. 1 mm × 1 mm when 10× objective lens is selected) and smaller (proportional to the magnification of the objective lens). It is safe to assume that no objective lens would have a large enough FOV to measure the entire filter at once. It is necessary to record and stitch multiple FOVs to image the whole filter. The higher the magnification, the smaller the FOV, the more FOVs required to image the filter, and the longer it takes to acquire and stitch the whole filter as depicted in Figure 15.13.

High-speed analysis is favorable, and it would be best to select the lowest magnification available when selecting the objective lens considering the speed of analysis. However, a low magnification objective lens may not offer sufficient resolving power for a small size fraction such as 1–20 μm. Without sufficient resolving power, the analysis results (i.e. counts, sizes, shapes, etc.) would be unreliable. The objective lens selection must take sufficient resolving power for each size fraction into account as well.

Resolution of a microscope is defined "as the shortest distance between two points on a specimen that can still be distinguished by the observer or camera system as separate entities" [26]. Airy disk can be defined as the "diffraction pattern of a circular aperture, as seen in the image plane of a well-corrected lens. The radius of the Airy disk is a measure of the resolution limit of a well-corrected lens" [27].

The equations for the Airy disk diameter, d, and the resolution, r, are below.

$$d = \frac{1.21 \times \lambda}{NA} \tag{15.3}$$

$$r = \frac{d}{2} = \frac{0.61 \times \lambda}{NA} \tag{15.4}$$

$$NA = n \times \sin\theta \tag{15.5}$$

where λ is the wavelength of the light, NA is the numerical aperture of the objective lens, n is the refractive index of the medium (e.g. air), and θ is half of the angular aperture of the objective lens.

Equation (15.4) calculates the theoretical resolution for ideal targets (= points of infinitesimal size) and ideal optics (e.g. 100% reflectance in the mirror and 100% transmittance in the lens). Empirical resolution depends on many factors, including the samples' characteristics (e.g. opaque vs. transparent, smooth vs. rough surface). It is often twice to thrice the best-case estimates and roughly proportional to $1/NA$. Working estimates for the best-case scenario are $r \approx 0.5$ μm and $d \approx 1$ μm with short wavelength (e.g. 500 nm), high numerical aperture (e.g. $NA = 0.9$) objective lens, and well-behaving samples (e.g. clean, flat, reflective, and opaque).

Another aspect of objective lens selection is the depth of field (DOF), defined as "the depth of the specimen layer which is in sharp focus at the same time, even if the distance between the objective lens and the specimen plane is changed when observing and shooting the specimen plane by microscope" [28]. The definition is written for multi-layer specimens but can be applied to samples of rough surfaces or irregular heights, as illustrated in Figure 15.14.

Equation (15.6) describes a modified DOF equation for an optical microscope with a digital camera.

$$DOF = \frac{r \times 250{,}000}{NA \times M} + \frac{\lambda}{2 \times NA^2} \tag{15.6}$$

where r is resolution of the microscope; M is total magnification = magnification of the objective lens × magnification of the lens for the camera; and λ is wavelength in μm.

Assuming the magnification of the lens for the camera is 10×, Table 15.5 provides the calculated DOFs for several common objective lenses (Olympus, Tokyo, Japan) used in this analysis. The manufacturer's specifications for magnification (M) and numerical apertures (NAs) are also provided [29–32].

The practical implication is that targets (in this chapter, microplastics) whose height variances are smaller than the DOF are all in focus at once. In other words, there is no need for auto-focusing of individual targets when their height variances are smaller than DOF, and the analysis speed improves accordingly. Given that auto-focusing is a time-consuming step, finding the focal plane by scanning the distance between the objective lens and the target and selecting the objective lens whose DOF is large enough for the height variances can significantly improve the speed of analysis.

Considering both resolution and DOF, the recommended objective lens selection for the size of microplastics is provided in Table 15.6. This is a best-effort recommendation based on rough estimation and extensive experience. Optimizing each case may result in different objective lens selections.

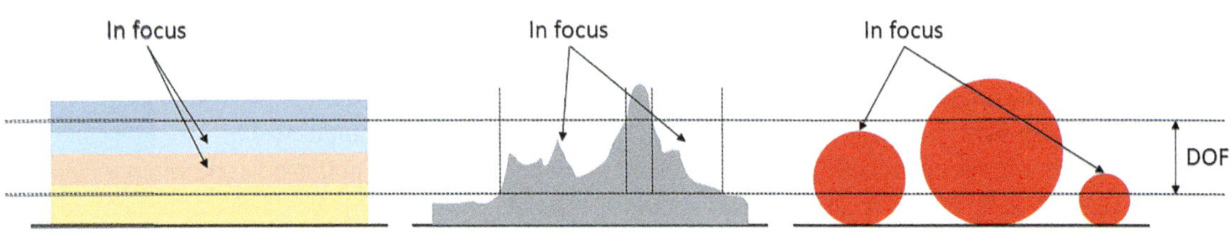

Figure 15.14 Illustration of depth of field for different types of samples. From the left, multi-layer specimen, rough surface, and discrete particles. *Source*: Eunah Lee, author.

Table 15.5 Depths of fields (DOFs) for common objective lenses used for this analysis were calculated with Eq. (15.6).

Manufacturer's objective lens annotation	M	NA	DOF (μm)
MPLN 10×, MPLN 10× BD, LMPLFLN 10×, LMPLNFLN 10× BD	100	0.25	18.26
MPLN 20×, MPLN 20× BD, LMPLNFLN 20×, LMPLNFLN 20× BD	200	0.4	6.04
LMPLFLN 50×, LMPLFLN 50× BD	500	0.5	2.46
MPLN 50×, MPLN 50× BD	500	0.75	1.41
MPLN 100×, MPLN 100× BD	1000	0.9	0.72

Table 15.6 Best-case recommendations for objective lens selection per microplastic sizes.

Objective lens NA	0.1	0.25	0.5	0.75	0.9
Minimum microplastics size	~20 μm	~10 μm	~5 μm	~2 μm	~1 μm

Please note that the recommendation is based on numerical apertures, not on magnifications, as both resolution and DOF are functions of numerical apertures. Considering CA Waterboard and ASTM procedures, 10× (e.g. $NA = 0.25$) and 20× (e.g. $NA = 0.4$) are the best choices for the size fractioning ranges of (>500 μm, 500 μm to 212 μm, and 212 μm to 20 μm). The determination of microplastics smaller than 20 μm and nanoplastics is rapidly increasing [33], and thus, for these particle size ranges, 50× (e.g. $NA = 0.75$) and 100× (e.g. $NA = 0.9$) objectives would be the best choices.

15.5.2 Digital Camera Selection

Digital camera specifications have no theoretical impact on the resolution. The practical effect, however, is significant enough to influence the measurement (i.e. size) results. To illustrate the impact of digital camera specifications, especially the number of pixels and, conversely, the size of pixels, we performed simulations below.

Figure 15.15 considers two targets, separated by one-tenth diameter. Imagine recording a micrograph of these two targets with an optical microscope equipped with a digital camera. The resolution of the microscope is fixed once the objective lens is selected.

We draw each of these two targets with 150 pixels along the diameter and 15 pixels apart. They are clearly defined and resolved. The size and count measurements are accurate. We then reduced the number of pixels and thus increased the size of pixels, using the "Resize" function in Microsoft® Paint, in succession to simulate the impact of digital cameras specifications. We then applied threshold and measured the areas diameters using ImageJ (National Institutes of Health, Bethesda, MD, USA).

The simulation in Table 15.7 shows the worsening image quality as the number of pixels decreases and the size of pixels increases. So why not always get the camera of the small pixels and large number of pixels? The readout time increases as the number of pixels increases, and it can take a while to stitch all FOVs to cover the entire filter. Then what is the best choice? The simulation indicates at least 10 pixels per microplastic to render it satisfactorily.

Table 15.8 further describes the parameters for several commercially available digital cameras. The image pixel size is based on images acquired with an XploRA Raman Microscope (HORIBA, Piscataway, NJ, USA) and LMPLFLN100 × BD (Olympus) objective lens.

If you select an objective lens other than 100×, you can estimate the pixel size in the image based on the inversely proportional relationship. For example, if the 20× objective (one-fifth times magnification) is

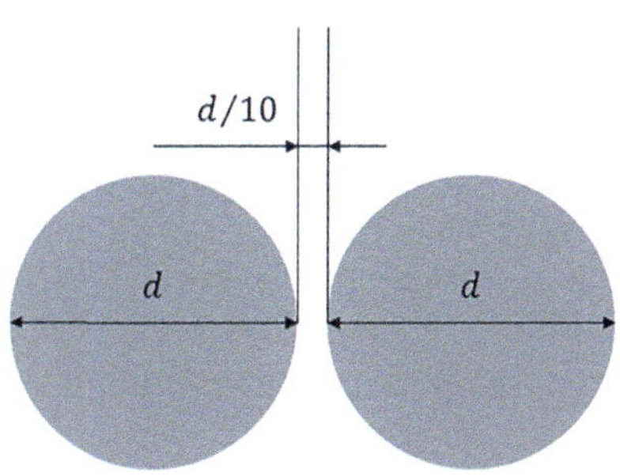

Figure 15.15 Illustration of two targets separated by a short distance. *Source*: Eunah Lee, author.

Table 15.7 Demonstration of data degradation as a function of the pixel counts/sizes of the digital camera.

Number of pixels along the diameter	150 pixels 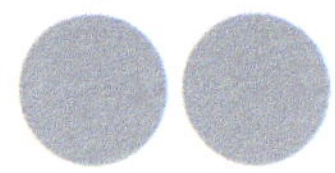	75 pixels 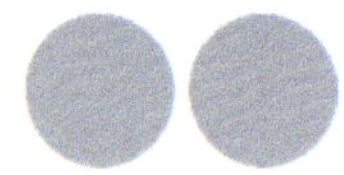	50 pixels 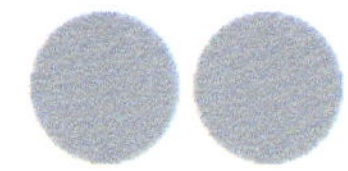
Counts:	2	2	2
Areas:	17,584	17,572.5	17,667
Relative areas:	100%	99.9%	100.5%
Number of pixels along the diameter	25 pixels 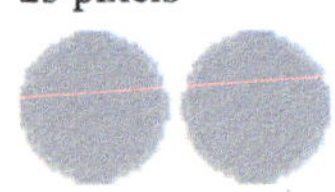	10 pixels 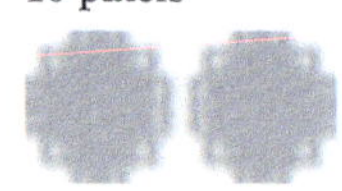	5 pixels 
Counts:	2	2	1
Areas:	17,532	18,000	37,800
Relative areas:	99.7%	102.4%	215.0%

Source: Eunah Lee, author.

Table 15.8 Numbers and sizes of pixels of commercial digital cameras (IDS Imaging, Obersulm, Germany).

	Number of pixels	**Optical area**	**Pixel size Physical vs. image**
UI-1220LE-C-HQ Rev.2	0.36 Mpix = 752 × 480	4.512 mm × 2.880 mm	6 μm vs. **0.10 μm**
U3-3880LE Rev.1.2	6.41 MPix = 3088 × 2076	7.411 mm × 4.982 mm	2.40 μm vs. **0.05 μm**
UI-3590LE-C-HQ Rev.2	18.10 MPix = 4912 × 3684	6.140 mm × 4.605 mm	1.25 μm vs. **0.025 μm**

selected, the pixel size in the image, when recorded by the UI-1220LE-C-HQ Rev.2 camera would be 0.5 μm (five times the pixel size in the image).

As an exercise, let's consider recording images of microplastics that are 20 μm (and larger) using 20× objective lens for visual examination. Which of the three cameras in Table 15.8 would be a good choice? The simulation in Table 15.7 indicates that 10 (or more) pixels across the target, 20 μm in this exercise, are necessary for good image quality and reliable measurement. The pixel sizes in the image (highlighted in bold) recorded with 20× objective lens and using the three cameras listed in Table 15.8 are 0.5, 0.25, and 0.125 μm from the top. The numbers of pixels across 20 μm are 40, 80, and 160, respectively, all of which are larger than 10. All three cameras are good choices, with the third one (UI-3590LE-C-HQ Rev.2) being the best choice.

15.6 Chemical Identification

Advantages of Raman spectroscopy as a chemical identification method include its non-destructive nature (compared to pyrolysis–gas chromatography–mass spectrometry), high spatial resolution (compared to mid-infrared spectroscopy), and high chemical specificity compared to other spectroscopy technologies. Raman microscopy combines optical microscopy with Raman spectroscopy. The workflow can be automated to capture an image of microplastics on a filter (or a substrate) and then analyze the image to identify locations, counts, sizes, shapes, and colors of the microplastic particles and then acquire Raman spectra from them.

Microplastics are identified by searching and matching their Raman spectra using reference spectral libraries. Figure 15.16 provides a real-world example of data collected with a Raman microscope. We

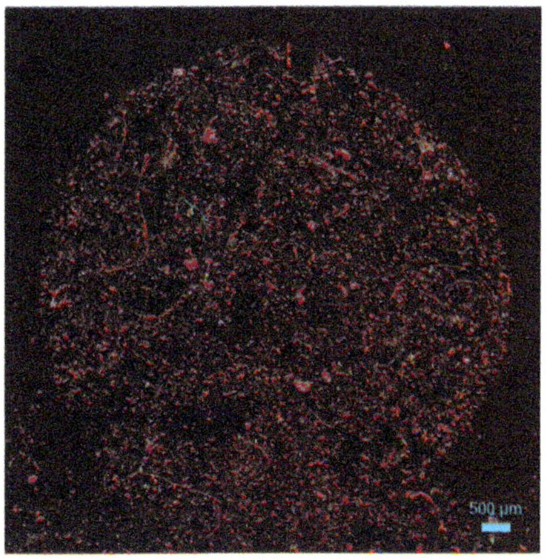

(a)

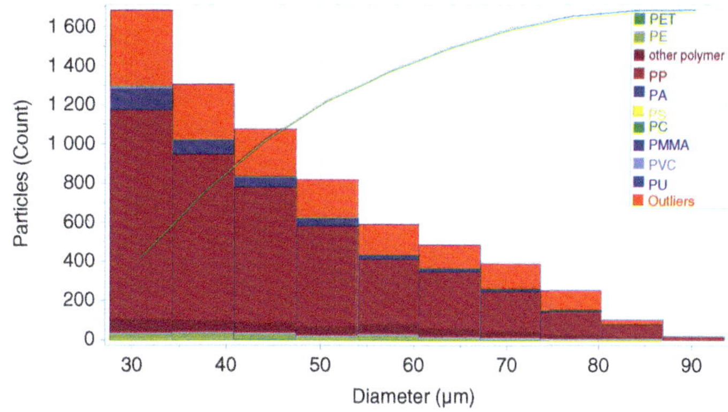

(b)

Figure 15.16 (a) A micrograph of microplastics retained on a Si filter after vacuum suction filtration. Microplastics are colored in red, while the filter appears in black. (b) Histogram of the number of particulates as a function of area equivalent diameter. Please note that each bin is colored proportional to each polymer's number ratio. *Source*: Author: Eunah Lee.

prepared the sample by putting pellets of polymers (Scientific Polymer Products Inc., Ontario, NY, USA) in a clean glass bottle, shaking it vigorously to induce break-outs, removing (bulk) pellets, adding RO water to the bottle, shaking it gently to separate microplastics created from the inner surface, and then filtered the suspension to retain microplastics on the filter.

The data matrix includes Raman spectra from all microplastics (a Raman spectrum per microplastic particulate, Total Model from Section 15.4.1, no subsampling), based on chemical identification made by searching and matching reference spectra of microplastics in the Raman found in the library [34]. Table 15.9 includes chemical identification and quantification by size using Raman Microscopy. This is a different presentation of the same data used in Figure 15.16

Table 15.9 Size distributions of individual polymers with customized size fractions.

	Diameter ranges (μm)				
Polymers	**20...40**	**40...60**	**60...80**	**>80**	**Counts per polymer**
PET	2	11	5	0	18
PE	75	77	28	0	180
PP	2011	1613	517	21	4162
PA	101	57	19	0	177
PS	2	1	0	0	3
PC	0	0	1	0	1
PMMA	81	59	20	1	161
PVC	24	15	5	0	44
PU	0	1	0	0	1
Other polymers	129	163	65	1	358
Unidentified	695	620	313	7	1635
Counts per diameter range	3120	2617	973	30	Total = 6740

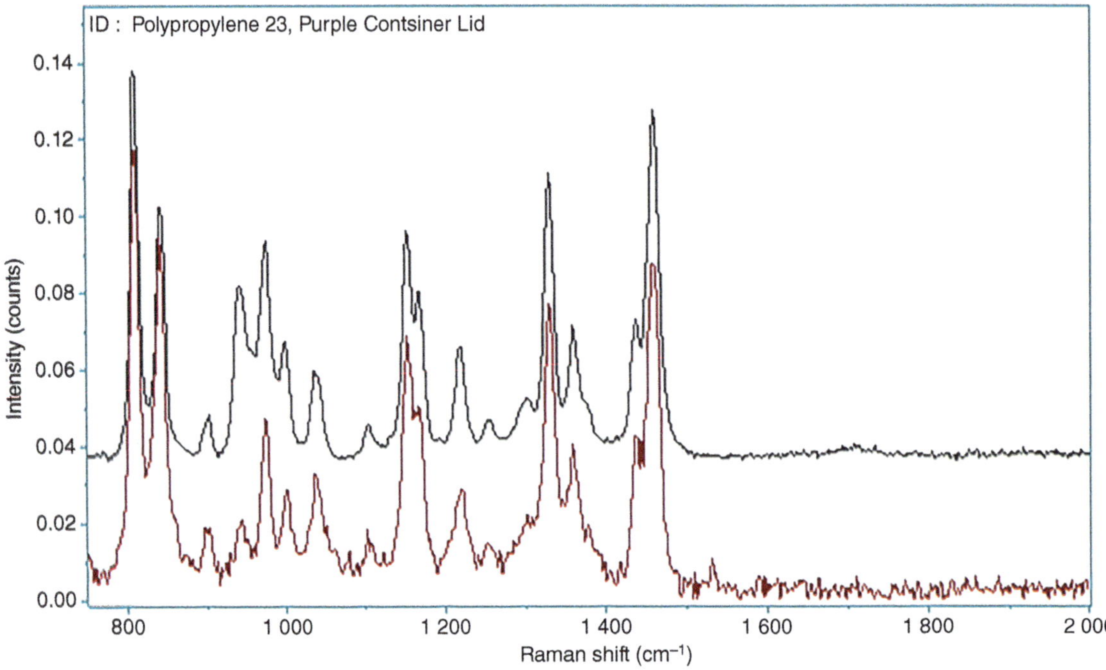

(a)

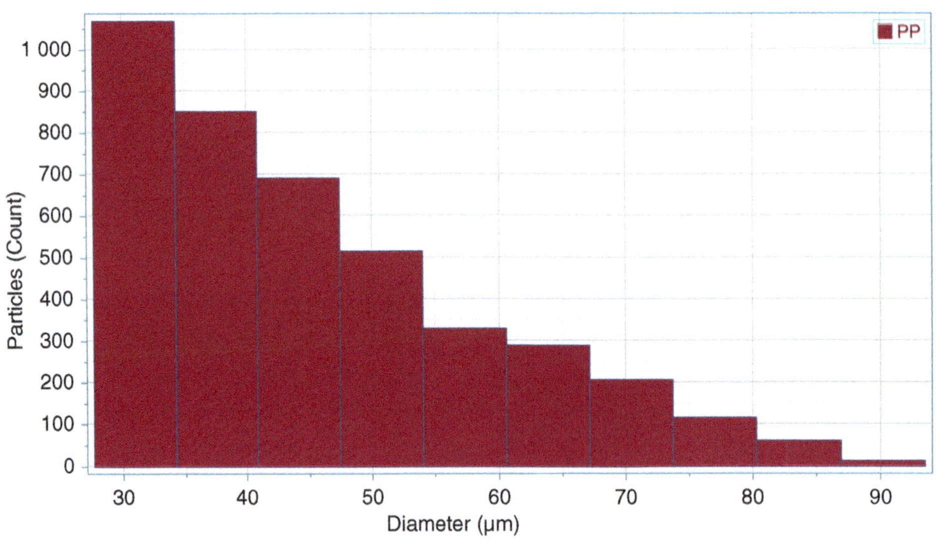

(b)

Figure 15.17 (a) A microplastic spectrum (black) overlaid of a reference spectrum of polyethylene (dark red). The microplastic is identified as PE. (b) Size distribution of all microplastics identified as PE. *Source*: Eunah Lee, author.

We selected polyethylene (PE) and polyethylene terephthalate (PET) as examples to present the individual results in Figures 15.17 and 15.18, respectively. The particle spectra for PE and PET show good agreement with reference spectra in the Raman library, and the size distribution (counts vs. size) for each type of microplastic identified is also good.

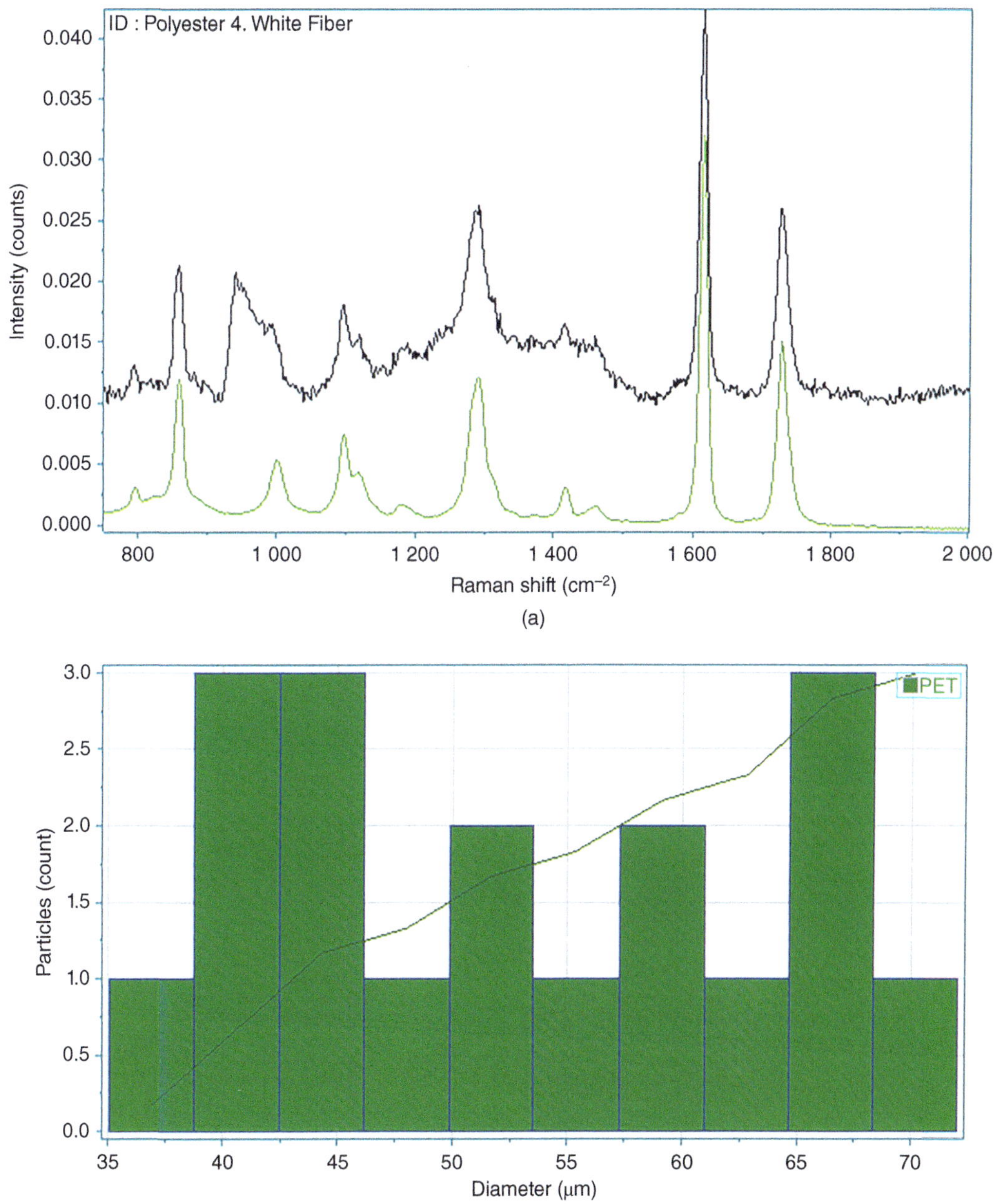

Figure 15.18 (a) A microplastic spectrum (black) overlaid with a reference spectrum of polyethylene terephthalate (dark green). The microplastic is identified as PET. (b) Size distribution of all microplastics identified as PET. *Source*: Author Eunah Lee.

15.7 Conclusion

This chapter reviewed regulatory agency methods to screen and select microplastics for analysis. This review included procedures for sample preparation, models to subsample based on filter areas at sample presentation, and parameters to consider when selecting objective lenses and digital cameras at the beginning of sample analysis. Based on extensive research in this field by the authors and the simulated data provided for analyzing subsampling protocols, the following conclusions are presented.

- Selective fluorescence staining (and other optical or spectroscopic technologies) is an effective way to increase the (overall) speed of analysis by selecting samples that contain high concentrations of microplastics only for analysis.
- The Random subsampling model seems to be the best, holding statistical representativeness even when the sampling ratio (the number of analyzed particles/all particles collected) is reduced to 2%.
- Numerical aperture, as well as magnification, is a critical criterion when selecting the most effective objective lens. For example, the most efficient objective lens is $NA \sim 0.25$ (e.g. 10× objective lens) when analyzing microplastics down to 20 μm.
- If the numerical aperture is the same (or close) for two objective lenses, the lower magnification is preferable to the higher magnification because the former has a larger FOV. For example, to analyze microplastics down to 5 μm, the best objective lens is $NA \sim 0.4$ or 0.5, which presents two choices: 20× and 50× long working distance (LWD). Between the two, 20× is preferable because of its larger FOV. It will be faster to record the image of the entire sample area.
- Raman microscopy is a well-established technology used for chemical identification. Modern Raman microscopes equipped with a motorized stage and digital camera can be programmed and automated to efficiently and effectively analyze microplastics for quantitative count information and chemical identification.

Acknowledgments

The authors wish to thank Ms. Yujeong Park and Dr. Andrew Whitely for their support, wisdom, and contribution. The authors also wish to thank HORBIA Instruments for the funding and instrument uses.

References

1 Geyer, R., Jambeck, J., and Law, K. (2017). Production, use, and fate of all plastics ever made. *Science Advances* 19 (3): e1700782.
2 Haederle, M/University of New Mexico (2024). Microplastics in Every Human Placenta, New UNM Health Sciences Research Discovers [Internet]. Albuquerque, NM: University of New Mexico Health Sciences Center Newsroom; Feb 20, [cited 2024 Dec 2]. Available from: https://hsc.unm.edu/news/2024/02/hsc-newsroom-post-microplastics.html (accessed 15 December 2024).
3 ASTM Fact Sheet [Internet]. West Conshohocken, PA: ASTM International; [cited 2024 Dec 4]. Available from https://www.astm.org/about/overview/fact-sheet.html.
4 About ISO [Internet]. Geneva, CH: ISO; [cited 2024 Dec 4]. Available from https://www.iso.org/about.
5 Microplastics [Internet]. Sacramento, CA: California State Water Resources Control Board; [updated 2024 Aug 20; cited 2024 Dec 4]. Available from: https://www.waterboards.ca.gov/drinking_water/certlic/drinkingwater/microplastics.html.
6 State Water Resources Control Board Resolution No. 2020-0021 (2020). Adoption of Definition of 'Microplastics in Drinking Water' [Internet]. Sacramento, CA: California State Water Resources Control Board; [posted 2020 Jun 16; cited 2024 Dec 4]. Available from: https://www.waterboards.ca.gov/board_decisions/adopted_orders/resolutions/2020/rs2020_0021.pdf.

7 Standard Practice for Collection of Water Samples with High, Medium, or Low Suspended Solids for Identification and Quantification of Microplastic Particles and Fibers [Internet]. West Conshohocken, PA: ASTM International; [updated 2020 Aug 14; cited 2024 Dec 4]. Available for purchase from: https://www.astm.org/d8332-20.html.

8 Wong C/Southern California Coastal Water Research Project, Coffin S/California State Water Resources Control Board. SWB-MP2-rev1 Standard Operating Procedures for Extraction and Measurement by Raman Spectroscopy of Microplastic Particles in Drinking Water [Internet]. Sacramento, CA: California State Water Resources Control Board; [revised 2022 May 27; cited 2024 Dec 4]. Available from: https://www.waterboards.ca.gov/drinking_water/certlic/drinkingwater/documents/microplastics/swb-mp2-rev1.pdf.

9 Wong C/Southern California Coastal Water Research Project, Coffin S/California State Water Resources Control Board. SWB-MP1-rev1 Standard Operating Procedures for Extraction and Measurement by Infrared Spectroscopy of Microplastic Particles in Drinking Water [Internet]. Sacramento, CA: California State Water Resources Control Board; [revised 2022 May 27; cited 2024 Dec 4]. Available from: https://www.waterboards.ca.gov/drinking_water/certlic/drinkingwater/documents/microplastics/swb-mp1-rev1.pdf.

10 About Southern California Coastal Water Research Project [Internet]. Costa Mesa, CA: Southern California Coastal Water Research Project; [cited 2024 Dec 4]. Available from: https://www.sccwrp.org/about/.

11 Resources for Microplastics Research [Internet]. Toronto, ON: University of Toronto, Rochman Lab; [cited 2024 Dec 4]. Available from: https://rochmanlab.wordpress.com/resources-for-microplastics-research/.

12 Standard Practice for Preparation of Water Samples with High, Medium, or Low Suspended Solids for Identification and Quantification of Microplastic Particles and Fibers Using Raman Spectroscopy, IR Spectroscopy, or Pyrolysis-GC/MS [Internet]. West Conshohocken, PA: ASTM International; [updated 2020 Aug 14; cited 2024 Dec 4]. Available for purchase from: https://www.astm.org/d8333-20.html.

13 Shanmugasundaram, M. and Lee, E. HORIBA Instruments Inc. Quantitative Analysis Using Raman Spectroscopy in Pharmaceutical Applications [Internet]. Piscataway, NJ: HORIBA Instruments Inc.; [cited 2024 Dec 10]. Available from: https://www.horiba.com/int/scientific/applications/pharmaceutical/pages/quantitative-analysis-using-raman-spectroscopy-in-pharmaceutical-applications/.

14 Johansson, J., Sparén, A., Svensson, O. et al. (2007). Quantitative transmission Raman spectroscopy of pharmaceutical tablets and capsules. *Applied Spectroscopy* 61 (11): 1211–1218.

15 Hampton, L., Brander, S., Coffin, S. et al. (2022). Characterizing microplastic hazards: Which concentration metrics and particle characteristics are most informative for understanding toxicity in aquatic organisms? *Microplastics and Nanoplastics* 2: 20.

16 ISO/DIS 16094-2 (2023). *Water Quality – Analysis of Microplastic in Water - Part 2: Vibrational spectroscopy methods for waters with low content of suspended solids including drinking water [Internet].* Geneva, CH: ISO [cited 2024 Dec 10]. Available for purchase from: https://www.iso.org/standard/84460.html.

17 Microplastics Analysis Workflow [Internet]. Palaiseau, FR: HORIBA Instruments Inc.; [cited 2024 Dec 12]. Available from: https://www.horiba.com/usa/scientific/applications/environment/pages/microplastics/#microplastics-analysis-workflow.

18 Nanoparticle Tracking Analysis [Internet]. Kyoto, JP: HORIBA Instruments Inc.; [cited 2024 Dec 12]. Available from: https://www.horiba.com/usa/scientific/technologies/nanoparticle-tracking-analysis/nanoparticle-tracking-analysis/.

19 No, Microplastics Are Not The "New Acid Rain." Not Even Close [Internet]. Washington, DC: American Chemical Council; 2020 Jul 27 [cited 2024 Dec 12]. Available from: https://www.americanchemistry.com/

chemistry-in-america/news-trends/blog-post/2020/no-microplastics-are-not-the-new-acid-rain-not-even-close.

20 Maes, T., Jessop, R., Wellner, N. et al. (2017). A rapid-screening approach to detect and quantify microplastics based on fluorescent tagging with Nile Red. *Scientific Reports* 7: 44501.

21 Karukstis, K. (2001). *Handbook of Surfaces and Interfaces of Materials. Volume 3: Nanostructured Materials, Micelles, and Colloids. Chapter 13 Encapsulation of Fluorophores in Multiple microenvironments in Surfactant-Based Supramolecular Assemblies. Section 5.3.1. Fluorescence Properties of Nile Red.* Cambridge, MA: Academic Press.

22 Zhang, L., Cao, X., Cai, W., and Li, Y. (2011). Observations of the effect of confined space on fluorescence and diffusion properties of molecules in single conical nanopore channels. *Journal of Fluorescence* 221: 1865–1870.

23 Dresselhaus, M., Dresselhaus, G., Saito, R., and Jorio, A. (2005). Raman spectroscopy of carbon nanotubes. *Physics Reports* 409 (2): 47–99.

24 Cheusheva, S. (2023). How to Get Random Sample in Excel Without Duplicates [Internet]. Lomianki, PL: Ablebits Docs; Mar 14 [cited 2024 Dec 20]. Available from: https://www.ablebits.com/office-addins-blog/excel-random-sample-no-duplicates/.

25 ParticleFinder [Internet]. Palaiseau, FR: HORIBA Instruments Inc.; [cited 2024 Dec 20]. Available from: https://www.horiba.com/usa/scientific/products/detail/action/show/Product/particle-finder-1657/.

26 Resolution [Internet]. Minato City, JP: Nikon MicroscopyU; [cited 2024 Dec 20]. Available from: https://www.microscopyu.com/microscopy-basics/resolution.

27 Sheppard, C. (2005). *Encyclopedia of Modern Optics, Microscopy – Overview.* Amsterdam, NL: Elsevier.

28 Depth of Field [Internet]. Waltham, MA: Evident Scientific Inc.; [cited 2024 Dec 20]. Available from: https://www.olympus-ims.com/en/microscope/terms/focal_depth/.

29 MPLN [Internet]. Waltham, MA: Evident Scientific Inc.; [cited 2024 Dec 20]. Available from: https://www.olympus-ims.com/en/microscope/mpln/.

30 MPLN-BD [Internet]. Waltham, MA: Evident Scientific Inc.; [cited 2024 Dec 20. Available from: https://www.olympus-ims.com/en/microscope/mpln-bd/.

31 LMPLFLN [Internet]. Waltham, MA: Evident Scientific Inc.; [cited 2024 Dec 20. Available from: https://www.olympus-ims.com/en/microscope/lmplfln/.

32 LMPLFLN-BD [Internet]. Waltham, MA: Evident Scientific Inc.; [cited 2024 Dec 20. Available from: https://www.olympus-ims.com/en/microscope/lmplfln-bd/.

33 Qian, N., Gao, X., Lang, X. et al. (2024). Rapid single-particle chemical imaging of nanoplastics by SRS microscopy. *Proceedings of the National Academy of Sciences of the United States of America* 121 (3): e2300582121.

34 Munno, K., De Frond, H., O'Donnell, B., and Rochman, C. (2020). Increasing the accessibility for characterizing microplastics: Introducing new application-based and spectral libraries of plastic particles (SLoPP and SLoPP-E). *Analytical Chemistry* 92 (3): 2443–2451.

16

Material-Sparing Approaches for Spectroscopic PAT Method Development in the Pharmaceutical Industry

Zhenqi Shi[1] *and Benoît Igne*[2]

[1] *Synthetic Molecule Pharmaceutical Science, Genentech Research and Early Development, Genentech, A Member of Roche Company, South San Francisco, CA, USA*
[2] *Manufacturing Science and Technology, Vertex Pharmaceutical Inc, Boston, MA, USA*

16.1 Introduction

Traditional approaches to developing spectroscopic process analytical technology (PAT) methods often require data collection at the manufacturing scale intended for commercial manufacturing. Spectroscopic PAT sensors, particularly Near-infrared (NIR) spectroscopy, are sensitive to sample presentation, particularly driven by changes in density. Thus, to avoid scale-up activities for the PAT methods, in-line and online applications are typically developed at the manufacturing scale.

This situation is challenging in multiple aspects. First, the amount of active pharmaceutical ingredient (API) available throughout development may be limited until late development activities (registration and quality by design activities) or commercial manufacturing. Also, the cost of API may be prohibitive. Finally, for accelerated clinical programs, there may be a very limited number of campaigns that can be used to develop and test PAT method/model performance ahead of process validation.

Spectroscopic PAT method development will require the collection of spectral data representing the range of variability expected during commercial manufacturing. In addition to variability in API(s) and excipient concentrations, spectral data collection at the representative scale of the process and equipment (sample presentation, variability in light sources, probes, instruments, etc.) is necessary to ensure that the model is fit for the intended purpose.

While method validation activities are routinely performed at the commercial scale, incorporating all the necessary spectral variability at commercial scale for method development is typically impossible, particularly for batch manufacturing. With respect to PAT method development, continuous manufacturing (CM) presents significant advantages compared to batch manufacturing. With method development performed directly at manufacturing scale, significant efficiencies are gained in terms of material needs. Process interfaces developed in the lab are directly implemented on the factory floor, and methods no longer need to be scaled up. However, CM represents only a small portion of the pharmaceutical manufacturing equipment currently used by the industry. PAT scientists have thus had to develop approaches to reduce the use of material to enable spectroscopic PAT as a tool for generating product and process understanding during development for process monitoring and potentially for process control during commercial manufacturing.

This chapter will review material-sparing approaches reported in the pharmaceutical industry. They include the use of new algorithms based on pure component spectra, the selection of efficient designs of experiments (DoE), and sampling interfaces minimizing powder needs. Sample preparation approaches

Practical Quantitative Vibrational and Electronic Spectroscopy: A Guide for Developing, Optimizing, and Validating Procedures, First Edition. Edited by Deborah A. Peru.

Companion website: www.wiley.com/go/peruelectronicspectroscopy

are also discussed, leveraging manufacturing equipment dynamics to generate calibration samples and process-scale-independent models. Finally, the use of synthetic calibration standards is reviewed.

16.2 Sampling Interfaces

A significant amount of work has been performed on developing robust interfaces for in-line sensors on moving powder. In this section, efficient process-sampling interfaces will be discussed as well as scale-independent sampling approaches.

16.2.1 Efficient Process-Sampling Interfaces

The concept of a material-sparing approach for pharmaceutical process development is not a new concept. Many scale-down models have been developed using small-scale, but representative, equipment/apparatus with the main driver to reduce material consumption that would otherwise have been used if deployed at scale. A well-known example in the pharmaceutical chemistry, manufacturing, and control (CMC) environment is the *in vitro* dissolution test commonly used to mimic what could happen *in vivo* without the expensive *in vivo* test. A similar concept has also been applied in spectroscopic method development for the intended purpose of real-time process monitoring to improve process understanding.

Content uniformity is one of the commonly investigated quality attributes for oral solid dosage form manufacturing on small molecules. It is routinely conducted through measurement on the tablet itself. Recently, a trend that has gained popularity is moving the measurement to the feed frame of a tablet press. This location has the advantage of being the location closest to the final tablet compression where a PAT measurement can be performed on powder and thus provides a representative measure of the API concentration in the compressed tablets. Qualitative trending was initially reported to capture the API concentration temporal profile in the feed frame [1–3]. With more and more experience accumulated on process monitoring through feed frame, researchers started to realize that the feed frame actually represents a challenging location for spectroscopic measurements due to the very dynamic nature of the sample (i.e., moving powder) at that location, such as changes in working distance, measurement angle, and powder physical properties (e.g., density). Finally, the paddle speed inside the feed frame is also sometimes used in close-loop control as a parameter to adjust the die fill to ensure consistent tablet weight. Thus, unless properly accounted for in a spectroscopic calibration, factors such as mass flow rate and feed frame paddle speed are expected to impact spectra [4, 5]. Calibrating spectral signals on a commercial tablet press across varied mass flow rates and paddle speeds is expected to consume a lot of API. That is when material-sparing approaches started to gain traction.

An NIR method was developed at Eli Lilly and Company in an offline "feed frame table" and successfully tested on a full-scale CM line [6]. The offline "feed frame table" was designed to mimic the full-scale line and required a minimal amount of raw materials to incorporate variations from process, formulation, instruments, and raw material lots. The use of such an offline "feed frame table" was able to consume only 5 kg of API for full-scale process monitoring, which would otherwise have consumed over 100 kg of API if a similar design of experiment had been conducted at full scale. The material-sparing NIR PAT method was demonstrated to provide accuracy, precision, and robustness in predicting blend potency during tablet manufacturing.

Similar to the approaches adopted by researchers at Lilly, researchers at Pfizer developed a benchtop feed frame system for the same purpose [7]. A 15-point DoE was created using the benchtop unit to develop a calibration set. The resulting NIR method on a surrogate API was successfully implemented during five test batches across process scales in two different manufacturing sites across different potency levels on a CM platform for direct compression. This study shows that a benchtop feed frame can be used for material-sparing calibration method development without operating a full-scale process line and applied across multiple sites and instruments at different potency levels.

The same approaches were also adopted by academic researchers. A group at the University of Puerto Rico at Mayaguez reported the use of NIR spectroscopy in a homemade feed frame device to successfully monitor API concentrations at both high (15% w/w) and low (5% w/w) levels [8]. Both calibration models developed using partial least-squares (PLS) on five blends provided highly accurate predictions of the test sets with low bias (0.03% w/w and – 0.06% w/w) and relative standard error of prediction (1.9% and 3.7%), respectively. It was also found that recirculation within the closed feed frame did not change the physical properties of the blends studied. The prediction of blends flowing through the open feed frame was possible with a calibration model developed in the closed feed frame. The closed feed frame would allow the recycling of blends before collecting data at various mass flow rates and paddle speeds, which could reduce the materials needed to develop calibration models by more than 90%.

Additionally, Raman spectroscopy was also reported by researchers at Purdue University to investigate the effects on Raman signals by fluorescence interference, photobleaching, feed frame rotation speed, and material particle size [9]. With the careful selection of effective preprocessing to remove fluorescence interferences, the study showed that Raman measurements were robust; that is, they were not significantly affected by feed frame rotation speed and excipient particle size. Additionally, the calibration via stationary powder was proved effective in monitoring API concentration on flowing powder. This implies that the nature of Raman spectroscopy, with effective spectral preprocessing, could be more forgiving toward its sensitivity to non-chemical interferences, which further reduces the material necessary for a method calibration intended for API concentrations.

Although using an offline feed frame to serve as a surrogate apparatus saves a significant amount of calibration material, it still consumes kilogram-scale material. For a high-dose drug formula, the consumption of API is not trivial, especially for early-phase programs. Thus, a group of researchers at GSK collaborated with an academic institute to develop an in-house moving powder set-up to simulate the environment of the feed frame [10]. In their study, the model developed offline via Raman spectroscopy could predict the active ingredient content after a bias correction and used only a fraction of the material (i.e., 20 g). The established Raman method was demonstrated as suitable for the determination of bulk assay of the active ingredient in powders inside the feed frame for use during batch and CM processes.

As mentioned above, many reported case studies in the feed frame of the tablet press originated from their applications in CM. Given the nature of CM on dispensing and processing material constantly, the typical time and the amount of material needed to reach state-of-control is significantly reduced compared to the batch process [11, 12]. This minimal resource requirement has enabled the early use of PAT during process and formulation development and clinical trial material production, when the raw material availability may be limited. Upon successful technology transfer, the PAT method can be transferred from the laboratory scale to the full commercial scale whenever necessary. Given the at-scale transfer from the laboratory to a commercial CM process, typical spectroscopic calibration method transfer efforts across scales for a batch process were also significantly reduced. An example of using CM to streamline PAT method calibration is the unique approach adopted by Eli Lilly and Company to generate blends at various API concentrations via the use of a decay curve in their continuous direct compression line [13]. It refers to compensating the loss of API via excipients by loss-in-weight feeders at the end of steady-state operations. This not only consumes zero API during the decay curve, but also generates multiple API concentrations for the purpose of spectroscopic method calibration.

16.2.2 Scale-Independent Sampling

As previously mentioned, most spectroscopic PAT methods are impacted by changes in the density of the material. The resulting variability in pathlength can result in significantly different spectra that can appear to fall outside the model space. Some changes in density are expected due to variability in raw materials and processes, and methods should be made robust to that source of variability to ensure method robustness. However, demonstrating robustness when the sample presentation is responsible for the main differences in sample density can be difficult. This scenario is typical of two common issues for PAT scientists: (1) scale-up activities from laboratory or pilot scales to commercial scale(s), and (2) uncontrolled sample presentation in cases where the sampling interface does not ensure reproducible sample presentation. Section 16.2.1 discussed how sampling interfaces can be developed to mimic commercial-scale sample flow. In this section, scale-independent approaches will be discussed.

Process development typically takes place with processing units at the gram or low kilogram levels since access to API may be limited. As such, unit operations such as bin blenders, and batch granulator processes can be scaled up to equipment multiple orders of magnitude larger for commercial manufacturing. From a spectroscopic standpoint, this typically means that PAT methods developed to monitor blend uniformity or water content, for instance, can struggle to accommodate the various scales in a single model.

Talwar et al. used a small number of pilot-scale blend experiments to complement a comprehensive DoE performed at a small scale (data collected in static and dynamic conditions) [14]. The authors determined that including the density variability from static, dynamic small scale, and dynamic pilot scales allowed the development of fit-for-purpose models to monitor blending dynamics and determine endpoints while minimizing data collection at larger scales.

A similar issue exists with tablets. Running a commercial press consumes a lot of material. Generating representative tablets from blends with varying API levels can be difficult on a commercial tablet press. Compaction simulators can generate tablets from gram-scale blends representative of the commercial-scale press. However, when a PAT method is developed on tablet presses of various scales, spectra may need to be standardized across scales to accommodate for the differences in compression mechanism. An example of this approach was published by Farrell et al. where they used a regularization approach (Tikhonov regularization) to update a model developed in the laboratory so it would be fit for purpose when deployed for monitoring tablets compressed with a commercial-scale press [15].

Another strategy to address this problem is to collect spectra representative of large-scale powder density at a small scale. Mohan et al. presented such an approach for bin blending [16]. Using an Instron system to apply a precise force and generate known powder density, the authors collected NIR spectra of the powder of small powder quantities (<50 g) and developed a model to monitor a 1 kg scale blender. The authors achieved an equivalent model performance to that of a model developed at the 1 kg scale. In addition, by collecting spectra during the powder consolidation process (not just when the target density is reached), these authors also demonstrated increased model robustness when applied to a test 1 kg blending experiment.

Velez-Silva et al. presented an approach where powder feed rates and feed chute angles were modified to develop a comprehensive variance space including chemical variability and density variability (associated with feed chute angles) [17]. The resulting models demonstrated that the density-independent model obtained through varying the chute angles performed as well as models developed by varying powder rates and significantly reduced the amount of powder necessary for method development.

To simplify in-line sampling for PAT, hardware solutions have also been created. Powder streams can be sampled using tools that control powder flow such as the "spider wheel" and TASO/RADO probe holder systems commercialized by Expo Process Analytics (www.expoprocessanalytics.com). With these hardware solutions, powder flow is precisely controlled over the PAT probe to ensure consistent density to the target application environments, such as the feed frame of a tablet press and granulator across scales, etc. Because they allow powder development outside of the manufacturing process, models can be developed with small amounts of powder and directly integrated with the commercial-scale manufacturing equipment without requiring the need for model update or redevelopment.

Another approach is to control powder density on the manufacturing line through buffer (or hold) sections where the powder level is carefully controlled by feeders or powder dosing valves. These solutions allow powder density and flow rate to be controlled within the range used during development, thus ensuring the representativity of powder during online sampling. These solutions are particularly useful when coupled with feed chute or feed frame measurements. In these situations, varying powder levels in the tablet press hopper can impact powder density at the probe level, potentially impacting the PAT measurements.

Finally, a hardware solution that combines powder sampling, cleaning, and referencing should be mentioned. GEA is marketing the Lighthouse Probe® (www.gea.com), which is designed to mitigate powder density variability through the use of baffles above and below the probe, clean the probe in place to limit fouling, and if required, present the probe to referencing material for periodic hardware calibration. The incorporation of such a probe interface to the powder flow is expected to enable a robust PAT method development across scales.

16.3 Design of Experiments and Synthetic Calibration

In an industry as risk-averse as the pharmaceutical industry, development activities often follow a rigorous process driven by risk assessments and risk mitigation strategies. After a demonstration of feasibility, all reasonable sources of variability that could have an impact on the PAT measurement system are typically identified and DoEs are often put in place to generate spectral data representing these sources of variability. As stated in this chapter, they can be from the equipment (probe changes, lamp changes, instrument response to manufacturing conditions, etc.), sampling interfaces, and process variability (excipient variability, manufacturing variability, etc.).

The DoEs vary from full factorial where each source of variability is varied independently of others to ensure orthogonality of the various factors tested in the resulting spectra to efficient designs where a reduced number of levels are used to capture multiple sources of variability at once. Bondi et al. studied the impact of various experimental designs (five-level full factorial, three-level full factorial, central composite, I-optimal, and D-optimal) on the prediction performance of an API via the use of NIR spectroscopy [18]. Results showed that an I-optimal design performed similarly to the five-level full factorial but with 16 fewer conditions. In another study, Scheibelhofer et al. determined that evenly spaced levels in a DoE improved performance and that when non-linear relationships are observed, more levels in the design are required [19]. Bogomolov introduced diagonal designs as an alternative solution to more exhaustive approaches for generating evenly spaced designs with fewer levels [20]. The author suggested that the designs' simplicity and interpretability were compounded by their competitive performance compared to traditional DoEs.

Examples of efficient designs used for calibration have been published for pharmaceutical PAT applications. A D-Optimal design was used to develop a material-sparing NIR method for a CM process [21]. The factors of the D-Optimal design were selected to incorporate the maximum amount of variation while reducing the correlation with the analyte of interest. In total, a set of 11 blends was designed for developing the material-sparing method. A supersaturated design was also used to develop a calibration design to incorporate a larger amount of variation using a minimal number of experiments [6]. The design was created via a nearly orthogonal array using a specific combination of factors to get as much balance as possible for pairs of factors, with a high priority on API concentration and lot-to-lot variability. A 7-point calibration design was used to develop the material-sparing method. Additionally, the generalized reduced gradient method was also used to develop a 9-point calibration set for in-line monitoring of blend potency by NIR [22]. The gradient method was used with constraints to minimize the correlation between formulation components and generate an appropriate range of API concentrations. The true density of the blends was also used as a constraint in the gradient method to optimize blend properties during the experiments.

While these aforementioned examples demonstrate the suitability of efficient designs, they carry a level of risk that should be evaluated. If the possibility of collecting additional samples is limited in the future due to short development timelines or high costs, more comprehensive designs may be better suited to ensure all the relevant variance is initially captured. On the other hand, if it is not necessary to meet all performance criteria, the number of design points could be reduced during model optimization.

Another approach that has yet to gain significant traction is augmenting existing calibration data with synthetic (or virtual) spectra that fill in gaps in the data, thus reducing the number of experiments to conduct. Cogdill et al. [23] generated synthetic spectra by estimating the structure and unstructured error signals in Eq. (16.1):

$$\hat{X} = C \cdot S^T + E + e \tag{16.1}$$

where $\hat{X}$ is the resulting synthetic spectra, C is the concentration matrix, S is a matrix of pure component spectra, E is a matrix of structured noise, and e is a matrix of unstructured noise. E was estimated by generating noise representing baseline curvature and offset, batch-to-batch variation, instrument drift (long-term), instrument noise (short-term), and sampling repeatability. e was estimated from linear combinations of short-term synthetic instrument noise spectra representing unstructured white noise. The influence of model rank, the intensity of the noise, and the number of spectra on the model performance were evaluated. The authors successfully developed NIR models from synthetic spectra (generated pure component spectra, parallel testing data, and instrument noise information) to predict the API content in tablets. The resulting model was directly comparable to a model built on 500 NIR spectra and reference values. A similar study was also reported on powder blends using a similar methodology [24].

Finally, despite the best efforts put into collecting high-quality data, sampling errors and laboratory errors can still result in samples that are not well predicted by the model. This can be due to a mismatch between when the spectra are collected and where the sample is withdrawn from the process for reference testing. Or it can also come from the inherent reference method variability. To overcome these issues during early-phase development and enable the use of PAT without collecting any additional samples, empirical target distribution optimization approaches were developed [25]. The basic premise of the methodology is that reference values for samples presenting poor prediction performance can be optimized to increase the performance of the model. One of the approaches was used by Pedersen et al. [26]. Samples presenting high prediction residual values were excluded from the model and predicted by a model without these high-error samples. The predicted values for these high error samples were then used to replace the reference values and a new model was built. The authors found that optimizing the reference values increased the robustness of models based on ternary mixture models by reducing non-linearity. However, it should be noted that while useful in early-phase development activities, practitioners should not rely on this approach for final model development due to the risk of overfitting and reduction in model robustness. Indeed, by self-correcting the values that are deemed to not fit within the model, important variability could be ignored and thus affect model performance on the long term.

16.4 Efficient Modeling Algorithms and Approaches

16.4.1 Calibration-Free or Lean Calibration Algorithms

Besides the use of scale-down and scale-independent experimental setup for material-sparing spectroscopy method development and efficient design of experiments, the modeling algorithm is another critical component of the material-sparing spectroscopic method development. The resource requirement of the PAT/spectroscopic method may vary depending on the modeling algorithm used. PLS is a widely known modeling approach used for developing PAT/spectroscopic methods. A rigorous calibration dataset encompassing many representative sources of variabilities is often required to successfully develop and deploy PLS models [6, 7]. By contrast, calibration-free or lean calibration approaches are an alternative modeling strategy to minimize the material requirement for model development. This could be especially attractive in applying process monitoring for early-phase development when API supply is very limited.

The concept of lean calibration has a long-standing history in the field of chemometrics. It is commonly acknowledged that a spectrum is composed of three component signals, including the analyte of interest, structured noise, such as both chemical and physical interference, and white noise. A rigorous spectroscopic calibration method development process often focuses more on collecting representative variability of those structured noises than on the signal from the analyte of interest. For a pharmaceutical example, calibrating a content uniformity method, the API concentration often ranges monotonically from 70% to 130% of label claim, but most of the calibration data is collected to represent sources of variability such as excipient composition, excipient lot variability, API lot variability, tablet compaction force, and tablet weight. Thus, if the structured noise can be characterized ahead of time followed by proper mitigation during method development, lean method calibration naturally becomes feasible, not only to reduce calibration efforts but also to make the model potentially more parsimonious and more robust to maintain during its lifecycle management.

In the literature, a couple of methodologies on the topic of lean calibration or "lean chemometrics" are reported. The first category is to leverage advanced spectral preprocessing to mitigate/remove the impact of both chemical and physical interferences on a pre-existing model that is unknown to these interferences, such as net analyte signal (NAS) [27] and extended multiplicative scattering correction (EMSC) [28, 29]. The second category is to leverage spectral vectors representative of structured noise and apply those with the signal of the analyte of interest simultaneously at the step of estimating concentrations for unknown samples. Great examples of this category originate from classical least-squares (CLS) and many of its derivative approaches, such as prediction-augmented CLS (PACLS) [30, 31], and iterative optimization technique (IOT) [32]. The last category uses virtual spectra representative of calibration samples. Information on synthetic calibration can be found in Section 16.3.

Lorber et al. were the first group of researchers to report the conceptual use of NAS as a preprocessing step to project a sample spectrum onto a null space spanned by the interference vectors [27]. Cogdill et al. reported the first pharmaceutical application of NAS on a tablet dataset [33]. Although three algorithms reported in the paper, including PLS, NAS, and pure component projection (PCP), reached equivalent performance, the potential of deploying NAS and PCP while using limited reference values was highlighted. Given both approaches (NAS and PCP) are capable of preprocessing and projecting a sample spectrum to a unit scalar value that is linearly correlated to the API concentration, the number of necessary reference values could be reduced to a set as small as two samples. Its potential for being leveraged as lean calibration approaches has been further reported by other researchers in pharmaceutical applications [34–38].

Additionally, NAS preprocessing could also be used as a spectral pre-treatment before PLS modeling [24, 39]. Given the number of interference vectors removed at the NAS preprocessing step, the number of latent variables required to reach model convergence is usually small, which makes a parsimonious model potentially more robust during lifecycle management. Moreover, such a NAS-based preprocessing was reported to be similar to external parameter orthogonalization (EPO) for the case study of calibration transfer [40]. Zhao et al. successfully extended the concept of NAS and EPO to a continuum regression (CR) filter to transfer a powder blend model collected offline to inline applications with the intended purpose of saving material necessary for the in-line application [41]. The main idea was to use EPO/CR to mitigate the difference between off-line and in-line data so that the PLS modeling, after such preprocessing, can focus on the API concentration in the blend.

CLS is one of the most commonly reported ordinary least-squares (OLS) approaches compared to PLS as the most popular example of inverse least-squares (ILS). The main difference between CLS and PLS is that CLS requires a full-ranked data matrix, while PLS can handle the ill-conditioned matrix. That difference has been leveraged by many researchers on the use of such a unique requirement by CLS to reduce the number of calibration samples needed in the training set. Typically, two approaches have been reported in the pharmaceutical applications of CLS model calibration. If chemical compositions are the analytes of interest, pure component spectra of interested species by themselves can serve as the pure component spectra matrix, without preparing any calibration mixture samples. Great examples of using CLS are found for analyzing chemical imaging data [42, 43]. Similarly, a lean calibration data set only containing chemical concentration variability can also be used to estimate the pure component spectral matrix by preparing a very small number of calibration mixture samples [44]. When CLS is being applied to unknown samples, any spectral vectors representative of new interferences that have not been seen by the initial model can be added to the original pure component spectral matrix to augment the prediction via attributing that interference at the last step of the regression. The use of spectral vectors to represent those interferences, instead of adding new samples with reference values to recalibrate CLS model, has proven to be an effective way to generate robust predictions against those unknown sources of variability, such as calibration transfer from pharmaceutical blends to tablets [45] and when sample matrices were changing [46].

Recently, CLS was also reported to monitor batch-blending process using a lean two-point calibration model, including a placebo blend and nominal blend at 100% label claim [44]. As long as a placebo blend (as simple as pure microcrystalline cellulose) is collected under the same blending condition as the nominal blend, the same CLS approach can be applied across many different APIs. This enables pharmaceutical researchers to start monitoring batch blending from the very first at-scale blending operation without submitting any thief samples for reference testing. When enough blending runs are monitored in development, a robust CLS model can be further developed before technology transfer to commercial manufacturing. Thus, such an approach is expected to be fit for purpose for early-phase development when API supply is very limited. Especially considering that more and more small-scale blenders are becoming available (from 10 g of material with the SentroBlender® by Sentronic, Inc. [www.sentronic.eu]), the use of near-infrared spectroscopy (NIRS) combined with CLS is expected to improve process understanding from the very first bench-scaled blending experiment.

Leveraging a similar concept as CLS but within an optimization framework, IOT was also reported as a minimal calibration technique using pure component spectra to develop a PAT method for both drug substance and drug product manufacturing processes [32]. The unique difference of IOT compared to CLS is the ability to apply constraints during the concentration estimate. Two common constraints are (1) the estimated concentrations between zero and one; (2) the summation of estimated concentration across chemical components equals to one. Multiple case studies were reported in the initial publication, such as solvent swap and continuous blending. Recently, a variable selection step was added to the IOT algorithm via the use of variable importance plot (VIP-IOT) to demonstrate comparable model performance with PLS algorithm [47]. Such a VIP-IOT PAT method was successfully applied to the commercial-scale data to predict blending potency during tablet manufacturing. Additionally, the same group of researchers reported the equivalent performance between CLS and IOT in the same application at the feed frame of a tablet press on a continuous direct compression process [48].

A common question arising when deploying these types of lean approaches is whether such a lean model is robust enough during lifecycle management. Given what has been reported so far, lean approaches present a reasonably robust performance upon deployment in commercial manufacturing. An extended version of the IOT algorithm was presented to develop a material-sparing PAT method for CM [13, 49]. The extended term(s) are added into the model to account for non-chemical interferences, a similar concept to what has been reported for CLS. The extended term of interferences can be further divided into two categories. Interference of known origin could be estimated in a least-squares fashion, hence defined as "supervised extended iterative optimization technology (EIOT)" algorithm. While the "unsupervised EIOT" is designed to handle those interferences unknown to the experimentalists, that still impacts the model prediction. So far, the extended IOT has been demonstrated to be robust against different raw material lots, different NIR spectrometers, and across different manufacturing sites. Finally, the use of these interference vectors can be further integrated within the optimization framework to improve robustness. If a library of historical interference vectors is available, a binary optimization variable can be implemented in the optimization routine to allow the optimizer to not only attribute those unknown variabilities per the interference vectors in the library, but also pick the most relevant historical interference vector for the unknown sample, given the underlying similarity between those two. This is a concept theoretically feasible, but waiting to be demonstrated in commercial settings, and it is expected to streamline the model maintenance and reduce the burden of deploying such models for in-process control and even real-time release testing.

16.4.2 Efficient Modeling Approaches

Most of the aforementioned modeling algorithms have been applied to measure the API concentration in pharmaceutical samples, given the limited supply of API in early-phase product development. However, when the analyte of interest is no longer API concentration, pharmaceutical researchers have also been developing formulation and/or matrix-independent approaches to enable efficient PAT method development. Two commonly reported examples are monitoring organic solvent content in drug substance manufacturing and monitoring ribbon solid fraction (SF) during drug product manufacturing.

When manufacturing drug substances, organic solvents play significant roles. They are involved in bond-forming reactions, impurity rejection during work-up procedures, crystallization to ensure the desired polymorph form is achieved, and filtration drying to enable acceptable residual solvent levels according to the International Council of Harmonization (ICH) guidelines. Thus, it is necessary to understand the solvent kinetics to ensure desirable process outcomes. The use of vibrational spectroscopy to monitor solvent content during drug substance processing has been heavily reported [50–55]. Given that temperature is a common process parameter and the known sensitivity of vibrational spectroscopy to temperature, spectroscopic calibration for a given solvent within a range of operation temperatures in a given API matrix is expected to be time and labor-intensive. Thus, multiple efficient modeling approaches have been developed to enable the material-sparing PAT method development across different temperatures, such as spectral space transformation [53], loading space standardization [51], and systematic prediction error correction [50].

Since the first report on the sensitivity of spectral slope in NIRS to density change, researchers have been using this mechanistic understanding to develop formulation-independent PAT method/model for physical properties throughout the manufacturing process of solid oral dosage form [56]. Ribbon SF is one of those reported key quality attributes. Talwar et al. developed an NIR method for predicting the SF of dry granulated ribbons manufactured across formulation matrices [57]. The study utilized an excipient-only formulation system to develop calibration compacts to predict SF of compacts containing an API at various concentrations. Both PLS and spectral-slope algorithms were evaluated to model compact SF. The risk associated with unmodeled chemical variation (i.e., the introduction of API) manifested itself through the generation of new peaks and decreased baseline absorbance in the NIR spectra. Thus, the spectral slope was found to better manage such a risk, as demonstrated by relatively higher robustness to the increasing load of the API. The reduced robustness of the PLS approach was attributed to the impact of chemical variability on both spectral baseline and peak absorbance. A prediction error of approximately 5% was observed at 10% drug load using the spectral-slope approach. Therefore, an understanding of the risk associated with unmodeled variability is expected to enable spectroscopic method development on vibrational spectroscopy as an API-sparing technique, at least for low-dose formulation where the excipients mainly dominate the spectral slope.

16.5 Chemical Imaging for Pharmaceutical Product Development

Since the introduction of hyperspectral imaging, the sheer volume of data provided has enabled pharmaceutical product development in multiple dimensions, from development to manufacturing and from process understanding to troubleshooting. Given such an advantage on data volume, a single hyperspectral imaging measurement has often been deemed as "two birds, one stone" compared to traditional single-point spectroscopy measurement. That is also why pharmaceutical researchers considered chemical imaging as a lean process development tool by its design.

A group of researchers at Pfizer leveraged the combinational use of both NIR chemical imaging and Raman spectroscopy to select the manufacturing platform of a solid oral dosage form [58]. In this study, NIR Chemical Imaging and single-point Raman spectroscopy were used simultaneously to understand the impact of manufacturing processes on two quality characteristics of a formulation under development, namely polymorphism purity and drug distribution in the dosage form. These studies illustrated a rapid material-sparing approach via sensor combination to develop an optimal manufacturing platform selection for the formulation.

In another study, a one-dimensional Raman line-mapping approach was developed and used to extract critical information assessing two performance-indicating parameters of a drug in polymer formulation: sedimentation of the API within a capsule formulation and phase separation of individual components [59]. The use of Raman hyperspectral imaging to assess homogeneity in solid dosage forms has traditionally been limited by the time required to acquire images from a statistically representative sample area. This novel approach enabled less than half an hour per capsule for data collection and processing to assess drug and polymer domain sizes, specifically the degree of drug and polymer segregation. Additionally, the Raman mapping data was compared with both the dissolution profiles and processing parameters of the samples tested, and a strong correlation was shown between formulation homogeneity and dissolution behavior.

16.6 Conclusion

In this chapter, a review of material-sparing approaches used to develop spectroscopic PAT methods in the pharmaceutical industry was performed. A wide range of techniques are available for practitioners. Some standardize sample presentations to the PAT system to avoid discrepancies across scales. Others model the scale-to-scale differences to make models robust. Many calibration-free or lean-calibration approaches have been developed to overcome burdensome traditional method calibration approaches. Finally, approaches aimed at reducing the number of samples collected have also been proposed. Through the literature surveyed, the use of efficient calibration algorithms and "at-scale" sample presentation systems seems to have been the most favored approaches, particularly with the adoption of CM. However, these approaches are all valid for batch manufacturing, and with their success increasingly reported in the literature, batch manufacturing is expected to significantly benefit from simplified workflows to support early-phase development activities.

However, it remains to be seen how all these algorithm-based material-sparing approaches will perform when deployed during commercial manufacturing activities. PLS-based approaches are still used in the majority of methods submitted to regulatory agencies due to their acceptance and description in available regulatory guidance documents. However, their development and deployment require significant amounts of time and material and, thus, are not well suited for early-phase activities. Thus, the material-sparing methods described in this chapter have the possibility to completely revolutionize the use of spectroscopic PAT in the pharmaceutical industry by enabling the use of NIR, Raman, etc., from the first formulation development experiment.

References

1 Liu, Y. and Blackwood, D. (2017). Sample presentation in rotary TabletPress feed frame monitoring by near infrared spectroscopy. *American Pharmaceutical Review* 15: 1.

2 Sasic, S., Blackwood, D., Liu, A. et al. (2015). Detailed analysis of the online near-infrared spectra of pharmaceutical blend in a rotary tablet press feed frame. *Journal of Pharmaceutical and Biomedical Analysis* 103: 73–79.

3 Ward, H.W., Blackwood, D.O., Polizzi, M., and Clarke, H. (2013). Monitoring blend potency in a tablet press feed frame using near infrared spectroscopy. *Journal of Pharmaceutical and Biomedical Analysis* 80: 18–23.

4 Mateo-Ortiz, D., Colon, Y., Romanach, R.J., and Mendez, R. (2014). Analysis of powder phenomena inside a Fette 3090 feed frame using in-line NIR spectroscopy. *Journal of Pharmaceutical and Biomedical Analysis* 100: 40–49.

5 Peeters, M., Peeters, E., Van Hauwermeiren, D. et al. (2022). Determination and understanding of lead-lag between in-line NIR tablet press feed frame and off-line NIR tablet measurements. *International Journal of Pharmaceutics* 611: 121328.

6 Hetrick, E.M., Shi, Z., Barnes, L.E. et al. (2017). Development of near infrared spectroscopy-based process monitoring methodology for pharmaceutical continuous manufacturing using an offline calibration approach. *Analytical Chemistry* 89 (17): 9175–9183.

7 Alam, M.A., Liu, Y.A., Dolph, S. et al. (2021). Benchtop NIR method development for continuous manufacturing scale to enable efficient PAT application for solid oral dosage form. *International Journal of Pharmaceutics* 601: 120581.

8 Movilla-Meza, N.A., Sierra-Vega, N.O., Alvarado-Hernandez, B.B. et al. (2023). The use of a closed feed frame for the development of near-infrared spectroscopic calibration model to determine drug concentration. *Pharmaceutical Research* 40 (12): 2903–2916.

9 Kumar, S., Nagy, Z., Reklaitis, G.V., and Gonzalez, M. (2023). Considerations in Raman spectroscopy for real-time API concentration measurement at tablet press feed frame. *Journal of Spectroscopy* 2023 (1): 8631288.

10 Li, Y., Anderson, C.A., Drennen, J.K. 3rd et al. (2018). Method development and validation of an inline process analytical technology method for blend monitoring in the tablet feed frame using Raman spectroscopy. *Analytical Chemistry* 90 (14): 8436–8444.

11 Almaya, A., De Belder, L., Meyer, R. et al. (2017). Control strategies for drug product continuous direct compression-state of control, product collection strategies, and startup/shutdown operations for the production of clinical trial materials and commercial products. *Journal of Pharmaceutical Sciences* 106 (4): 930–943.

12 Conway, S.L., Rosas, J.G., Overton, P. et al. (2023). Implementation of a fully integrated continuous manufacturing line for direct compression and coating at a commercial pharmaceutical facility – part 1: operational considerations and control strategy. *International Journal of Pharmaceutics* 642: 122820.

13 Shi, Z.Q., Hermiller, J., and Munoz, S.G. (2018). Estimation of mass-based composition in powder mixtures using Extended Iterative Optimization Technology (EIOT). *AICHE Journal* 65 (1): 87–98.

14 Talwar, S., Pawar, P., Wu, H. et al. (2022). NIR spectroscopy as an online PAT tool for a narrow therapeutic index drug: toward a platform approach across lab and pilot scales for development of a powder blending monitoring method and endpoint determination. *The AAPS Journal* 24 (6): 103.

15 Farrell, J.A., Higgins, K., and Kalivas, J.H. (2012). Updating a near-infrared multivariate calibration model formed with lab-prepared pharmaceutical tablet types to new tablet types in full production. *Journal of Pharmaceutical and Biomedical Analysis* 61: 114–121.

16 Mohan, S., Momose, W., Katz, J.M. et al. (2018). A robust quantitative near infrared modeling approach for blend monitoring. *Journal of Pharmaceutical and Biomedical Analysis* 148: 51–57.

17 Velez-Silva, N.L., Drennen, J.K. 3rd, and Anderson, C.A. (2024). Continuous manufacturing of pharmaceutical products: a density-insensitive near infrared method for the in-line monitoring of continuous powder streams. *International Journal of Pharmaceutics* 650: 123699.

18 Bondi, R.W. Jr., Igne, B., Drennen, J.K. 3rd, and Anderson, C.A. (2012). Effect of experimental design on the prediction performance of calibration models based on near-infrared spectroscopy for pharmaceutical applications. *Applied Spectroscopy* 66 (12): 1442–1453.

19 Scheibelhofer, O., Grabner, B., Bondi, R.W. Jr. et al. (2015). Designed blending for near infrared calibration. *Journal of Pharmaceutical Sciences* 104 (7): 2312–2322.

20 Bogomolov, A. (2017). Diagonal designs for a multi-component calibration experiment. *Analytica Chimica Acta* 951: 46–57.

21 Alam, M.A. and Liu, Y.A. (2021). An agile and robust in-line NIR potency deviation detection method for monitoring and control of a continuous direct compression process. *International Journal of Pharmaceutics* 601: 120521.

22 Roman-Ospino, A.D., Baranwal, Y., Li, J. et al. (2021). Sampling optimization for blend monitoring of a low dose formulation in a tablet press feed frame using spatially resolved near-infrared spectroscopy. *International Journal of Pharmaceutics* 602: 120594.

23 Cogdill, R.P., Herkert, T., Anderson, C.A., and Drennen, J.K. (2007). Synthetic calibration for efficient method development: analysis of tablet API concentration by near-infrared spectroscopy. *Journal of Pharmaceutical Innovation* 2 (3): 93–105.

24 Shi, Z.Q., Cogdill, R.P., Martens, H., and Anderson, C.A. (2010). Optical coefficient-based multivariate calibration on near-infrared spectroscopy. *Journal of Chemometrics* 24 (5): 288–299.

25 Igne, B., Talwar, S., Drennen, J.K., and Anderson, C.A. (2011). Improvement of pharmaceuticalnear infrared calibrations by empirical target distribution optimisation. *Journal of Near Infrared Spectroscopy* 19 (1): 1–14.

26 Pedersen, T., Rantanen, J., Naelapaa, K., and Skibsted, E. (2020). Near infrared analysis of pharmaceutical powders with empirical target distribution optimization (ETDO). *Journal of Pharmaceutical and Biomedical Analysis* 181: 113059.

27 Lorber, A. (1986). Error propagation and figures of merit for quantification by solving matrix equations. *Analytical Chemistry* 58 (6): 1167–1172.

28 Martens, H., Nielsen, J.P., and Engelsen, S.B. (2003). Light scattering and light absorbance separated by extended multiplicative signal correction. Application to near-infrared transmission analysis of powder mixtures. *Analytical Chemistry* 75: 394–404.

29 Martens, H. and Stark, E. (1991). Extended multiplicative signal correction and spectral interference subtraction: new preprocessing methods for near infrared spectroscopy. *Journal of Pharmaceutical and Biomedical Analysis* 9 (8): 625–635.

30 Haaland, D.M. and Melgaard, D.K. (2000). New prediction-augmented classical least-square (PACLS) methods: application to unmodeled interferents. *Applied Spectroscopy* 54 (9): 1303–1312.

31 Haaland, D.M. and Melgaard, D.K. (2002). New augmented classical least squares methods for improved quantitative spectral analyses. *Vibrational Spectroscopy* 29: 171–175.

32 Muteki, K., Blackwood, D.O., Maranzano, B. et al. (2013). Mixture component prediction using iterative optimization technology (calibration-free/minimum approach). *Industrial and Engineering Chemistry Research* 52 (35): 12258–12268.

33 Cogdill, R.P. and Anderson, C.A. (2005). Efficient spectroscopic calibration using net analyte signal and pure component projection methods. *Journal of Near Infrared Spectroscopy* 13 (3): 119–131.

34 Shi, Z., Hermiller, J.G., Gunter, T.Z. et al. (2012). A novel sample selection strategy by near-infrared spectroscopy-based high throughput tablet tester for content uniformity in early-phase pharmaceutical product development. *Journal of Pharmaceutical Sciences* 101 (7): 2502–2511.

35 Medendorp, J., Buice, R.G., and Lodder, R.A. (2006). Acoustic-resonance spectrometry as a process analytical technology for the quantification of active pharmaceutical ingredient in semi-solids. *AAPS PharmSciTech* 7 (3): E22–E29.

36 Medendorp, J.P., Paudel, K.S., Lodder, R.A., and Stinchcomb, A.L. (2007). Near infrared spectrometry for the quantification of human dermal absorption of econazole nitrate and estradiol. *Pharmaceutical Research* 24 (1): 186–193.

37 Naseri, A., Ghorbani-Kalhor, E., Vallipour, J. et al. (2009). Simultaneous spectrophotometric determination of benzoic acid, sorbic acid, and ascorbic acid using a net analyte signal-based method. *Journal of AOAC International* 92 (6): 1807–1814.

38 Skibsted, E.T., Boelens, H.F., Westerhuis, J.A. et al. (2006). Simple assessment of homogeneity in pharmaceutical mixing processes using a near-infrared reflectance probe and control charts. *Journal of Pharmaceutical and Biomedical Analysis* 41 (1): 26–35.

39 Shi, Z.Q. and Anderson, C.A. (2009). Scattering orthogonalization of near-infrared spectra for analysis of pharmaceutical tablets. *Analytical Chemistry* 81 (4): 1389–1396.

40 Roger, J.M., Chauchard, F., and Bellon-Maurel, V. (2003). EPO–PLS external parameter orthogonalisation of PLS application to temperature-independent measurement of sugar content of intact fruits. *Chemometrics and Intelligent Laboratory Systems* 66: 191–204.

41 Zhao, Y., Mohan, S., Dave, K. et al. (2022). NIR spectroscopic methods for monitoring blend potency in a feed frame – calibration transfer between offline and inline using a continuum regression filter. *International Journal of Pharmaceutics* 614: 121363.

42 Amigo, J.M., Ravn, C., Gallagher, N.B., and Bro, R. (2009). A comparison of a common approach to partial least squares-discriminant analysis and classical least squares in hyperspectral imaging. *International Journal of Pharmaceutics* 373 (1–2): 179–182.

43 Ravn, C., Skibsted, E., and Bro, R. (2008). Near-infrared chemical imaging (NIR-CI) on pharmaceutical solid dosage forms-comparing common calibration approaches. *Journal of Pharmaceutical and Biomedical Analysis* 48 (3): 554–561.

44 Shi, Z., Rao, K.S., Thool, P. et al. (2022). Development of a near-infrared spectroscopy (NIRS)-based characterization approach for inherent powder blend heterogeneity in direct compression formulations. *The AAPS Journal* 25 (1): 9.

45 Shi, Z.Q., Igne, B., Bondi, R.W. et al. (2012). Calibration transfer from pharmaceutical powder mixtures to compacts using the prediction augmented classical least squares (PACLS) method. *Applied Spectroscopy* 66 (9): 1075–1081.

46 Bondi, R.W., Igne, B., Shi, Z.Q. et al. (2012). Classical least squares for quantitative analysis of pharmaceutical solid dosage forms: part 2—prediction across a changing sample matrix. *NIR News* 23 (5): 12.

47 Rish, A.J., Henson, S.R., Anik Alam, M. et al. (2022). Development of calibration-free/minimal calibration wavelength selection for iterative optimization technology algorithms toward process analytical technology application. *International Journal of Pharmaceutics* 614: 121463.

48 Rish, A.J., Henson, S.R., Alam, M.A. et al. (2022). Comparison between pure component modeling approaches for monitoring pharmaceutical powder blends with near-infrared spectroscopy in continuous manufacturing schemes. *The AAPS Journal* 24 (4): 82.

49 Munoz, S.G. and Torres, E.H. (2020). Supervised extended iterative optimization technology for estimation of powder compositions in pharmaceutical applications: method and lifecycle management. *Industrial and Engineering Chemistry Research* 59 (21): 10072–10081.

50 Chen, Z.P., Li, L.M., Yu, R.Q. et al. (2011). Systematic prediction error correction: a novel strategy for maintaining the predictive abilities of multivariate calibration models. *Analyst* 136 (1): 98–106.

51 Chen, Z.P., Morris, J., and Martin, E. (2005). Correction of temperature-induced spectral variations by loading space standardization. *Analytical Chemistry* 77 (5): 1376–1384.

52 Chong, M.W.S., McGlone, T., Chai, C.Y. et al. (2022). Temperature correction of spectra to improve solute concentration monitoring by in situ ultraviolet and mid-infrared spectrometries toward isothermal local model performance. *Organic Process Research and Development* 26 (11): 3096–3105.

53 Du, W., Chen, Z.P., Zhong, L.J. et al. (2011). Maintaining the predictive abilities of multivariate calibration models by spectral space transformation. *Analytica Chimica Acta* 690 (1): 64–70.

54 Parrott, A.J., McIntyre, A.C., Holden, M. et al. (2022). Calibration model transfer in mid-infrared process analysis with in situ attenuated total reflectance immersion probes. *Analytical Methods* 14 (19): 1889–1896.

55 Chen, Z.P., Morris, J., and Martin, E. (2004). *Modelung Temperatljre-Induced Spectral Variations in chemical process Monitoring*. IFAC Dynamics and Control of Process Systems.

56 Kirsch, J.D. and Drennen, J.K. (1999). Nondestructive tablet hardness testing by near-infrared spectroscopy: a new and robust spectral best-fit algorithm. *Journal of Pharmaceutical and Biomedical Analysis* 19 (3–4): 351–362.

57 Talwar, S., Nunes, C., Stevens, T. et al. (2017). Understanding the impact of chemical variability and calibration algorithms on prediction of solid fraction of roller compacted ribbons using near-infrared (NIR) spectroscopy. *Applied Spectroscopy* 71 (6): 1209–1221.

58 Romero-Torres, S., Chokshi, R., Huang, J., and Krishnan, M. (2011). Concomitant use of near-infrared chemical imaging and Raman spectroscopy during the selection of a solid dosage manufacturing platform. *American Pharmaceutical Review* 14 (7).

59 Brown, S.C., Claybourn, M., Sievwright, D. et al. (2010). Lean Raman imaging for rapid assessment of homogeneity in pharmaceutical formulations. *Applied Spectroscopy* 64 (4): 442–447.

17

Development and Implementation of an NIR Calibration Model to Determine an Active Pharmaceutical Ingredient in a Commercial Film-Coating Process

Maritza Reyna Liriano[1], Luis Torres[2], and Rodolfo J. Romañach[1]

[1] *University of Puerto Rico at Mayagüez, Mayagüez, Puerto Rico*
[2] *Quality Control Analyst, Avara Pharmaceuticals, Arecibo, Puerto Rico*

17.1 Introduction

This chapter describes the application of near-infrared (NIR) spectroscopy to determine the content of an active pharmaceutical ingredient (API) in the film coating of a tablet. The method seeks to quantify an API added as a suspension spray during the film-coating process. The NIR spectra will contain bands related to the multiple components of the tablet. The determination of the added API in the presence of bands from the multiple components of the formulation makes this a challenging application.

The interaction between light and particles has been the subject of study for hundreds of years and still remains a challenging area of study for spectroscopists today. Light scattering occurs as NIR radiation encounters a particle. The radiation may be absorbed by the particle. However, some of the radiation may be elastically scattered by the front surface of the particle. In addition, the radiation may be backscattered by the tablet after some of the radiation is absorbed. Diffuse transmission may also occur as light is forward-scattered through the tablet. The light transmitted through the tablet will be considered "absorbed" in a diffuse reflection experiment where the detection and source of radiation are on the same side. The detector does not measure the light absorbed; it measures the radiation that returns to it, and the remaining radiation is considered "absorbed" although it may have been transmitted. NIR spectra do not result from light interacting with a single particle but from multiple interactions between light and particles. The light that reaches the detector will likely have interacted with different particles at different distances from the detector. Therefore, an NIR spectrum is a complex pattern that includes information from the sample's chemical composition and the material's physical properties, both affecting light scattering. This information is highly convolved, and separating the chemical composition from the physical property information is challenging even for even the most experienced and knowledgeable practitioners of NIR spectroscopy [1, 2].

Several previous studies have elucidated the interaction between light and particles, an understanding that is essential for the application described in this chapter. Multiple theories that explain the interaction between light and particles have been thoroughly reviewed [2]. The effect of compression pressure on both diffuse and transmission spectra has been investigated [3, 4]. In diffuse reflection, the baseline and slope of NIR spectra increase as tablet compression increases [4]. The increased compaction pressure reduces the space between particles and facilitates the transmission of light through the tablet away from the diffuse reflection detector. Light-scattering studies have also shown that radiation travels through many paths within a tablet [5–7].

Practical Quantitative Vibrational and Electronic Spectroscopy: A Guide for Developing, Optimizing, and Validating Procedures, First Edition. Edited by Deborah A. Peru.

Companion website: www.wiley.com/go/peruelectronicspectroscopy

The determination of an API in the film coating is a special case where the interest is in analyzing a thin layer instead of analyzing the entire tablet. Nevertheless, the radiation is transmitted through the film coating and includes spectral features from the tablet. A previous study involved tablets coated with a hydroxypropyl methylcellulose (HPMC) film of only 28 μm [8]. Another previous study estimated the thickness of the film coating to be approximately 50 μm [9]. These previous studies have provided valuable information for the current implementation in an industrial environment.

In this chapter, the API content in a tablet's film coating is determined using an NIR spectrometer at a commercial pharmaceutical manufacturing site. The chapter goes beyond the usual description of developing an NIR spectroscopic calibration model within a research laboratory. NIR predictions are used as inputs in the manufacturing process to determine the optimal end point of the coating process. The NIR spectroscopic results have provided consistent endpoints for the film-coating process. The chapter also presents a series of model acceptance criteria that must be met every time the calibration model is updated. These model acceptance criteria were established through careful evaluation by the team that developed the process more than 10 years ago. These same criteria are still considered valid despite the many advances that have occurred in the understanding and validation of NIR spectroscopic methods. The chapter also presents a model update as an example of how the NIR method is developed while considering the model acceptance criteria.

17.2 Materials and Methods

17.2.1 Materials (Tablets from the Manufacturing Process)

Compressed tablets containing an API are first manufactured. A second different API is added to the tablets as a suspension spray during the film coating. Four Hi-Coaters from Freund Vector Corporation (Iowa, USA) are used to spray the suspension that holds the second API and then dry it. The second API must be added accurately to provide the correct dose for patients.

The film-coated tablets are withdrawn from the Hi-coaters after adding 40, 50, 60, 70, and 80% of the second API. The tablets are analyzed by NIR spectroscopy to determine the content of the second API. The NIR results are then used to determine the amount of suspension spray needed to reach the target content (100% of the second API).

17.2.2 Near-Infrared Spectral Acquisition

Near-infrared spectra were obtained using two Antaris FT-NIR spectrometers from Thermo Fisher Scientific (168 Third Avenue, Waltham, MA, USA) controlled by Thermo Results software. The NIR spectrometers are located within the commercial production area where film coating occurs. Diffuse reflectance NIR spectra were obtained with a resolution of 8 cm^{-1} and 32 scans per tablet over the 10,000–4000 cm^{-1} spectral range. The tablets were placed on the rotating carousel of the Antaris FT-NIR spectrometer, which holds up to 40 tablets. The data is processed and reported by Process Pulse II version 5.38. Process Pulse II stores spectral data, including the manufacturing date, batch ID, sample process points, formulation description, and instrument ID.

Two hundred spectra were acquired per in-process batch using the NIR spectrometer. Forty spectra are collected per sample process point at approximately 40, 50, 60, 70, and 80% of the targeted drug content. Each spectrum acquired is stored in the database. A total of five NIR spectra were averaged since the HPLC procedure is a composite method that analyzes five tablets simultaneously. Even though 40 spectra were obtained per sample point, only the first 20 spectra were used. The remaining 20 spectra are not predicted as they are reserve samples that can be used if problems occur during the analysis process.

17.2.3 High-Performance Liquid Chromatography Analysis

The API content in the film coating was determined through high-performance liquid chromatography (HPLC) using the Waters Alliance 2695E Separation module/2998 PDA detector. The HPLC analyses were conducted after approximately 40, 50, 60, 70, and 80% of the film-coating process was completed. The in-process HPLC composite assay analyzes five tablets per sample placed in a 1-L Pyrex Reagent Bottle. The sample preparation for the HPLC method requires HPLC-grade water, HPLC-grade acetonitrile, HPLC-grade o-phosphoric Acid (85%), and Potassium Phosphate Monobasic ACS grade. The HPLC analysis time for each sampling point is approximately an hour and a half, including the time needed to calculate, double-check, and report the results. The analysis of the five sampling points would take about 7.5 hours after the samples are moved to the HPLC laboratory.

17.2.4 Multivariate Data Analysis

Partial least squares (PLS) regression calibration models were developed using Aspen Technologies software. The Non-Iterative Partial Alternating Least Squares (NIPALS) algorithm was used to develop all calibration models. The reference values for the PLS model were obtained through HPLC, as described in Section 17.2.3. The API film-coating content from the production batches was determined using PLS models constructed from various calibration sets and variable selections using Aspen Technology's chemometric software.

The calibration model was constructed using first-derivative Savitzky–Golay with a 15-point moving window and a second polynomial order. The first-derivative pretreatment was effective in removing the variation commonly observed in the baseline of NIR spectra [10].

17.2.5 Model Acceptance Criteria

The quality of the models was evaluated based on the model acceptance criteria, shown in Table 17.1. These model acceptance criteria were established by the team that originally developed the calibration model and the process. The calibration model must comply with the calibration range; accuracy and precision established in the model acceptance criteria are shown in Table 17.1. The model acceptance criteria are a unique feature of this challenging application and are necessary in industrial applications. While many applications of NIR spectroscopy have been described in the scientific literature [10], the establishment of acceptance criteria for pharmaceutical NIR spectroscopic methods is discussed in very few studies [11]. The model acceptance criteria have not required change or modification in over a decade.

The model acceptance criteria also require the root mean square of calibration (RMSEC) to be ≤4.0 (% target API/tablet). The RMSEC is calculated by comparing the $\hat{c}_i$ (i.e., film-coating API content) values from the NIR prediction directly with the c_i values of the HPLC-determined composite assay value. The RMSEC equation also includes n, the number of samples in the calibration, and f, the PLS model rank or the number of PLS latent variables used in the model.

The RMSEC provides an initial assessment of the calibration model, which is validated by predicting an external validation set that spans the entire suitable range of operation conditions during commercial production [12]. The RMSEC is calculated as shown below:

$$\text{RMSEC:} \left[\frac{\sum_{i=1}^{n} (\hat{c}i - ci)^2}{n - f - 1} \right]^{1/2} \tag{17.1}$$

Per the acceptance criteria, the correlation coefficient (R) of the HPLC results vs. NIR predictions must be ≥0.990 which was calculated with the same results as in the RMSEC. The acceptance criteria for the slope (best-fit line) from 0.975 ≤ slope ≤ 1.025 and the intercept ≤2.0.

Table 17.1 Model acceptance criteria for determination of API added in film-coating suspension.

Parameter	Acceptance criteria	Results
Model acceptance criteria		
Calibration range (% target API/tablet)	40–80%	Calibration: 37.2–91.6%
RMSEC (% target API/tablet)	<4.0%	1.5%
Correlation coefficient (R) of HPLC vs. NIR	≥0.990	0.995
Slope of HPLC vs. NIR best-fit line (% target API/tablet)	0.975 < slope < 1.025	0.996
Intercept of HPLC vs. NIR best-fit line (% target API/tablet)	<2.0%	0.3%
Precision (RSD) – Method	<2.5%	0.48%
Precision (RSD) – Measurement	<1.0%	0.39%
Relative difference of projected endpoint (HPLC vs. NIR)	0.970 < relative endpoint < 1.030	Min: 0.976, Max: 1.025
REP results (cross validation) in model development		
Coater A (REP range)	N/A	Min: 0.980, Max: 1.001
Coater B (REP range)	N/A	Min: 0.993, Max: 1.019
Coater C (REP range)	N/A	Min: 0.983, Max: 1.018
Coater D (REP range)	N/A	Min: 0.990, Max: 1.002
Coater E (REP range)	N/A	Min: 1.000, Max: 1.000
Calibration model description		
Spectral range (cm^{-1})	N/A	5619–5789, 6032–6201, 6869–6954, 8593–8735
Spectral preprocessing	N/A	Savitzky–Golay 1st derivative, mean centering
Number of samples used for calibration	N/A	580
Number of PLS factors (Model rank)	N/A	4
Calibration correlation coefficient (R)	≥ 0.990	0.995
Slope of HPLC vs. NIR linear regression	0.975 < slope < 1.025	0.995
Intercept of HPLC vs. NIR linear regression	<2.0%	0.30%
Validation set for at-line NIR API-coating method		
Correlation coefficient (R)	≥ 0.990	0.995
Slope of HPLC vs. NIR prediction	0.975 < slope < 1.025	0.996
Intercept of HPLC vs. NIR prediction	<2.0%	0.30%
RMSEP (% target API/tablet)	<4.0%	1.50%
Validation range (% target API/tablet)	40–80%	39.2–91.4%
Relative end point	0.970 < relative endpoint < 1.030	Min: 0.976, Max: 1.025

The precision is evaluated through the prediction of 120 stored spectra that were obtained when the method was first implemented. This initial study included the placement of tablets at each of the 40 different positions, and three replicate studies were obtained at each position. A fully nested ANOVA is conducted at the manufacturing plant to compare the initial predictions to those obtained with the new calibration model. Method precision was evaluated by obtaining spectra of tablets at different positions in the tablet carousel. The measurement precision is obtained through repeated spectra of a tablet without changing its position in the tablet carousel. The method precision should be less than 2.5%, and the measurement precision should be ≤1.0%.

The relative end point (REP) values for the calibration set must meet the acceptance criteria of 0.970 ≤ REP ≤1.030; see Table 17.1. The REP is calculated by the ratio of the sum of HPLC results to NIR predictions from the 20 composite spectra per batch. A REP of 1.000 signifies a perfect agreement between HPLC and NIR results.

$$\text{REP:}\ \frac{\sum \text{HPLC values}}{\sum \text{NIR predicted value}} \tag{17.2}$$

The RMSEP (root mean square error of prediction) is determined for the external validation batches.

$$\text{RMSEP} = \sqrt{\sum_{i=1}^{N_p} (\widehat{c}_i - c_i)^2 / N_p} \tag{17.3}$$

where N_p is the number of samples used in external validation in this work; a single "sample" (i) refers to a matching pair of data.

The bias was also determined. Bias is calculated as the average of the residuals, calculated as shown below:

$$\text{Bias} = \sum_{i=1}^{N_p} (\widehat{c}_i - c_i) / N_p \tag{17.4}$$

The bias is based on the difference between the predicted and reference values. In contrast, the REP is based on the ratio of the sum of predicted and reference values. The two parameters measure the closeness to an accepted HPLC value.

17.2.6 Model Implementation and Monitoring in Manufacturing

The PLS calibration model developed in the Unscrambler X software is saved into the Process Pulse II configuration method. Process Pulse II is installed on the company server and computer/client, which the operator controls. The system is a client–server that requests a service or resource from another program (the server). The Thermo Results Operation is the Thermo Antaris II program used to control the spectrometer and acquire the NIR spectra. Process Pulse makes a copy of the NIR spectra obtained and predicts the content of the spray-dried API. Process Pulse is also used to obtain a report that describes the predicted results.

Tablets from an in-process batch are analyzed monthly using the NIR and HPLC methods. These are called monitoring batches, which evaluate the NIR method's predictive ability. The REP is calculated for the monitoring batches. A REP value between 0.970 and 1.030 indicates that the calibration model performs adequately (Figure 17.1).

17.3 Development of Calibration Model

17.3.1 Overview of Near-Infrared Spectra

Figure 17.2 shows a spectrum of the API that is applied through the film coating, as well as the principal formulation excipients. Broad and overlapping bands, typical of NIR spectra, are observed. However, narrow

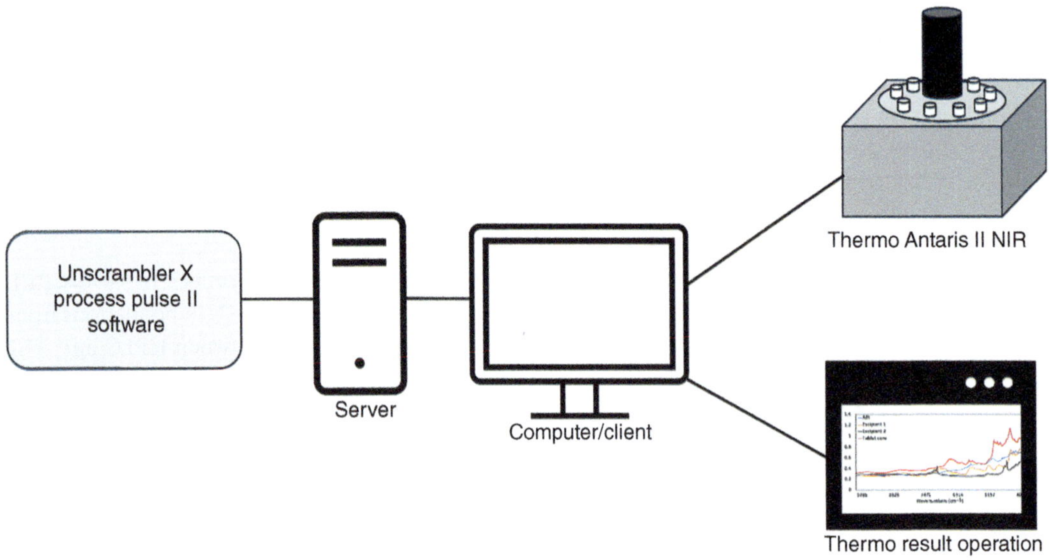

Figure 17.1 Scheme describing the acquisition of NIR spectra and the prediction of the API content by the NIR calibration model. *Source*: Maritza Reyna Liriano.

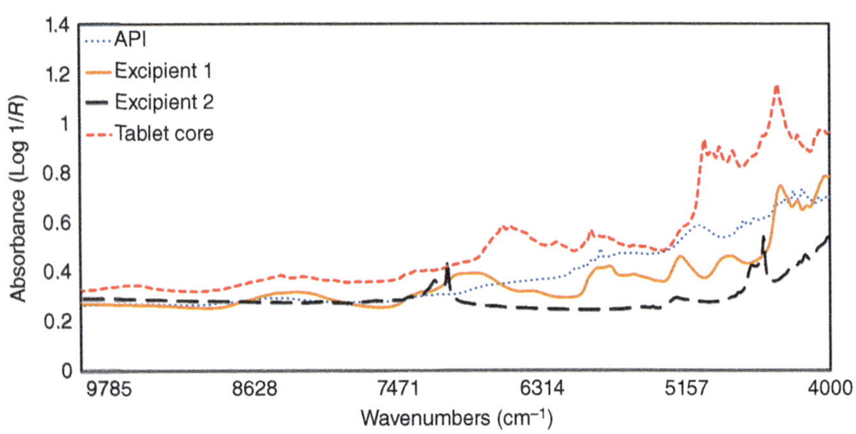

Figure 17.2 Shows spectra of the API applied through film coating, the tablet core, and two of the significant formulation excipients. *Source*: Maritza Reyna Liriano.

bands are observed around 7100 cm^{-1} for the first overtone of the O–H bands in one of the excipients. The identity of the API and excipients is proprietary information that cannot be revealed.

Figure 17.3 shows the spectra of the film-coated tablets at 40% (black line), 60% (black line), and 80% (red line) of the targeted drug content and the spectral regions used to develop the calibration model. The baseline increases throughout the entire spectral range during the film coating. The NIR diffuse reflectance spectra baseline depends on the fraction of diffusely reflected radiation received by the detector. When the radiation that returns to the detector I_{refl} is similar to that received from the reference (I_0), log (I_{refl}/I_0) is log 1, and the baseline is 0. The baseline increases when only a fraction ($1/c$) of the diffusely reflected radiation returns to the detector. For example, if 1/10th of the radiation returns to the detector,

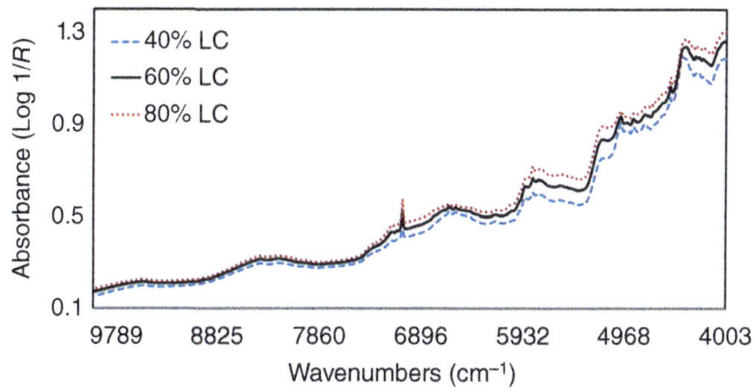

Figure 17.3 NIR spectra of the coated tablets of a monitoring batch from the first 40% target (black), 60% target (red), and final 80% (red) from sample process points of the coating process. The bands used in the calibration model are shaded in gray. *Source*: Maritza Reyna Liriano.

the baseline will be 1 (−log 0.1) (Eq. 17.5). Baselines over 0 are frequently observed in NIR spectroscopy. An increase in baseline has been observed in real-time spectra of powder flowing from a funnel. A high baseline was observed when few particles interacted with the NIR radiation [12, 13]. The baseline change (or offset) is described as log c in Eq. (17.5).

$$A_{\text{meas}} = -\log\left(1/c \times I_{\text{refl}}\right)/I_0 = \log c + A \tag{17.5}$$

As the film coating progresses, the film absorbs more radiation, and correspondingly, the fraction of radiation that returns to the detector decreases. The decrease in the fraction of radiation that reaches the detector has been observed in previous studies for both in-line [12], and off-line measurements [14]. These changes in the fraction of radiation that reaches the detector are the reason why differences are observed in the baseline of NIR spectra [13].

The baseline changes are observed throughout the entire spectrum as the film coating progresses from 40% (black line), 60% (black line), and 80% (red line) of the targeted drug content. The NIR spectra show bands from the film coating and the tablet core. The depth of penetration of the NIR radiation using polymer film layers has been studied with results of 2.95–3.12 mm [14]. The film coating applied to the tablets has a much smaller diameter, and therefore, the diffuse reflectance spectra contain bands related to the tablet core. The bands from the tablet core are present in the spectra obtained at all stages of the film-coating process, and they decrease as the film coating progresses.

Figure 17.4 shows the first-derivative Savitzky–Golay (15 pts) spectra at 40%, 60%, and 80% of the addition of the film coating. The spectral bands related to the film coating, such as the 7080 cm^{-1} and 5250 cm^{-1} bands shown in Figure 17.4 inserts, grow as the process progresses. However, the spectral areas related to the tablet core became proportionally lower as the thickness of the film increased. The inserts in Figure 17.4 also show the decrease observed in the 6784–6630 cm^{-1} and 5365–5172 cm^{-1} spectral regions related to the tablet core.

Figure 17.5 shows batches projected at the top (batch #1), middle (batch #2), and bottom along PC-2 (batch #3). The changes along PC-1 describe the addition of the API suspension as the film coating progresses. The most negative scores along PC-1 are for the spectra obtained at 40% target, the first stage of the coating process where samples are collected and measured by NIR spectroscopy. The most positive scores along PC-1 (far right) correspond to adding 80% target of the spray-dried API. The vertical lines in Figure 17.5 indicate the five stages at which samples are collected. The changes along PC-2 are related to the moisture content of the tablet film coating. The tablets with lower residual moisture will project in the score plot's bottom part (PC-2 negative).

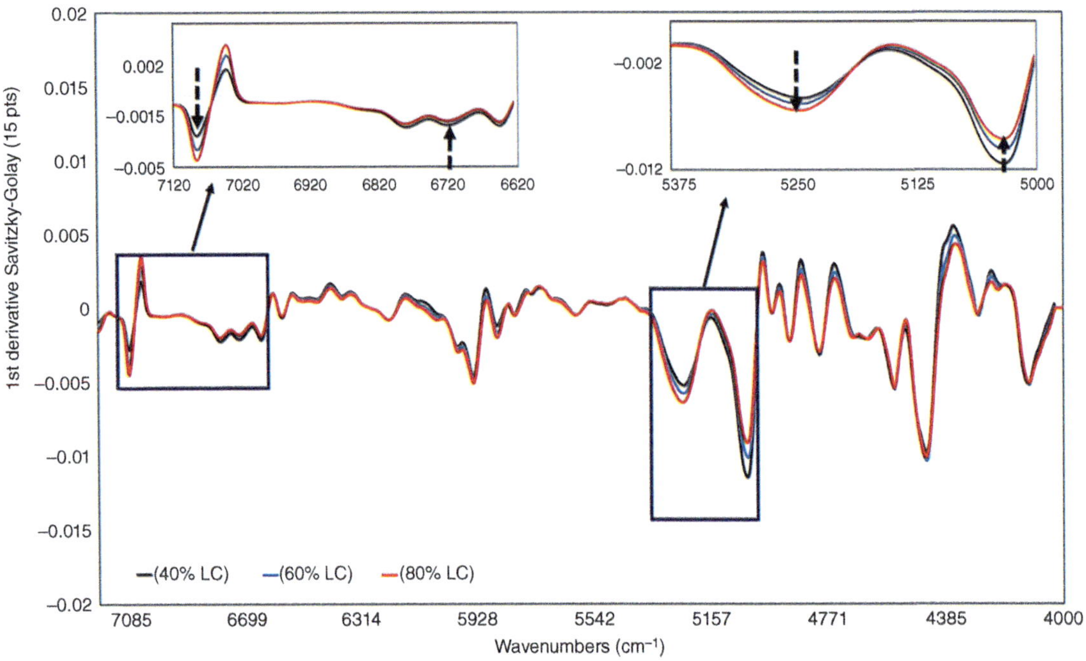

Figure 17.4 First-derivative spectra at 40%, 60%, and 80% targeted API/tablet. The insert shows spectral areas that grow as film coating (~7100 and 5250 cm^{-1}) progresses and a spectral area that decreases (~6900–6572 cm^{-1}). *Source*: Maritza Reyna Liriano.

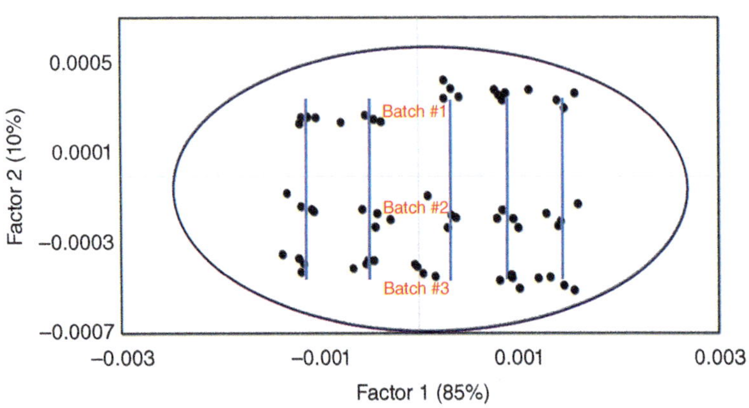

Figure 17.5 PCA score plot for three in-process batches, which differ along PC-2. *Source*: Maritza Reyna Liriano.

17.3.2 Selection of Spectral Range for Model Development

The calibration model is built with 20 spectra from each in-process batch. This comprises four averaged spectra from tablets collected at approximately 40, 50, 60, 70, and 80% of the API added during the film-coating process.

Table 17.1 offers a comprehensive overview of the calibration set used in this study. A total of 29 batches, comprising 580 spectra, were selected for the calibration set. The calibration set includes spectra from

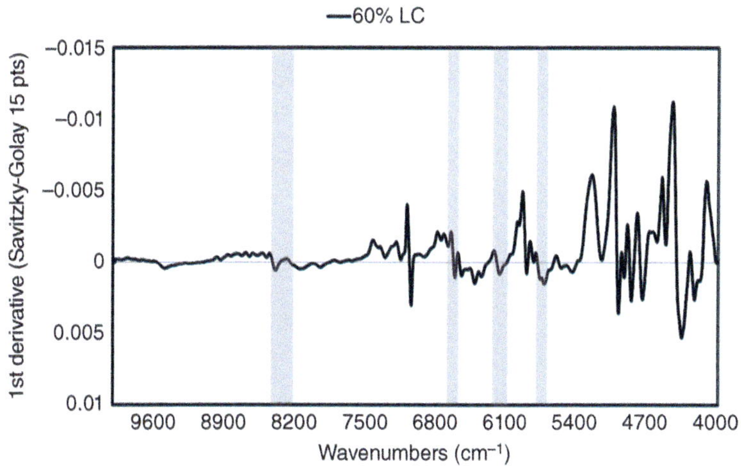

Figure 17.6 The first-derivative spectrum with the spectral areas used for the calibration model is marked in gray. *Source*: Maritza Reyna Liriano.

four coaters (A–D) used by the company and two NIR instruments NIR01 and NIR02 used to collect the in-process spectra. The objective was to ensure an equal number of batches using the two NIR instruments and incorporate all NIR spectrometers and coater combinations in the calibration model. The tablets included in the calibration model varied in API content from 37.2 to 91.6% of the target API/tablet.

Figure 17.6 shows the first-derivative spectrum and the areas included in the calibration model. Four spectral ranges were included in this calibration model: 8735–8593, 6954–6869, 6201–6032, and 5789–5619 cm^{-1}. These spectral ranges include several bands associated with the spray-coated API. All the bands in the spectral range chosen are weak or moderate intensity. Based on previous studies, the NIR radiation should penetrate the entire film coating. Previous studies found a penetration depth of 2.95–3.12 mm for polymer films, while tablet film coatings are typically less than 100 μm thick.

The REP results for the calibration set are obtained from the cross-validation method on the predictions of the NIR spectra using the same set indicated in the model. A cross-validation was performed, leaving out all the spectra from a batch and predicting the left-out batch as if they were unknown samples. This calculation determines whether the calibration model meets the acceptance criteria for the REP values in Table 17.1. The data matrix is organized so that each of the 20 rows corresponds to a different batch. The REP values in Table 17.2 are obtained from cross-validation results corresponding to PLS models. These results satisfy the REP range of 0.970–1.030 described in Table 17.1. The lowest REP result was 0.977 from batch A05 and coater C. The highest REP result was 1.029 from batch A14 and coater B, within the REP limit criteria.

The value of the root mean square error of calibration (RMSEC) obtained was 1.5, which is <4.0, and the calibration correlation coefficient (R) was $0.995 \geq 0.990$; these parameters are part of the model acceptance criteria described in Table 17.1.

Table 17.1 describes the model parameters used to build the calibration model and includes the acceptance criteria, where applicable. The selected model parameters use the same spectral regions marked in Figure 17.6 and the preprocessing described in Table 17.1. A total of 580 spectra (20/per batch) were selected to build the calibration set ranging from 37.2 to 91.6 (% targeted API/tablet) using 4 PLS factors. The correlation and slope were 0.995, respectively. The intercept of 0.3 and an RMSEC of 1.5 (% targeted API/tablet) conform with the model acceptance criteria for the determination of API added in film-coating suspension provided in Table 17.1.

Table 17.2 Summary of 29 batches used to build the calibration model.

NIR	Batch ID	Coater	REP
NIR01			
	A01	B	1.010
	A02	D	0.990
	A03	E	1.000
	A04	D	1.002
	A05	C	0.977
	A06	A	0.980
	A07	C	0.998
	A08	B	1.019
	A09	C	1.006
	A10	C	1.018
NIR02			
	A11	C	1.013
	A12	C	1.019
	A13	A	0.991
	A14	B	1.029
	A15	A	0.991
	A16	D	0.994
	A17	A	0.993
	A18	D	0.992
	A19	C	0.994
	A20	B	0.993
	A21	C	1.011
	A22	B	1.001
	A23	D	0.996
	A24	B	1.018
	A25	B	1.007
	A26	C	0.988
	A27	A	0.990
	A28	A	1.001
	A29	C	0.983

17.4 Validation of Calibration Model

The new model was validated with an external set of batches not included in the calibration model. The validation set contains 940 validation spectra from NIR01 and NIR02 (averaged), representing 47 batches obtained during monitoring batches. The spectral projections in the PLS score plot of the NIR model are

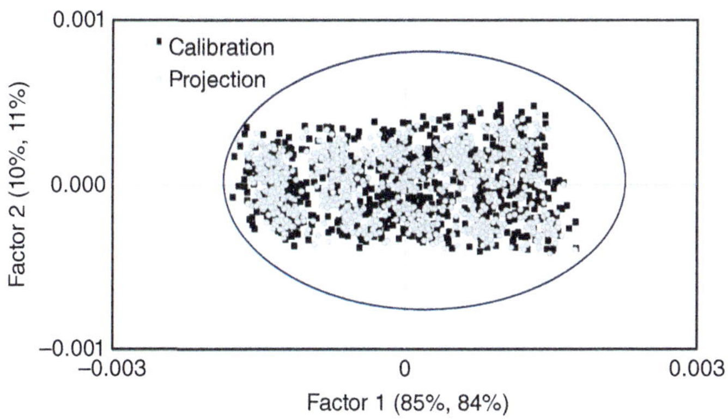

Figure 17.7 PLS score plot (99.0% CL ellipse) for spectra projection from monitoring batches into the model.
Source: Maritza Reyna Liriano.

shown in Figure 17.7, and a summary of prediction results is provided in Table 17.3. Validation batches have a collection period of over two years and were manufactured using multiple batches of API and excipients. These validation batches are independent of the calibration set.

The validation spectra are within the variation summarized by the first two principal components of the model. The lighter gray color in Figure 17.7 denotes the prediction sample scores. All the scores from NIR01 and NIR02 and coaters 1 to 4 are within the variation observed in the model. The PLS score plot summarizes 95% of the spectral variation in the calibration samples and 95% in the projected validation samples.

The validation statistics presented in Table 17.1 include the REP, bias, and RMSEP performed to evaluate model performance. The lowest REP value obtained was 0.973, and the highest was 1.023, within the model acceptance criteria. The lowest bias was −1.868 (% target) and the highest was 1.851 (% target).

17.5 Compliance with Model Acceptance Criteria

17.5.1 Range and Linearity

The range of the at-line NIR model is defined by the samples containing the minimum and maximum % target API/tablet in the validation data set. The NIR model contains coating content ranging from 39.2% to 91.6% of the target. Figure 17.8 shows the linearity between the NIR-predicted API/tablet values and the matching HPLC results for all the unknown samples in the validation set. The correlation coefficient obtained was 0.995. The predicted values agree with the reference values obtained from the HPLC method.

Two additional linearity metrics, namely slope and intercept of the regression of the in-process HPLC result vs. the corresponding NIR prediction, are described in Table 17.1. The slope and intercept of the updated model, which successfully meets the acceptance criteria are listed in Table 17.1 for the API dosage.

17.5.2 Specificity

Specificity is generally defined as the ability to unequivocally determine the analyte in the presence of components in a formulation (e.g., impurities, degradants, or excipients). Specificity for an API is conventionally verified if the loading plots' main features [15] correspond to the NIR spectrum of the analyte of interest [16]. The NIR spectrum should be pretreated in the same way as the spectra used in the model. Figure 17.9 shows the [X]-loadings weight vector for the first PLS factor vs. the API spectrum using the same pretreatment and

Table 17.3 Summary of predictions of 47 batches of validation set.

NIR	Batch ID	Coater	REP	Bias (% target)	RMSEP (target)
NIR01	V01	A	0.988	0.763	1.214
	V02	A	1.020	1.391	1.834
	V03	B	0.997	−1.331	2.053
	V04	A	0.978	0.177	0.980
	V05	A	1.001	−0.050	1.174
	V06	B	1.000	0.016	0.887
	V07	C	1.023	−1.444	1.722
	V08	A	0.999	0.139	1.127
NIR02					
	V09	C	0.981	0.837	1.326
	V10	B	0.990	−1.868	2.192
	V11	B	0.974	1.275	1.910
	V12	C	0.994	1.007	1.330
	V13	C	0.988	1.739	2.043
	V14	B	0.982	0.388	1.018
	V15	B	0.984	0.774	1.657
	V16	B	0.986	1.135	1.950
	V17	C	0.981	0.993	1.699
	V18	C	0.995	0.952	1.508
	V19	B	0.990	1.289	1.621
	V20	A	1.008	0.291	1.089
	V21	B	1.012	0.663	1.152
	V22	A	0.982	−0.498	1.763
	V23	C	1.008	−0.772	1.406
	V24	B	1.005	1.153	1.519
	V25	C	0.991	−0.484	1.421
	V26	C	1.010	−0.320	1.276
	V27	C	1.015	0.603	1.424
	V28	C	0.998	−0.577	1.348
	V29	A	0.994	−0.932	1.336
	V30	C	0.987	0.384	1.969
	V31	C	1.008	−0.521	1.134
	V32	A	0.995	0.303	1.794
	V33	B	0.995	0.335	1.252
	V34	C	1.007	−0.442	1.185
	V35	C	1.030	−0.007	1.116
	V36	A	1.000	−0.446	1.579

Table 17.3 (Continued)

NIR	Batch ID	Coater	REP	Bias (% target)	RMSEP (target)
	V37	B	1.007	−0.919	1.502
	V38	C	1.014	−0.270	1.436
	V39	A	1.004	−0.071	1.939
	V40	A	1.001	0.038	0.929
	V41	C	0.992	−0.503	1.185
	V42	B	0.973	0.554	1.273
	V43	B	1.001	1.851	2.122
	V44	A	1.002	0.059	0.820
	V45	D	1.007	−0.109	2.054
	V46	A	0.996	0.241	1.492
	V47	B	1.008	−0.441	1.144

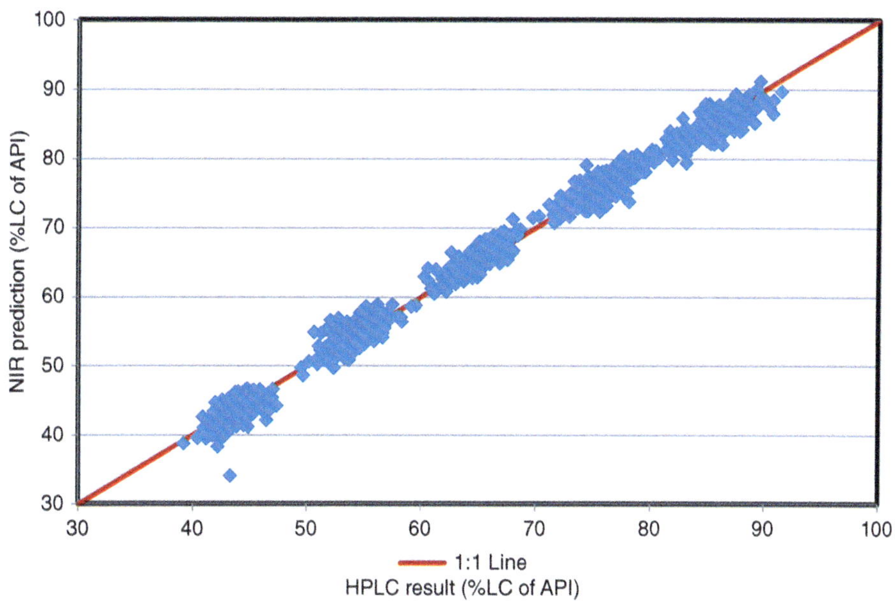

Figure 17.8 The plot of in-process HPLC result vs. NIR predictions % target API/tablet regression results for the NIR model. The red line represents a 1:1 fit. *Source*: Maritza Reyna Liriano.

spectral region of the selected model (5619–5789, 6032–6201, 6869–6954, and 8593–8735 cm^{-1}). Figure 17.9 demonstrates good spectral similarity between the loadings weight vector for factor 1 and the API spectrum in the spectral region of the selected model. The correlation coefficient was calculated, yielding a value of 0.60, indicating that 60% of the variability explained by the first factor is related to API content. The loading vectors indicate the importance of the spectral variables related to the API [17].

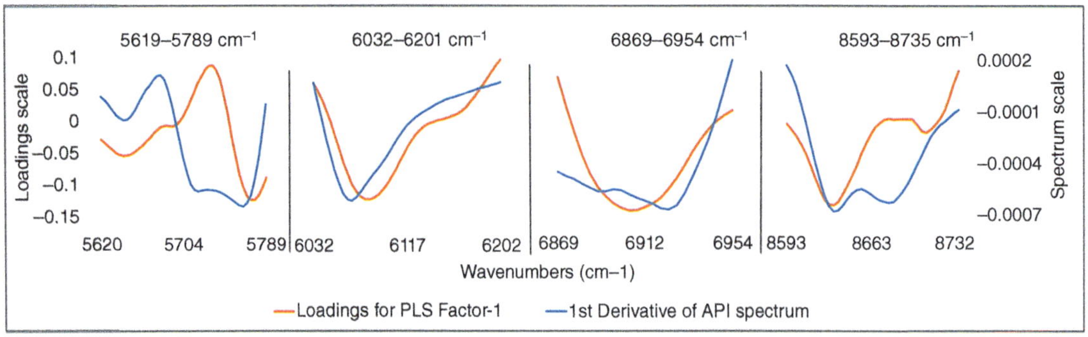

Figure 17.9 Loading weight vector from NIR calibration model developed using 1st derivative (15 points) in the selected model spectral regions from the 1st PLS factor. *Source*: Maritza Reyna Liriano.

17.5.3 Accuracy

The accuracy of an analytical procedure "is the closeness of test results obtained by that procedure to the true value" [18]. Model accuracy is defined using two parameters: REP and RMSEP, as described in Table 17.1. The REP (Eq. 17.2) is a metric of accuracy since it measures the closeness of the agreement of the predicted targeted API/tablet to an accepted reference HPLC value. The RMSEP in Table 17.3 summarizes the bias and standard deviation of the validation prediction set.

17.5.4 Precision

Method and measurement precision were estimated from the variance estimates of a fully nested ANOVA experimental design. Measurement precision is obtained with three consecutive spectra without changing the position of a tablet in the carousel. Method precision involves removing and placing the tablets into four (4) different positions within the table carousel. The spectra used in the precision studies were collected years ago and have been predicted repeatedly in the model updates. The spectra are predicted in the model update, and NIR predictions using the same number of factors from the model are analyzed in the fully nested ANOVA. Table 17.4 contains the precision metric results using the updated NIR calibration model.

17.5.5 Robustness

The robustness of an analytical procedure is a measure of its capacity to meet the expected performance criteria during normal use [16]. The validation batches include several batches of the API and excipients received at the commercial manufacturing site. The validation also includes variation in processing conditions, as spectra were obtained from tablets manufactured in all coaters.

Table 17.4 Precision metrics obtained using the NIR procedure.

Metric	Mean (% target API/tablet)	Standard deviation (% target API/tablet)	% RSD	Acceptance criteria (% RSD)	Pass/Fail
Method precision	57.092	0.272	0.48%	≤2.5%	Pass
Measurement precision		0.222	0.39%	≤1.0%	Pass

The REP validation metric was calculated batch-wise, where a single pair of NIR- and HPLC-determined spray coating endpoints was determined for each batch. Using the updated NIR model, the REPs for the forty-seven (47) commercial batches in the validation set ranged from 0.976 to 1.025 (see Table 17.3), and these results meet the acceptance criteria in Table 17.1. The acceptance criteria are ±3% of target (i.e., 0.970 to 1.030 acceptance range). As part of method validation, the conformance to REP acceptance criteria demonstrates the robustness of the NIR model to typical batch, coater, and instrument variation.

17.6 Outlier Diagnostics

In addition to predicting % target API/tablet values, the multivariate model provides two outlier metrics to ensure method predictability. The first metric is Hotelling T^2 (Eq. 17.6), which evaluates the new NIR spectral data project relative to the center of the model space within the modeled multivariate space. The higher the T^2 value, the greater the deviation from the population mean. Hotelling T^2 is widely used in quality control and multivariate process monitoring [17, 19].

$$T^2 = n(\bar{x} - \mu)^{\mathrm{T}} \boldsymbol{S}^{-1} (\bar{x} - \mu) \tag{17.6}$$

where n is the sample size; $\bar{x}$ is the sample mean vector; μ is the population mean vector; S is the sample covariance matrix, and T is the transpose of a matrix.

The second metric is Q-residuals (Eq. 17.7), which evaluates whether the new spectral data are adequate, like those used to develop the calibration model. The Q-residual quantifies the amount of variance in the data that is not explained by the principal components included in the model. The larger Q residual indicates that the sample has significant variation not captured by the model, suggesting potential outliers or model inadequacy [20].

$$Q = e^{\mathrm{T}} e \tag{17.7}$$

where e is a residual vector (difference between the original data point and its projection onto the principal component subspace) and T is the transpose of a matrix. As with other statistical metrics, alert limits can be applied to these metrics during the model's routine operation to indicate an outlier condition. The specific values of these alert limits for the two-outlier metrics are listed in Table 17.5.

Figure 17.10 helps to visualize Hotelling vs. Q-residuals in two dimensions for the calibration set (black circle) and validation set (red circles). In this case, all Hotelling T^2 and Q-residuals are within the established manufacturing limit criteria in Table 17.5. The insert in Figure 17.10 shows that the T^2 and Q-residual values are similar for the calibration and validation spectra.

The Hotelling vs. Q-residuals plot is useful for identifying samples that do not conform to the model, potentially signaling deviations in the spectra. The points near the origin are samples with low Hotelling T^2 and Q residuals, suggesting they fit the model well in terms of latent variable space and residual variance. Points with high T^2 but low Q residuals would lie far along the x-axis. High Q residuals with low T^2 would be observed distant along the y-axis. Those samples would be close to the model center in latent space but have significant unexplained variance.

Table 17.5 Outlier diagnostic alert limits the NIR model.

Outlier diagnostic	99.0% limit	99.9% limit
Hotelling T^2	64.919	105.964
Q-residuals	4.53×10^{-6}	7.39×10^{-6}

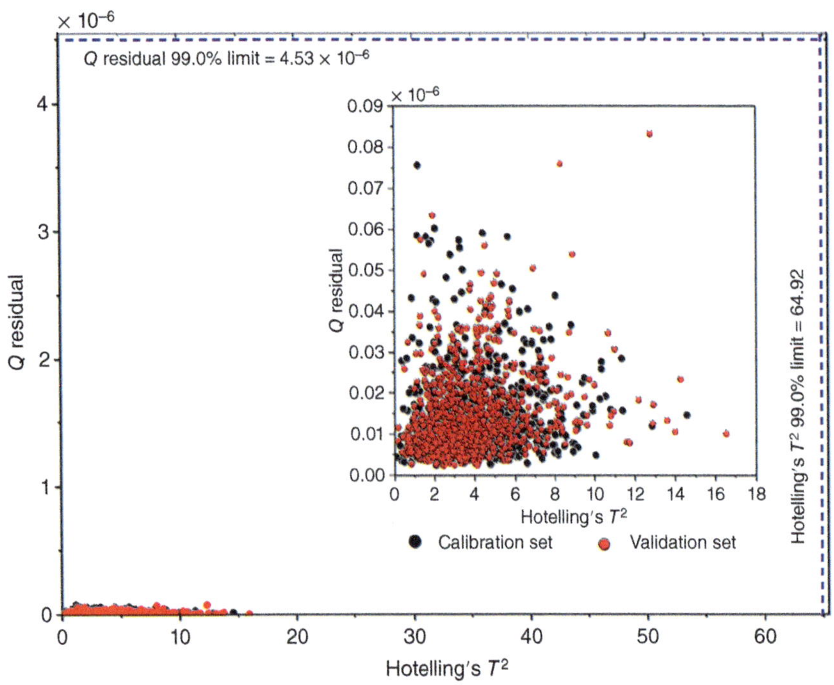

Figure 17.10 Q-residuals vs. Hotelling T^2 for calibration and validation. *Source*: Maritza Reyna Liriano.

17.7 Industry Final Assessment of Model

After reviewing the method's linearity, range, accuracy, and robustness metrics using the updated NIR model, it can be concluded that it meets the acceptance criteria (see Table 17.1) and is suitable for its intended use.

17.8 Conclusion

This chapter has presented the application of NIR spectroscopy to accurately add an API within the film coating of a tablet. A set of model acceptance criteria has guided the implementation of the method. A data management system was implemented to obtain the NIR spectra and determine the API content. A procedure was also developed to monitor the performance of the calibration model.

Acknowledgment

Graduate student Dhavalkumar Patel is thanked for his help with figures.

References

1 Talwar, S., Nunes, C., Stevens, T. et al. (2017). Understanding the impact of chemical variability and calibration algorithms on prediction of solid fraction of roller compacted ribbons using near-infrared (NIR) spectroscopy. *Applied Spectroscopy* 71: 1209–1221. https://doi.org/10.1177/0003702816671960.

2 Dahm, D.J., Dahm, K.D., and Norris, K.H. (2000). Test of the representative layer theory of diffuse reflectance using plane parallel samples. *Journal of Near Infrared Spectroscopy* 8: 171–181. https://doi.org/10.1255/jnirs.276.

3 Alcalà, M., Ropero, J., Vázquez, R., and Romañach, R.J. (2009). Deconvolution of chemical physical information from intact tablets NIR spectra: two-three-way multivariate calibration strategies for drug quantitation. *Journal of Pharmaceutical Sciences* 98: 2747–2758. https://doi.org/10.1002/jps.21634.

4 Short, S.M., Cogdill, R.P., Wildfong, P.L.D. et al. (2009). A near-infrared spectroscopic investigation of relative density and crushing strength in four-component compacts. *Journal of Pharmaceutical Sciences* 98: 1095–1109. https://doi.org/10.1002/jps.21473.

5 Shi, Z. and Anderson, C.A. (2009). Scattering orthogonalization of near-infrared spectra for analysis of pharmaceutical tablets. *Analytical Chemistry* 81: 1389–1396. https://doi.org/10.1021/ac802105v.

6 Johansson, J., Folestad, S., Josefson, M. et al. (2002). Time-resolved NIR/vis spectroscopy for analysis of solids: pharmaceutical tablets. *Applied Spectroscopy* 56: 725–731. https://doi.org/10.1366/000370202760077676.

7 Aldama, J., Shi, Z., Ortega-Zúñiga, C. et al. (2021). Fractal and polarization properties of light scattering using microcrystalline pharmaceutical aggregates. *Applied Spectroscopy* 75: 94–106. https://doi.org/10.1177/0003702820949272.

8 Möltgen, C.-V., Puchert, T., Menezes, J.C. et al. (2012). A novel in-line NIR spectroscopy application for the monitoring of tablet film coating in an industrial scale process. *Talanta* 92: 26–37. https://doi.org/10.1016/j.talanta.2011.12.034.

9 Andersson, M., Folestad, S., Gottfries, J. et al. (2000). Quantitative analysis of film coating in a fluidized bed process by in-line NIR spectrometry and multivariate batch calibration. *Analytical Chemistry* 72: 2099–2108. https://doi.org/10.1021/ac990256r.

10 Ciurczak, E.W., Igne, B., Workman, J., and Burns, D.A. (ed.) (2021). *Handbook of Near-Infrared Analysis*. Boca Raton, FL: CRC Press.

11 Goodwin, D.J., Van Den Ban, S., Denham, M., and Barylski, I. (2018). Real time release testing of tablet content and content uniformity. *International Journal of Pharmaceutics* 537: 183–192. https://doi.org/10.1016/j.ijpharm.2017.12.011.

12 Ropero, J., Beach, L., Alcalà, M. et al. (2009). Near-infrared spectroscopy for the in-line characterization of powder voiding Part I: development of the methodology. *Journal of Pharmaceutical Innovation* 4: 187–197. https://doi.org/10.1007/s12247-009-9069-z.

13 Næs, T. (ed.) (2004). *A User-Friendly Guide to Multivariate Calibration and Classification*. Chichester: NIR Publications.

14 Ortega-Zuñiga, C., Reyes-Maldonado, K., Méndez, R., and Romañach, R.J. (2017). Study of near infrared chemometric models with low heterogeneity films: the role of optical sampling and spectral preprocessing on partial least squares errors. *Journal of Near Infrared Spectroscopy* 25: 103–115. https://doi.org/10.1177/0967033516686653.

15 ICH Q2(R2) Guideline on validation of analytical procedures – Step 5 (n.d.).

16 U.S. Department of Health and Human Services FDA Center for Drug Evaluation and Research, U.S (2006). Department of Health and Human Services FDA Center for Biologics Evaluation and Research, U.S. Department of Health and Human Services FDA Center for Devices and Radiological Health. Guidance for industry: patient-reported outcome measures: use in medical product development to support labeling claims: draft guidance. *Health and Quality of Life Outcomes* 4: 79. https://doi.org/10.1186/1477-7525-4-79.

17 Esbensen, K. (2018). Fundamentals of principal component analysis. In: *Multivariate Data Analysis*, 69–105. Camo Software, AS.

18 United States Pharmacopeial Convention (2013). Validation of compendial procedures. In: *United States Pharmacopeia 36*, 1225–1241. Rockville, MD: United States Pharmacopeial Convention.

19 Bu, D., Wan, B., and McGeorge, G. (2013). A discussion on the use of prediction uncertainty estimation of NIR data in partial least squares for quantitative pharmaceutical tablet assay methods. *Chemometrics and Intelligent Laboratory Systems* 120: 84–91. https://doi.org/10.1016/j.chemolab.2012.11.005.

20 Jolliffe, I.T. (2002). *Principal Component Analysis*, Springer Series in Statistics, 2e. New York: Springer https://doi.org/10.1007/b98835.

Part III

Real-World Implementation Vignettes

18

Automating the Spectrometer Calibration and Maintenance Processes

Brian G. Rohrback

Infometrix, Inc., Bothell, WA, USA

18.1 Introduction

Much of this book is focused on a practitioner's guide to managing the development of optical spectroscopy procedures for use in research and routine analysis. This section is devoted to the next phase of implementation: maintaining the ongoing relationship between the information in the spectra and the property of interest. We propose that the process of instrument calibration can be automated using concepts like those used in artificial intelligence (AI) and machine learning, with the benefits of efficiency, objectivity, and performance reliability.

As is well documented in this book, spectroscopy is the most practical mechanism for investigating the chemistry of a reaction process in real time. It is critical to make these chemistry measurements both to ensure product quality and to flag situations tied to process upsets and unexpected production pathways. We have spent the last half-century investigating the primary optical spectroscopy technologies spanning much of the electromagnetic spectrum, and through this experience, investigators have identified a set of best practices for calibrating and maintaining calibration quality in routine analytical settings. Purchasing a quality (Fourier transform infrared (FTIR), near-infrared (NIR), or Raman instrument is not enough. The quality of the final predictions is a function of both that instrument and the calibration that extracts the information from spectra and delivers content to a control system.

For most of its history, the process of calibrating optical spectrometers to perform in an application setting has fallen to those versed in the art of chemometrics. There is considerable variability in the training and the skill of those charged with initiating and maintaining calibration quality. Over the last 20 years, the increased deployment of spectrometers for routine quality chores has resulted in an even larger gap in model quality across the industry. The calibration work largely falls to technicians who may not have been exposed to adequate training and who perform the calibration work sporadically, often in a rush. The revolving door of company personnel is exacerbating the situation, frequently creating a yearly swap of personnel charged with this work, and thus, quality suffers.

Assume that our goal is to automate completely the installation and maintenance of an optical spectrometer for a specific application. The idea has been presented in numerous publications, including Workman et al. [1]. We can break down the chemometrics tasks, assigning best practices procedures for each. The first is to optimize the method to be employed, identify the best preprocessing, perform wavelength selection, and choose the best algorithmic approach. Secondly, we find the primary breakdown in industrial use is the maintenance of a model as conditions, such as the spectrometer, ingredients, and unit operations, change.

Practical Quantitative Vibrational and Electronic Spectroscopy: A Guide for Developing, Optimizing, and Validating Procedures, First Edition. Edited by Deborah A. Peru.

Companion website: www.wiley.com/go/peruelectronicspectroscopy

The main thrust of model maintenance involves two tasks: finding the optimal samples to include in the calibration and defining model complexity (number of factors or rank).

The current discussion outlines the steps to mimic the results that an experienced chemometrician would achieve if charged to produce an optimized model for a specified application.

18.2 Review of the Calibration Process

Deploying multivariate analysis has significant power; the task is to build a calibration model for the analyte concentration or physical property of interest. Regression establishes a functional relationship between some quantitative sample property, the dependent variable, and one or more independent variables. In analytical chemistry, the independent variables are often chromatographic or spectroscopic measurements.

Compared to univariate regression, multivariate methods offer improved precision, more sophisticated outlier detection, and, in the case of factor-based algorithms, the possibility of compensating for interferences. By describing the correlations among the variables, these models can point out phenomena that give rise to the data structure. If more data becomes available later, it is either added to the existing data pool or modeled separately.

In generating predictive systems, modeling dependent variables in terms of independent variables is only the first step. Later, the model is applied to independent variable measurements on new samples (sometimes called unknowns) to predict their dependent variable properties. The data used to create a predictive model are called the training set, and the model-building phase is often referred to as calibration. The term calibration model emphasizes that prediction, not description, is the ultimate objective.

After a model is created, the entire procedure should be validated before it is used. Validation, which establishes the reliability of a predictive model, entails making predictions on samples with already-known dependent variable settings. A model may describe the training set very well but yield poor predictions when applied to future samples. Thus, validation is as important as the model-building process itself because it tells us how well a model should perform.

Because factor-based methods characterize variance, they produce combinations of independent variables that account for variation associated with unquantified components even without explicit information about dependent variable settings for those components. These factor models will work on unknowns whose component values exceed the range of values in the model but may have additional prediction errors due to the extrapolation.

Two demands offset the advantages of factor-based methods. First, a relatively large number of training set samples are required to characterize the variance structure adequately. Second, the user must decide how many factors are necessary to model the properties to be predicted, a task that may not be perfectly clear-cut.

18.3 Automating the Calibration Procedure

Selecting a method for processing future data is the first step when setting up a spectrometer for routine analysis. This process can be automated and, in some cases, performed completely intervention-free. The goal is to identify the best combinations of preprocessing (scaling and transformations) and to select the optimal wavelengths to use. The steps for multivariate calibration method automation involving spectral data must be reliable across different spectroscopies and diverse applications. The system must avoid unnecessary complications of the model as this tends to reduce prediction performance. The steps that could be incorporated into an automated procedure for building and selecting models are described in Figure 18.1:

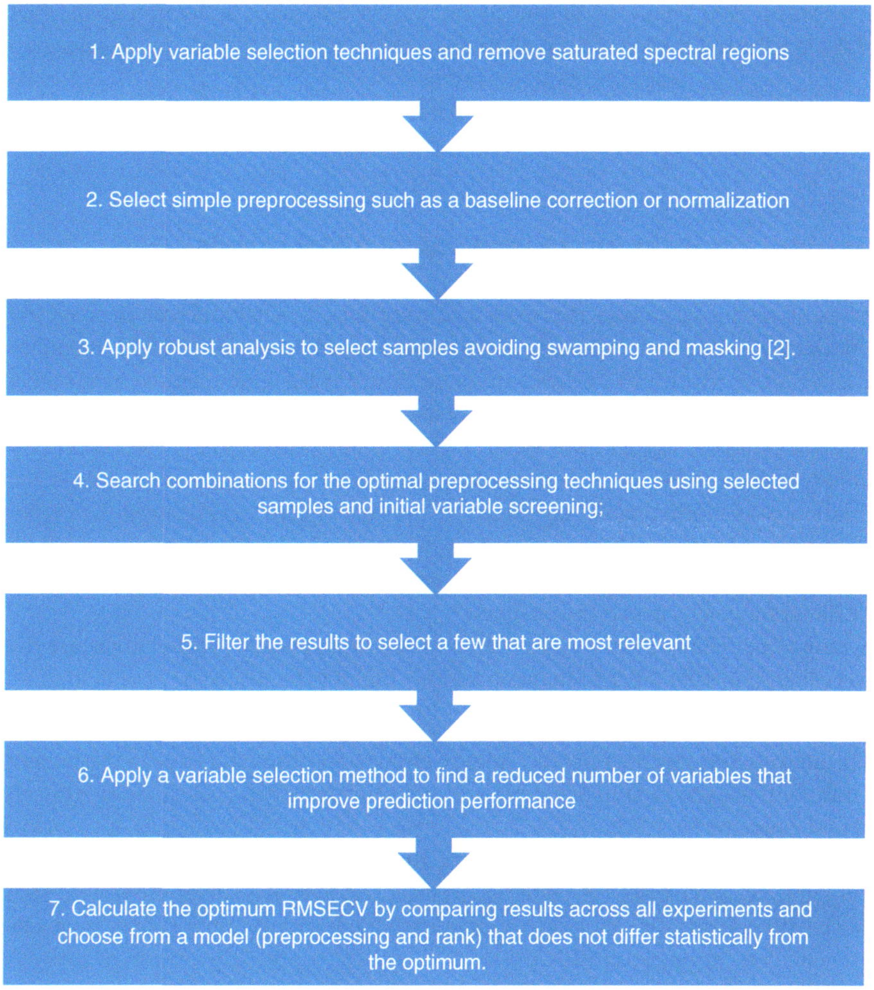

Figure 18.1 Automated model building and selection procedure. *Source*: Brian Rohrback.

The purpose of preprocessing is to remove variability from the data that would otherwise degrade prediction performance. Variable selection is designed to find a reduced number of variables that improve prediction performance. Of course, this could mean selecting variables that are the most chemically relevant, but note that often, for inverse modeling algorithms like partial least squares (PLS), the selected variables may not be chemically relevant.

Assume there is systematic baseline variation in the data. Without first preprocessing to remove the random baseline offset, a variable selection method with an inverse model can select some baseline effects to incorporate into the model to account for this variability. This can make the model unnecessarily complicated and reduces prediction performance. Some preprocessing approaches, such as derivatives, should not be applied to discontinuous wavelength ranges because the gaps in otherwise continuous profiles would greatly affect the resulting transformed data.

Although few software packages support locally weighted options, PLS can be applied globally and in a locally weighted approach. The automated locally weighted regression (LWR) system is more completely documented in Pell et al. [2]. It is best to compare predictions from LWR model to those from a global

PLS to see if the improvement is statistically significant. If there is no difference between a global PLS model's predictions and the locally weighted model's performance, the automated system should select the global PLS.

Once the method development stage is completed, it likely will not need to be addressed again for a routine application during the spectrometer's lifetime. However, maintaining the calibration remains an issue.

18.4 Practical Considerations for Automated Model Maintenance

When looking to automate a maintenance process, the issues to consider are efficiency and cost. In preparing for calibration work, efficiency is linked to minimizing error and the speed of assembling and maintaining data. Cost considerations are partly driven by direct measures, such as the time required to update a model by hand, and indirect factors, such as the operator's skill and experience in calibration modeling. A single, coordinated automated calibration system will speed response time to changes in process unit operation or ingredient chemistry and lower the cost of calibration by removing a time consuming, manual task. A fully automated calibration system also enables continuity in the face of personnel changes. In corporate settings, there are often several locations where spectroscopy is being used, and automation offers to scale multi-instrument installations under a common computational and visualization umbrella, enabling a common baseline for calibration quality across the company.

A succession of steps should be followed for the automated model maintenance process to ensure the dataset is free of spurious calibration samples. The stepwise manner found to be successful is as follows:

1) It is critical to eliminate extreme outliers as they can have undue influence on later screening steps. These extremes can result from bad spectra, reference values, or both. Spectral errors due to instrument or sampling system transients are removed in this first stage, adjusting for differences in the length and covariance structure of the spectra.
2) The spectra's property information may also contain errors, particularly if reported manually. Transposing numbers or slipping a decimal place are common occurrences. Even if a laboratory information management system (LIMS) system is in the loop, the reference methodology often suffers from low precision, leading to errant data logging.
3) Screening the combined spectra and associated reference values for extreme values, where samples exceeding a threshold set by several median absolute deviations from the median (MADs), mimics best practices that would be done manually.
4) Automation should anticipate clustering in the data and adjust so that a large population does not confer outlier status on a smaller valid cluster.
5) Robust measures of location and spread are used to identify aberrant reference values and vector lengths and detect anomalies in the correlation of spectra (see Egan and Morgan [3]).
6) After eliminating obvious mistakes, the next phase of analysis is designed to identify the optimal set of samples to use for calibration by screening the spectra and property values using robust PLS, as implemented in the RSIMPLS program Hubert et al. [4] used to detect concentration and out-of-model space outliers. This generates a matrix of possible experiments across a series of ranks and robustness metrics, qualifying the individual results and selecting the optimal model with the lowest root mean square error of calibration (RMSEC).
7) Finally, one or more standard PLS analyses can be run to polish the results and set the model rank.

Three outlier diagnostics are usually computed in PLS that are useful in deciding whether a sample is better to include in a model or if an outlier should be discarded. In Figure 18.2, points deemed acceptable to include are marked in blue, and red dots are spectra determined to be outliers. While only two of the

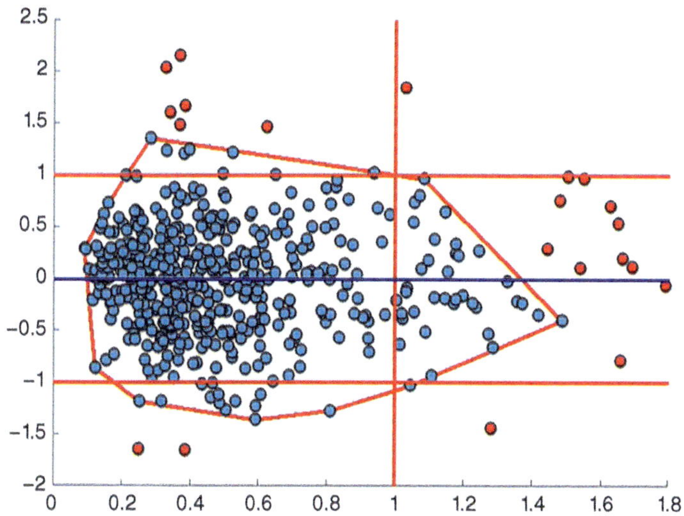

Figure 18.2 Robust evaluation of calibration samples. *Source*: Brian Rohrback.

three diagnostics used are displayed in this plot, all three are considered in the same manner. Here, the concentration residual is on the Y-axis, and Mahalanobis distance (an in-model diagnostic) is on the X-axis. The diagnostic not shown is the out-of-model diagnostic (F-ratio). The vertical and horizontal red lines mark the two-sigma limits, so we would expect 95% of all good samples to fall within those bounds. The system evaluates the density profile for the "acceptable" spectra within the two-sigma bounds, then uses set-specific information to judge those falling outside rigid boundaries. The final inlier sample set is outlined in a red polygon.

With an optimal calibration sample set in hand, the last step is to decide on the model complexity by choosing the optimal number of factors. Figure 18.3 shows how this is done with a standard PLS model

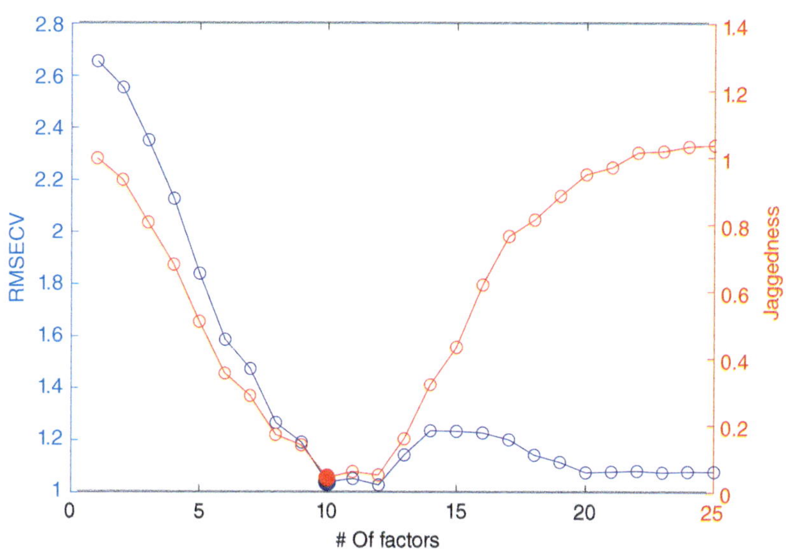

Figure 18.3 Automating the selection of "Rank" (# of factors). *Source*: Brian Rohrback.

constructed from the cumulative inliers found using the previous steps and is used to select model rank based on the root mean square error during cross validation (RMSECV) values and a jaggedness criterion [5].

18.5 Performance of Automated Models

Of course, all this automation would be pointless if the performance, as measured by the model's quality and delivery speed, is unsatisfactory. Several comparison tests have been run, matching the automated system against practicing chemometricians in academia and industry. One test was run as a blind data set supplied as part of the 2019 Paris Conference on Chemometrics. Here, 18 chemometricians were given the same data set, and an implementation of an automated calibration system was added for comparison. The results are shown in Figure 18.4.

When presented with a blind dataset, about a quarter of the chemometricians did not find an acceptable model, a quarter did an acceptable job, and the remaining half, including the automated system, built optimized models that were not significantly different from one another.

In a separate set of runs, six data sets were supplied to expert chemometricians to compare the time required to build optimized models. The six data sets were of various sizes, roughly 300 for two sets, 650 for two sets, and 2200 spectra in the last two sets. As shown in Figure 18.5, the time required for the automated process to optimize models for all six sets was 15 minutes; the experts needed between 2 and 3 hours to perform the same task. As expected, an automated system can be more responsive than an analyst. Because automation is available anytime, the response to modeling needs can be 10 to 100 times faster.

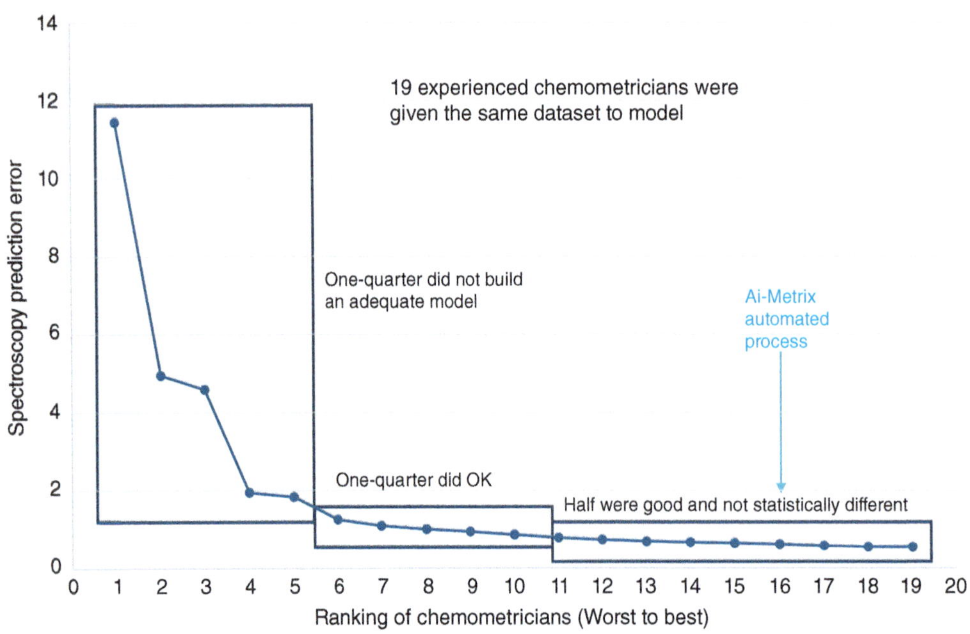

Figure 18.4 Prediction error for experienced chemometricians. *Source*: Brian Rohrback.

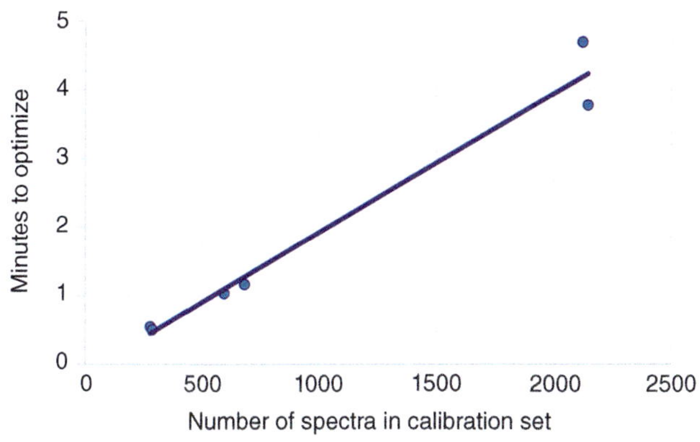

Figure 18.5 Minutes to develop optimized models for six data sets using the automated modeling software system. *Source*: Brian Rohrback.

18.6 Conclusion

Optical spectroscopy is the most useful technique for measuring chemistry in or near the process. In many industries, an increasing number of spectrometers require application-specific calibrations. However, most operators/users have little time to assess the quality of models objectively, and few skilled operators are available to build and maintain calibration models regularly. Frequent employee turnover exacerbates the ability to manage instrument calibration and model maintenance activities properly.

Because the automated approach follows the same best practices established by chemometric experts, model quality will not be degraded, results will be consistently documented, and regularly scheduled model updates will be more responsive to changes in the process chemistry.

References

1 Workman, J., Kowalski, B., Mobley, P. (1995). The design of an expert calibration system for spectroscopic based process analytical chemistry. In: Proceedings of the 40th Annual ISA Analysis Division Symposium. Toronto, ON, April 25–27, p. 97.

2 Pell, R.J., Ramos, L.S., and Rohrback, B.G. (2025). Automation of local regression model building for spectroscopic data. *Journal of Chemometrics* 389 (1): e3637.

3 Egan, W.J. and Morgan, S.L. (1998). Outlier detection in multivariate analytical chemical data. *Analytical Chemistry* 70 (11): 2372–2379. https://doi.org/10.1021/ac970763d.

4 Hubert, M. and Branden, K.V. (2003). Robust methods for partial least squares regression. *Journal of Chemometrics* 17: 537–549. https://doi.org/10.1002/cem.822.

5 Gowen, A.A., Downey, G., Esquerre, C., and O'Donnell, C.P. (2011). Preventing over-fitting in PLS calibration models of near-infrared (NIR) spectroscopy data using regression coefficients. *Journal of Chemometrics* 25 (7): 375–381. https://doi.org/10.1002/cem.1349.

19

Considerations When Selecting a Sample Interface

Ryan Lerud

Process Insights Inc., Sacramento, CA, USA

19.1 Introduction

The sample interface is arguably the most critical decision any process engineer can make for a process analyzer. An analyzer with the most precise X-axis and the most repeatable Y-axis will be rendered useless if the sample interface is inadequate for the application. A non-operating analyzer provides little benefit nor contributes to process quality or control if the system is not optimized for its intended purpose. The selection of the sampling interface needs careful attention to ensure proper analyzer utilization for either optimization or ongoing process monitoring and control purposes.

The focus of the sample interfaces in this vignette will be limited to the fiber optic-based process spectroscopy instruments of semitransparent liquids suitable for transmission measurements with UV/Vis or near-infrared (NIR). By moving light rather than process materials, the analyzer can be installed in a general-purpose environment, and an intrinsically safe sample interface can be installed in situ in a potentially hazardous (Class 1 Division 1 or zone 1) environment.

Applications can be found in petrochemical, polymer, food, semiconductor, and other industries. Extractive liquid sample conditioning used in transmission Fourier transform infrared (FT-IR) will not be discussed. Additionally, reflectance probes for solids or powders such as feed and forage or highly reflective slurry such as 3D printer ink, which may require an insertion attenuated total reflectance (ATR) probe, are excluded from this discussion.

With advanced engineering tools, careful planning is needed to select the analyzer type. Process parameters (e.g., pressure, temperature, and flow rate) and the chemical compatibility of the sample streams with the analyzer hardware must be evaluated. Considering these parameters and their implications on the analyzer hardware ensures a fit-for-purpose measurement. Furthermore, the proper location for interfacing and monitoring the process must be identified to ensure it is consistent with the defined measurement goals. For example, the location where a quality control (QC) "grab sample" is taken may be in a transfer line downstream from the reactor. Measuring after the reactor may limit the measurement to an automated post-reaction QC inspection. Meanwhile, by installing an insertion probe through the sidewall of the reactor, the chemical reactions can be monitored directly. The in situ measurement allows process spectroscopy to provide real-time actionable information for process trim control. Either location may be ideal, depending on the desired measurement goal.

The following discussion provides selection considerations when working with process spectroscopy instrument providers.

Practical Quantitative Vibrational and Electronic Spectroscopy: A Guide for Developing, Optimizing, and Validating Procedures, First Edition. Edited by Deborah A. Peru.

Companion website: www.wiley.com/go/peruelectronicspectroscopy

19.2 Commercial Analyzer Interfacing and Design Considerations

Every analyzer vendor has an application worksheet that collects information necessary for a technical sales proposal. Vendors have their selection criteria based on the types of products they offer. In the case of process spectroscopy, the site engineer needs to work with the sales team to select the sample interface. For some applications, buying the analyzer and probes from different vendors may be better.

When considering the sample interface, care must be taken to ensure the material compatibility of wetted parts, temperature and pressure ratings, serviceability, flow requirements (Reynolds number and density), direction of flow relative to the optical path, and other parameters. These considerations can be used to select the type of optical sample interface best suited to the process. The following sections get more granular regarding selecting the sample interface.

19.2.1 Material Compatibility

Selecting the proper materials for constructing the sample interface is vital for stable, long-term operation. A sample interface is only fit for purpose if it is compatible with the process to be monitored. If measuring trace water in an acid stream, a Teflon flowcell body may last longer than 316 L stainless steel. Durability also depends on the types of seals associated with the optics in a sample interface to prevent process infiltration. Typically, this is achieved with synthetic O-rings such as Viton™ or Kalrez™. A more permanent design will use a thin metal braze such as GEORO™. O-Ring seals require scheduled changeout. Sealing technology can be of critical importance. The following questions should be considered: What is your occupational health and safety risk if the primary seal fails? Does the probe have a secondary backup seal in case the primary fails?

19.2.2 Temperature, Pressure, and Fluid Density

Likely, the most straightforward parameters to consider when selecting a sample interface are the temperature and pressure of the process stream within the design limits. Some additional considerations include: If the pressure is over 500 psi, should a hydrostatic test be performed to verify the workmanship of the sample interface and reduce the risk of a leak? Will the probe hold up to the thermal cycling of the process, or will the optics or seal fail due to incompatibility with a material coefficient of expansion?

Process fluid density is often an overlooked process parameter in sample interface selection. All materials have leak rates. This is why balloons filled with helium are usually deflated after a few days. For a metal braze seal on an insertion probe, the vendor will typically conduct a helium leak test and then rate the probe as water-tight, fuel-tight, or virus-tight [1]. Leak rate tests are typically performed at room temperature. If the process stream has low density, high temperature, and high pressure, it is more likely that the process may find a pinhole defect in the seal, creating a leak. Many vendors offer dual seals to reduce the risk of the process escaping into the environment. It is essential to understand if the leak rate of the probe is sufficient to avoid an unexpected leak for your process conditions.

19.2.3 Serviceability

A sample interface that is fit for purpose will also be designed to allow easy routine service or consider other occupational health and safety risks. For example, a cleanout port in a flowcell can be immensely useful, but if the process is pyrophoric, the cleanout port may become a risk that should be avoided. The frequency of how often you have to clean the probe, change the O-rings, or take a new reference (background correction) is a critical factor for determining your serviceability requirements. These considerations will be process-specific. A flowcell installed in a sample conditioning system for measuring the cetane value of diesel fuel will require less frequent maintenance than a probe installed in a transfer line measuring the polyol content of a polymer or paint.

Some questions when considering serviceability requirements: Is the process likely to plug the sample interface? What will the procedure be for removing the sample interface from service for cleanout maintenance? How often will the O-rings need to be replaced? Should you consider a bypass line for a flowcell, an insertion probe on a gear-driven extractor, or a flanged probe that can only be removed when the whole line is depressurized? If the sample interface needs to be factory serviced, how far away is the factory, and what is their typical turnaround time for an O-ring service?

19.2.4 Flow Characteristics

Most sample interfaces require consistent and fully developed process flow through the light beam or optical path. Either laminar and/or turbulent flow is typically recommended. Bubbles or other types of two-phase sample flow disturbances (separation of solvent and solutes) can occur with Reynolds numbers between 2000 and 4000 and are not ideal for most sample interfaces. Figure 19.1 demonstrates the flow-induced noise vs. Reynolds number. Interfacing regions to avoid are where transitions occur between laminar and turbulent flow. If these types of flow-induced noise occur, the process stream may violate the assumptions of Beer–Lamberts law and cause erroneous readings to be reported by the analyzer.

In fluid dynamics, there is a concept of an "entrance limit," which refers to the distance required for the flow to be fully developed [2]. The requirement of fully developed flow is not limited to spectroscopic measurements; many types of flow instrumentation, such as flow meters, require a fully developed flow to function properly [3]. This implies that many of the same rules for where to place a flowmeter in a process pipe apply to an inline optical flowcell. In the author's experience, the general rule is to select a sample

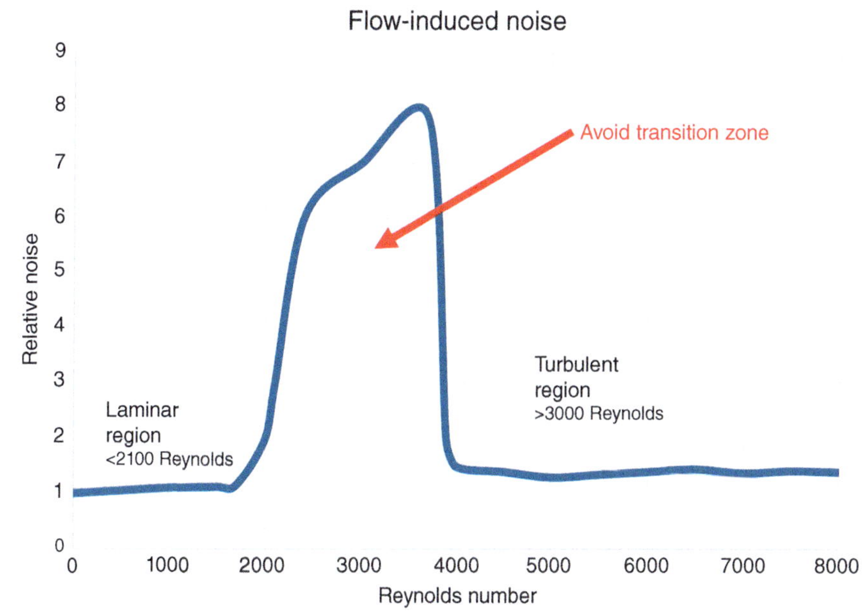

Figure 19.1 Example chart of relative noise vs. Reynolds number. The transition zone where flow irregularities may cause measurement noise occurs between 2000 and 4000 Reynold numbers. *Source*: Ryan Lerud, Process Insights-Guided Wave Inc.

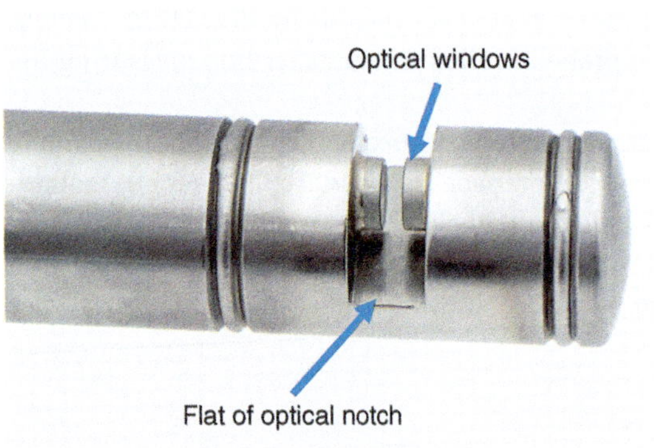

Figure 19.2 Anatomy of a transmission probe. Optical windows control the spot size. The distance between the windows is the optical pathlength. The flat of the optical notch is a slot cut into the probe that directs the process to flow between the windows. The process flange is user-defined and seals the probe to the process pipe. *Source*: Ryan Lerud, author Process Insights-Guided Wave Inc.

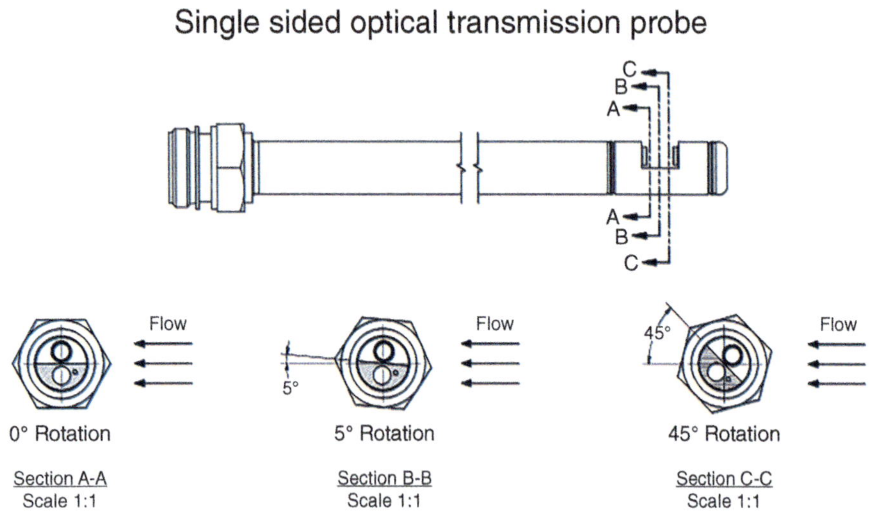

Figure 19.3 Sections lines A, B, and C illustrate fluid flow through the flat in the optical path of a transmission probe at various angles. *Source*: Ryan Lerud, Process Insights-Guided Wave Inc.

installation point 10 times the pipe diameter downstream from pipe elbows and avoid pre-pump suction zones. In practice, if you have a 90° bend in a ½ inch, there must be at least 5 in. of straight pipe before and after the flowcell to ensure a fully developed flow maintained in the optical path.

In the case of optical transmission insertion probes, the angle of the flow relative to the flat of the optical path or optical notch is also an important consideration, see Figures 19.2 and 19.3. If the flat is parallel to the flow, then cavitation, due to flow over a sharp edge, can occur, causing bubbles. If the flat is perpendicular to the flow, process fluid flow between the optical windows will be sporadic. The flow angle is typically recommended to be between 5 and 45°. It is the author's experience that 3–7° of optical cavity presentation, or "bite" relative to the process flow, provides for best measurement practice. This technique provides a "micro-zone compression" that helps push the sample through the optical notch and, therefore, fully developed flow between the optical windows. A simple fluid flow computational model can be used to optimize the angle of an insertion probe for any specific process stream.

19.2.5 Pathlength Reproducibility

When designing an optical sampling system, be certain to ask the vendor for information regarding the reproducibility of the optical pathlength. Some vendors offer adjustable optical pathlength sample interfaces.

These are suitable only for proof-of-concept benchtop studies or R&D at the beaker level. However, a fixed pathlength is essential for process applications to achieve precision needed for "trim control" process adjustments. Some analyzer vendors include a pathlength correction within the chemometric preprocessing tools used for calibration model development. However, experienced vendors will offer precise fixed pathlengths. Reducing errors at the point of data collection will improve precision and accuracy compared to mathematical correction for measurement variances such as pathlength deviations.

In transmission optical spectroscopy, a pathlength variance of no more than ±0.2% allows sample interfaces to be interchangeable without impacting calibration performance. The change in absorbance due to small pathlength variance is typically less than the diurnal thermal noise in the fiber optic cables connecting the sample interface to the analyzer.

Table 19.1 Commercially available remote insertion probes.

(a) 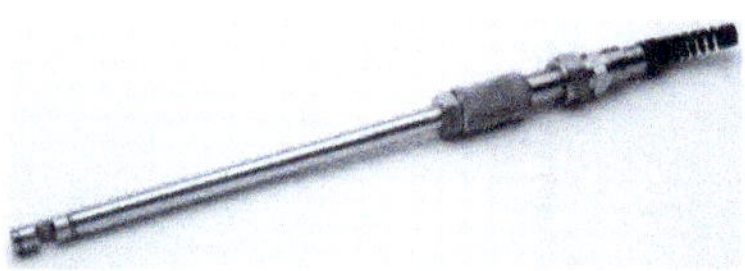	**Single-sided transmission insertion probe** • Standard continuous process measurement probe • Braze or O-ring seal • O-ring seal recommended for corrosive process streams containing strong acids, bases, or cyanides • Process flange can be welded to the body to enable the probe to be installed in reactor vessels • Compatible with gear-driven extractor.
(b) 	**Transflectance transmission insertion probe** • Internal mirror enables longer path lengths in shorter optical notch • Recommended for light duty or lab use.
(c) 	**Dual-sided transmission shuttle process probe** • Installed in the process pipe • Two transmission probes connected with a couple allow the optical path to be serviced by pulling one end out of the process • Often used in the polymer industry to allow 24/7 process uptime and probe downtime.
(d) 	**Turbidity single-sided transmission insertion probe** • Measures 90° scattering with additional fiber optic cable • Typically, O-ring seals • Typical for fuel color or cloudiness or other scattering applications.
(e) 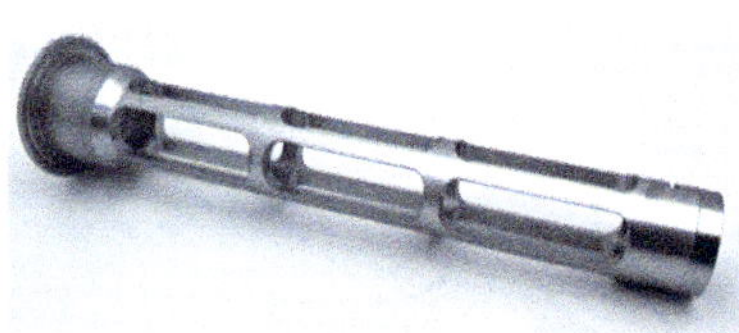	**Vapor probe for vacuum applications** • Vapor probe designed to be installed in glove box and handle deep vacuum applications • NIR can be used to measure sterilization concentration per ISO 22441:2022.

Source: Ryan Lerud, Process Insights – Guided Wave Inc.

Table 19.2 Commercially available remote flowcells.

(a) 	**Multipurpose flow cell** • Available in multiple pathlengths • Ideal for use inside loops or small-diameter process lines • Maintenance video on YouTube.
(b) 	**Axial flow cell** • For applications that require long optical pathlengths • May be installed at a slight angle to prevent bubble formation if this is an issue • An example application is phenol breakthrough in bromide plant • Sufficient pathlength for identification of water and acetone by APHA color.
(c) 	**Gas flow cell** • Longer pathlength up to 1 m for monitoring gases • Many materials of construction are possible.
(d) 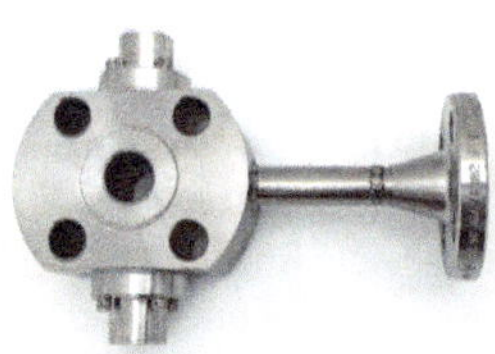	**High safety flow cell HSFC** • Contains sniffer port for early warning leak detection • For applications involving extremely hazardous materials • Patented double O-ring seals for added protection • Built into a ½″ class 300 flange • Standard optical pathlength is 20 and 30 mm.
(e) 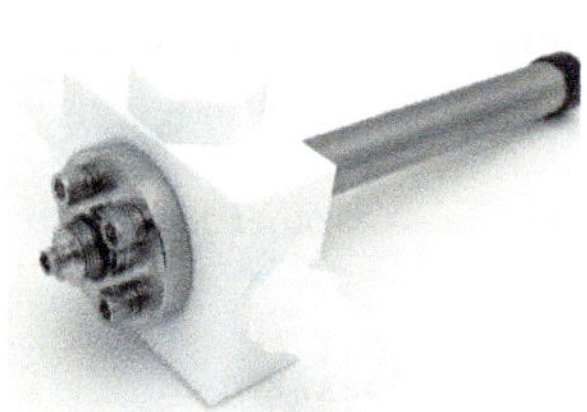	**Teflon flow cell** • For applications not compatible with metals • Performs well in corrosive streams containing strong acids, bases, peroxides, or halogenated compounds • Ideal for processes where even parts per billion levels of metals contamination can create serious problems, as in semiconductor fab etching and cleaning steps • Teflon-to-Teflon seal reduces the chance of leaks.

Source: Ryan Lerud, Process Insights – Guided Wave Inc.

Table 19.3 Commercially available remote flowcells with unique design features.

(a)		**Teflon PFA/PEEK** • Designed for hydrofluoric applications • For applications where Teflon and metal are not compatible • Typically used in semiconductor applications.
(b)	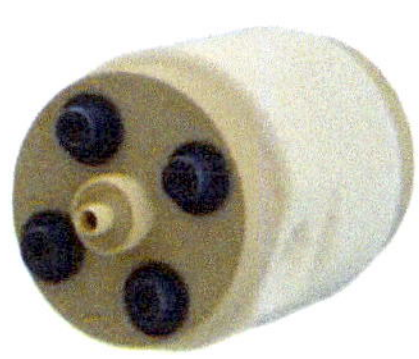	**Teflon axial** • For applications where pathlength is greater than 25 mm and small flow, volumes are required • For applications not compatible with metal • Great for applications where trace metal sensitivity is required.
(c)		**Turbidity flow cell** • For fuel color and turbidity application • Collects 90° scatter light • Possible use in fluorescence measurements with a 90° connection.
(d)		**Heated flow cell** • For applications such as polymers where temperature control is critical • Can be used to maintain the temperature of a preconditioned sample with heated fluid • For use with process temperatures up to 300 °C • Temperature limit depends on O-ring selection.

Source: Ryan Lerud, Process Insights – Guided Wave Inc.

The following tables provide examples of various sampling interfaces commercially available. Table 19.1 provides examples of remote insertion probes, Table 19.2 provides samples of remote flowcells, and Table 19.3 provides additional examples of remote flowcells with unique designs for special circumstances. Each example was engineered to meet different measurement needs in different markets and applications.

19.3 Conclusion

As an optical spectroscopy practitioner, it is important to ask the right questions and ensure that the complete analytical solution (spectrometer and sample interface) is fit for purpose. This requires considering that the analyzer and probe are correctly designed for making the measurement and that the physical installation in the analyzer shed or process pipe does not add noise sources or negatively impact serviceability. If the implementation of PAT is poorly thought through, then the value added by the technology cannot be realized.

References

1 Guide To Good Leak Testing (2009). Institute of Refrigeration (IoR). Guide to Good Leak Testing. https://www.epa.gov/sites/default/files/documents/RealZeroGuidetoGoodLeakTesting.pdf (accessed 12 January 2025).

2 Morega, A. (2001). 7 – Principles of heat transfer. In: *Mechanical Engineer's Handbook* (ed. D.B. Marghitu), 521. San Diego, CA: Elsevier Academic Press.

3 Çengel, Y.A. and Cimbala, J.M. (2018). *Fluid Mechanics: Fundamentals and Applications*, 4e, 391. New York, NY: McGraw-Hill Education.

20

Continuous Performance Qualification Systems for On-line Near-Infrared (NIR) Analyzers Used in Fuels Blending Operations

Ryan Lerud

Process Insights Inc., Sacramento, CA, USA

20.1 Introduction

In oil refineries and petrochemical plants, process analytical technology (PAT), such as NIR spectroscopy, is used to monitor chemical reactions to ensure the recipe or manufacturing procedure is being followed. The output of these process analyzers is sent back to a control room. When the reported process conditions deviate beyond allowable tolerances, the plant personnel must troubleshoot the issue and correct the deviation to maintain reaction or process yield. In the field, false alarms can occur, disrupting the smooth operation of the plant. Often, the operational integrity of the process analyzer reporting the out-of-specification condition is the first thing to be questioned as the source of the reported deviation. Additionally, if data from the quality control laboratory disagrees with the online measurement, the operational integrity of the process analyzer is often questioned. Continual hardware validation, an optional approach used for many process analyzers, can quickly rule out a malfunction within the instrument.

In the case of NIR spectroscopy, the process analyzer, at its most basic description, consists of the light source, spectrometer, the calibration model, and the sample interface. Each component could be the source of a bad reading being reported to the control room, and each has its own troubleshooting steps. Continual hardware validation is often the best option to identify invalid measurements and speed up the troubleshooting process. For NIR process analyzers, continual validation is addressed in ASTM D6122 [1]. The recommendation cited is referenced below:

> Verification of adequate instrument performance, verification of the applicability of the calibration model to the spectrum of the sample under test, and verification of equivalence between the result calculated from the infrared measurements and the result produced by the primary test method used for the development of the calibration model

The tests defined in ASTM D6122 are split into three levels of priority and complication (Level 0, Level A, or Level B) and are conducted on control or test samples. Level 0 tests directly relate to the analyzer hardware's capability to generate a consistent spectrum. This is accomplished by measuring photometric noise (root mean square at a specific wavelength), baseline stability, spectral resolution, photometric linearity, and wavelength stability. Level A is sensitive to all the Level 0 parameters. Additional tests compare key parts of the spectral data with historical data to identify deviations, such as peak size, shape, or position. This is typically achieved by comparing the current scan with a saved benchmark scan. Level A tests do not identify specific failure modes but do indicate if the instrument performance is within historical bounds. Level B

Practical Quantitative Vibrational and Electronic Spectroscopy: A Guide for Developing, Optimizing, and Validating Procedures, First Edition. Edited by Deborah A. Peru.

Companion website: www.wiley.com/go/peruelectronicspectroscopy

tests monitor the instrument performance for deviations the system calibration models are sensitive to. This is achieved by applying online calibration models to determine the concentration of a standard reference material (SRM) or control sample, such as reagent-grade toluene. The results are compared to historical values to detect any change in the analyzer's performance. Level B tests can temporarily produce a false alarm if the calibration model has been significantly updated and is now outputting values that are sufficiently different than the historical trend. The false alarm often clears once enough new data is collected and incorporated as the new historical validation. Online calibration models are developed and validated in the petrochemical industry following ASTM E1655 [2].

20.2 Continual Validation Procedures

Continual validation procedures are implemented differently depending on how the sample is presented for measurement: inline, outline, or extractive with a sample conditioning system (SCS). If the NIR analyzer can multiplex, it can switch between the process interface(s) and a dedicated ASTM D6122 reference cell, as shown in Figure 20.1.

If the analyzer is extractive with an SCS, a wash and validation skid, as shown in Figure 20.2, will introduce a control sample into the measurement cell. Both methods have trade-offs, as described in Table 20.1, and follow the general order of operations provided in Figure 20.3.

A dedicated reference cell allows for near real-time on-demand measurement of the control sample and reduces the downtime from measuring the process to under one minute. Thus, spectral collection of the control sample can be conducted automatically in the background with no human intervention and little to no interruption of process monitoring. This allows the plant operation to continuously measure the quality of the process and the analyzer's performance. The downside to a dedicated reference cell is that the complete measurement loop, source, fiber optic cables, remote sample interface, and spectrometer are not tested. The dedicated reference cell limits hardware validation to the reference measurement loop, source, reference cell, and spectrometer. By quickly validating the performance of the critical hardware components, the plant personnel can move on to inspecting the process fiber optic cables for damage, the sample interface for fouling, and the actual process for deviations.

In contrast, using an SCS with a wash and validation skid ensures complete testing of the entire sample measurement loop (i.e., source, fiber optic cables, measurement cell, and spectrometer). If all the components operate as expected, the plant personnel can safely rule out the instrument as producing a bad result. Not only does this build confidence in process analyzer results, but it also more quickly alerts process operators that an actual process upset is occurring, and remedial steps to correct it can happen more quickly.

Figure 20.1 The stability monitoring reference cell is installed inside the environmentally controlled analyzer enclosure. Inside the sample holder is a polymeric crystal that provides a repeatable spectral fingerprint. *Source*: Ryan Lerud / Process Insights-Guided Wave Inc.

Figure 20.2 Wash and validation skid provides wash solvent (methylcyclohexane) for the sample cells and instrument standardization solvent (spectrophotometric-grade Toluene) for periodic background and validation. Contains a 3–6 month supply of reagent. *Source*: Author: Ryan Lerud / Process Insights-Guided Wave Inc.

Table 20.1 Tradeoffs between real-time continual validation methods.

Continual validation method	Pros	Cons
Dedicated reference cell	• Near real-time and on-demand • Automatic, no human intervention or interruption of the process • One-time upfront cost • Small footprint • No consumables/maintenance	• Complete measurement loop not tested
SCS with wash & validation skid	• Validates complete measurement loop • Can rule out instrument false alarm	• Longer downtime • Uses consumables • Larger footprint • Maintenance required

Source: Ryan Lerud, Process Insights – Guided Wave Inc.

The downside to a wash and validation skid is the increased downtime. Depending on process conditions, up to 30 minutes may be needed to flush the measurement cell with a wash solvent, introduce the control sample, collect the validation spectra, wash the measurement cell, and then begin measuring the process again. Typically, a validation measurement using an extractive measurement with an SCS will be performed at the beginning and end of a measurement campaign. Plant personnel would consider running additional validation tests only in the event of a process deviation.

Additional considerations when choosing between an extractive SCS and a dedicated reference cell are the installation and ongoing maintenance costs. The dedicated reference cell has only a one-time upfront cost. The SCS with a wash and validation skid has a larger footprint within the plant, an ongoing cost of consumable fluids, and the inconvenience of periodic maintenance of the various valves and pumps. Therefore, the cost of ownership is much lower for a dedicated reference cell.

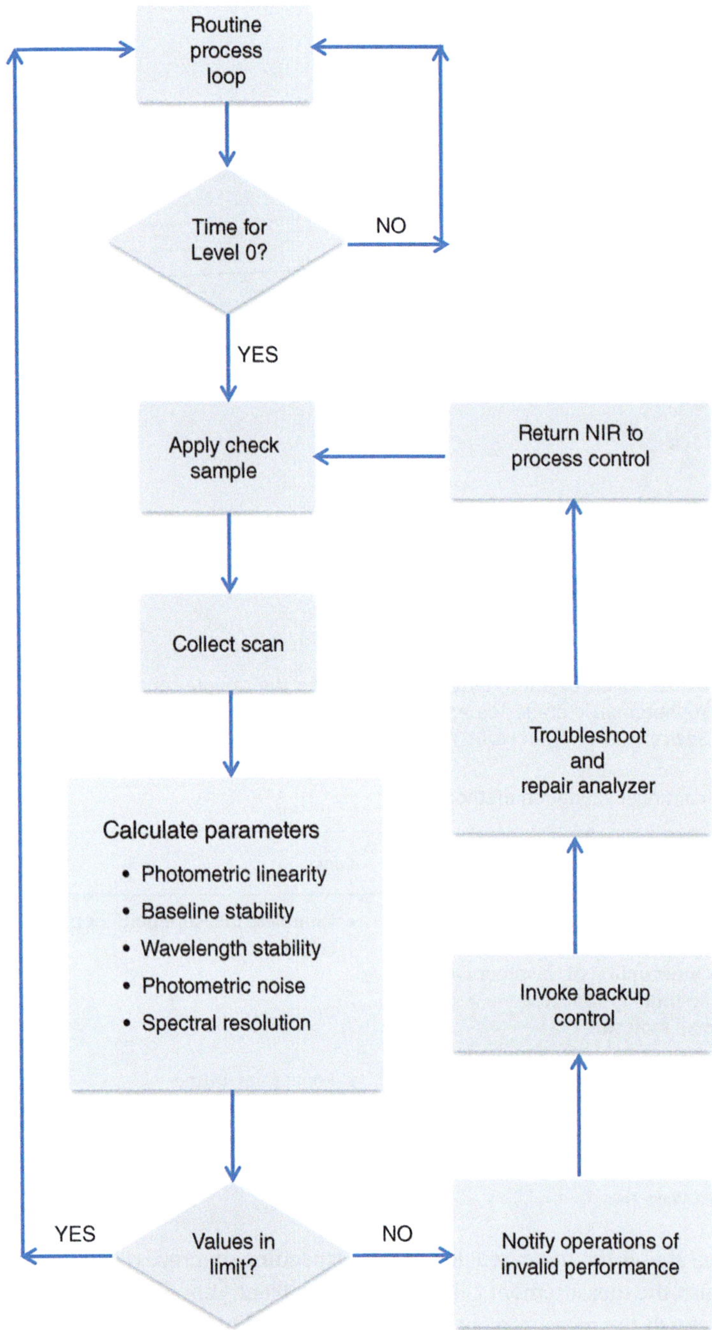

Figure 20.3 Flowchart of level 0 tests used in an NIR-O Full Spectrum Analyzer©. Courtesy of Susan Foulk, Process Insights-Guided Wave Inc [3]. *Source*: Ryan Lerud-Process Insights-Guided Wave Inc.

Real-time hardware validation can also be used as a diagnostic tool for outlier detection and model maintenance. If the NIR analyzer detects an outlier, a grab sample can be collected for further analysis (see Chapter 18). If the NIR analyzer passes validation and the outlier fits within the model, the sample is a true outlier and can be considered for inclusion in the model. Conversely, if the hardware is validated, and the outlier spectrum is not similar to the model, then the NIR analyzer correctly alerts the engineers of a process deviation.

20.3 Conclusion

Process optimization and yield are the main goals of any chemical plant investing in PAT. The real-time hardware validation checks outlined in ASTM D6122 have been developed to support this goal. By ensuring that the NIR analyzer is functioning within expected bounds, the plant operation team can trust the online analyzer's output and make informed decisions to maximize profits.

References

1 American Standard for Testing and Materials (2020). *Validation of the Performance of Multivariate Online, AtLine, Field and Laboratory Infrared Spectrophotometer, and Raman Spectrometer Based Analyzer Systems* (ASTM D6122-20). Pennsylvania: ASTM International. ASTM International – Standards Worldwide.

2 American Standard for Testing and Materials (2024). *Standard Practices for Infrared Multivariate Quantitative Analysis* (ASTM E1655-17). Pennsylvania: ASTM International. ASTM International – Standards Worldwide.

3 Foulk, S.J. and Lerud, R. (2019). Automated near-infrared analyzer validation (ASTM D6122) for the oil and gas industry. *NIR News* 30 (5–6): 39–41.

Index

Practical Quantitative Vibrational and Electronic Spectroscopy: A Guide for Developing, Optimizing, and Validating Procedures, First Edition. Edited by Deborah A. Peru.

Published 2026 by John Wiley & Sons, Inc.
Companion website: www.wiley.com/go/peruelectronicspectroscopy

j

k

l

m

p

q